EDITED BY
ROBERT J. KOESTLER
VICTORIA H. KOESTLER
A. ELENA CHAROLA
FERNANDO E. NIETO-FERNANDEZ

ART, BIOLOGY, AND CONSERVATION: BIODETERIORATION OF WORKS OF ART

The Metropolitan Museum of Art, New York

The papers in this volume were presented at the conference "Art, Biology, and Conservation: Biodeterioration of Works of Art," held at the Metropolitan Museum of Art, June 13–15, 2002.

The conference and accompanying publication were made possible by an anonymous donor and the Samuel H. Kress Foundation.

Published by The Metropolitan Museum of Art
ISBN 1-58839-107-8

Book design, composition, and production by Binocular, New York
Printed and bound in Canada

Cover illustration: Detail from *Angel Appearing to Three Marys at the Tomb*, Louis Comfort Tiffany, Tiffany Studios 1892–1918, Metropolitan Museum of Art, accession number 67.654.206. Reproduced in full on page 87.

CONTENTS

9 DIRECTOR'S INTRODUCTION
Philippe de Montebello, Director, The Metropolitan Museum of Art

10 ACKNOWLEDGMENTS
Robert J. Koestler, Research Scientist, The Metropolitan Museum of Art, Symposium Chair

SPECIAL TOPICS

14 THE EVALUATION OF BIODETERIORATION PROCESSES ON CULTURAL OBJECTS AND APPROACHES FOR THEIR EFFECTIVE CONTROL
Th. Warscheid

28 PRESERVING THE TIMBERS OF THE TUDOR WARSHIP MARY ROSE
A.M. Jones, R. Mouzouras, A.J. Pitman, and S.B. Pointing

50 BIODETERIORATION STUDIES ON PASTELS AND OIL-BASED PAINTINGS
M. Berovič

60 CHEMICAL AND MICROBIOLOGICAL CAUSES OF DETERIORATION OF TOOTH-BRUSHES THAT USED TO BELONG TO PRISONERS OF AUSCHWITZ-BIRKENAU CONCENTRATION CAMP: RESEARCH AND METHODS OF THEIR CONSERVATION
A. Strzelczyk and H.E. Rosa

72 COLLATERAL DAMAGE: ANTHRAX, GAS, AND RADIATION
D. Erhardt, C.S. Tumosa, D. von Endt, and A.-S.M. El-Easely

PAPER

82 AN INTRODUCTION TO TIFFANY STUDIES
A.M. Baldwin

94 STUDIES OF FUNGAL INFESTATIONS OF TIFFANY'S DRAWINGS: LIMITS AND ADVANTAGES OF CLASSICAL AND MOLECULAR TECHNIQUES
M.P. Di Bonaventura, R. DeSalle, D.E. Eveleigh, A.M. Baldwin, and R.J. Koestler

110 ENZYMATIC APPROACH TO REMOVAL OF FUNGAL SPOTS FROM DRAWINGS ON PAPER
F.E. Nieto-Fernandez, S.A. Centeno, M.T. Wypyski, M.P. Di Bonaventura, A.M. Baldwin, and R.J. Koestler

128 ARTWORKS, DRAWINGS, PRINTS, AND DOCUMENTS–FUNGI EAT THEM ALL!
H.M. Szczepanowska and A.R. Cavaliere

152 PRACTICAL APPLICATIONS OF ENZYMES IN PAPER CONSERVATION
Y. Van Dyke

172 FUNGAL-PROBLEM MONITORING FOR HERITAGE COLLECTIONS: THE NEED FOR BASELINE REFERENCE LEVELS FOR FUNGAL STRUCTURES AND BETA-GLUCANS
M.-L.E. Florian

TEXTILES

190 CHARACTERIZATION OF BACTERIA ISOLATED FROM NATURALLY AGED SILK FIBROIN
O. Ciferri, M.E. Becker, and E. De Rossi

STONE AND MURAL PAINTINGS

210 MICROBIAL COMMUNITIES IN CAVES: ECOLOGY, PHYSIOLOGY, AND EFFECTS ON PALEOLITHIC PAINTINGS
L. Laiz, J.M. Gonzalez, and C. Saiz-Jimenez

226 RED STAINS ON CARRARA MARBLE: A CASE STUDY OF THE CERTOSA OF PAVIA, ITALY
E. Zanardini, P. Abbruscato, L. Scaramelli, E. Onelli, M. Realini, G. Patrignani, and C. Sorlini

248 MICROBIAL PROCESSES IN THE DETERIORATION OF MAYA ARCHAEOLOGICAL BUILDINGS IN SOUTHERN MEXICO
C. McNamara, T.D. Perry, M. Zinn, M. Breuker, R. Müller, G. Hernandez-Duque, and R. Mitchell

266 MICROBIODETERIORATION OF MURAL PAINTINGS: A REVIEW
J. Karbowska-Berent

302 METHODOLOGICAL APPROACHES TO THE STUDY OF STONE ALTERATION CAUSED BY CYANOBACTERIAL BIOFILMS IN HYPOGEAN ENVIRONMENTS
P. Albertano

316 STUDY OF MICROBIAL COMMUNITIES COLONIZING HYPOGEAN MONUMENT SURFACES USING NONDESTRUCTIVE AND DESTRUCTIVE SAMPLING METHODS
C. Urzì, F. De Leo, P. Donato, and V. La Cono

WOOD AND ARCHAEOLOGICAL MATERIALS

328 DETERIORATION IN HISTORIC AND ARCHAEOLOGICAL WOODS FROM TERRESTRIAL SITES
R.A. Blanchette

348 DEGRADATION PATTERNS IN WATERLOGGED WOOD AND THE TWO-STEP PEG TREATMENT FOR LARGE ARCHAEOLOGICAL FINDS: THE CASE OF THE BREMEN COG
P. Hoffmann

358 THE CONSERVATION OF THE WOODEN OBJECTS FROM GORDION, TURKEY: METHODS FOR THE TREATMENT OF DRY ARCHAEOLOGICAL WOOD
E. Simpson

370 DETERIORATION AND CONSERVATION ISSUES ASSOCIATED WITH ANTARCTICA'S HISTORIC HUTS
B.W. Held, R.A. Blanchette, J.A. Jurgens, S. Duncan, and R.L. Farrell

390 EVALUATING THE WOODEN REMNANTS OF THE TEKTAŞ BURNU SHIPWRECK
J.A. Jurgens, R.A. Blanchette, and D.N. Carlson

TREATMENT AND PREVENTION

410 VISUAL ASSESSMENT OF BIOCIDE EFFECTS ON JAPANESE PAINT MATERIALS
J. Suzuki and R.J. Koestler

426 ANOXIC ERADICATION OF FUNGI IN WOODEN OBJECTS
C. Tavzes, J. Pohleven, F. Pohleven, and R.J. Koestler

440 MICROBIOLOGICAL IMPURITY OF CONSERVATION TREATMENT WATERS
V. Bullock

452 CONSEQUENCES OF MICROBE-BIOFILM-SALT INTERACTIONS FOR STONE INTEGRITY IN MONUMENTS
E. May, S. Papida, and H. Abdulla

472 MECHANISMS OF MICROBIAL CALCIUM CARBONATE PRECIPITATION
C. Barabesi, E. Tamburini, G. Mastromei, and B. Perito

486 BIOMEDIATED CALCITE PRECIPITATION FOR THE REINFORCEMENT OF MONUMENTAL STONES
P. Tiano, S. Bracci, and S. Rescic

498 BIOLOGICAL MORTARS: A SOLUTION FOR STONE SCULPTURE CONSERVATION
G. Orial, Th. Vieweger, and J.-F. Loubiere

518 BIOCIDES AND TREATMENT OF STONE: LIMITATIONS AND FUTURE PROSPECTS
M.P. Nugari and O. Salvadori

536 THE USE OF METALLIC OXIDES IN CONTROL OF BIOLOGICAL GROWTH ON OUTDOOR MONUMENTS
D.P. Wessel

ENDNOTE
554 IN DEFENSE OF THE FUNGI
A.R. Cavaliere

566 AUTHOR INDEX
568 CONTACT LIST

DIRECTOR'S INTRODUCTION

The interface of art, science, and conservation is an exciting place. That's what we at the Metropolitan Museum of Art discovered in June 2002 when we hosted a three-day symposium that brought these areas together. The first such conference ever held at the Met, the Art, Biology, and Conservation 2002 meeting provided an opportunity for close to 40 scientists and conservators to present their latest research on the interactions of microbes and art, and to offer approaches to counter the effects of biodeterioration.

As I noted at the time, no work of art is immune to microbial attack. Microbes, being cosmopolitan, are a threat to art collections worldwide, as they are always present in the environment, lying dormant, and waiting for the right conditions to occur so they can flourish. Such conditions include subtle environmental changes, such as a temporary increase in moisture level, as well as major events, such as a water leak; both can foster an infestation within just hours. Whether this leads to serious damage depends on many factors, including the material composition of the art, the amount of water present, the species of microbe on the art, and how long conditions are favorable for growth. Once such mishaps occur, the work of scientists and conservators becomes vital, as they strive to protect our often fragile cultural heritage.

The papers in this volume detail the work of the select group of biologists and conservators that came together at ABC2002 to show how they are addressing these concerns. A wonderfully diverse group, they had in common a desire to preserve the historic and the beautiful, the skills to do so, and an excitement about the whole process. I think that excitement comes through in this volume as well.

Philippe de Montebello
Director
The Metropolitan Museum of Art

ACKNOWLEDGMENTS

The papers in this volume were presented at the conference "Art, Biology, and Conservation: Biodeterioration of Works of Art," held at the Metropolitan Museum of Art, June 13–15, 2002. They give a timely overview of many of the issues and directions of research in biodeterioration of cultural property today.

Many individuals assisted in making possible this first-ever art, science, and conservation symposium at the Metropolitan Museum of Art. Museum Director Philippe de Montebello encouraged this endeavor and mobilized the efforts of the Metropolitan's Development Office under Christine Scornavacca to find the necessary funding to host the event. He also guided Education, under Kent Lydecker, to provide the essential structural support for the meeting. Elizabeth Hammer and Tara Diamond ably handled all of the many arrangements necessary for this program to happen; this included accommodations for the speakers, contracts, registration, and all aspects of the social program. Many other members of Education gave vital assistance: Masha Turchinsky and Lena Howansky helped with many details of the poster and program brochure, which was designed by Kevin Park and expertly edited by Merantine Hens. Thanks also go to Karen Ohland and Nathaniel Wilcox for their financial expertise, and to Stella Paul, head of General Programs.

I want to give a special thank you to my department head, James H. Frantz, Conservator-in-Charge of the Sherman Fairchild Center for Objects Conservation, for his many years of support for a strong science program in conservation. His behind-the-scenes efforts and encouragement ensured that this symposium became a reality. Without his support we would not have been able to have this meeting.

There are many other people who assisted in various ways: A. Elena Charola was an excellent sounding board and an able organizer, helping me develop the details of the program. Ann Baldwin was always there to give advice on the poster, the extended abstract proceedings cover, and many other facets of this project. Bruce Schwarz produced a wonderful photograph of a fungal-damaged Tiffany drawing that we used for the poster art and abstract booklet. Rebecca Rushfield spent many hours promoting the program to the conservation community. Dan Koestler deftly assisted the speakers in their PowerPoint presentations. Joseph Cho and Stefanie Lew of Binocular, New York, are to be thanked for their expert and artful book design.

To each and every one involved with this program, thank you.

Finally, this symposium was made possible in part by the generous support of an anonymous donor and the Samuel H. Kress Foundation.

Robert J. Koestler, Symposium Chair

ART, BIOLOGY, AND CONSERVATION

SPECIAL TOPICS

TH. WARSCHEID

THE EVALUATION OF BIODETERIORATION PROCESSES ON CULTURAL OBJECTS AND APPROACHES FOR THEIR EFFECTIVE CONTROL

LBW – Bioconsult, Wiefelstede, Germany

ABSTRACT The importance of biodeterioration processes for built heritage has reached the attention of those in charge of its restoration and conservation. The microbial impact on the deterioration process has already been acknowledged and its analysis has been integrated into anamnesis programs within conservation practice. In order to meet conservators' and restorers' needs as they tackle biodeterioration problems and their prevention, it will be necessary to focus on future research and case studies. This paper highlights some aspects of the evaluation process and its relevance to effective control of biodeterioration processes within art objects.

BIODETERIORATION OF BUILT HERITAGE Recent research has elucidated the basic mechanisms of biodeterioration of cultural heritage objects—for this paper this includes architectural heritage—and a wide range of analytical techniques to describe microbial impacts on materials are available. Since microbially induced deterioration differs widely with various climatic and expositional conditions, each problem needs to be addressed individually, with tailored analytical approaches and countermeasure strategies (Warscheid, 2000).

Exposed monuments in moderate climates of Northern Europe and comparable regions in the U.S., with regular rainfall and moderate temperatures, bear mainly mixed consortia of microorganisms that settle in the uppermost layers of stones and other building materials, depending on their porosity. The impact of this colonization is mainly aesthetic (e.g., discoloration) and in most cases leads to detrimental effects for the material only when it is in conjunction with other damage factors, such as those that lead to crust formation, salt crystallization, and mineral oxidation. Any measures taken against biocolonization need to be evaluated with regard to their relative importance within the entire damage process. These usually rely on adequate cleaning procedures, protcction from water, and careful biocidal treatments.

In semiarid areas of the Mediterranean as well as similar climatic zones in South America, with less rainfall and comparable higher temperatures, the microbial population is dominated mostly by specially adapted microorganisms, such as cyanobacteria, black yeast (*Dermaticiaceae*), and/or lichens. The biodeterioration processes here are surface-bound and dominated by mechanical influences and thermo-hygric alterations. Biocorrosive activity is only critical in particular cases.

The specialization of microorganisms to the environment reaches its highest degree in arid zones, whereas the biodeteriorating effects can be mostly neglected, despite the phenomenon of "rock-varnish" formation caused by cyanobacteria and mineral-oxidizing fungi. Due to the specialization of the respective microflora, effective protective treatments are quite difficult to achieve and a thorough analysis of the object under restoration is required

It is in tropical climates that the impact of microorganisms on biodeterioration is greatest. [FIG. 1] High humidity and temperatures lead to highly developed consortia of microorganisms with extensive capabilities in terms of biocorrosion and biofouling. The microflora are extremely stable and resistant to any kind of environmental changes, as well as to conservation interventions such as cleaning or biocidal treatments. Therefore, any countermeasures against microbial contamination in tropical areas need to be carefully planned, carried out with precision, and regularly monitored in order to achieve long-term efficacy (Warscheid and Braams, 2000).

Indoor biodeterioration processes are determined by many factors. Exposed objects in churches, caverns, archives, castles and archaeological sites, to name but a few locations, contain materials ranging from the mineral or metallic to the organic. The latter are particularly subject to the attack of highly specialized microorganisms, given that environmental conditions (humidity, light, temperature) occur at appropriate levels, an important issue that will be discussed in more detail further on.

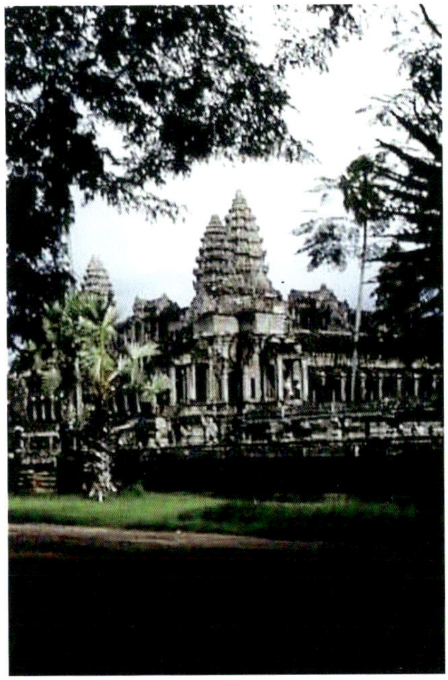

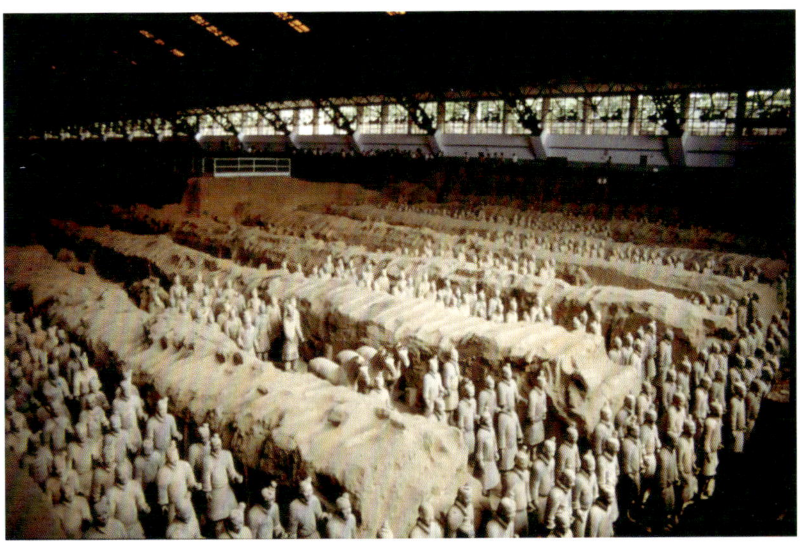

[1] Biodeterioration in subtropical climates: Temple of Ankgor Wat (Cambodia). [2] Microbial contamination of an archaeological excavation buried in loamy soil: Terracotta Army in Lintong (China).

Mural paintings offer a wide range of organic binders (e.g., casein, oil, egg yolk) and, in the course of restoration treatments, consolidants and fixatives based on polymeric compounds (e.g., cellulose acetate, PVA, PMA). If sufficient water is available, biodeterioration will result in detrimental discoloration and/or decomposition, unless pigments present (e.g., copper-containing malachite) limit the microbial activity. If complete climatic control can be achieved for these objects and their surrounding environments, biodeterioration processes may be limited and controlled with minimal use of biocides, or even without them.

Archaeological sites can also be covered by organic, sometimes colorful coatings, but these materials should not be considered as potential nutrient sources, since there might be a good reason why they have survived for centuries. In this respect, microbiological investigations on the polychromy on the warriors of the terracotta army in Lintong, China, have shown that the oriental lacquer layer with its phenolic compounds is barely attractive to the contaminating fungi found in the overall excavation. [FIG. 2] The microorganisms were mainly responsible for the mechanical detachment of the superficial paint layers, rather than "eating them up" (Warscheid and Rudolph, 2002).

The recovery of archeological objects implies considerable changes in the environmental conditions that the material is exposed to. Oxygen appears to be the most important factor; it might cause a start or even an acceleration of biodeteriorating processes on metals, wood, leather, or textiles. On the other hand, if objects should remain in their found environment, i.e., in sediments, soils, and closed caverns, biodeterioration could also continue due to the activity of anaerobic microorganisms, such as sulfate-reducing bacteria or fermentative microorganisms. The possibility of in-situ preservation is currently being studied in various Viking excavations in Denmark (Soerensen and Gregory, 1998).

The intensity of the microbial attack on historical glass objects and paintings depends mainly on the composition of the silica material. Especially high concentrations of potassium will make the glass increasingly susceptible to microorganisms and their biocorrosive attack. Additives of manganese- and iron-containing minerals to the glass will increase microbial oxidation processes. Nevertheless, the biodeterioration processes on glass are mainly secondary with respect to the initial corrosion processes caused by atmospheric pollutants. However, they can also be initiated and supported by organic oil and wax varnish residues deriving from former historical restoration treatments. In this concern, careful cleaning and sufficient aeration in the latter will help to control the biodeterioration processes (Krumbein et al., 1993).

In archives, the impact of microorganisms on the deterioration of paper, parchment, leather, and textiles constitutes the classical case of biodeterioration of cultural objects. Nevertheless, it is to be remembered that, besides the nutritious organic material, considerable amounts of moisture and humidity are needed to keep the biodeterioration going. Defined ranges of humidity levels that could be regarded as favorable for microbial growth on archival material are lacking and difficult to assess, since the climatic properties of the building and building materials (e.g., building physics, absorbency), the maintenance of objects (e.g., cleaning) and the types of materials affected differ from case to case and are currently not fully understood (Padfield, 2002). Over and above that, our knowledge of the physiological behavior of the respective microorganisms (e.g., sporulation and germination) under these specific environmental conditions is poor (Sedlbaur, 2002). Once a better understanding of the prevailing growth conditions of microorganisms in archives is achieved, the formulation of effective countermeasure strategies would be far easier and based on environment-friendly approaches.

Studying biodeterioration processes worldwide we have to consider that microbial impacts on objects of cultural heritage are expressed in many forms and to different extents. Nevertheless, any action to limit or control possible biodeteriorating effects requires, first, a careful analysis and evaluation of the biogenic impact in relation to other damage functions present on the particular object under investigation.

THE EVALUATION OF BIOGENIC DAMAGE FUNCTIONS The importance of microbial impacts on materials and their relevance to the entire deterioration process has to be evaluated very carefully. "Biofouling" (i.e., the presence of colloidal microbial biofilms on or inside materials) leads to an aesthetic impairment of the original surfaces and potentially causes alterations in the physicochemical characteristics of the material matrix, [FIG. 3A] while "biocorrosion" (i.e., microbially induced or influenced corrosion of materials) alters the structure and stability of materials. [FIG. 3B] Therefore all potential deteriorating properties of the microflora and possibly related biodeterioration phenomena have to be proven to be able to take real qualitative and, more important, quantitative measures to counteract these changes in material properties (Warscheid, 2000).

While on organic materials (e.g., paper, parchment, leather, textile, binders, and polymers) the proof of microbially induced destruction is generally easy to establish because the enzymatic and metabolic pathways are generally known, it is quite difficult to distinguish the microbial impact on inorganic materials (e.g., stone, plaster, glass) from other deterioration factors and to determine its relative importance in cases of damage (Koestler et al., 1997). Thus, interdisciplinary studies are desperately needed on the effects of microbial biofilms on thermal-hygric behavior (Dornieden et al., 2000), moisture transport (Warscheid, 1996a), microbial disintegration (Mitchell, this volume), the enhanced deposition of aerosols and pollutants (Warscheid et al., 1991 [FIG. 3C]), and the modification of crystallization patterns of harmful masonry salts (May et al., this volume) on mineral materials, in order to establish the importance of biodeterioration processes for the conservation effort (Warscheid, 1996b).

The health risks posed by microbial cells and metabolites (e.g., allergenic spores, toxins, microbial volatile compounds, pathogenic microorganisms) are being increasingly noted by conservators and restorers, especially in the field of archaeology and archive conservation. [FIG. 3D] While the potential risks are known, and standardized analytical setups and evaluation guidelines have been formulated (Gabrio, 2002), so far no reliable information exists on the real danger to the employees involved. It would be advantageous to apply modern molecular-biological techniques in order to target even nonculturable, but possibly allergenic, toxic, or even pathogenic microbial strains as well as their hazardous metabolites, since the latter are only detectable by high-resolution analytical techniques (Gabrio, 2002). Nevertheless, those evaluating data should always consider whether the possible microbial endangerment is only theoretically possible or whether it is actually probable under the prevailing climatic conditions (Warscheid, 2000).

EXPOSURE CONDITIONS AND MICROBIAL ACTIVITY Microbial contamination in and on materials is basically determined by the availability of water in its various forms—humidity, moisture, etc. The growth and biodeteriorating activity of microorganisms is furthermore defined by respective material properties (e.g., porosity, roughness, inner surface, absorbency) and the available nutrient supply (e.g., organic material components and aerosols) to microorganisms (Warscheid, 2000).

The water activity around a given object determines the danger threshold for biodeterioration processes. In order to understand the limits of microbial growth and biodeteriorating activity, there is an urgent need to reliably define the moisture or humidity level ranges for the specific environments to which historical objects are exposed. Here lies the key for controlling biodeterioration processes on valuable objects, whether these are wall-paintings, glasses, caverns, archives, excavation sites, and, particularly, restoration laboratories.

Much data exists on the limiting relative humidity for biological growth. However, some of it is unfortunately rendered useless by the lack of care in controlling

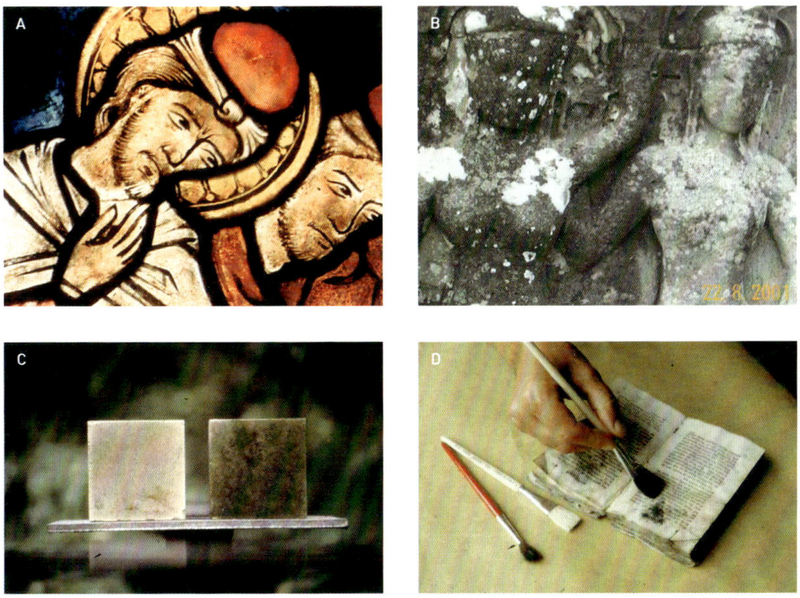

[3] Biodeterioration mechanisms and environmental impacts: a) Color alterations by fungal melanins on medieval glasses ("Cosmas und Damian" in Goslar, Germany); b) Biocorrosion by lichens and fungi on natural stone (Angkor Wat, Cambodia); c) Biofilms as precursor for patina and crust formation on natural stone (Exposure field study in Duisburg, Germany); d) Health-related impacts of fungal spores and mycotoxins on archive objects (Stadtbibliothek Trier, Germany).

the RH and by the failure to distinguish water available in the culture medium from water in the air (Senkspiel et al., 1994). There is also some uncertainty about the effect of ventilation rate over the organism (Padfield, 2002). Nevertheless, it can be agreed upon that the limit for germination is seldom less than 80% RH and that the limit for growth of established microorganisms might be as low as 55% RH. These xerophilic organisms are mainly fungi. Bacteria require much higher RH for survival, although some salt-dependent bacteria actually need a moderately low RH. Many researchers have used culture gels and have thus bypassed the very relevant question of whether water activity alone is a deciding variable. In other words, we need more systematic studies on the effect of water activity as expressed by the RH over an essentially "dry" substrate compared with water activity as defined by a salt or gycerol solution on a petri dish (Padfield, 2002).

There are several ways in which microorganisms gather water in "dry" environments. Some achieve osmotic equilibrium with air at a moderate RH by concentrating salts and soluble organic substances, such as sugars, in their cell fluid; other organisms use active, energy-consuming methods for pumping water through their cell wall. In this context, it is important to stress the fact that a material-specific microflora is usually embedded in a colloidal slime layer, called biofilm. The biofilm protects the microorganisms by balancing out changes in humidity and temperature as well as osmotic changes, as well as pH-relevant influences due to the presence of colloidal polymeric substances (Warscheid, 1996a, 2000).

Science has only begun to study biofilms and their significance in the biodeterioration of cultural property. The chemical nature of these biofilms, where microorganisms are embedded in a polymeric slime, is almost unknown, and their relevance to the growth and survival of the respective microorganisms under given expositional conditions remains unclear. All we know is that on contaminated biofilm-covered materials, probably originating in a single episode of wetness, the microflora resist environmental changes and conservation treatments more easily than on unaltered objects, which they would only be able to attack under quite favorable growth conditions, defined by laboratory data.

In addition, we have to realize that the hygroscopicity of materials themselves will determine the availability of water to the respective microflora and therefore their metabolic activity. Even building materials in an object can support or control the microbial growth depending on how the object is designed and its resulting moisture-retaining capacity. Absorbent walls and ceilings will significantly improve the climatic stability of indoor spaces that are inadequately ventilated. This is an as yet untapped resource for museum control (Padfield, 1999).

In conclusion, we do not really understand the ecology of microbial growth on and in materials well enough, and more attention should be paid to microclimatology, to the material, and to building physics in order to evaluate the potential threat of biodeterioration processes, since these have been studied only for positively wet environments so far.

APPROACHES FOR EFFECTIVE BIODETERIORATION CONTROL In order to control biodeterioration processes on cultural objects, the first step is an entire anamnesis and survey of physical and chemical damage factors to describe and evaluate the importance of microbial impact and to define adequate conservation strategies.

The control of biodeterioration processes will be basically achieved by effective measures to limit and restrict the aforementioned growth conditions for microflora. Therefore, "good housekeeping" and climate control, as well application of appropriate microbe-resistant conservation treatments, is preferable to applying ecotoxicologically questionable and unhealthy biocides (Warscheid and Braams, 2000).

As discussed above, the protection of cultural artifacts against biodeterioration will be primarily achieved by the reduction of moisture in and around the endangered object in the course of drying, drainage, ventilation, and climatized storage. Due to the moisture-conserving effects of microbial biofilms, contaminated artifacts should be kept at humidity levels under 55% RH, whereas noncontaminated objects can even be kept up to 65% RH, depending on the type of material and its sorption isotherm. Further research is also desperately needed in this area.

The growth-controlling effect of desiccation will be intensified by a subsequent soft cleaning (i.e., vacuum cleaner, brushes) and, if necessary, careful disinfection (i.e., medical alcohol eventually combined with conservation agents) of the material surface removing moisture-absorbing dust, attached particles, crusts, or biogenic slimes. [FIGS. 4A–D] In this context, it has to be emphasized that during cleaning the input of additional moisture should be limited as much as possible (e.g., by wrung-out cloths, fine-part dry cleaning); otherwise the cleaning effect will remain only temporarily until the remaining microflora recover in the moist condition (Warscheid et al., 1988).

The long-term effect of these measures will be extended as long as further contamination of the objects by aerosols, particles, and microorganisms can be avoided. This will require the formulation and realization of technical guidelines for "good housekeeping" (i.e., regular cleaning and repair, optimal heating and ventilation). Modifications in the illumination, temperature conditions, pH-range or redox potential of the affected materials and their environment will have only

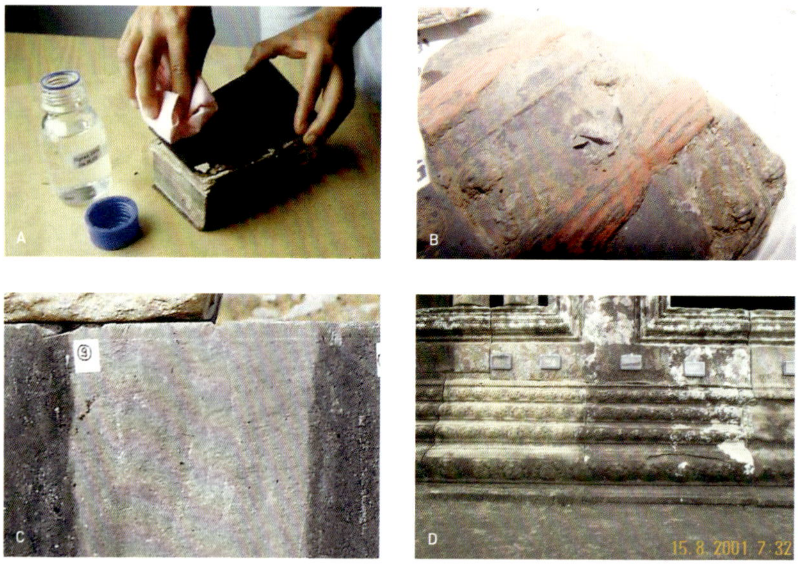

[4] Examples of approaches for effective biodeterioration control: a) Decontamination of an archival object: after vapour cleaning, application of a disinfecting, biostatic solution of parabens (Stadtbibliothek, Trier); b) Due to an effective climate control in the excavation, the application of biocides is mainly restricted during the conservation and fixation treatment of detaching oriental lacquer layers (Terracotta Army, Lintong, China); c) Successful discoloration of fungal contaminated marble in the Mediterranean area by the application of an effective cleaning solution with biostatic properties (Milet, Turkey); d) Long-term control of microbial contamination and biocorrosion processes by a newly designed algicide applied on the left side of the stone test field (Angkor Wat, Cambodia).

slight inhibiting effects against the mostly highly adaptable microflora, not to mention the practical problems of their implementation.

Nevertheless, these practical steps will provide sufficient relief in most cases of biodeterioration of cultural artifacts, even when microbially sensitive materials are present, and the individual steps can be easily integrated into the general conservation work scheme. [FIG. 4B] After all, the importance of regular maintenance and care should not be neglected (Warscheid and Braams, 2000).

Biocidal treatments are advised only when all of the aforementioned measures show limited effect. While cleaning and adequate storage of artifacts, books, and other archival material may or may not prevent biodeterioration processes, archaeological sites probably need short-term conservation by biocides right after excavation in order to avoid microbially induced corrosion during restoration. In contrast, openly exposed monuments, especially in semi-arid or (sub-)tropical climates where biodeteriorating effects are considerably more important, need more intense treatments and long-term care to reduce irretrievable losses of our cultural heritage. [FIGS. 4C, D]

CONCLUSION Cultural objects, whether originating from soil and anoxic excavations, kept in dark and humid environments, or openly exposed to corrosive and nutritive atmospheric pollutants, need, first of all, a profound anamnesis, constant care, and conservative maintenance.

With regard to biodeterioration processes, more detailed information is needed on the real growth conditions and nutrient requirements of the different biodeteriorating microflora: whether they are alive and active, or whether they are simply present without any measurable deteriorating impact. This will allow a clearer determination of the necessity and intensity of cleaning processes, and whether these need to be combined with subsequent disinfection treatments. Over and above that, knowledge of microbial growth patterns will help in formulating guidelines for effective "good housekeeping" and adequate hygiene control and monitoring. In this way, biocide use could be kept to a minimum out of concern for ecological and health implications, while long-term effectiveness of biocides, in cases where they are desperately needed, might be improved in the future by synergistic formulations in specially designed hygienic coatings. Bioremediation techniques in the conservation of materials, such as biocalcification, biodesalination, and enzymatic removal of biofilm and stains, may in the future open a new chapter on the bioprotective—as opposed to the deteriorative—impacts of microorganisms on cultural objects.

REFERENCES

DORNIEDEN, T., GORBUSHINA, A. and KRUMBEIN, W.E. 2000. Patina. In: Ciferri, O., Tiano, P. and Mastromei G. (Eds.), *Of Microbes and Art—The Role of Microbial Communities in the Degradation and Protection of Cultural Heritage.* Kluwer Academic, New York. pp. 105–120.

GABRIO, T. 2002. Schimmelpilze in Innenräumen—Nachweis, Bewertung, Qualitätsmanagement. Bericht der AK "Qualitätssicherung—Schimmelpilze in Innenräumen—" am Landesgesundheitsamt Baden Württemberg, Stuttgart.

KOESTLER, R.J., WARSCHEID, TH., and NIETO, F. 1997. Biodeterioration: Risk Factors and their Management. In: Snethlage, R., and Baer, N.S. (Eds.), *Saving our Cultural Heritage: The Conservation of Historic Stone Structures.* John Wiley and Sons, New York. pp. 25–36.

KRUMBEIN, W.E., GORBUSHINA, A., RUDOLPH, C., URZI, C., and WARSCHEID, TH. 1993. Biological investigations on the question of organic and inorganic eutrophication induced biocorrosion and biogenic deposits on late medieval church windows of the cathedral of Tours and St. Katharina in Oppenheim. In: von Welck, S.F. (Ed.), *Gemeinsames Erbe gemeinsam erhalten—* 1. Statuskolloquium des Deutsch-Französischen Forschungsprogramms für die Erhaltung von Baudenkmälern, Karlsruhe. pp. 269–278.

MAY, E., PAPIDA, S., and ABDULLA, H. 2003. Consequences of Microbe-Biofilm-Salt Interactions for Stone Integrity in Monuments. In: Koestler, R.J., Koestler, V.R., Charola, A.E., and Nieto-Fernandez, F.E. (Eds.), *Art, Biology, and Conservation: Biodeterioration of Works of Art.* The Metropolitan Museum of Art, New York. 452–471.

MCNAMARA, C., PERRY, T.D., ZINN, M., BREUKER, M., MUELLER, R., HERNANDEZ-DUQUE, G., and MITCHELL, R. 2002. Microbial Processes in the Deterioration of Maya Archaeological Buildings in Southern Mexico. In: Koestler, R.J., Koestler, V.R., Charola, A.E., and Nieto-Fernandez, F.E. (Eds.), *Art, Biology, and Conservation: Biodeterioration of Works of Art.* The Metropolitan Museum of Art, New York. 248–265.

PADFIELD, T. 1999. Humidity buffering by absorbent materials in walls. www.natmus.dk/cons/tp/wallbuff.htm.

PADFIELD, T. 2002. The window in context: The influence of the other components of a building on the function of the window. *Progress Report to Energiestyrelsens "Projekt Vindue"*, 10 pp.

SEDLBAUR, K. 2002. Beurteilung von Schimmelpilzbildung auf und in Bauteilen. *Bauphysik* 24 (3), 176–176.

SENKSPIEL, K., OHGKE, H., and BECKERT, J. 1994. Kinetik der Auskeimung von Schimmelpilzsporen auf Baustoffen in Abhängigkeit von deren Gleichgewichts (a_w)-, Material ($:_m$)- und Oberflächen(n_o)-Feuchte. *Gesundheitsingenieur* 115 (2), 77–85.

SOERENSEN, B. and GREGORY, D. 1998. *In situ* preservation of artifacts in Nydam Mose. In: Mourey, W. and Robbiola, L. (Eds.), *Metal* 98—Proceedings of the International Conference on Metal Conservation. James & James (Science Publishers, Ltd.), 94–99.

WARSCHEID, TH., PETERSEN, K., and KRUMBEIN, W. E., 1988. Effects of cleaning on the distribution of microorganisms on rock surfaces. In: Houghton, D.R., Smith, R.N., and Eggins, H.O.W. (Eds.), *Biodeterioration* 7. Elsevier Applied Science, London. 455–460.

WARSCHEID, TH., OELTING, M., and KRUMBEIN, W.E. 1991. Physico-chemical aspects of biodeterioration processes on rocks with special regard to organic pollution. *International Biodeterioration* 28, 37–48.

WARSCHEID, TH. 1996a. Biodeterioration of Stones: Analysis, Quantification and Evaluation. In: *Proceedings of the 10th International Biodeterioration and Biodegradation Symposium* (Dechema-Monograph No. 133, Frankfurt). 115–120.

WARSCHEID, TH. 1996b. Impacts of microbial biofilms in the deterioration of inorganic building materials and their relevance for the conservation practice. *Internationale Zeitschrift für Bauinstandsetzen* 2 (6), 493–504.

WARSCHEID, TH. 2000. Integrated Concepts for the Protection of Cultural Artifacts against Biodeterioration. In: Ciferri, O., Tiano, P., and Mastromei G. (Eds.), *Of Microbes and Art—The role of Microbial Communities in the Degradation and Protection of Cultural Heritage.* Kluwer Academic, New York. 185–202.

WARSCHEID, TH. and BRAAMS, J. 2000. Biodeterioration of Stone: A review. *International Biodeterioration and Biodegradation* 46, 343–368.

WARSCHEID, TH. and RUDOLPH, C. 2002. Microbiological examinations for the preservation of Chinese Polychromy from the Qin and Han Dynasty. In: Blänsdorf, C., Emmerling, E., and Petzet, M. (Eds.), *Qin Shihuang—The Terracotta Army of the First Chinese Emperor.* Arbeitshefte des Bayerischen Landesamtes für Denkmalpflege, München. pp. 702–724.

A.M. JONES
R. MOUZOURAS
A.J. PITMAN
S.B. POINTING

PRESERVING THE TIMBERS OF THE TUDOR WARSHIP MARY ROSE

The Mary Rose Trust, HM Naval Base, Portsmouth, UK

ABSTRACT The Tudor warship Mary Rose was built in Portsmouth, England, between 1509 and 1511 and served in Henry VIII's navy until she sank on July 19, 1545. Over the many years that followed, the hull and her contents were covered by a deposition of sediments. Within this environment, the hull and objects remained protected from further deterioration. Upon recovery from the protective sediments and re-exposure to aerobic conditions, chemical, physical, and biological degradation ensued. Experiments were undertaken to evaluate the effectiveness of passive conservation methods with the view of controlling the activity of wood decay organisms in hull timbers. Storage in CO_2 atmosphere, cold storage, reburial, and gamma irradiation were evaluated for their effectiveness in controlling the bacterial, fungal, and insect activity in Mary Rose archaeological timbers. Of these methods, cold storage was shown to reduce microbial activity, although soft-rot fungi were identified growing on hull timbers. The suitability of reburial within anoxic sediments was assessed. Physicochemical conditions were characterized by a near neutral pH, low Eh, and absence of dissolved oxygen. However, cellulolytic activity in recent and Tudor sediments was demonstrated. Gamma irradiation was found to be a highly effective passive conservation method against bacteria, fungi, and insects. A dose of 15 KGy is required for inactivation of these organisms. At doses in excess of 100 KGy radiolytic damage resulted in increased hygroscopicity, increased cracking and warping, loss of surface texture, reduction in compression and bending strength, and chemical alteration and degradation of cell wall components. This work concluded that cold temperature and gamma irradiation are suitable methods of storing waterlogged archaeological wood for long periods.

Following long-term storage of the hull (1982–1994), an active conservation technique involving a two-stage polyethylene glycol treatment method has been employed to retain the structural integrity of the waterlogged hull timbers. Active conservation of the hull commenced in 1994 and will be completed by 2008. Controlled air-drying will then dry the hull timbers.

INTRODUCTION The decision to proceed with the recovery of material from a wreck site must be made with the full realization of the implications involved in the storage and treatment of waterlogged material (Grattan and Clarke, 1987). Many problems and loss of potential archaeological information have occurred to recovered waterlogged wood owing to inadequate and unsuitable storage conditions.

The storage of archaeological wood should be discussed at the planning stage with input from a range of specialists including conservators, wood technologists, microbiologists, and chemists. Advice from these people may prove useful for determining whether a storage method chosen prevents or minimizes further deterioration.

Ideally, storage facilities must be established and ready for use before the excavation begins. As previously discussed, waterlogged archaeological wood will deteriorate immediately on exposure to air. Therefore safe and immediate transfer of recovered archaeological wood into an ideal storage environment is essential. Archaeological wood recovered from the Mary Rose site was placed into temporary storage soon after reaching the surface.

Once the wood is in storage, it is essential that ideal storage conditions be maintained. During a passive storage phase of variable duration, information relating to post-excavation recording, cleaning, and reconstruction work can be carried out with relative ease. In addition, information regarding the physical and chemical properties of recovered archaeological material is usually determined during the passive holding phase. Ideally, any conservation method adopted to treat an artifact should be determined according to the condition and nature of the material to be treated (Christensen, 1970; van der Heide, 1972; Florian et al., 1978). Therefore, it is essential that archaeological wood be maintained in ideal storage environments until such determinations can be made. Equally important, financial support for active conservation may not be available during excavation and recovery periods. Long or extended storage is inevitable in cases where large wrecks and their contents are recovered (Rule, 1983). If suitable storage facilities are not adequately supported, then the recovery of wreck material is hard to justify. For the Mary Rose project, suitable storage facilities were set up prior to the recovery of archaeological material.

PASSIVE STORAGE The fundamental problem if waterlogged archaeological wood is allowed to dry in an uncontrolled manner is that it may collapse, shrink, distort, split, and even, in severe cases, disintegrate completely (Grattan, 1987). The situation becomes a particular problem if the wood is stored under unfavorable conditions. This post-excavation damage will continue unless suitable storage environments are maintained. In an ideal world, it is probably worth aiming for a holding period of less than a year (English Heritage Guidelines, 1995). This is not always possible when dealing with a well-preserved marine wreck site, which can produce many tons of waterlogged wood. In such cases, passive holding periods of up to 20 years are not uncommon.

For the storage of the Mary Rose Hull and individual timber elements, the Mary Rose Trust adopted a number of methods. Three basic requirements for the ideal storage of waterlogged wood were suggested by a panel of experts and proved useful as a guide to good practice. The requirements were the ability to:

- Keep the wood at maximum moisture content;
- Exclude light to reduce the growth of algae; and
- Prevent or minimize the activity of bacteria, fungi, and insects.

These requirements were met in a number of ways:

COLD WATER SPRAYING OF THE MARY ROSE HULL Problems arise when complete or partially complete large wooden structures have to be held passively without suffering serious drying. Tanking large complete or incomplete hulls is expensive and often impractical. In such instances, direct spraying of large waterlogged structures is employed. The spraying of ships and other large structures with fresh water has an added advantage in that the mud, silt and salt deposits are washed out. The purpose of passive holding of the Mary Rose hull, as with any waterlogged or decayed artifact, is to maintain water saturation and prevent further deterioration. It also provided an opportunity to determine the active conservation method best suited to treat and conserve the hull of the Mary Rose. A major problem with the passive storage of the Mary Rose hull is one of size. Many of the storage methods described in the literature (Van der Heide, 1972; Barkman, 1975; Baynes-Cope, 1975; Dawson et al., 1982; Jespersen, 1985) have been developed for the storage of small finds, and most are not capable of expansion to a larger scale. The only suitable method to passively hold the hull of the Mary Rose is by water spraying. A research program was then set up to compare a warm (20°c) and cold water (5°c) spraying regime using original ship timbers. The results were predictable, showing significant differences in bacterial activity between warm water and cold water spraying regimes. Maintaining a water spray temperature of below 5°c substantially reduced the degree of microbial activity within hull timbers.

Based upon these findings, a low-temperature water-spraying regime was chosen to passively store the hull of the Mary Rose. The reasons for this were as follows:

- Water saturation of the hull timbers would be maintained by the constant flow of water;
- Temperature of the spray water remained between 2 and 5°c. At these temperatures, microbial activity is well below the optimum growth temperature of many marine and terrestrial fungi and bacteria that cause wood decay in archaeological timbers;
- The method is safe and nontoxic; and
- The method is compatible with future conservation treatment processes.

From October 1982 to September 1994, the Mary Rose hull was sprayed continuously with cold water. [FIG. 1] During this long period of passive storage, post-excavation recording, cleaning, and reconstruction work of the hull was completed. In addition to these important tasks, the passive holding period provided the Trust's conservators with an opportunity to investigate the nature and condition of the hull timbers.

Throughout the period of passive holding, the hull was displayed to the general public in a specially constructed building within Dry Dock Number 3, HM Naval Base, Portsmouth. The building itself protects the hull from the external environment by providing appropriate thermal and vapor insulation properties. The fabric of the building consists of a double layer of Treveira (polyester cloth covered with PVC). Inside this building a spray system ensured that a film of recycled, refrigerated, filtered freshwater (water temperature of 2–5°c) covered the entire hull.

The spray system operated continuously during nonworking hours, being turned off only for a maximum of four 1-hour periods each weekday to allow staff time to reconstruct the various decks. The water spray system consists of an interconnecting system of various-sized PVC pipes with full jet spray nozzles positioned at 1-m intervals. Filtered, chilled freshwater was furnished to each nozzle,

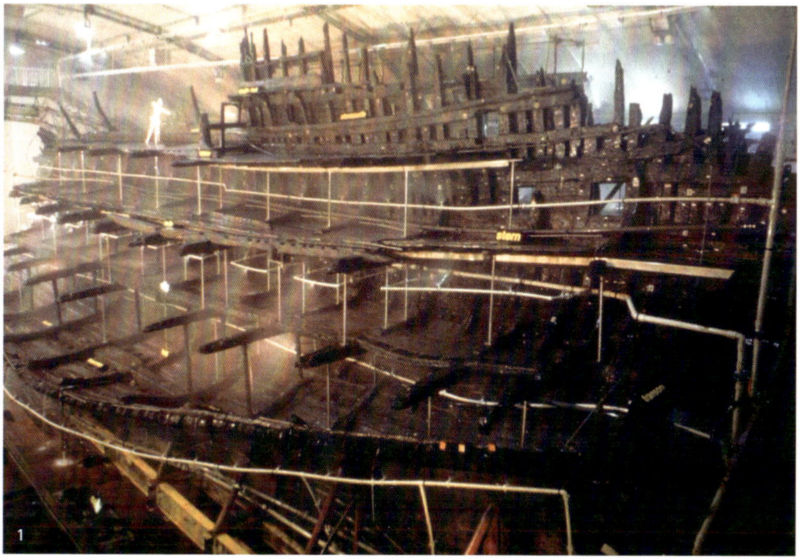

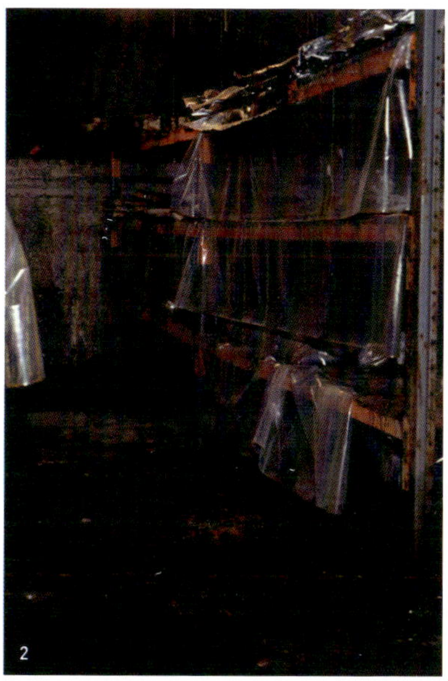

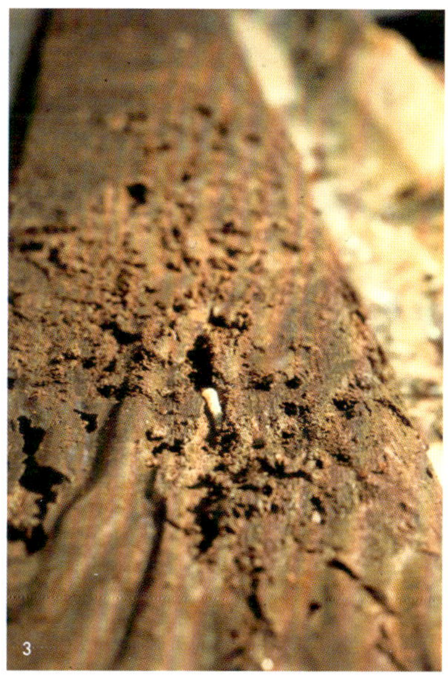

[1] Passive storage of the Mary Rose hull. [2] Wrapped timbers stored on shelves within the Block Mill cellars. [3] Larval tunnels at the surface of stored archaeological timber. Note exposed larval stage within tunnel.

where it was atomized into small droplets (250-micron diameters). In total, 567 nozzles were positioned around the hull, ensuring complete coverage.

Reducing the temperature of the spray water was achieved by passing filtered water through four industrial-sized refrigeration units. These units reduced the temperature of the water to between 1 and 3°C. Two submersible pumps situated in the water reservoir pumped cold water to the spray nozzles. Returning spray water was collected in a catchment tank situated beneath the hull of the Mary Rose. The water then passed into the settlement tank, where it was filtered and pumped into the spray water reservoir. Re-chilling occurred and the filtered, cold water was once again furnished to the spray system.

EXPERIMENTAL After preliminary studies, a standard monitoring procedure was adopted by Trust scientists and subsequently used throughout the passive holding period. This is summarized here:

Environmental Data Ship-hall environment and spray water temperature were monitored continuously using a multi-channel data logger. External temperatures were also recorded.

Chemical Analysis of Spray Water Total chemical analysis of spray water and raw mains water was carried out on a monthly basis to monitor diffusion of sea salts from the ships' timbers.

Monitoring of Bacterial Populations in the Spray Water Weekly measurements were taken to quantify the number of colony forming units per millilliter (CFU/ml) of spray water. This was achieved by the use of Tillomed nutrient agar dipslides to measure total aerobic bacteria. The dipslide tongue, which is coated with a layer of nutrient agar, is removed from its bottle and immersed into the spray water for about 10 seconds so that the agar surfaces are covered. The slide is then removed and allowed to drain for 2 seconds and replaced into its bottle. Dipslides were then incubated at 20°C for 2–7 days. Growth of bacterial colonies was then compared with charts provided to estimate level of contamination (in CFU; 10^3—very slight growth; 10^4—slight growth; 10^5—moderate growth; 10^6—heavy growth; 10^7—very heavy growth).

Macroscopic Examination of Hull Timbers The hull timbers were examined macroscopically for the following features:

- Presence of slime
- Presence of fungal sporulating structures

This was achieved by direct examination of the timber surfaces with the aid of an illuminated magnifier. In addition, slivers of wood bearing fungal sporulating structures were removed by means of a razor blade and examined under a light microscope. To isolate fungi that did not sporulate on the ships' timbers, core samples taken with an increment borer were inoculated onto antibiotic containing seawater cornmeal agar and seawater agar plates. These plates were then incubated at 20°c and examined regularly for fungal growth. Techniques and procedures for the examination and identification of marine fungi are those described by Kohlmeyer and Kohlmeyer (1979).

Monitoring Microbial Activity Within Hull Timbers Catalase is an enzyme common to all microbes and can be used to determine a microbial activity through timbers from surface layers to the inner sound core. Catalase activity (Line, 1983) expressed as dissolved oxygen in parts per million was recorded every 10 seconds over a period of 1 minute. The amount of dissolved oxygen after 1 minute was then used as a measure of catalase (microbial) activity within hull timbers.

GENERAL OBSERVATIONS Throughout the 12 years of passive holding the water temperature was maintained between 2 and 5°c. During this period the system functioned well, although some localized biocide (polybor) treatment of hull timbers was necessary to control microbial slimes (bacteria, cyanobacteria, diatoms, and green algae) and fungal fruiting structures. Both were found to be symptoms of poor water flow over certain regions of the hull. Accelerating flow rate of the spray water prevented surface fouling by microorganisms. However, excessive flow can result in the removal of the decayed outer surface of timbers to a depth of a few millimeters. Such destruction of important surface layers is unacceptable since they contain unique information, e.g., details of fabrication marks. Further evidence that suggested the success of the cold water system relates to data obtained from microbiological analysis of the spray water. Maintaining spray water temperature below 5°c substantially reduced the degree of microbial growth. Although microorganisms can survive low temperatures (Atlas and Bartha, 1987), their metabolic activities and growth rates decline below the optimal temperature. Differences in the number of viable aerobic bacteria for spray water temperatures of below 5 and at 20°c are 10^3 and 10^8, respectively. This reduction in viable bacterial numbers was clear evidence that a cold water spraying regime constituted an effective passive holding system. The optimal growth for bacterial and fungal isolates found inhabiting hull timbers was found to occur between 20 and 30°c. This supports the view that suppression of microbial activity is temperature dependent. Although bacteria and marine fungi [TABLE 1] were found to grow superficially on hull timbers, little active decay of the wood was observed by microscopic examination of wood samples removed from hull timbers. Bacterial and fungal attack of the wood cell walls was only observed in a few cases.

Throughout the passive holding period, there was a real risk of bacterial infection to staff working in the ship hall. Many common aquatic bacteria are able to colonize manmade water systems, and poorly maintained systems provide ideal sites for their growth. Bacteria such as *Legionella pneumophila* were considered a threat. The risk of Legionnaire's disease from the Mary Rose spray system was controlled by maintaining low water temperatures, maintaining a clean spray system, carefully monitoring microbiological populations, and selective use of biocides to reduce the growth of biofilms (slime). Adhering to these simple guidelines removed the risk of infection to both public and staff.

WRAPPING ARTIFACTS AND INDIVIDUAL TIMBERS IN POLYETHYLENE BAGS The majority of Mary Rose ships' timbers were carefully packaged in polyethylene bags of various sizes. To create suitable bags 500-gauge polyethylene sheeting was used, and all edges were hermetically sealed. Timbers were often kept within buildings [FIG. 2] that had no temperature control. To control bacterial and fungal activity, artifacts and timbers were pre-treated with biocides and then wrapped in polyethylene bags. All individual timbers were kept in the dark. Random sampling of polyethylene-wrapped objects was undertaken every three months to assess stability. Polyethylene acts as a good barrier to water movement, provided that the bags are well sealed. Temperature values within the wet storage sites were more dependent on the external environments. Temperatures ranged from 5°c, in the winter, to 21°c, in the summer. The polyethylene wrapping did not act as a temperature barrier, and thus the temperature within the wrapping was very close to that of the storage site.

Maintaining the waterlogged state of timbers is an important criterion and should not be underestimated. Desiccation of wood leads to collapse and distortion, which is normally an irreversible process. If the wood is maintained in a waterlogged condition, the growth of wood-rotting fungi may be restricted since water saturation will limit the amount of available oxygen, which is required by the fungi

BACTERIA	FUNGI
Aeromonas sp.	*Neriospora cristata*
Alcaligenes sp.	*Lulworthia sp.*
Pseudomonas sp.	*Humicola alopallonella*
Vibrio sp.	*Monodictys pelagica*
	Cirrenalia macrocephala
	Trichocladium sp.

[TABLE 1] Microbial isolates recorded during the passive holding period

SPECIES	DECAY TYPE
Ascomycotina	
Ceriospora halima	Soft rot (type I)
Corollospora maritima	Soft rot (type I)
Lulworthia sp.	Soft rot (type I)
Neriospora cristata	Soft rot (type I)
Zopfiella sp.	Unknown
Deuteromycotina	
Cirrenalia macrocephala	Soft rot (type I)
Dictyosporium toruloides	Unknown
Gliomastix murorum	Unknown
Humicola alopallonella	Soft rot (type I)
Monodictys pelagical	Soft rot (type I)
*Penicillium sp.**	Staining
*Stachybotrys atra**	Staining
*Trichocladium achrasporum**	Soft rot (type I)
Zalerion maritimum	Soft rot (type I)
Zalerion varium	Soft rot (type I)
Basidiomycotina	
Digitatispora maritima	White rot
Nia vibrissa	White rot

* Terrestrial species

[TABLE 2] Fungal species observed on the surface of stored timbers (Mourzouras, 1987)

(Scheffer, 1973). However, low oxygen tensions may not totally prevent the growth of certain fungi, especially soft-rot fungi (Cartwright and Findlay, 1958; Duncan, 1961; Scheffer, 1973). These fungi are better adapted to decay wood in conditions of low oxygen tensions, since they preferentially degrade cellulose through a non-oxygen process (Levi and Preston, 1965).

Examination of stored Mary Rose timbers for fungal activity has been demonstrated by Mouzouras (1987). Fruiting structures of 17 fungi were found on the surface of timbers. Three of these were terrestrial and 14 were marine fungi. [TABLE 2] The most frequent species recorded was *Monodictys pelagica*, which was recorded on 23% of the timbers examined.

Despite the large number of fungi found fruiting on the surface of timbers, few signs of recent attack were observed (Mouzouras, 1987). Soft-rot cavities were seen, but were not numerous. Such decay patterns were only observed after extensive examination of a large sample of timbers, indicating that fungal attack of stored polyethylene-wrapped timbers was very localized. The absence of brown-rot fungi and the isolation of two white-rot fungi were not surprising. This can be attributed to the vulnerability of their enzymes to dilution in the marine environment. As a result, brown-rot fungi were not found in stored Mary Rose artifacts and timbers. The marine white-rot fungi *Digitatispora marina* and *Nia vibrissa* were found to be inefficient degraders of stored timbers.

Bacterial decay of stored wooden artifacts and ships' timbers was more frequently observed. Once again this form of decay was also very localized. Erosion and tunnelling decay patterns were observed in a number of timbers examined. The erosion troughs were observed below the decayed layers of timbers, while tunnelling or burrowing bacteria were observed closer to the outer cell layers of the timber. Tunnelling is restricted to aerobic bacteria and is usually a characteristic of an advanced stage of decay. If, as suggested by Mouzouras (1987), anaerobic bacteria can only cause erosion of the wood cell wall, this indicates that the core of waterlogged archaeological timbers may be anaerobic. A low oxygen tension, which exists in the core of Mary Rose timbers, may also explain the absence of fungi deeper in the wood.

The possible explanation for the lack of active decay in polyethylene-wrapped stored timbers and artifacts may be the unfavorable conditions existing in timbers encapsulated in polyethylene sheeting. Oxygen levels within the polyethylene bags are very low, with little gaseous exchange across the polyethylene barrier. Also, with little or no water vapor exchange between the polyethylene sheets, a buildup of toxic substances and secondary metabolites takes place, thereby inhibiting the growth of fungi and aerobic bacteria. The combined action of low oxygen and the presence of these substances may explain the lack of recent decay by aerobic microorganisms in stored timbers.

Biocides have been employed in attempts to control microbial growth (Jones et al., 1986). However, their limited efficacy over time makes re-treatment necessary. This is often time consuming and creates future problems for conservators in handling the wood, and in the safe disposal of the biocide solution. It must also be noted that biocides may interfere with radiocarbon dating and future active conservation treatment processes.

Cold storage of polyethylene-wrapped Mary Rose artifacts and timbers revealed that the activity of wood-degrading microorganisms is arrested. However, viability of the mixed microbial flora (bacteria and fungi) supported by a range of wrapped timbers was not affected by cold storage. The threat of renewed biodeterioration under more favorable conditions therefore exists. Cold storage and biocide treatments are therefore only suitable for the short-term storage of wrapped timbers where the risk of biodeterioration is considered low.

The fruiting structures of fungi present on the wood surface must be considered a health hazard to the handlers of stored timbers. The fungus *Stachybotrys atra* was found to be present on a number of stored timbers. This fungus causes a mycotoxicosis known as stachy-botryotoxicosis (Eppley and Bailey, 1973), which can be fatal in humans. It must be stressed that archaeological timbers in storage should be carefully monitored for microbial growths and that control of these organisms should be undertaken.

A number of Mary Rose timbers have also been stored in a carbon dioxide atmosphere. This method was ineffective in preventing microbial wood decay activity, or in reducing the viability of the mixed microbial flora found growing within stored timbers. During a study to evaluate the efficacy of this method, the anaerobic environment created was found to support the growth of anaerobic human pathogens, presenting a serious health risk to conservators.

During the study of stored polyethylene-wrapped wet wood it became apparent that this method was efficient in preventing desiccation but ineffective in preventing microbial decay activity or viability of the mixed microflora identified. Cold storage and biocide treatments did, however, help reduce microbial activity of stored timbers.

More recently another type of wood decay organism was found living in stored timbers wrapped in polyethylene sheeting. A survey of 1568 stored timbers revealed that the larvae of the wharf-borer beetle, *Nacerdes melanura*, infested 2% of these timbers. Attack of the timbers is not always apparent from examination of

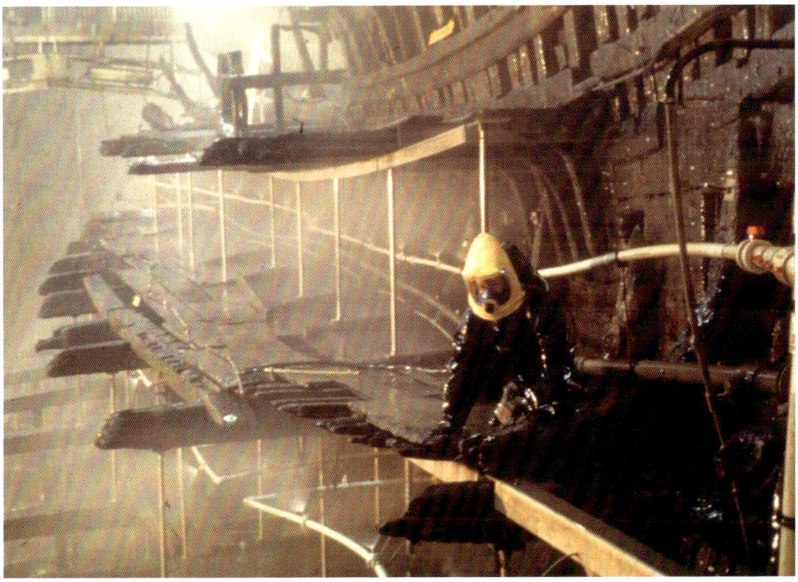

[4] Spraying of the Mary Rose hull with a solution of PEG 200.

their outer surfaces. The presence of exit holes or small piles of frass was often the only visible indication that the timbers were infested.

Damage caused by the wharf-borer larvae is often extensive. They produce tunnels throughout the softer decayed regions of timbers. The tunnels vary in size depending on the size of the larva, and tunnelling was found on both upper and lower surfaces. The extent of damage within stored timbers varied, and depended upon the degree of larval infestation. Timbers with high larval numbers (18–20) showed extensive attack and destruction. This attack was restricted to the outer surfaces of the timbers to a depth of 10 mm. Timbers attacked included oak, poplar, and pine, indicating that the larvae of *Nacerdes melanura* [FIG. 3] were capable of infesting both softwoods and hardwoods. The infested timbers surveyed had a moisture content ranging between 131 and 670%.

The survey of stored timbers also indicated that the polyethylene sheeting used to wrap Mary Rose timbers could be penetrated by emerging adults, and that seals on some packaging were inadequate. Studies involving different types of packaging material indicated a need to replace polyethylene sheeting with a more suitable material. Technical data presented in Table 3 show the barrier properties of a number of wrapping materials tested. [TABLE 3] Data show that all materials tested were more impermeable to oxygen and moisture than the polyethylene originally used to wrap stored timbers.

The material chosen to replace polyethylene was a barrier foil, T1-19v. This material possesses low oxygen and water transmission properties, making it a suitable candidate for the storage of archaeological timbers. Lower oxygen tensions that existed within T1-19v-wrapped timbers reduced microbial activity and prevented further infestation by the wharf-borer beetle. T1-19v is now used to wrap all stored archaeological material.

STORAGE OF LARGE INDIVIDUAL TIMBERS IN WATER TANKS Large individual ships' timbers were stored in large tanks containing water. [FIG. 4] The size of individual objects dictated the type of water tanks used. Plastic tanks with lids to exclude light were often used for the storage of smaller material. The advantage of this type of container is that it is stackable and can be used where space is an issue.

For the many large wooden objects and timbers recovered from the Mary Rose, purpose-built tanks were constructed from a variety of materials such as polypropylene and polyethylene and supported by a framework of steel tubing. These tanks were covered with black polyethylene to prevent light penetration. Tank water was changed on a regular basis to prevent heavy buildup of slime.

WRAPPING MATERIAL	OXYGEN TRANSMISSION RATE*	WATER VAPOR TRANSMISSION RATE**
60 Surlyn/100I Surlyn	9.5	1.1
70 Nylon/125I Surlyn	9.5	0.9
T1-19v	<0.01	<0.05
T1-25x	<0.01	<0.05
Polyethylene (500, gauge)	4.0	1500

* In $cm^3/m^2/24$ h
** In $g/m^2/24$ h

[TABLE 3] Properties of wrapping material

ORGANISM	SOURCE	MINIMUM LETHAL DOSE (KGy)
Aerobic bacteria	Mary Rose timbers	2.4
Anaerobic bacteria	National collection of bacteria	2.5
Invertebrates	Mary Rose timbers	3.2
Marine fungi	Mary Rose timbers	15.0
Terrestrial fungi	Mary Rose timbers	13.0

[TABLE 4] Radiosenstivity of wood decay organisms

All tanks, apart from one extremely large one, were kept inside a building to prevent the occurrence of temperature extremes. Freezing of waterlogged wood will cause damage, and high temperatures will accelerate the decay process. Although this method proved successful in preventing loss of moisture from the timbers, growth of slime and wood-degrading organisms was not arrested. From time to time, it was necessary to control the growth of microorganisms within storage tanks. Biocontrol using snails (*Physa sp.*) proved most effective in controlling bacterial numbers, while ampholytic surfactants and ultraviolet light performed relatively poorly.

For a limited time after recovery, the Trust stored a large number of individual ships' timbers in a local pond. The pond was found to support large numbers of the common pond snail, which was well suited for the removal of bacterial slime found on the timber surface. No grazing damage to the timber surface was observed and there was little evidence of surface fouling.

Like most projects, the temporary storage of wood after excavation all too frequently turns into longer-term storage while conservation issues such as funding are debated or deferred. Extending passive storage indefinitely may lead to the degradation of the wood over a period of several years until the wood is in too poor a condition for conservation. To prevent this situation from developing, Mary Rose conservators have been involved in the search for an improved passive holding method, one that fulfills all the criteria laid down by Dawson et al. (1982).

GAMMA RADIATION STERILIZATION OF WATERLOGGED ARCHAEOLOGICAL WOOD The biodeterioration of irreplaceable wooden timbers has been demonstrated in the various methods employed by the Mary Rose Trust, and the need for a passive conservation method that will deter biodeteriogenic organisms has resulted in the use of gamma irradiation to control bacteria, fungi, and insects living in stored archaeological wet wood. Research by Pointing (1995) has shown that gamma irradiation is a superior alternative to cold and wet storage. This method was evaluated by Pointing (1995) as a novel storage treatment and has been adopted by the Trust since 1998.

Screening a representative range of organisms [TABLE 4] revealed a dose of 15 KGy is required for inactivation. Gamma irradiation may also be desirable as a treatment for wood excavated from polluted archaeological sites. Applying U.K. medical sterilization standards of 25 KGy (Anon., 1994) would ensure inactivation of any human pathogens in addition to wood-decay organisms.

A dose level of 15–25 KGy had no adverse effects on the chemical and physical properties of waterlogged archaeological wood (Pointing, 1995). However, at dose levels in excess of 100 KGy radiolytic damage was observed, resulting in increased hygroscopicity, increased warping and splitting during drying, loss of surface texture, reductions in compressive and bending strength, and chemical alteration and degradation of wood cell wall components (Pointing, 1995; Pointing et al., 1996). At very high dose levels, losses in cellulose crystallinity are observed, with alkane and cycloalkane breakdown products formed. Lignin damage was found to be restricted to altered linkages.

ADVICE TO CONSERVATORS A range of wrapped waterlogged timbers from the Mary Rose hull supporting a mixed biodeteriogenic population have been successfully treated using gamma irradiation. More than three years after their treatment the timbers have remained sterile, with no evidence of microbial or insect activity. This level of efficacy cannot be achieved by other passive holding techniques (wrapping, cold storage, use of biocides, UV sterilization, and water storage at low temperature).

Gamma irradiation is suitable for the treatment of both slightly and heavily degraded waterlogged wood. An adsorbed dose of 100 KGy is acceptable for any one timber, and this dose level should not be exceeded in single or repeated treatments since higher irradiation doses may adversely affect the physical and chemical structure of the wood. An additional drawback of this process is that gamma irradiated timbers are unsuitable for radio carbon dating and any analytical process invalidated by exposure to radiation.

Gamma irradiation treatment involves wrapping timbers and transporting them to an irradiation facility. Exceptionally large objects that cannot be transported easily by hand will be unsuitable for this treatment. Smaller objects such as artifacts and individual ships' timbers were sterilized by this technique. For Mary Rose timbers inactivation of wood biodeteriogens is achieved by a single dose of 20 KGy for slightly degraded timbers not exceeding 75 mm in thickness and for heavily degraded timbers up to 150 mm in thickness. For much larger timbers a dose must be recalculated based on size and density of the timber, but the dose must not exceed 100 KGy.

Timbers selected for gamma irradiation are heat-sealed in a radiation-stable wrapping material (volumetric barrier foil T1-19v). Further wrapping with a protective material such as "bubble pack" is advised to avoid damage during transport. The wrapped object should then be clearly labelled. Wrapping, transportation, and treatment should always be supervised by a conservator to minimize the risk of accidental damage.

Following irradiation treatment, it is important to maintain the integrity of the wrapping material surrounding the sterile object. Any breach will allow contamination of the sterile environment within the wrapping. Conservators at the Trust periodically check wrapping material for damage. No specialized storage environment (cold storage) is required. Once the timber is unwrapped, timber sterility will no longer be maintained. Gamma irradiated timbers and artifacts can help reduce biological activity during active conservation.

One of the advantages of gamma irradiation treatment is that there is no residual radioactivity in the treated object, which is therefore completely safe for handling by conservators and other specialists. This presents a clear advantage over biocide treated objects, which often present toxicity problems to conservators. Gamma irradiation is also a low-cost alternative to cold storage, with far greater efficacy in preventing ongoing wood decay. Finally, gamma irradiation may be compared favorably to the criteria for an ideal passive holding method as laid out by Dawson et al. (1982). [TABLE 5]

CRITERIA FOR IDEAL STORAGE*	EVALUATION OF GAMMA IRRADIATION
Prevent or minimize further decay	Complete sterility of artifacts and ships' timbers
Compatibility with future conservation treatments	At recommended dose levels (below 100 KGy), effects unlikely but require further evaluation
Ease of maintenance	No specialized environment needed, but wrapping material must remain intact
Cost-effectiveness	Initial costs involve wrapping and irradiation, but low storage and maintenance costs
Health and safety issues	Gamma irradiation does not result in any residual activity in the exposed wood

[TABLE 5] Gamma irradiation 'an ideal passive holding method.' *After Dawson et al., 1982.

ACTIVE CONSERVATION To conserve the Mary Rose, a two-step polyethylene glycol (PEG) treatment program is currently underway to stabilize the size and shape of hull timbers. Phase I, incorporating the PEG 200 spraying program, is designed to prevent cell wall shrinkage in the wood cells within the sound core of hull timbers. PEG 200 is utilized to replace water in the cell walls, keeping the wood cells swollen during the drying process and dramatically reducing shrinkage. This first phase began in 1994 and is due to be completed in 2003.

Following the completion of Phase I, the hull timbers will be sprayed with a high-grade PEG (possibly PEG 4000 or PEG 1500). This grade of PEG will only penetrate the outer surface of hull timbers, where degraded cell wall layers will allow it to pass through. Once drying of the hull commences, this high-grade PEG will actually solidify in the outer timber layer, providing mechanical support to the fragile wall remnants. Because PEG 4000 or PEG 1500 solutions are very viscous, higher temperatures will need to be maintained within the ship hall environment during Phase II (30–35°c), so that the high-grade PEG solution can be more easily sprayed. An entirely new spray system will be needed for this, the design of which is currently under discussion. It is anticipated that Phase II will be completed by 2008 and the spraying system turned off.

On completion of Phase II, the PEG-treated timbers of the *Mary Rose* hull will be dried (Phase III). An important research program is currently underway aimed at identifying the best temperature and humidity conditions to dry the hull. Results will be used in the design of an air conditioning system that will help dry the PEG-treated timbers of the *Mary Rose*. This system will be fully adjustable between 50 and 95% relative humidity. The drying program will take approximately 4 years.

ACKNOWLEDGMENTS The authors wish to dedicate this paper to Dr. S.T. Moss, of the University of Portsmouth. It is also a pleasure to acknowledge financial support from the Heritage Lottery Fund, U.K., and The Metropolitan Museum of Art, New York.

REFERENCES

ANON., 1994. A guide to designing for radiation sterilization. Isotron plc, 13 pp.

ATLAS, R.M., and BARTHA, R., 1987. *Microbial Ecology, Fundamentals and Applications*, 2nd Editions, B. Cumins Publishers, 533 pp.

BARKMAN, L., 1975. The preservation of the Vasa. In: Oddy, W.A., (Ed.), *Problems of the conservation of waterlogged wood. Maritime monographs and reports* No.16, National Maritime Museum, Greenwich, pp. 65–106.

BAYNES-COPE, A.P., 1975. Fungicides and the preservation of waterlogged wood. In: Oddy, W.A., (Ed.), *Problems of the conservation of waterlogged wood. Maritime monographs and reports* No.16, National Maritime Museum, Greenwich, pp. 31–33.

CARTWRIGHT, K.ST.G., and FINDLEY, W.P.K., 1958. *Decay of timber and its preservation*, H.M.S.O., London, 332 pp.

CHRISTENSEN, B.B., 1970. The conservation of Waterlogged Wood, In: *Conservation of waterlogged wood in the National Museum of Denmark*, Copenhagen, p. 98.

DAWSON, J.E, RAVINDRA, R., and LAFONTAINE, R.H. 1982. A review of storage methods for waterlogged wood. In: Grattan, D.W., and McCawley, J.C., (Eds.), *Proceedings of the ICOM waterlogged wood working group conference*, ICOM, Ottawa, pp. 227–235.

DUNCAN, G.C., 1961. Relative aeration requirements by soft rot basidiomycete wood-destroying fungi, U.S. Forest Service., *Forest Products Laboratory Report* No. , p. 2218.

English Heritage Guidelines on the recording, sampling, conservation and curation of waterlogged wood, 1995. pp. 1–27.

EPPLEY, R.M., and BAILEY, W.J., 1973. 12, 13-epoxy-trichothecenes as the probable mycotoxins responsible for stachybotryosis, *Science*, 181, 758–760.

FLORIAN, M.L.E., SECCOMBE-HETT, C.E., and MCCAWLEY, J.C., 1978. The physical, chemical and morphological condition of marine archaeological wood should dictate the conservation process. In: *Papers from the First Southern Hemisphere Conference on Maritime Archaeology*, Oceans Society of Australia, Melbourne, pp. 128–144.

GRATTAN, D.W., 1987. Conservation of waterlogged wood. In: Pearson, C., (Ed.), *Conservation of Marine Archaeological Objects*, Butterworths, pp. 164–200.

GRATTAN, D.W., and CLARKE, R.W., 1987. Conservation of waterlogged wood, In: Pearson, C., (Ed.), *Conservation of marine archaeological Objects*, Butterworths, pp. 200–206.

JESPERSEN, K., 1985. Extended storage of waterlogged wood in nature. In: Grattan, D.W., and McCawley, J.C. (Eds.), *Proceedings of the ICOM waterlogged wood working group conference*, ICOM, Ottowa, pp. 39–54.

JONES, A.M., RULE, M.H., and JONES, E.G.B., 1986. Conservation of timbers of the Tudor ship Mary Rose. In: Houghton, B.S., Llewellyn, G.G., and O' Rear, C. (Eds.), *Biodeterioration* 6, The Biodeterioration Society, London, pp. 354–362.

KOHLMEYER, J. and KOHLMEYER, E., 1979. *Marine Mycology: The Higher Fungi*, Academic Press, London, 690 pp.

LEVI, M.P., and PRESTON, R.D., 1965. A chemical and microscopic examination of the action of the soft rot fungus *Chaetomium globosum* on beechwood (*Fagus sylvatica*). *Holzforschung*, 19, 83–190.

LINE, M.A., 1983. Catalase activity as an indicator of microbial colonization of wood. In: Oxley, T.A., and Barry, S., (Eds.), *Biodeterioration* 5. pp. 38–43.

MOUZOURAS, R., 1987. Microbiological aspects of stored timbers from the Mary Rose and the decay of wood by marine fungi. PhD thesis, Portsmouth Polytechnic.

POINTING, S.B., 1995. Gamma irradiation and reburial studies as a novel passive conservation treatments for waterlogged archaeological timbers of the Mary Rose. PhD thesis, University of Portsmouth.

POINTING, S.B., JONES, E.B.G., and JONES, A.M., 1996. Radiosensitivity of fungi isolated from waterlogged archaeological wood. *Mycoscience*, 37, 455–458.

RULE, M.H., 1983. *The Mary Rose; The Excavation and Raising of Henry VIII's Flagship*, Windward, Leicester.

SCHEFFER, T.C., 1973. Microbial degradation and the causal organisms. In: . Nicholas, D.D., (Ed), *Wood deterioration and its prevention by preservative treatments*. Syracuse University Press, New York, pp. 31–106.

VAN DER HEIDE, G.D., 1972. Problems of ship-archaeology and the preservation of ancient ship remnants. In: Walters, A.H., and Huek-Van der Plas, (Eds.), *Biodeterioration of Materials*, Vol. 2, Applied Science, Essex, pp. 376–380.

M. BEROVIČ

BIODETERIORATION STUDIES ON PASTELS AND OIL-BASED PAINTINGS

University of Ljubljana, Slovenia

ABSTRACT The biodeterioration of pastels and oil-based paintings was studied, including the influence of ambient temperature and humidity. The mechanisms of filamentous fungi infestation and the different ways it manifests on pastels and oil-based paintings are discussed. Preservation and conservation procedures, including the use of benz-isothiazolene in ethanol (1:4), are presented.

INTRODUCTION Biodeterioration of art monuments is defined as a natural phenomenon in which microorganisms, at different life-cycle stages, deteriorate various organic substances as well as natural resins and cellulose fibers (Cruger and Cruger, 1990). In this process organic compounds, carbohydrate fibers, proteins, or other kinds of natural polymers are decomposed by microbial enzymes into various metabolites, including organic acids and amino acids. These products, together with cellulolytic enzymes released from various filamentous fungi, accelerate further hydrolysis of cellulose fibers from paper or canvas ground. A pH from 5 to 7, and high relative humidity in combination with temperatures ranging from 20 to 30°c, significantly promote this process. These phenomena, on the one hand, represent the life cycle of various microorganisms present on art monuments, but on the other hand they could constitute a serious problem for the restorer (Larpent-

Gourgaud and Sanglier, 1992). In many cases, mixed fungal infestation is simply cleaned or mechanically removed from the affected surface, without any removal of the rest of the fungal spores and mycelia, sanitizing and further preservation of the ground (Wallhäusser, 1982, 1985; Bader 1986).

MATERIALS AND METHODS Samples from infested areas were transferred in a laminar flow chamber onto agar plates and incubated for 248 hours at 30°c and 90% RH in an incubator. For identification of fungal mycelia standard taxonomic methods were used.

For inactivation of fungal mycelia and spores from the picture area, from the front and back side and the frame and underframe, sterilization by gaseous formaldehyde was used. The time of treatment was 12 hours.

For conservation of art surfaces benz-isothiazolone in ethanol (1:4) was used. The addition of 10 ppm to agar was found to be the optimal concentration.

A JEOL JSM-T220 (Japan) scanning electron microscope was used, with an acceleration voltage of 20 kV. Organic compounds related to fungal spores and filamentous fungi biomass on the surface of pastel and oil grounds were checked by X-ray diffraction analysis. For optical observation, a Wild-M 20 stereo microscope was used, as well as a Wild 1.25× (Switzerland) optical microscope.

RESULTS AND DISCUSSION

Pastel Ground The biodeterioration of pastel ground and flaxen canvas base are actually two different mechanisms. Pastel ground is often represented by cardboard or starch pasted rough paper on cardboard ground. In this case fungal infestation develops on the pigment-covered porous painted surface. The main source of fungal growth is starch glue and some organic wastes on the painted surface. A dry atmosphere and good air circulation in exhibition halls inhibits the growth of fungal mycelia, so that these factors are naturally protective (Wallhäusser, 1982). Conversely, factors that strongly promote development of fungal infestation are relative humidity over 50% combined with temperatures from 20 to 30°c. The presence of some pigments based on copper, tin, and lead oxides and salts usually blocks local infestation, while the other pigments play a neutral role (Larpent-Gourgaud and Sanglier, 1992). [FIG. 1]

Fungal mycelia on the pastel surface [FIG. 2] exploit a glucose hydrolyzed from a starch glue or gum tricocanth or the other polysaccharides contained in pastel pigments. In the presence of micro amounts of water in a solid matrix, microbial growth develops as a function of time and temperature. In very dry climates and

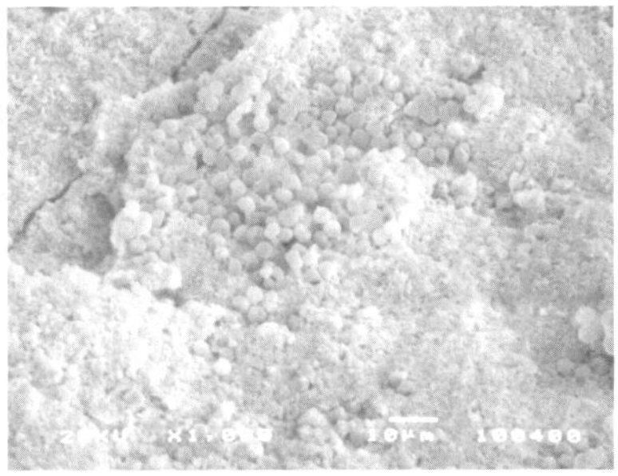

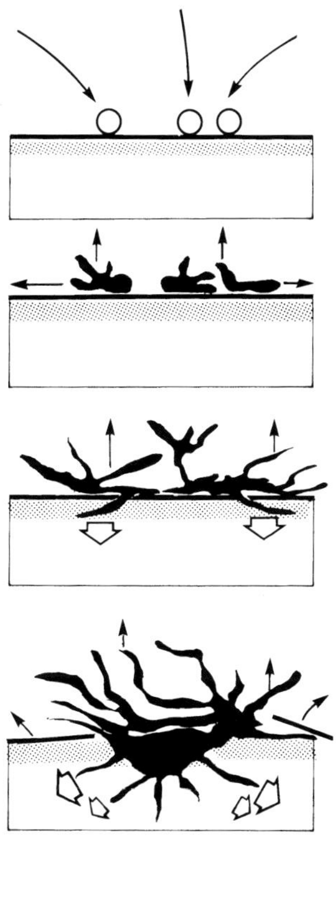

[1] Filamentous fungi spores on pastel surface. Scanning electron microscope magnification 1000×.
[2] Mechanism of fungal growth on pastel surface.

after exhausting all substrate and water sources, microbial growth stops, and fungal spores patiently wait for the next suitable opportunity (Larpent-Gourgaud and Sanglier, 1992).

This process also includes the presence of various metabolic products and their action on the painted surface. Various organic acids, mostly citric, oxalic, malic, or fumaric acids secreted in the metabolic TCA cycle, can react with a pastel pigment, changing it to its salts—this results in decolorization and changed optical properties.

The effect of secreted organic acids on paper fibers of various drawings or printed pages of graphics or books stored in humid and nonaerated depositories often results in brownish circles and spots of carbonized cellulose fibers. This effect on pastel painted surfaces results in the formation of various funnel forms usually filled with fungal mycelia. If the microbial infestation is linked with wet starch in the under layers, further growth may increase rapidly. [FIGS. 3, 4]

To prevent this process, rapid intervention measures have to be taken. [FIG. 5] The first essential step is drying of the whole painted surface, the ground, as well as the backing and frame. The next step is mechanical cleaning of fungal mycelia from the painted surface followed by sterilization of the whole picture using gaseous formaldehyde for denaturation of fungal mycelia and spores (Richards, 1968; Wallhäusser, 1985; Bader, 1986).

Sterilization with gas is the most effective method; use of gas ensures that all painted surfaces and porous layers will be disinfected.

For conservation of pastel painted surfaces a solution of benz-isothiazolone in ethanol (1:4) is sprayed on the inner side of the protecting glass. A thin layer of this preservative enables indirect contact of its vapors with the pastel surface to a distance of a few millimeters.

Oil-Based Paintings Contrary to the situation with pastels, fungal infestations on oil paintings start from the back of the painting. They actually start as a result of microdroplets of water that condense on the back surface; this is followed by microtransport of water from the back surface toward the painted surface.

As humidity levels increase, and temperatures stay in the range of 20 to 30°c, particularly in the spring or autumn, fungal infestation of the sized canvas commences on the back side of the oil painting. The first sizing layer is often a glue that penetrates the canvas from front to back. It is this layer that is attacked first. After consuming the available organic compounds from the surface of sized areas, fungal growth penetrates into the lime-chalk ground. The increase of fungal biomass and the activity of its metabolic products, enzymes, and organic acids result

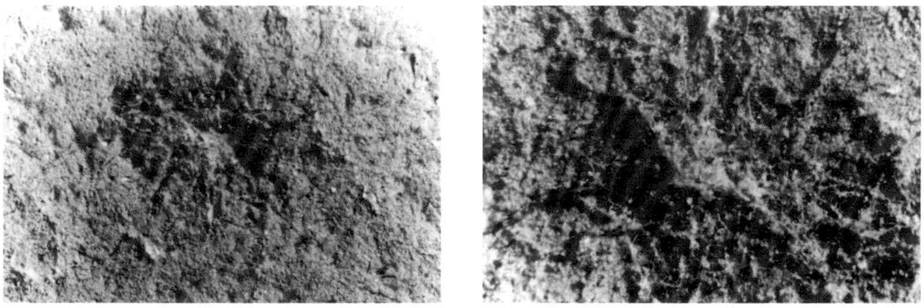

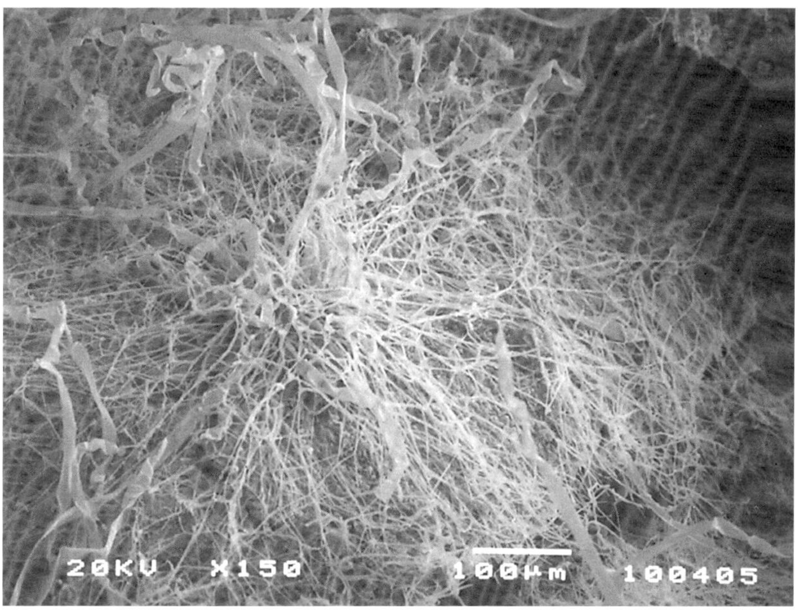

[3] *Mucor michei* mycelia on the pastel surface. Left, fungal biomass; right, cavities on pastel surface. (Wild stereo microscope, 12 and 25× magnification.) [4] Filamentous fungus hyphae on pastel surface. Scanning electron microscope magnification 150×.

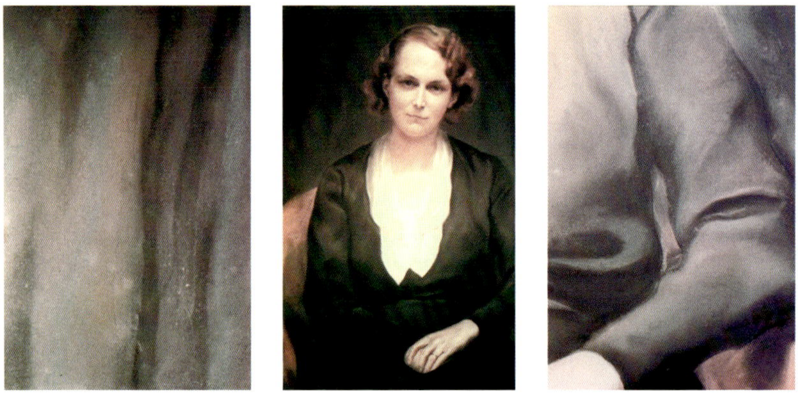

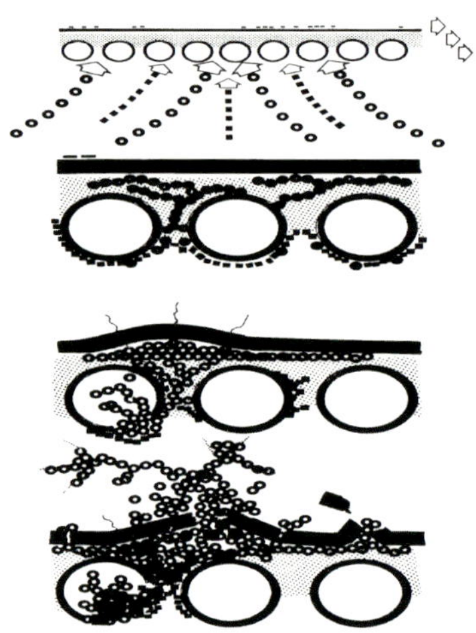

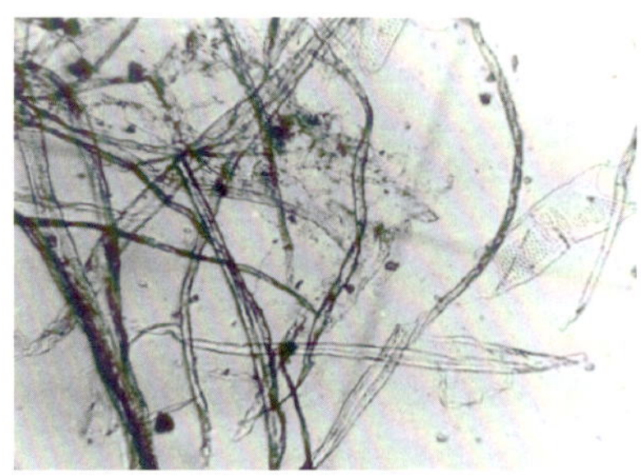

[5] Božidar Jakac, pastel, 50 × 70 cm, 1955: Portrait of Miss Lunaček. Samples of microbial infestation are shown in a and c; b shows pastel after restoration and conservation.
[6] Mechanism of fungal infestation of flaxen canvas.
[7] *Trichoderma viridae* (a) and *Aspergillus niger* (b) growth on cellulose fibers. Wild–20, optical, 100× magnification.

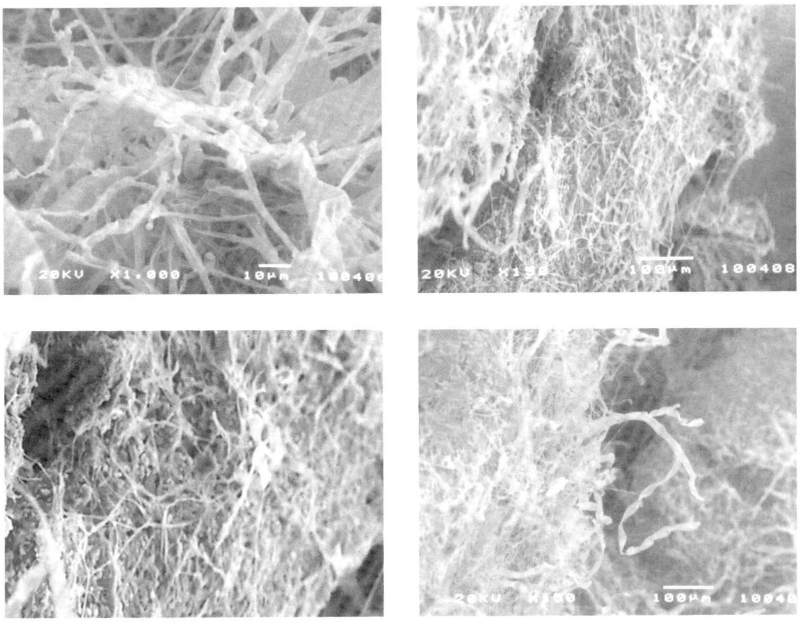

[8] Filamentous fungi and canvas fibers. Scanning electron microscope magnification 150–1000×.

in stretching of the painted surface layer. This stretching finally results in cracking and bubbling of the surface. In later stages, fungal biomass penetrates completely from the back side through the painted surface until fungal mycelia become apparent on the painted surface. [FIG. 6]

Part of the growth of the fungal mycelia usually stops when the organic compounds from the size and ground are exhausted. However, some aspects of the infestation, especially cellulose degradation, will still continue. [FIG. 7] Cellulose degradation represents the most problematic and dangerous part of biodeterioration since this group of fungi could hydrolyze cellulose fibers from flaxen canvas. [FIG. 8] Under attack by fungal xylanases and organic acids, cellulose fibers start crumbling and, eventually, the whole of the canvas background finally disappears. The rest of the painted surface and the chalk-lime ground are then left without support, so they break at the first mechanical shock, at which point holes appear in the picture. Restoration and conservation in such a case start with mechanical cleaning of fungal mycelia from the painted surface, followed by sterilization of the whole picture—including the frame, underframe, and unframed picture—with gaseous formaldehyde, which enables denaturation of fungal mycelia and spores.

After this step a new canvas layer has to be added to support the damaged canvas. This operation is performed on a vacuum table with the fungally caused holes filled with a wax-damar-natual-resin mixture. The last step is to use classical restoration procedure. For conservation of the back of oil paintings and the wooden parts of frames, a solution of benz-isothiazolone in ethanol (1:4) in spray form is recommended.

CONCLUSIONS For pigments and binders of pastels and oil-based paintings, sterilization by gaseous formaldehyde is a suitable and nondestructive method for inactivation of microbial spores and developed filamentous fungi mycelia.

Benz-isothiazolone in ethanol (1:4) in a film form, on the inner side of protective glass in the case of pastel pictures, or in a spray form in the case of oil-based paintings, is an effective conservation measure against fungal mycelia. Although our previous experiments that were conducted with tin tetrabutilate showed excellent results, this substance has been removed from practice because of its carcinogenicity. Results using benz-isothiazolone in ethanol (1:4) were excellent, so this substance is recommended for future use.

REFERENCES

BADER, F.G., 1986. Sterilisation: prevention of contamination. In: Demain, A., and Solomon. N.A., (Eds.), *Manual of Industrial Microbiology and Biotechnology*. American Society of Microbiology, Washington, DC, pp. 345–361.

CRUGER, W., and CRUGER, A., 1990. *Biotechnology—A Textbook of Industrial Microbiology*, 2nd edition. Editor of English Edition, Thomas D. Brock, Sinauer Associates, Inc., Sunderland, MA. Assoc. Inc. Press, 1982. Sunderland, Madison, WI, pp. 84–86.

LARPENT-GOURGAUD, M., and SANGLIER, J.J., 1992. *Biotechnologies*, Principes et Methodes, Doin Editeurs, Paris.

RICHARDS, J.W., 1968. *Introduction to Industrial Sterilisation*, Academic Press, London.

WALLHÄUSSER, K.H., 1982b. Securité programmée lors la filtration sterilisante, *Inf. Chim.* 230, 201–216.

WALLHÄUSSER, K.H., 1985. In: Rehm, H.J., and Reed, G., (Eds.), *Sterilisation*, vol.2, VCH, Weinheim, pp. 700–716.

A.B. STRZELCZYK

H.E. ROSA

CHEMICAL AND MICROBIOLOGICAL CAUSES OF DETERIORATION OF TOOTHBRUSHES THAT USED TO BELONG TO PRISONERS OF AUSCHWITZ-BIRKENAU CONCENTRATION CAMP: RESEARCH AND METHODS OF THEIR CONSERVATION

Nicholas Copernicus University, Toruń, Poland

ABSTRACT A collection of toothbrushes (nearly 5000 specimens), originally owned by Auschwitz-Birkenau concentration camp prisoners, after having been stored for over 60 years, has undergone a high degree of chemical deterioration. The first step of this research was to establish the chemical nature of the material used to make them. This was celluloid. Results of natural aging included evaporation of the plasticizer—camphor, as well as liberation of highly concentrated nitric acid. These phenomena caused the crumbling of the toothbrushes' material. In this work, methods of deacidification were developed, and reinforcement of the structure of the toothbrushes was carried out.

INTRODUCTION The museum in question is not a building where valuable objects are collected and stored. It is the area of the Auschwitz-Birkenau concentration camp set up by the Nazis close to the Polish historical capital, Cracow. After the war the whole terrain of the camp was turned into a museum. Everything that remained after the prisoners and torturers were gone was collected; many objects that were gathered had belonged to former prisoners. The museum objects nowadays consist, among others, of: documents of camp life, buildings and barracks where people were imprisoned, their equipment (furniture, vessels), prisoners' clothes (striped uniforms, plain clothes, shoes), glasses, dentures, brushes, and bowls, as well as toothbrushes—about 5000 of them. [FIG. 1] In the late 80s a number of toothbrushes were deposited in the Holocaust Memorial Museum in Washington.

The storage of the objects presents many difficulties. In the course of the nearly 60 years of storage, the structure of the materials has been irreversibly changed by aging, with the poorer raw materials undergoing the most intense changes. These processes drastically affected the collection of toothbrushes. Throughout the post-war period they were stored in open-work baskets made of synthetic material and lined with paper, and they were protected from dust by being covered with paper or flannel cloths. At first they were stored in an unheated storeroom; then they were transferred into an air-conditioned hall.

In the 80s the first changes in the look and structure of the toothbrushes were observed. On the bottom of the containers in which they were kept appeared a sticky liquid, which soaked into the paper and caused it to completely disintegrate. The handles of some of the toothbrushes showed traces of cracks or even crumbled. [FIG. 2] Others were covered with a light deposit and tended to stick together. [FIG. 3] The report of 1994 specifies those changes. The conservators of the Holocaust Memorial Museum also noticed some disturbing changes in the structure of toothbrushes. Samples were sent to the Smithsonian Conservation Analytical Laboratory with a request for help.

Toward the end of 2000 the whole collection of toothbrushes was transported from Auschwitz to the Department of Paper and Leather Conservation of Nicholas Copernicus University in Toruń, Poland, in order to find out the causes of the deterioration and treat the collection.

CONDITION OF THE TOOTHBRUSHES The condition of the toothbrushes at that time was actually much worse than what has been described above.

1. A large number of handles had disintegrated [FIG. 4] to so great a degree that we named it "the crumble." The weight of this mass was estimated at about 5 kg. The crumble tended to stick together. This tendency was, however, less marked than in the case of toothbrushes made of ivory-colored material.

2. The latter were stuck together and covered with a thick layer of a white-gray sticky encrustation. The heads of toothbrushes were deformed, split along the lines of attachment of the bristles; the bristles were darkened and very brittle. As a result of chemical changes the wires used for the attachment of the bristles had given off blue-green dye. This was the copper rust called verdigris.

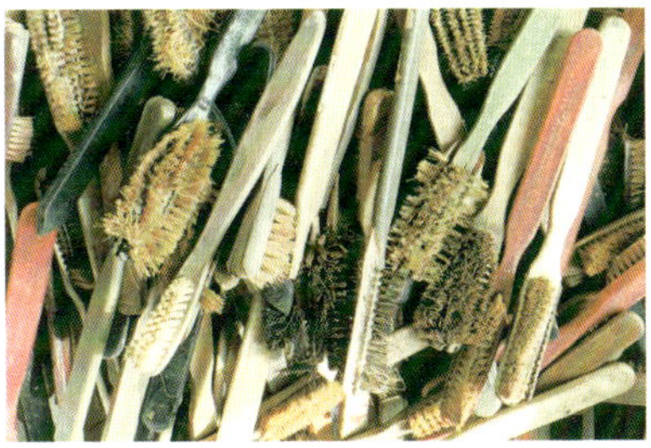

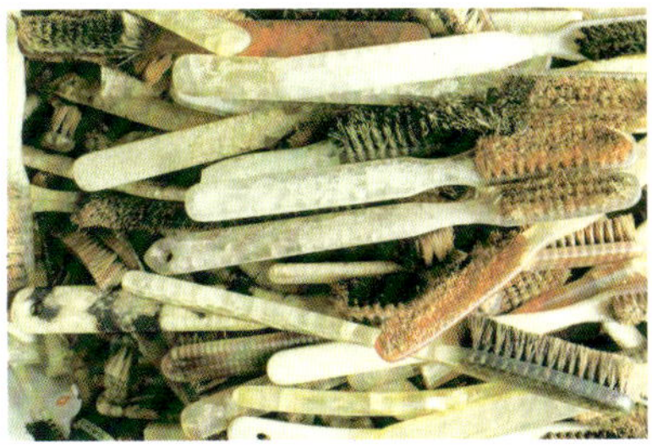

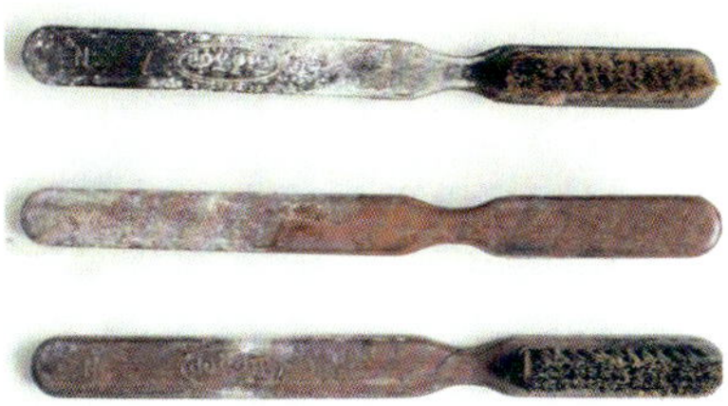

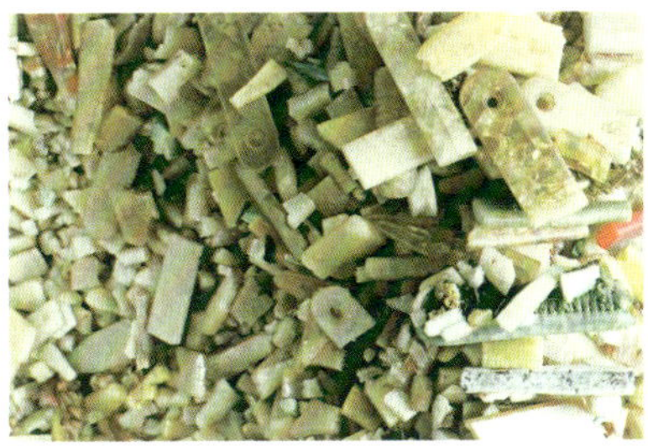

[1] Diversity of toothbrushes before sorting and conservation. [2] Cracks on toothbrush handles. [3] Slight light deposit on intact toothbrushes. [4] The so-called "crumble," i.e., crumbled handles.

3. The cloths with which the baskets containing the toothbrushes were covered had disintegrated into powder. Also, the papers lining the bottoms were completely decayed. No stains were found on the cloths or on the papers. On the contrary, their color had lightened.

PRELIMINARY STUDIES OF THE PROPERTIES OF THE DETERIORATED TOOTHBRUSHES

Preliminary studies looked at fragments of the flannel cloths, the white-gray encrustation on the handles of toothbrushes made of ivory-colored plastic, the cracked majority of handles, and the crumble formed by their disintegration. The subject of study was the reaction to water of those elements, their acidity, and the microorganisms inhabiting them. It turned out that all the deteriorated samples either swelled or even dissolved in water. This resulted in deep degradation of the cellulose rings in the celluloid material. That process was most evident in the severely disintegrated elements that had previously formed the toothbrushes' handles. Those made of the ivory-colored material had hard cores, which did not react to water, and the superficial water-sensitive thick encrustation. Upon being washed in water the crust washed away and the thinner solid inner part of the handles was exposed. The fact that the basic material of the major part of the toothbrushes was sensitive to water considerably influenced the methods of conservation treatment. Both the intact toothbrushes and the damaged ones, as well as the fragments of flannel cloth, were strongly acidic, their pH reaching 2–3. The most highly acidic material was the sticky substance coating the toothbrushes. It was most abundant in the bottom of the baskets. It seems certain that the interaction of the water-soluble portion of the toothbrushes with the resistant core stimulated their deterioration.

CHEMICAL STUDIES Chemical studies were carried out on representative samples of the deteriorated toothbrushes by Dr. Wojciech Czerwiński and Dr. Jacek Nowaczyk of the Department of Physical Chemistry at Nicholas Copernicus University in Toruń, Poland. Methods used were FTIR and the magnetic resonance method C^{13} NMR CPAS. What was found was that the badly cracked and disintegrating toothbrush handles were made of cellulose nitrate with camphor as a plasticizer. This material, popular in the past, was commonly called celluloid. The toothbrushes with ivory-colored handles covered with encrustation were made of a mixture of nitro and acetylcellulose. They did not contain the camphor plasticizer.

The causes of the deterioration have been specified as follows:

• The plasticizer gradually evaporated, a process that was facilitated by the easy access of humid air to the exhibits while they were stored in plastic baskets.

• Decrease of moll mass caused by the disappearance of the plasticizer, as well as by hydrolysis, resulted in the release of large quantities of NO_2 ions; these combined with moisture in the air to form nitric acid. Those two processes caused the reduction in volume of the material, and its cracking.

• Further acidification of the toothbrushes led to the oxidation of carbon C-6 bonds in the cellulose ring and the formation of carboxylic groups. This resulted in a drastic increase in hydroscopicity of the toothbrush material.

The results of these studies are in agreement with those presented by Eileen Blakenbaker, Mary T. Baker, Melanie E. Feather, et al., who have studied the collection of toothbrushes at the United States Holocaust Memorial Museum.

MICROBIOLOGICAL STUDIES As has already been mentioned, no traces of the development of microorganisms were found on the handles of crumbled toothbrushes, on the handles and heads of the better-preserved specimens, or on the covering cloth. There were no discolorations characteristic of microbial colonies, or points of more intensive deterioration. That, however, does not mean that the surfaces of those objects were completely free from microorganisms. They certainly had been settled with dust for many years. That is why routine studies were carried out to find out the extent of microbial infestation on the toothbrushes. The result of those studies would help determine whether the whole collection should be subjected to disinfection.

The studies were carried out separately on completely disintegrated toothbrush handles and on fragments of bristles that had separated from the heads.

The Microflora on Toothbrush Handles A suspension of the crumble from the toothbrushes was obtained by mixing 1 g of the finest powder with 5 ml of sterile water. The pH of this suspension was 2–3. One-milliliter portions of that suspension were poured onto five petri dishes with the medium for *Streptomycetes* (Waksman's) as well as with malt extract agar (Difco). The inoculated petri dishes were incubated at room temperature. The developed colonies were used for further studies. The following 27 strains were isolated:

Aspergillus spp. 1–10
Penicillium spp. 1–7
Stemphylium spp. 1–5
Papularia spp. 1–2
Alternaria spp. 1–2
Streptomycete sp. 1

The results reveal what microorganisms were present on the surface of disintegrated toothbrush handles. The small number of colonies obtained from the suspensions is evidence that they developed only from the scattered colony forming units. Further, a simple method was used to check the cellulolytic abilities of 15 strains chosen from the isolated ones. They were inoculated onto an agar medium for cellulolytic fungi with acidic potassium phosphate and a disc of sterile filter paper as a unique carbon source on the surface of agar medium. The medium was inoculated at its center with suspensions of spores of fungi listed below. After three weeks of incubation at room temperature the intensity of growth, as well as the cellulose decomposition by the strains studied, were estimated.

The most intensely cellulolytic were *Alternaria spp.* 1 and 2 as well as *Penicillium spp.* 1 and 2. The remaining strains, though reaching fairly good growth, did not exhibit any well-marked abilities to decompose cellulose. That supports the view that microbiological decay was not the direct cause of the corrosive changes the Auschwitz toothbrushes were subject to.

The Microflora of the Toothbrush Bristles The bristles detached from the heads of the deteriorated toothbrushes were placed in petri dishes with drops of water supply ensured. Observations on the appearance of microflora were carried out during 10 weeks of incubation at room temperature. The growth of *Penicillium* as well as abundant development of *Actinomycetes* was observed in many places on the surface of the bristles. [FIG. 5] We suppose that the *Streptomyces sp.* isolated from the handles could have originated from the deteriorated bristles. In spite of the fact that no changes characteristic of microbial invasion were observed on the surface of toothbrush heads, in our opinion the bristles might have been attacked by actinomycetes, which are known to be ubiquitous and xerophilic, and have keratinolytic abilities. Our previous studies on the potential for their growth on PVA and PVAC membranes have confirmed the ability of actinomycetes to spread over highly acidified materials. The presence of those alkaliphilic bacteria in acid environments is evidence of nature's great flexibility.

METHODS OF CONSERVATION The collection of toothbrushes is a very peculiar collection, considering the fact that each toothbrush represents a very personal possession of one of the Nazis' victims. The condition of the toothbrushes, before conservation, was shocking. The process of their destruction was clearly visible to the naked eye. In the room the collection had been placed in, after a short time there was a clearly perceptible smell of nitric acid.

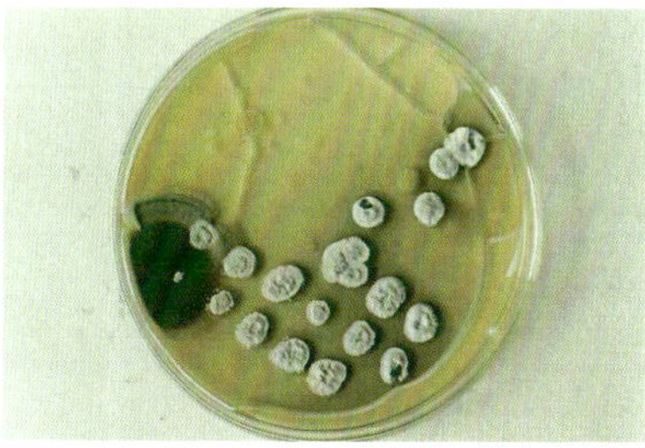

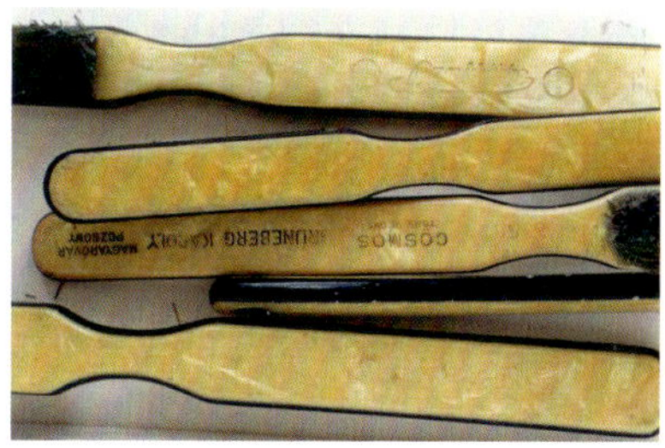

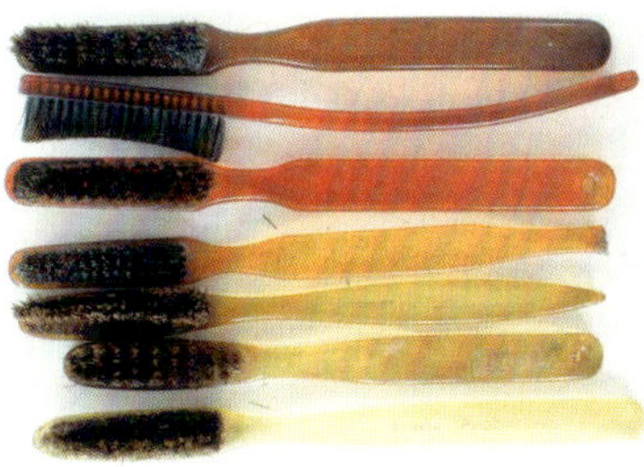

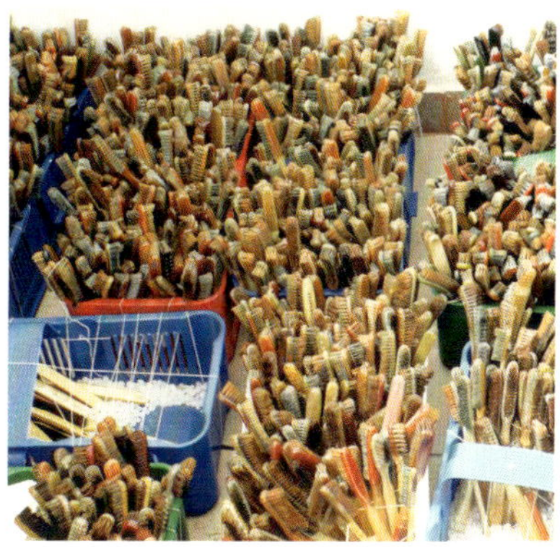

[5] *Actinomycete sp.* isolated. [6] The toothbrushes of group 4 after the conservation process.
[7] The toothbrushes of group 7 after conservation. [8] The collection of toothbrushes after conservation.

One of the mistakes made in the storage of the toothbrushes before conservation was that they had been placed in containers all mixed together, with no consideration given to their structure or the kind of material they were made of. As a result, nearly all the toothbrushes had contact with those most badly deteriorated and acidified, i.e., those with handles made of nitrocellulose and camphor.

Conservation of these unusual objects was a difficult problem. Studies of cellulose materials show that it is impossible to completely stop the chemical disintegration of celluloid, as it is a polymeric material. In drawing up the conservation program the aim was only to slow down the celluloid's rate of decay. The program included:

1. Neutralization of acid celluloid decay products;
2. Reinforcement of the deteriorated bristles of each kind of toothbrush;
3. Reinforcement of toothbrush handles for the whole collection; and
4. Establishment of conditions for further storage of the collection in the museum.

The collection included 4776 items. They were put into 11 groups depending on their structure, appearance, and preservation condition.

Group 1 consisted of 48 toothbrushes with strongly corroded surfaces, colored from white to dark beige.

Group 2 consisted of 173 pieces with three-layer decorated handles, often imitating mother-of-pearl. Most toothbrushes in this group had their makers' names on their handles, printed with paint or impressed.

Group 3 consisted of 85 pieces with handles comprised of several colored layers, ranging from transparent to cream colored, to a gray to tortoise-shell imitation or striped.

Group 4 consisted of 227 pieces with wooden handles coated with plastic in various colors, as well as toothbrushes with unpainted and uncoated wooden handles.

Group 5 consisted of 1002 pieces with transparent or semitransparent handles in various colors. On the handles were the logos of the toothbrush producers. A characteristic feature was their comparatively good condition, with damage occurring mainly at the attachment site of the bristles.

Group 6 consisted of 212 pieces with matte multicolored handles, many of them dark brown to black, showing a diversity of shape, bearing printed maker's signs.

Group 7 consisted of 493 pieces, 335 of which were very severely deteriorated. Their handles were transparent, diversely shaped, and covered with a network of cracks, often very dense. The handles were very fragile and brittle, breaking easily when touched.

Group 8 consisted of 314 pieces with matte single-colored white or light yellow handles. Many of them bore imprints of the maker's name.

Group 9 consisted of 123 pieces, characteristically smaller in size, probably children's toothbrushes.

Group 10 consisted of toothbrushes with severe mechanical damage, incomplete, consisting of only parts of the heads, without handles. That group included 2099 toothbrushes.

Group 11 consisted of about 5 kg of crumbled handles, ranging from larger fragments to powder form.

CONSERVATION PROCEDURES Initial studies were carried out in order to establish the methods of cleaning the toothbrush handles, reinforcing the bristles, and reinforcing the celluloid and wooden handles. Also, it was determined which solvents and binders could be safely used in the treatment and which methods of deacidification of the toothbrushes could be used. Based on observation of the toothbrushes' behavior in the course of initial conservation assessment, safe storage practices were established. The most important thing was to keep the toothbrushes in an upright position so that the bristles would be in contact with freely circulating air. That protected the bristles against moisture accumulation, and it prevented the acid celluloid decomposition products from penetrating inside the handles. Some silica gel was placed in the bottom of the containers in which the toothbrushes were stored.

Cleaning After all items were dried under dry air conditions, the handles were cleaned with cotton wool pads soaked with 1% Sterinol aqueous solution and isopropanol (1:1). Sterinol is a detergent and fungicide and, being an ammonium salt, it is of alkaline pH.

Reinforcement of the Bristles After drying, the bristles were weak and brittle. To reinforce them each toothbrush and fragments of heads without handles were soaked with a 2% Paraloid B-67 soluted in a small amount of xylene and diluted with White Spirit. The consolidating solution was poured into petri dishes, and the heads were placed in them so that only the bristles were immersed in this solution. The bristles readily absorbed the consolidant. After the treatment the toothbrushes were dried. The effect was satisfactory: The bristles were reinforced and acquired higher tolerance to changes in air humidity.

Deacidification Ammonia vapors were used in this process. Two neutralization methods were used:

1. The toothbrushes of group 7, the most highly acidified, fragile, and brittle, as well as other highly degraded ones, were dried and neutralized under vacuum conditions. A Laybold vacuum installation under a glass bell was used with NaOH tablets as moisture absorbent. The initial drying lasted 24 hours at reduced pressure. Neutralization under reduced pressure conditions lasted 24 hours.

2. The remaining toothbrushes, their fragments, and the crumble, after prolonged storage under relative humidity conditions of 45–50%, were subjected to a neutralization process in ammonia vapors in glass desiccators. A 3% ammonia solution was placed in the bottom of the desiccators. The toothbrushes were placed in the upper part of the desiccators, which were then tightly closed. The neutralization process lasted 48 hours. After that time the brushes were aerated under an extractor, and after one week's time they were soaked with imidazol in order to fix the pH acquired. The pH of the external layers of the handles was 9–10, that of the internal layers 7–8. The neutralization processes affected the colors of the toothbrushes in that both the handles and the bristles became slightly darker: The white and cream-colored handles turned beige; other handles became amber-colored.

Reinforcement of Toothbrush Handles To reinforce celluloid toothbrush handles, 0.5% acetylcellulose solution in acetone was used. This substance was singled out because of the fact that handles made of cellulose nitrate and cellulose acetate were generally in better condition than those made of celluloid with camphor. Destructive changes in them were found only on the surface. In both the preliminary tests and the actual treatment the acetylcellulose solution readily penetrated deep into the structure of the material of the handles. The acetylcellulose solution was also used for gluing together fragments of handles. The toothbrushes with wooden handles and those of group 5 were soaked with 2% Paraloid B-72 solution in xylene. [FIGS. 6, 7, AND 8]

Recommendation for Exhibit Storage The external conditions recommended for storing the toothbrush collection are: a temperature of 12°c, and 35% RH. The toothbrushes should not lie in a disorderly mess, but instead should be arranged in rows with their heads toward the external air and with their handles in dry conditions. The ideal solution would be storing them on perforated shelves in cabinets with the aforementioned parameters.

FUNGUS	COLONY DIAMETER, IN MILLIMETERS	CELLULOLYTIC ACTIVITY QUOTIENT*
Alternaria sp. 1	65	2.0
Alternaria sp. 2	65	2.0
Penicillium sp. 1	60	2.0
Penicillium sp. 2	60	2.0
Penicillium sp. 3	60	1.5
Aspergillus sp. 1	60	1.5
Aspergillus sp. 2	55	1.5
Penicillium sp. 4	65	1.5
Penicillium sp. 5	65	1.0
Aspergillus sp. 3	55	1.0
Aspergillus sp. 4	55	1.0
Papularia sp. 1	55	1.0
Penicillium sp. 6	65	1.0
Penicillium sp. 7	60	1.0
Aspergillus sp. 5	50	1.0

* Cellulolytic activity was determined as the quotient of colony diameter to the diameter of the filter paper dissolution zone.

[TABLE 1] Intensity of growth and cellulose decomposition by the strains studied

D. ERHARDT
C.S. TUMOSA
D. VON ENDT
A.-S.M. EL-EASELY

COLLATERAL DAMAGE: ANTHRAX, GAS, AND RADIATION

Smithsonian Center for Materials Research and Education, Suitland, MD

ABSTRACT In response to acts of bioterrorism, chemical and radiation treatments have been employed to kill suspected pathogens. These treatments are effective in killing the microorganisms, but are potentially damaging to surrounding objects and materials. The chemical treatments usually involve oxidizing agents such as the gas chlorine dioxide or solutions, gels, or foams of various peroxides or chlorine bleaching agents. Electron irradiation of mail produces damage both directly and by the heating induced during treatment. For example, polystyrene slide mounts have softened and warped, and stacks of papers have adhered together and transferred printing and images. Tensile measurements on irradiated paper show a substantial (up to 80%) loss in the ability of the paper to be deformed. Chemical analysis of the paper shows a large increase in hydrolytic and oxidative soluble degradation products compared to controls. The irradiated paper is both brittle and discolored.

INTRODUCTION The fall of 2001 saw terrorism on a scale not previously seen in the U.S. in modern times. Among the problems encountered were a series of anthrax-contaminated letters sent to a number of locations including the offices of two senators. This resulted in the contamination of the Hart Senate Office Building and its mail-handling center, as well as mail sorting locations starting from the point of origin of the letters. Because of the scale of contamination, gaseous fumigation with chlorine dioxide was proposed as the method of decontamination for the Senate building. However, because of the aggressive oxidizing nature of this gas, there was concern about the possible effects on artwork and sensitive technical equipment as well as personal items such as photographs in the building. Tests of exposure to a number of quickly prepared test samples showed that damage such as fading of inks, dyes, and photographs did indeed occur. Mitigation of the problem by such techniques as treatment with Oxone or hydrogen peroxide was also considered.

Chemical treatments to mitigate biohazards have include the traditional agents such as hypochlorite bleaches, peroxide-based agents such as that developed at Sandia National Laboratories for chemical and biological warfare agents, and Oxone, a DuPont product whose active ingredient is potassium peroxymonosulfate. All of these compounds and formulations are potentially damaging to many of the materials in the objects encountered in offices and museums.

Because of such concerns, it was ultimately decided that chlorine dioxide would be used to decontaminate only the most heavily contaminated areas of the Hart Building. Other surfaces throughout the building were decontaminated using other techniques, including a peroxide foam. Valuable materials, including artwork, were bagged, sealed, and taken to another site where methods such as vacuuming with HEPA filters were used to decontaminate them. However, none of these reagents and treatments are unsuitable for the treatment of mail.

Eventually, the decision was made to sterilize all mail that was present in the closed mail handling facility, as well as all subsequent mail that normally would have passed through it. Electron beam irradiation was chosen for this purpose. This process is routinely used for food items with no apparent problems. The results for mail were quite different, probably because of the higher total dose, 50–100 kGy, versus the 7–20 kGy typically used to kill spores or the even lower doses used to kill insects. The methods, dosage, and conditions chosen resulted in severe yellowing and embrittlement of papers, melting of plastics, and blocking of manuscripts. Chemical analysis and measurements of changes in color and mechanical and physical properties verify the damage, and confirm reports that the irradiated mail reaches temperatures well over 100°c. These effects have serious implications for museum objects and samples, and will affect decisions regarding the archival nature of irradiated documents.

METHODS

Tensile Tests and Paper Specimens Tensile tests were performed on screw driven tensile testers as described previously (Mecklenburg and Tumosa, 1991). Samples of paper were approximately 125 mm × 6 mm × 0.15 mm. Tests were performed on standard envelopes and paper used in the course of business (both irradiated and unirradiated samples) and on irradiated samples of Whatman paper that had been characterized previously.

Irradiation of Test Specimens Postal specimens were estimated to be irradiated at 10 MeV for a minimum dose of 50 kGy per pass and with at least two passes (McKnight, 2002). Some specimens were from routine mail routed through our own Washington, D.C., address. Other samples prepared specifically to evaluate the effects of irradiation were mailed to our Washington address from our Maryland facility and evaluated on their return. Other test specimens prepared before the use of irradiation for mailed specimens began were irradiated at an agricultural irradiation facility at 5.2 MeV at 250 microamps for a total dose of 257 kGy. This was anticipated to be two to five times the dose received at postal facilities.

Chemical Testing for Glucose Formation A sample of paper (about 1 g) was extracted with stirring in 25 ml of water for two hours. The extract was filtered, separated into 5-ml aliquots, and evaporated under vacuum. The residue was derivatized as follows: 0.1 ml STOX, a commercial reagent containing hydroxylamine hydrochloride and o-phenyl-β-D-glucopyranoside (internal standard) in pyridine, is added to the residue of one of the aliquots and heated at 70°c for 1 hour to convert carbonyl groups and any cyclic hemiacetals to the oximes. Then 0.1 ml hexamethyldisilazane and a drop of trifluoracetic acid are added. The *per*-trimethylsilylated supernatant is analyzed by gas chromatography on a DB-17HT column with FID detector starting at 50°c and immediately rising to 330°c at 10°c/min. Sugars are identified by comparison of retention times with standards, or by analysis by GC/MS. The basic method was described previously (Erhardt and Mecklenburg, 1995).

Thermal Treatments Plastics were heated in dry ovens at temperatures expected to be encountered in the irradiation of the mail between 80 and 130°c. Paper specimens were treated in both dry ovens and ovens with a controlled relative humidity. Control of relative humidity ensures a process more like that of natural aging (Erhardt and Mecklenburg, 1995).

Color Tests Color changes were calculated from standard L*a*b* measurements obtained with a HunterLab Ultrascan spectrophotometer.

RESULTS Plastics received in an altered state were identified by Fourier transform infrared spectroscopy and examined for heat damage. Test specimens were heated to determine the temperature ranges required for similar damage. The largest number of plastic specimens received through the mail was based on polystyrene. This included the clear windows on business envelopes, slide mounts for 35-mm slides, and the bodies of 3½-inch computer disks. Continuous heating of similar materials for up to 30 min at temperatures of up to 100°c showed only slight changes in dimension or stiffness. Above 100°c changes became more apparent and by 110°c changes were drastic. Deformation occurred due to softening and relaxation of casting strains and sometimes due to handling. Softening above 100°c is commonplace for polystyrene materials (Roff and Scott, 1971).

The most obvious effects on paper are changes in color. The changes are a distinct yellowing, which was found to be slight immediately after irradiation but increased in intensity with time. Color measurements, L*a*b*, for a typical paper showed distinct darkening and a color shift to the yellow in the irradiated sample. The decrease in L* (L* = −1.13) is indicative of a general darkening. The large increase in b* (b* = 4.68) is due to increased yellowness. There was little change in a* (−0.29).

Stress-strain testing of paper samples showed that irradiation greatly reduces the strength and extensibility of paper. Figure 1 shows stress-strain curves for paper from identical business envelopes, one of which was irradiated by the Postal Service. [FIG. 1] The paper is typical of business envelopes made from recycled fibers. The extensibility of the irradiated paper (the amount of strain, or stretching, required to break it) has been reduced from 6.2% to 1.3%. The strength of the paper (stress to break) has also been reduced, from 19 MPa to 11 MPa. Interestingly, the stiffness of the paper (slope of the initial stress-strain curve) is essentially unchanged—the paper is not stiffer, just weaker. In this case, it is the loss of extensibility that results in the impression that the paper is more brittle. Figure 2 shows stress-strain plots for Whatman paper samples, a control sample, and a sample irradiated to 257 kGy. [FIG. 2] While the loss in strength and extensibility is not as great as for the envelope paper, the overall effect is similar. It is expected that the Whatman paper would be more stable than the recycled paper of the envelope.

Gas chromatographic analysis of aqueous extracts of the envelope paper samples showed a significant increase in the amounts of extractable material in irra-

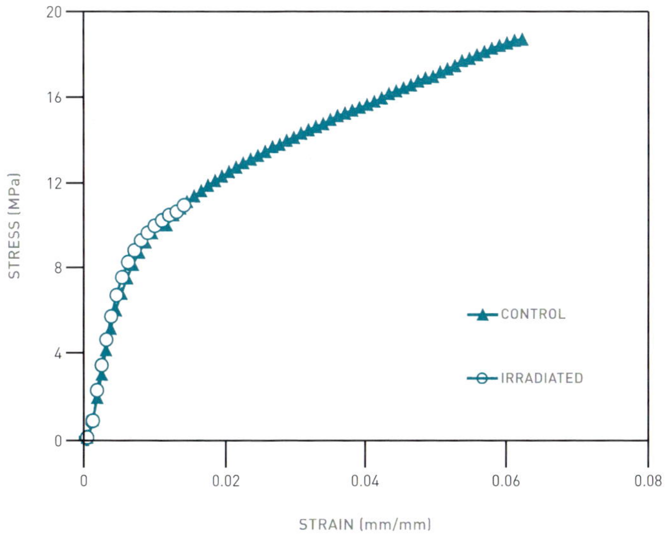

[1] Stress-strain curves for samples of initially identical mailed business envelopes, one unirradiated and one irradiated. The strength and extensibility are both reduced, but the stiffness (initial slope of the curve) is unaffected.

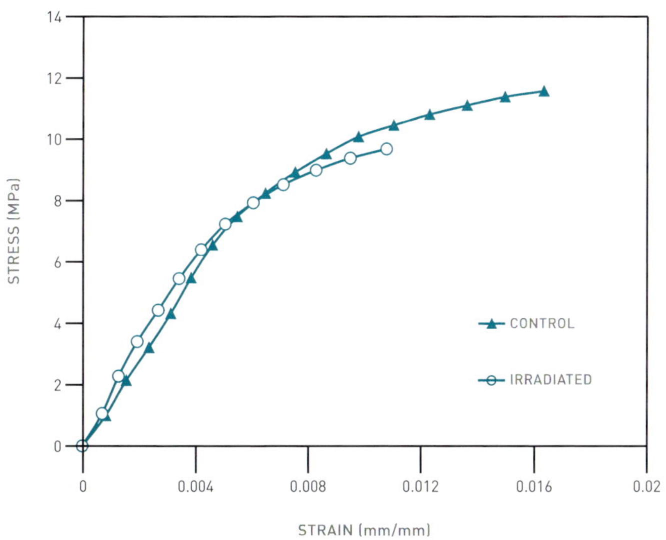

[2] Stress-strain curves for samples of Whatman paper, one unirradiated and one irradiated to 257 kGy. The effects are similar to those in Fig. 1, but less drastic.

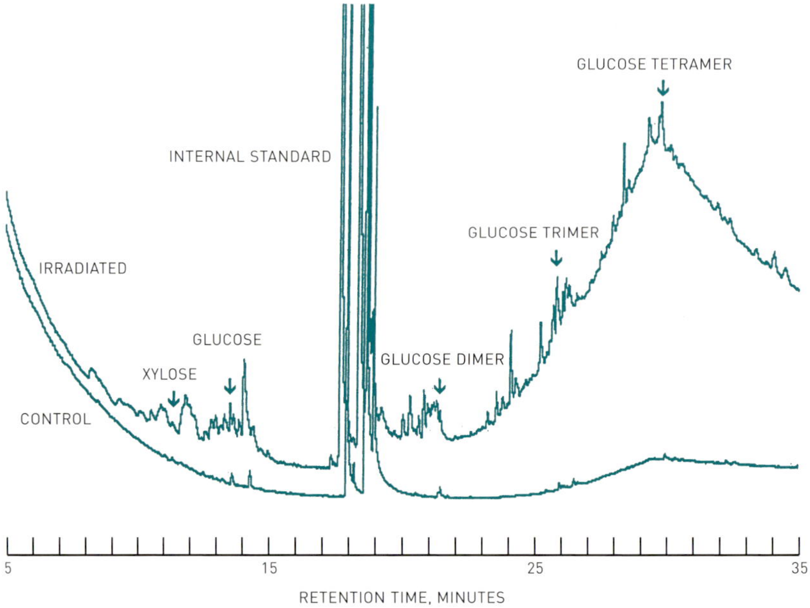

[3] Gas chromatograms of the *per*-trimethylsilylated oxime derivatives of the extracts of unirradiated and irradiated business envelopes. Small amounts of xylose, glucose, and glucose oligomers, compounds typically formed during natural aging, are seen. However, uncharacteristically large amounts of other compounds (the unlabelled peaks) are also formed.

diated samples. [FIG. 3] Unirradiated fresh paper contains only small amounts of soluble material. The increase in soluble material in the irradiated samples represents decomposition products. However, the mixture of products formed due to irradiation is quite different than what is found in typical naturally aged samples. The primary degradation process during natural aging is hydrolysis, and the typical product mixture includes xylose, glucose, and glucose oligomers (short chain polymers) resulting from hydrolysis at different points along the cellulose chain. Other compounds including products of oxidation and chain scission are also present, but in smaller amounts. The irradiated samples do show increased amounts of the hydrolysis products, but also significant amounts of other compounds. The hydrolysis products were probably formed before the water absorbed in the paper at ambient conditions was driven off by heating. The other products are formed either through scission reactions directly caused by irradiation, or by thermal scission or oxidation resulting from the greatly increased temperatures.

DISCUSSION The electron irradiation of mailed items causes both thermal and radiation effects that create damage not generally associated with the usual postal processes. In the procedures in use at the time of this work, mail up to 5 in. in thickness is placed in plastic bags, sealed, and irradiated, usually with two passes to ensure a minimum dose throughout the mass of mail. The changes are due to the irradiation combined with the resulting high temperatures and the effects of moisture being driven from the materials by the heating.

These effects of electron irradiation were often obvious. Paper is quite yellowed and brittle and plastics are melted or deformed. There were early reports of melted and burnt mail and even color changes in gems (Reese, 2001). Measured temperatures during irradiation have reached 130°c. This leads to various physical changes such as sticking and contraction of common materials (McKnight, 2002).

Because the heating of the packages of mail is uneven, moisture driven from the hottest regions of the blocks of mail by temperatures above 100°c can condense in cooler areas. The resulting liquid water can cause blocking or adhesion of paper, running of inks and dyes, and tide lines. In conjunction with the melting of resins in printed materials, this leads to inseparable masses of paper, particularly in the case of journals.

ACKNOWLEDGMENTS The authors would like to thank Walter Hopwood for technical assistance and Dr. Burrell J. Smittle of the Florida State Department of Agriculture for irradiation of some samples.

REFERENCES

ERHARDT, D. and MECKLENBURG, M.F. 1995. Accelerated vs natural aging: Effect of aging conditions on the aging process of cellulose. In: Vandiver, P.B. (Ed.), *Materials Issues in Art and Archaeology* IV, Materials Research Society, Pittsburgh, PA, pp. 247–270.

MCKNIGHT, D. 2002. Titan Corporation. Personal communication.

MECKLENBURG, M.F. and TUMOSA, C.S. 1991. An introduction into the mechanical behavior of paintings under rapid loading conditions. In: Mecklenburg, M.F. (Ed.), *Art in Transit: Studies in the Transport of Paintings*. The National Gallery of Art, Washington, pp. 137–171.

REESE, K.M. 2001. Mail irradiation discolors gems. *Chemical and Engineering News*. December 24, p. 64.

ROFF, W.J. and SCOTT, J.R. 1971. *Fibres, Films, Plastics and Rubbers*. Butterworths, London, pp. 48–49.

PAPER

A.M. BALDWIN

AN INTRODUCTION TO TIFFANY STUDIES

The Metropolitan Museum of Art, New York

ABSTRACT A noteworthy collection of drawings (ca. 1885–1930) produced by Louis Comfort Tiffany (1848–1933) and his studio artists sustained significant mold damage prior to its 1967 acquisition by the Metropolitan Museum of Art. In order to be exhibited and studied safely, the works required conservation intervention. Research was directed toward the identification of fungal contaminants to assess long-term health concerns and to determine the causes of dark fungal staining. Dematiaceous molds and fungal by-products were sampled from the works, which were water damaged sometime between 1933 and 1967. Scientific investigations included the use of classical and molecular techniques to identify the fungi, and the testing of oxido-reductases derived from the white-rot fungi *Phanerochaete chrysosporium* to reduce dark fungal staining on paper.

INTRODUCTION Louis Comfort Tiffany (1848–1933), the American painter, designer, and colorist, is perhaps best known for his innovations in glass. He was influenced by designers and artisans employed at his father's firm, Tiffany and Company, and studied informally with the American tonalist painter George Inness (1825–1894). His artistic sensibilities were further shaped by travels to Europe, northern Africa, and Egypt. From 1879 to 1933, Tiffany served as a principal in firms specializing in decorative arts, interior design, and glass. His distinctive work—parlor lamps, mantels, metalwork, and vases—would make Tiffany a household name. He revolutionized glass technology with a special type of glass called Favrile, patented in 1890. This process created novel optical effects—such as opalescence and iridescence—through the manipulation of molten glass and the addition of metallic oxides (Frelinghuysen, 1998). [FIG. 1]

Before design schemes were selected and ultimately realized, proposals took form in graphite pencil and watercolor drawings on supports of commercial artist papers and illustration boards. Drafting devices such as compasses were sometimes utilized, and photographic methods were employed as well. Finished presentation drawings were often shown in elaborately cut window mats, and bore pen and ink inscriptions. [FIG. 2]

In 1966, a collection of some 420 original preparatory sketches, renderings, and presentation drawings by Tiffany's studios was purchased by the Metropolitan Museum of Art. The collection offered a daunting challenge, for prior to acquisition it had sustained considerable water damage for a long enough period to foster extensive fungal growth. The effects of biodeterioration were apparent in mold growth, insect damage, and biofilm residues. Examination revealed that some works were infected significantly, and others, negligibly so, or not at all. Where water saturated the artwork, fungal growth and collateral loss and weakness were readily apparent.

In the late 1990s, Alice Cooney Frelinghuysen, the Anthony and Lulu Wang Curator of American Decorative Arts at the museum, established a preservation initiative so that these works, so obviously damaged, could be safely studied, exhibited, and prepared for publication. The project brought together a team of conservators, biologists, microbiologists, and chemists at the Metropolitan Museum of Art and local institutions to address these problems.

The following conservation priorities were established: (1) minimization of risks to the collection, to staff members, and to scholars handling the works; (2) adoption of appropriate measures to reduce or eliminate fungal contamination; (3) preservation of original materials; (4) consolidation of areas with weakness and loss; (5) reduction of dark hyphal staining; and (6) post-treatment segregation of works affected by and treated for mold.

Scientific research was directed along the following avenues: (1) classical culturing methods and genetic sequencing (PCR) to identify fungi and rule out toxic species; (2) Raman spectroscopy and scanning electron microscopy to characterize pigmented fungi and their chemical components; and (3) testing of enzymatic treatments to reduce pigmented fungal hyphae and conidia.

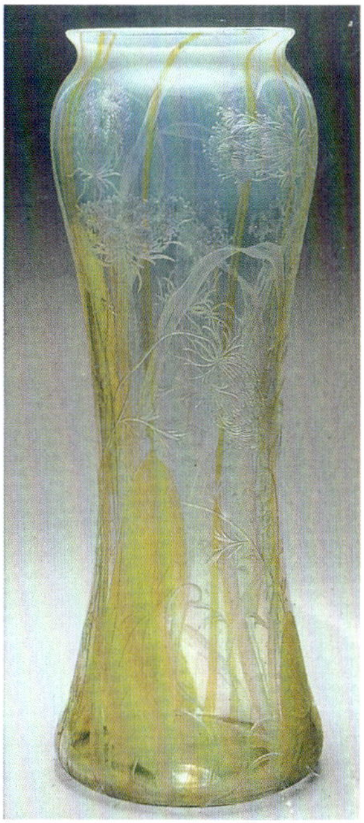

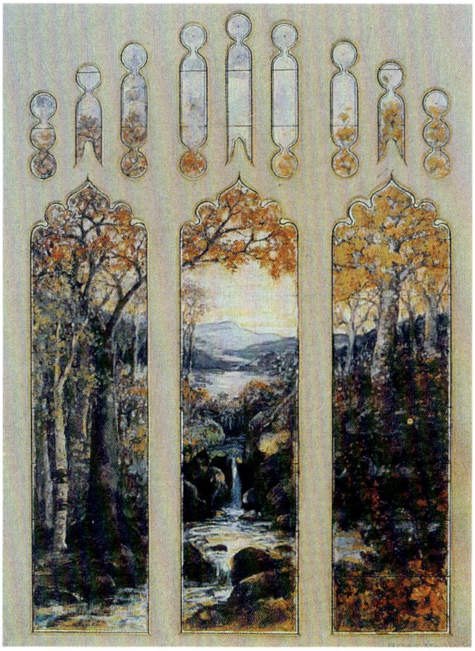

[1] *Favrile Glass Vase* by Louis Comfort Tiffany. Tiffany Glass and Decorating Company. Probably cut by Fredolin Kreischman (1853–1898) from Frelinghuysen, A.C., 1998. Louis Comfort Tiffany at the Metropolitan Museum of Art. Reprint of the *Metropolitan Museum of Art Bulletin* (Summer 1998), Metropolitan Museum of Art, New York, NY, p. 42. [2] *Design for Towle autumn landscape window*, 1923, by Louis Comfort Tiffany. Tiffany Studios. From Frelinghuysen, A.C., 1998. Louis Comfort Tiffany at the Metropolitan Museum of Art. Reprint of the *Metropolitan Museum of Art Bulletin* (Summer 1998), Metropolitan Museum of Art, New York, NY, p. 42.

FUNGAL SPECIES CULTURED AND IDENTIfiED Using both classical culturing methods and genetic replication techniques, the presence of the following genera of fungi was determined (Di Bonaventura et al., this volume): *Aspergillus, Cladosporium, Penicillium, Fusarium, Chaetomium,* and *Thielavia.* Many of these fungi are indigenous to soil. They are considered cosmopolitan, and are often found on cultural property (Florian, 2000; Dicus, 2000). Some of these strains were not able to be cultured, but, rather, were identified solely through genetic sequencing methods.

RISKS OF EXPOSURE Spores produced by many of the genera identified on the Tiffany drawings are truly microscopic, measuring about 2–5 μm in diameter. *Aspergillus, Cladosporium,* and *Penicillium,* for instance, are recognized allergens. Without filtration, they are readily inhaled, and can produce allergic responses in animals due to factors such as their morphology and chemical composition.

Some genera identified above are considered true human pathogens. Many strains of *Fusarium,* for instance, are known to produce mycotoxins—substances that are carcinogenic, teratogenic, and even lethal (University of Minnesota Environmental Health and Safety website, 2001). Other fungi found on the Tiffany drawings, such as *Aspergillus fumigatus, Penicillium chrysogenum,* and *Cladosporium sp.,* can cause serious illnesses in individuals with compromised immune systems (Carlile et al., 2001). As of 1998, only about six antifungal medications—which are not always effective—were available by prescription (Hadfield, 1998).

VISUAL APPEARANCE OF MOLD In general, fungal contamination of the artwork took one of three forms: (1) dense superficial mold growth; (2) palpably weakened paper supports with a mottled appearance and a grayish cast; and (3) localized, dark, circular spots. [FIGS. 3, 4, AND 5]

In works such as *Design for Square Mosaic Panel* (MMA accession # 67.654.053), patches of powdery, matte, or wooly mold, highly colored, white or dark brown, are visible to the naked eye. [FIG. 3] At low magnifications, these masses have a bushy, or fuzzy appearance. At high magnifications, innumerable microscopic spores, and their vegetative and reproductive hyphae, are discernible. [FIG. 4]

In a work such as *Angel Appearing to Three Marys at the Tomb* (MMA accession # 67.654.206), paper supports bear an amorphous, or difficult to discern, grayish cast. The paper looks and feels tangibly weakened, and there is a corresponding loss of paper constituents, indicating the presence of so-called "second colonizer" fungi (Szczepanowska, 1986). [FIG. 5] At magnifications of 90–225×, multiple twisting strands of black conidia entrenched within the fibrous paper are observed, and

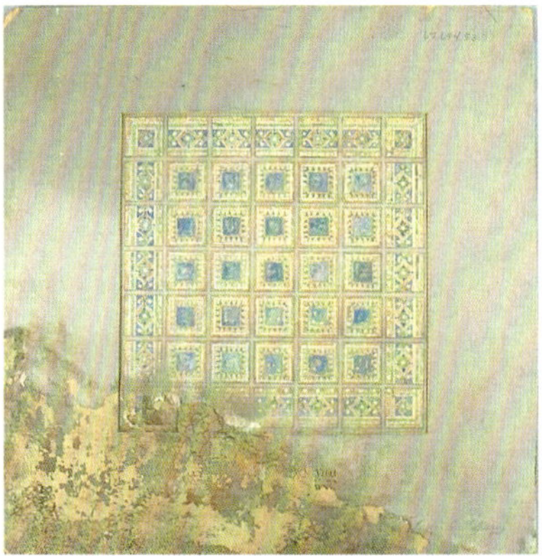

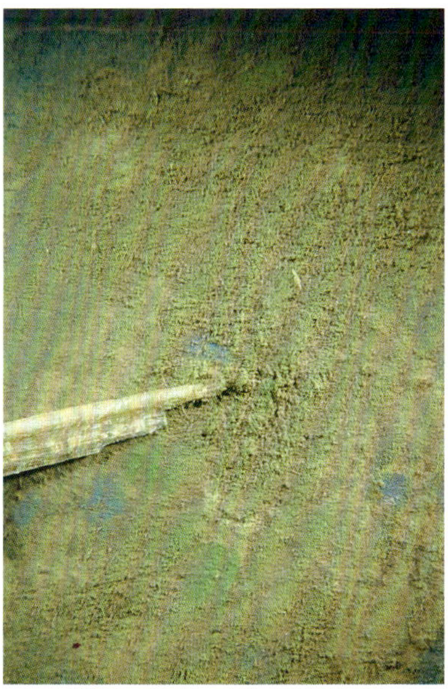

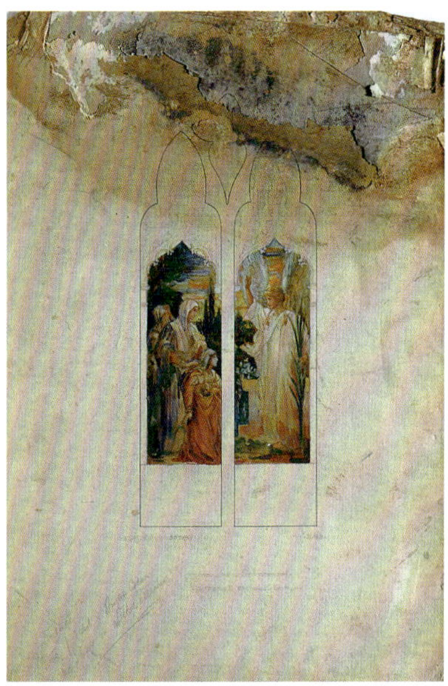

[3] *Design for square mosaic panel,* Tiffany Studios, c. 1892–1918, Metropolitan Museum of Art, accession number 67.654.053. [4] *Design for square mosaic panel,* Louis Comfort Tiffany, Tiffany Studios. Detail showing green fluffy superficial mold growth. [5] *Angel Appearing to Three Marys at the Tomb,* Louis Comfort Tiffany, Tiffany Studios 1892–1918, Metropolitan Museum of Art, accession number 67.654.206.

it becomes apparent that the gray staining is caused by the optical mixture of black conidia and white paper support. [FIG. 6]

A third type of disfiguration caused by darkly pigmented mycelia affects works such as *Design for Square Mosaic Panel*, MMA accession # 67.654.233. [FIG. 7] From ordinary viewing distances, these pigmented hyphae are visible as dark, nearly circular spots—often with dense centers—which extend radially through their host material. At higher magnifications, the very fine melanized fungal filaments are distinguishable from paper fibers. [FIG. 8]

MOLD REMEDIATION Aseptic techniques for the removal of superficial growth from cultural artifacts were adopted, based on practices outlined in the literature (Price, 1994; Nyberg, 1998; Dicus, 2000; Florian, 2000).

Treatment of areas weakened by digestive fungal enzymes is problematic, as entrenched conidia and hyphae cannot be extracted without disturbing the surrounding and underlying paper matrices. Denaturing of spores and/or their safe extraction, along with structural reintegration, are objectives of treatment.

POSSIBLE TREATMENTS FOR PIGMENTED HYPHAE Paper conservators recognize the need for the development of safe and practical treatments for the reduction of pigmented hyphal spots on cellulosic substrates.

Melanized cell walls have proven impervious to most conventional paper conservation reagents. Application of dilute solutions of aqueous sodium borohydride, a reducing agent, can lighten staining to a certain degree. However, the optical density of these dark, three-dimensional stains typically necessitates many repeat applications of more concentrated solutions of aqueous sodium borohydride, and a "melanin ghost" often remains.

The desire for specificity—for an agent or technique able to act only on fungal tissues, and not harm paper—has led to the testing of microbial enzymes to degrade fungal cell walls (Florian and Purinton, 1995). Chitinases have been found to break down unpigmented fungal cell walls; however, the presence of melanin appears to inhibit such penetration (Bloomfield and Alexander, 1967).

MELANIN Melanin is a natural pigment present in many types of living organisms. Depending on its chemical form, it can range from reddish-brown to brown to almost black.

The type of melanin found in many kinds of fungi is a polymer of 1,8-DHN, or dihydroxynaphthalene (Butler and Day, 1998b). It is usually bound with proteins and carbohydrates. An exact chemical structure has not been proposed.

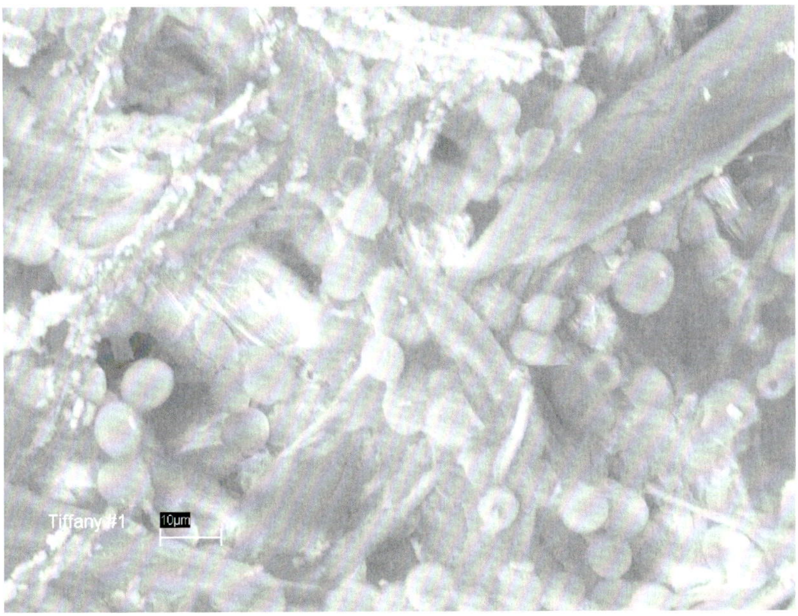

[6] Detail of *Angel and Three Marys at the Tomb*. Scanning electron micrograph 2,000×, 60 Pa, 20 kV (by Robert J. Koestler).

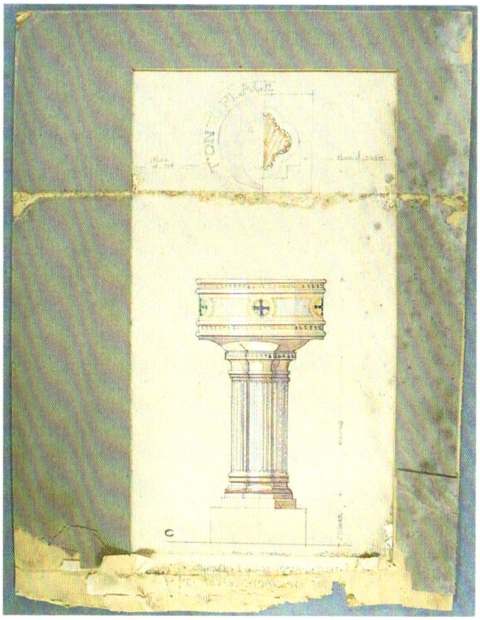

[7] *Design for a marble and glass mosaic font,* Louis Comfort Tiffany, Tiffany Studios. 1892–1918, Metropolitan Museum of Art, accession number 67.654.233. [8] *Design for a marble and glass mosaic font,* Louis Comfort Tiffany, Tiffany Studios. Detail showing dark fungal spots.

LIGNINASES AS A POTENTIAL MELANINASE Research has shown that certain oxido-reducto enzymes, such as the ligninases manganese peroxidase and laccase, produced by the white-rot fungi *Phanerochaete chrysosporium*, can act on melanized fungal walls and decolorize melanin (Butler and Day, 1998a). These findings have prompted experiments using ligninases as potential agents for the reduction or elimination of dark mold staining on paper artifacts. Our initial trials using oxido-reductases applied with dilute solutions of hydrogen peroxide have been successful, and indicate that bleaching of these spots is possible and that the resulting change is quantifiable (Nieto et al., this volume).

CONCLUSIONS Progress has been made in addressing problems relating to the preservation of mold-damaged Tiffany drawings in the collection of the Metropolitan Museum of Art, and aspects of this research may have a bearing on the conservation of other cellulosic-based collections subjected to fungal deterioration.

Fungal contaminants have been identified using both classical and genetic techniques (Di Bonaventura et al., this volume). Fungi—*Penicillium chrysogenum, Penicillium sp., Cladosporium sp., Aspergillus fumigatus,* and *Fusarium oxysporium*—were cultured, and genetic sequencing techniques made possible the identification of *Chaetomium sp., Cladosporium oxysporium,* and *Thielavia hyalocarpo,* which were otherwise not culturable.

From a preservation perspective, these findings suggest that a potential health risk can be posed by this collection, and emphasize the need for personal protective equipment during handling. The viability of some fungi and, conversely, the lack of viability of others, indicates a certain likelihood of future mold outbreaks, should environmental conditions favor mold growth.

Identification of the dematiaceous—or darkly pigmented—fungi found on the Tiffany drawings enables conservators and scientists specifically to tailor treatments to particular dark fungal stains on works of art (for instance, to the spores of *Chaetomium* or the hyphae of *Cladosporium*). Our work indicates that treatment of these spots with oxido-reductases derived from organisms such as *Phanerochaete chrysosporium* is possible and that color change is quantifiable (Nieto et al., this volume).

Further research is needed to test and assess the practicality and affordability of enzyme treatments used to remediate dark mold pigmentation on works of art. Of concern to the conservator are conditions favorable for enzyme activity, the ability to rinse enzymes from paper substrates, and assessment of long-term effects of such treatments. Recent developments in microbiology are providing art conservators with essential means to address problems of biodeterioration. The combination of research, practical intervention, and proper exhibition and storage will help to make these beautiful—if regrettably damaged—Tiffany drawings safe, structurally stable, and accessible to scholars and to the public.

ACKNOWLEDGMENTS The author gratefully acknowledges the following people for their guidance and support throughout this project: Marjorie Shelley, Alice Cooney Frelinghuysen, Margaret Lawson, Akiko Yamazaki-Kleps, Yana Van Dyke, Rachel Mustalish, Barbara Veith, Maria Pia Di Bonaventura, Mike Butler, Silvia Centeno, Douglas Eveleigh, Fernando Nieto-Fernandez, Mary Lou Florian, Paul Szaniszlo, and Robert J. Koestler.

REFERENCES

BELL, A.A. and WHEELER, M.H. 1986. Biosynthesis and functions of fungal melanins. *Annual Review of Phytopathology*, 24, 411–451.

BLOOMFIELD, B.J., and ALEXANDER, M. 1967. Melanins and resistance of fungi to lysis. *Journal of Bacteriology*, 93, 1276–1280.

BUTLER, M. and DAY, A.W. 1998a. Destruction of fungal melanins by ligninases of *Phanerochaete chrysosporium* and other white rot fungi. *International Journal of Plant Science*, 1596, 989–995.

BUTLER, M. and DAY, A.W. 1998b. Fungal melanins: a review. *Canadian Journal of Microbiology*, 44, 1115–1136.

CARLILE, M., WATKINSON, S., and GOODAY, G. 2001. *The Fungi*. 2nd edition. Academic Press. Harcourt Science and Technology Publication London, pp. 432–443.

DICUS, D.H. 2000. One response to a collection-wide mold outbreak: How bad can it be—how good can it get. *Journal of the American Institute of Conservation*, 39, 85–105.

FLORIAN, M.L. 1997. *Heritage eaters: Insects and fungi in heritage collections*. London James and James (Science Publishers) Ltd.

FLORIAN, M.L. 2000. Aseptic technique: A goal to strive for in collection recovery of moldy archival materials and artifacts. *Journal of the American Institute of Conservation*, 39, 107–115.

FLORIAN, M.L. and PURINTON, N. 1995. Determination of location of stains in fungal spots and enzymatic removal of pigmented hyphae in paper. Preprints, *3rd International Conference on Biodeterioration of Cultural Property*, 4–7 July, 1995, Bangkok, Thailand, 414–425.

FRELINGHUYSEN, A.C. 1998. Louis Comfort Tiffany at the Metropolitan Museum of Art. Reprint of the *Metropolitan Museum of Art Bulletin* (Summer 1998), Metropolitan Museum of Art, New York, NY.

HADFIELD, LT. COL. 1998. Health risks associated with molds and mildews. SCMRE Museum Support Center, C 9815—Preservation Fundamentals II: Mold and Mildew. Unpublished Typescript—Class Handout on Occupational Safety Lecture.

NYBERG, S. 1998. *Invasion of the giant mold spore*. Atlanta: Solinet.

PRICE, L. 1994. Mold: Managing a mold i nvasion: *Guidelines for disaster response technical series* 1. Philadelphia: Conservation Center for Art and Historic Artifacts.

SZCZEPANOWSKA, H. 1986. Biodeterioration of art objects on paper. *The Paper Conservator*, 10, 31–39.

UNIVERSITY OF MINNESOTA. 2001. UMN Environmental Health and Safety website: http://www.deh.umn.edu/iaq/fungus/glossary.html

M.P. DI BONAVENTURA[1*]
R. DESALLE[2]
D.E. EVELEIGH[3]
A.M. BALDWIN[1]
R.J. KOESTLER[1]

STUDIES OF FUNGAL INFESTATIONS OF TIFFANY'S DRAWINGS: LIMITS AND ADVANTAGES OF CLASSICAL AND MOLECULAR TECHNIQUES

1 The Metropolitan Museum of Art, New York
2 The American Museum of Natural History, New York
3 Cook College, Rutgers University, New Brunswick, NJ
* Corresponding author

ABSTRACT A noteworthy collection of drawings (c. 1885–1930) produced by Louis Comfort Tiffany (1848–1933) and his studio artists sustained significant mold damage prior to its acquisition by the Metropolitan Museum of Art in 1967. In order to be exhibited, published, and studied safely, the works required conservation intervention. The works, which were water damaged sometime between 1933 and 1967, were sampled for molds using both classical and molecular techniques to identify the fungi present. The range of molds isolated by culturing were identified using their ITS (internal transcribed spacer) region sequences via searching for similarity in GenBank. Five species were identified as belonging to: *Aspergillus fumigatus*, *Cladosporium sp.*, *Fusarium oxysporum*, *Penicillium chrysogenum*, and *Penicillium sp.* In contrast DNA extracted directly from dark, thick-walled spores present on the paintings showed a 98% similarity to *Chaetomium sp.* when compared with GenBank registries. Similarly from brown stained paintings, fungal samples were amplified. In their ITS1-5.8-ITS2 region 80% of the clones matched *Cladosporium* with a 99% similarity. Yet, the remaining 20% of the clones matched 100% of the 2 ITS regions of *Chaetomium sp.* with a 98% similarity. A similar trend occurred with samples of mixed brown fungal spots from another work, where 80% of the clones matched *Thielavia hyalocarpa*, and the remaining 20% of the clones corresponded to *Cladosporium oxysporum*.

The apparent disparities (80:20) at first sight could represent differences in the proportions of the two molds originally present, but they also may be related to the degradative processes that affect DNA, or to the order of colonization. For instance, *Chaetomium sp.* could be detected with sequences of a few hundred base pairs while *Cladosporium sp.* was identified with fragments of more than 500 bp. These overall results may indicate that *Chaetomium sp.* was an earlier colonizer or that *Cladosporium sp.*, because its DNA was still in better condition, could be thought of as a later colonizer of the drawings.

Cladosporium sp., *Chaetomium sp.* and *Thielavia* appear as the primary degradative culprits, while the other general isolates may be thought of as contaminants.

Unfortunately, the interpretations cannot be defined in a better manner from the historical background of the collection, since the disposition of Tiffany's drawings was unknown prior to 1967. Based on the analyses of the existing damage, it has been supposed that sometime between 1930 to around 1967 the collection sustained water damage. It is unclear how the handling of these 400 works, during the past decades, may have affected cross-contaminations among the drawings. In overview, the investigations indicate that the molecular techniques give a more definitive picture than do classical techniques of the fungal species composition on a work of art. This has implications in terms of restoration and conservation. In this instance, treatment methods should focus on removal of *Chaetomium sp.* and *Cladosporium sp.* or their by-products of metabolism, with lesser attention directed to *A. fumigatus*, *F. oxysporum*, *P. chrysogenum*, and *Penicillium sp.*

INTRODUCTION Objects of beauty, memorial windows for churches and synagogues, blown Favrile glass, and iridescent mosaics are among the many treasures created by Louis Comfort Tiffany (1848–1933) and his studio artists. These works spanned the years from 1885 to 1930. A formative part of these efforts was the design drawing, taking form as preparatory sketch, colored pencil rendering, and finished presentation drawing. Tiffany's artists were versed in representational depictions and architectural rendering, conveying translucence and opalescence with watercolor paints on artists' papers and illustration boards.

Of these works, some 420 drawings acquired by the Metropolitan Museum of Art in 1967 had sustained considerable mold damage prior to their acquisition. The mold problems were related to an accident, or accidents, that occurred sometime between 1933 and 1967, wherein the drawings received significant wetting. Removal of the mold and its by-products entails extensive conservation work.

The extreme fungal devastation of these works is in stark contrast to their delicacy, precision, and tonal harmony. Water damage created a number of condition problems, such as soil deposition, tide lines, delaminated substrates, and a population of long-since dead, and sometimes voracious, insects. The distribution of apparently dormant, multicolored, brown, and black fungal material found in and on as many as 20% of the objects is pronounced.

The storage history of this collection is unknown. The collection was found in a fourth-floor attic room of a marble supplier in Long Island. In 1967 the collection was acquired by the Metropolitan Museum of Art and, except for cataloguing and photography, the group has remained largely undisturbed in museum storage boxes.

Presumably between 1930 and 1967 the collection sustained water damage at least once and there was extensive fungal growth, both on the surface and through the cellulosic substrate of the stacked drawings. Once mold growth on paper is established, aesthetic and structural damage may occur. Stains usually arise from colored fungal bodies and from organic metabolic products such as pigments, while the extracellular enzymes released by fungi can cause the digestion of the cellulose fibers modifying the texture and strength of the paper. For instance, several species belonging to the genera *Penicillium, Fusarium, Alternaria,* and *Cladosporium* can cause stains on paper that range from light green to pink and brown to black. The presence of *Chaetomium globosum* or even *Stachybotrys atra* on paper is not surprising considering the cellulolytic activity displayed by these species (Szczepanowska and Cavaliere, 2000). Fungal-induced stains can sometimes be reversed through mechanical stain removal or extracted with solvents. Laser treatments have been successfully applied to some of the fungal stains, which also illustrates that their removal is dependent on the fungal species involved (Szczepanowska and Moomaw, 1994). Overall, however, removal of fungal-induced stains on works of art on paper remains problematic, and an understanding of their composition, both as individual fungal species and/or as mixed fungal communities, may contribute to their removal.

Culture-based methods have limitations in their ability to recover all infesting molds. This can be due to the age of the processed samples, many of the fungal species may be dead; plus there is the fact that not all the viable ones will grow under selected laboratory conditions. Conversely a single extraneous spore that floated by can yield an erroneous positive result. However, these limitations related to culture-based methods may be partially overcome by analysis of the total DNA extracted directly from a carefully selected environmental-molded sample, if the nucleic acid in a sample has not degraded entirely. The non-coding ITS region of the rDNA gene cluster has been widely used for studying fungal species profiles from uncultured fungal communities (Pennanen et al., 2001). The sequence variation of the ITS1 and ITS2 region in fungi especially has led to their use in systematic and phylogenetic analyses at lower taxonomic levels. Some fungal species complexes have, however, not yet been resolved using the ITS region, and many fungi may contain within-species variation in their ITS sequences (Skouboe et al., 1999; Travis et al., 2000). Such molecular techniques have mainly been applied directly for detection and identification of mycorrhizal fungi, phytopathogenic fungi from within plant tissues, endophytic fungal communities, or wood-decaying fungi and microbial diversity in soil samples (Guo et al., 2001; Vainio and Hantula, 2000). So far, to our knowledge, there have not been studies that have used molecular techniques on uncultured fungal communities colonizing art objects made with material such as paper.

In order to identify fungi based on genetic sequence analyses, DNA was extracted from scraped material and/or contaminated paper. When dealing with old fungal contaminations, for instance, the remains of fungal hyphae on paper, there are often problems in extracting DNA, and this may be related to processes that degrade DNA (Herrmann and Hummel, 1994). When samples are characterized by the presence of what can be identified as remains of the fungal mycelium, can DNA be found in it and what is the quality of the DNA? Further, the possibility of testing different extraction methods is, in many cases, limited by the small amount of starting material and, generally, the attempt to identify the fungal species is made on amplified DNA sequences of a few hundred base pairs. From older specimens, yields may be tentatively increased by individual amplification of the two ITS regions, ITS1 and ITS2, with the primer pairs ITS1/2 or ITS5/2 and ITS3/4 (Kretzer et al., 1996; Ristaino et al., 2001). In situations where drawings are infested by more than one species, other steps are required to obtain the fungal species profiles, such as cloning the PCR products, RFLP (restriction fragment length polymorphism) or DGGE (denaturing gradient gel electrophoresis).

Although an identification of the fungal species may be achieved once the DNA has been extracted from fungal spores (Zhou et al., 2000), it is possible that similar results may not be obtained from older fungal contaminations that are characterized by the remains of fungal hyphae.

MATERIALS AND METHODS

Selection of the Fungal Spots Preliminary visual and stereomicroscopical investigations were made in order to select the fungal stains from infected drawings. Particular focus was placed upon the brown, gray, and gray to black fungal spots, which are among the most difficult to treat. The drawings sampled included: *Mosaic Panel* (MMA Acc. No. 67.654.53), *Chancel Rail* (MMA Acc. No. 67.654.65), *Desk Items* (MMA Acc. No. 67.654.69), *Angel Appearing to Three Marys at the Tomb* (MMA Acc. No. 67.654.206), *Four Seasons* (MMA Acc. No. 67.654.317), *Adoration of the Magi* (MMA Acc. No. 67.654.319), and *Vassar College* (MMA Acc. No. 67.654.307); these are described in Baldwin (2003, this volume). Swabs, scraped fungal material, and, where possible, small pieces of contaminated paper were aseptically selected under stereomicroscopy and used as samples to be investigated with scanning electron microscopy, culturing tests, and molecular techniques.

Scanning Electron Microscopy Morphological observations on the remains of the fungal contamination and an evaluation of the existing damage to the cellulose fibers were conducted under SEM (LEO 1455VP). As the fungal spots had air-dried, no preparation of the specimens in terms of standard SEM procedures was considered necessary. A small amount of material from the selected specimens was directly mounted on aluminum stubs, coated with gold, and examined.

Culture-Based Methods An attempt to grow and isolate the fungal species was carried out by direct incubation of the scraped fungal material or small pieces of contaminated paper in a humidity chamber, and by culturing tests on selected media. A first set of samples was placed in petri dishes containing a filter paper moistened with sterile distilled water. In order to avoid contact with water, samples were located on the surface of glass rods (Arai, 2000). A second set of specimens, taken from the same fungal spots, was inoculated on 3% MEA (malt extract agar) plates. Swabs and scraped fungal material were, respectively, washed or resuspended in 0.01% Tween 80 and vortexed. Each suspension (200 µl), as well as pieces of contaminated paper, was inoculated in duplicate on the surface of the agar plates. The specimens were incubated in separate humidity chambers (airtight glass jars) for four weeks at laboratory conditions. Both of these tests were conducted under high relative humidity in order to simulate the environmental conditions under which the drawings could have been contaminated. The relative humidity was adjusted to 96–98% with a saturated aqueous solution of 150 gl^{-1} K_2SO_4 (Korpi et al., 1997).

FUNGAL DNA DIAGNOSTICS

DNA EXTRACTION

Fungal Isolates DNA extraction from pure fungal cultures, kept on PDA (potato dextrose agar) plates supplemented with cellophane membranes to facilitate the removal of the mycelial colony, was carried out using the DNeasy Plant Mini Kit (QIAGEN). Small pieces of the fresh mycelia (10 mm^2) were ground in sterile 1.5-ml microcentrifuge tubes with liquid nitrogen using a glass spatula. The tissue powder was then mixed with 400-µl lysis buffer and processed as suggested by the manufacturer's instructions.

Field Samples Total genomic DNA was directly extracted from scraped fungal material and/or small pieces of contaminated paper using the UltraClean Soil DNA Isolation Kit (MO BIO Laboratories, Inc.) (Pennanen et al., 2001). Lysis of the samples was performed on a flat bed vortex at maximum speed for 6 to 8 min., and the washing step was repeated with the samples that eluted brown. The concentration of the template was increased, if needed, by use of a speed vacuum, without heating, and the pellet was re-suspended in sterile distilled water.

PCR AMPLIFICATION OF ITS REGIONS The ITS1-5.8-ITS2 region was amplified by PCR using the primers ITS1 and ITS4. The two ITS regions were individually amplified with the primer pairs ITS1, ITS2 and ITS3, ITS4 (White et al., 1990). PCR amplifications were carried out in 50 µl of reaction mixtures containing 1 to 5 µl of the templates, 5 µl of 10 × PCR buffer (15 mM $MgCl_2$), 5 µl of 2.0 mM each of dNTP, 5 µl of 0.5 µM each of primer (ITS1 × ITS2, ITS3 × ITS4, ITS1 × ITS4), 5 µl of 20 mg/ml BSA (bovine serum albumin), and 0.2 µl of 5 U/µl of AmpliTaq DNA Polymerase (PE Applied Biosystems). Forty cycles of amplification were performed in a GeneAmp PCR System 9700 thermal cycler (PE Applied Biosystems) after an initial denaturation step at 94°c for 5 min. The temperature cycling parameters were denaturation at 94°c for 30 sec followed by primer annealing at 48°c for 30 sec and primer extension at 72°c for 45 sec to 1.30 min., with a final extension at 72°c for 10 min. following the last cycle. PCR products were visualized on 1.2% agarose gels run in 1x TAE buffer followed by ethidium bromide staining.

CLONING AND SEQUENCING OF ITS REGIONS PCR products were ligated into the pCR 2.1 plasmid vector using the TOPO TA Cloning kit (Invitrogen). Competent TOP10 One Shot cells were transformed and positive transformants were screened using

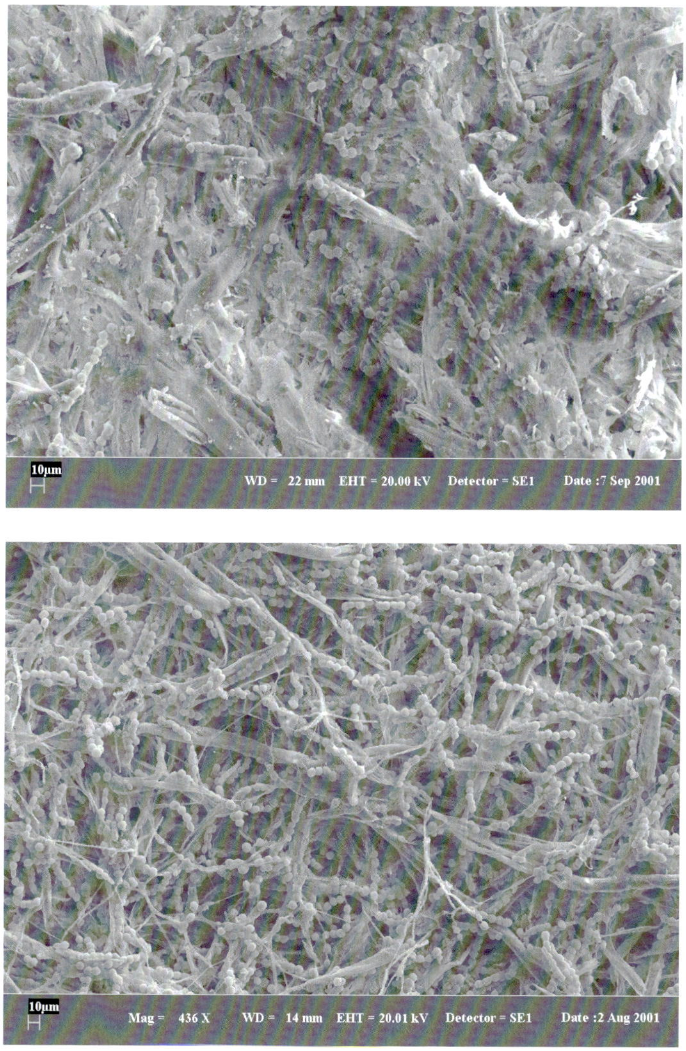

[1] Fragments of damaged paper contaminated by fungal spores (in *Angel and Three Marys*). Figure 1a shows damage to the paper fibers, while figure 1b shows many chains of mixed fungal spores and hyphae.

PCR amplification with universal M13 (–20) forward and reverse primers according to the manufacturer's instructions. PCR products were visualized on 1.2% agarose gels run in 1x TAE buffer followed by ethidium bromide staining. DNA sequencing was performed in an ABI 3700 DNA analyzer (PE Applied Biosystems). For the sequencing of the cloned fragments, both strands of the plasmid with the fungal insert were sequenced with M13 forward and reverse primers. ITS1, ITS2, ITS3, and ITS4 primers were used for direct sequencing of the PCR products in both directions. Cycle sequencing was carried out using the BigDye Terminator Cycle Sequencing (DNA sequencing kit, PE Applied Biosystems).

SEQUENCE ANALYSIS Sequences from both direct sequencing and cloning were aligned to complete or partial ITS sequences available in GenBank.

RESULTS AND DISCUSSION Using the stereomicroscope, selected spots were classified in one of three ways: fungal-induced stains caused by the presence of pigmented fungal hyphae, such as the green patina found in *Mosaic Panel* or the brown spots in *Desk Items*; staining that may have resulted from organic metabolic products such as the pigments, as observed in *Chancel Rail*; and the most problematic in terms of conservation, those caused by fungal spores and associated with cellulose digestion, as shown in *Angel and Three Marys*, *Four Seasons*, and *Vassar College*. This third group, when observed under SEM, was characterized by the presence of often-collapsed fungal hyphae and/or by single, groups, or chains of fungal spores. In some cases these fungal spores showed only conidiophore fragments and an absence of aerial structures that are required to identify the fungi at the genus level. Because of this no further work was undertaken to tentatively identify the fungal species using morphology. [FIGS. 1A,B]

The fungal culture isolates that were grown from the disfigured paintings, were identified based on comparison of similar ITS sequences in GenBank, namely: *Aspergillus fumigatus*, *Cladosporium sp.*, *Fusarium oxysporum*, and *Penicillium sp.* Few fungal cultures could be grown from the culturing tests. *Penicillium chrysogenum* was the only species that could be detected with direct incubation in the humidity chamber inside the petri dishes. This may be due to the fact that we were dealing with an old fungal contamination and the age of the processed specimens could affect the viability of the fungal species. Furthermore, the adequacy of the culture-based methods can be limited by the selectivity of the isolation procedures used.

Directly extracted DNA samples yielded PCR products in 33% of the total number of specimens processed. DNA extracted from thick-walled fungal spores in gray to black spots (in *Angel and Three Marys*, *Four Seasons*, and *Vassar College*) could be amplified and sequenced using the primer pairs ITS1 × 2, ITS3 × 4 and ITS1 × 4. The blast search of these sequences in GenBank showed that both the individuals' ITS regions and the entire ITS1-5.8-ITS2 region matched with *Chaetomium sp.* with 98% similarity. This result fits with the microscopical investigations made on the gray to dark fungal spots. These were characterized by the digestion of the cellulose fibers that can be related to *Chaetomium sp.* known for its cellulolytic activity. [FIGS. 2A,B] A *Cladosporium oxysporum* strain was identified with 99% similarity when both sequences from direct sequencing and cloned sequences of the entire ITS region were aligned for blast searches (in *Mosaic Panel*). In contrast, recovery from the fungal spots by direct sequencing yielded 2, and possibly a third, species compared to 5 species by culturing followed by sequence identification.

Thus, two different groups of cloned sequences were obtained from DNA extracted from brown stains (in *Angel and Three Marys*) when the sequences were blasted in GenBank. The alignment of 80% of the clones from the entire ITS1-5.8-ITS2 region matched with *Cladosporium sp.* with 99% similarity, while the remaining 20% of these clones and 100% of the smaller ITS1 and ITS2 regions matched with *Chaetomium sp.* with 98% similarity. [FIG. 2C]

The results of the lower number of species identified via direct PCR amplification compared to the greater numbers obtained by culturing should be set in general perspective. "Over 160 species of fungi have been described as commonly recovered from buildings. A careful assessment of these taxa shows that only a very small percentage of these fungi consistently grow on building materials within buildings." (Levetin et al., 2002). In library materials 234 species of fungi had been isolated, and reported on from 1919–1977 (Zyska, 1997). We agree that general culturing will tend to yield high species numbers, though many of the recovered isolates represent common aerial drifters. The single isolate of the potential toxin forming *Aspergillus fumigatus* falls into this category. In contrast to this over recovery of species, direct culturing could also yield a lack of representative species as the spores or hyphal fragments could have died through aging. The current samples are apparently at least 30 years old. However, caution is necessary as mushroom hyphae are recorded to have survived 83 years, mycelial fungi in a glacier for greater than 11 thousand years, and there was the amazing claim of 5 million years

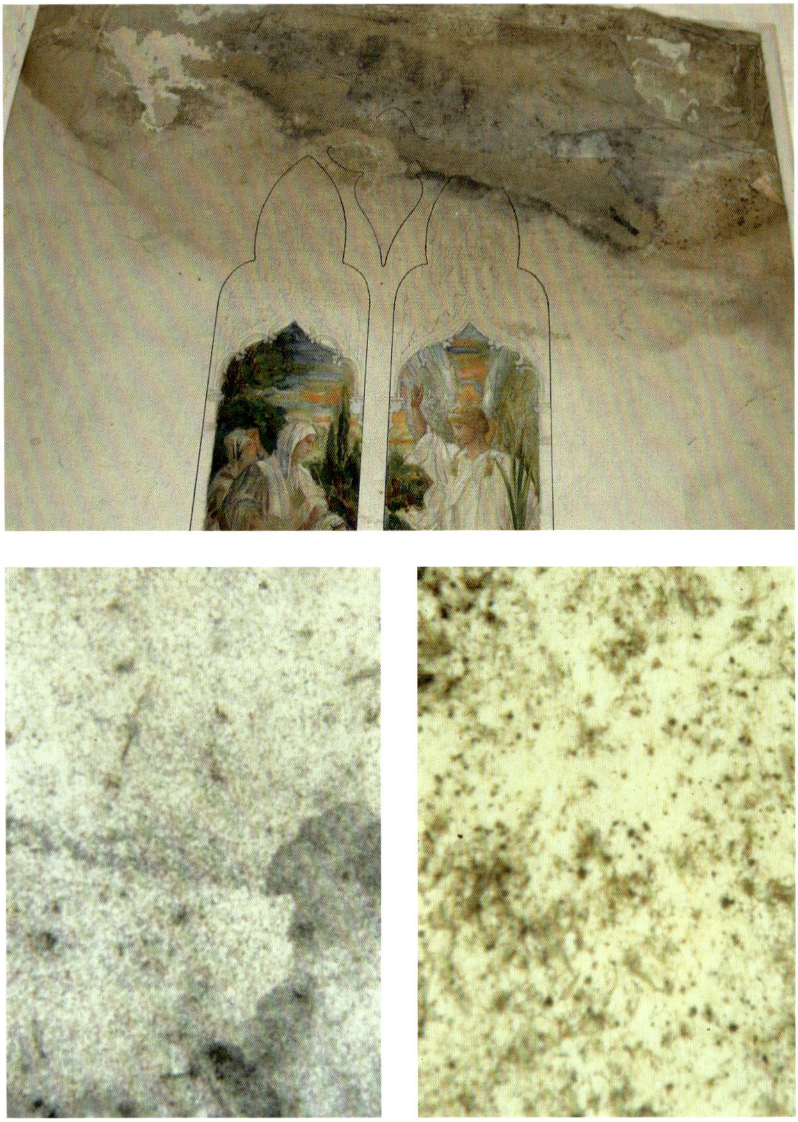

[2] Figure 2a, mixed fungal contaminations ranged from brown, and gray to dark in color (upper right corner) are shown in *Angel and Three Marys*. Figure 2b shows a gray to dark fungal stain caused by the presence of fungal spores, 6.6×. Figure 2c, brown fungal spots, 20×.

for *Aspergillus, Cladosporium* and *Penicillium* from Pliocene core samples (Kennedy et at., 1994). Even so, the storage in an attic for years could well have resulted in the death of a major portion of the fungi in the Tiffany art samples.

The species identity obtained through direct amplification of the ITS sequences directly from the drawings merits comment. These results may be related to the degradative processes that affect DNA, and the rationale for suggesting this is as follows. Older DNA samples from these drawings will probably be more severely degraded and will only amplify with PCR primers that are expected to produce small fragments of less than 200 base pairs (assuming an equivalence of all fungal DNAases). Therefore more recent colonizers of the drawings might be expected to have degraded less and hence could yield PCR products of larger size, or greater than 500 bp. In fact *Chaetomium sp.* ITS sequences could be detected only in PCR products with sequences of a few hundred base pairs, while *Cladosporium sp.* was identified in clones of more than 500 bp. This result seems to indicate that *Chaetomium sp.* could have been an earlier colonizer of the drawings. *Cladosporium sp.*, because its DNA was still in good enough condition to produce larger PCR products, could be thought of as a more recent contaminant of the drawings. On the other hand, the environmental conditions that the drawings were exposed to will affect the resistance of fungi and consequently the fungal species composition of the stained areas. A similar trend was shown in light to dark brown fungal stains (in *Mosaic Panel*). Cloned sequences of the ITS1 and ITS2 regions individually amplified matched with *Thielavia hyalocarpa* with 98% similarity while the alignment of 83% of the entire ITS region blasted with *Cladosporium oxysporum* with 99% similarity. [FIGS. 3A,B]

As shown above, differences in species composition were observed when the results from culturing tests were compared with those obtained from molecular investigations. The only fungal species identified from both culture-based methods and fungal DNA diagnostics was *Cladosporium sp.* In most of the directly extracted samples, *Cladosporium sp.* was identified doing the blast search with the entire ITS region amplified with ITS1 and ITS4 primers. In theory, the species composition of a mixed DNA sample can be revealed by PCR analysis, provided that the amplification efficiencies from different templates are similar. In our investigations it seems that the fungal cultures isolated and identified from the culturing tests, with the exception of *Cladosporium sp.*, may be thought of as contaminants and thus they are present in the samples at lower amounts. Unfortunately, these results cannot be supported by the historical background of the collection because its history was unknown prior to 1967, when it was acquired by the Metropolitan Museum

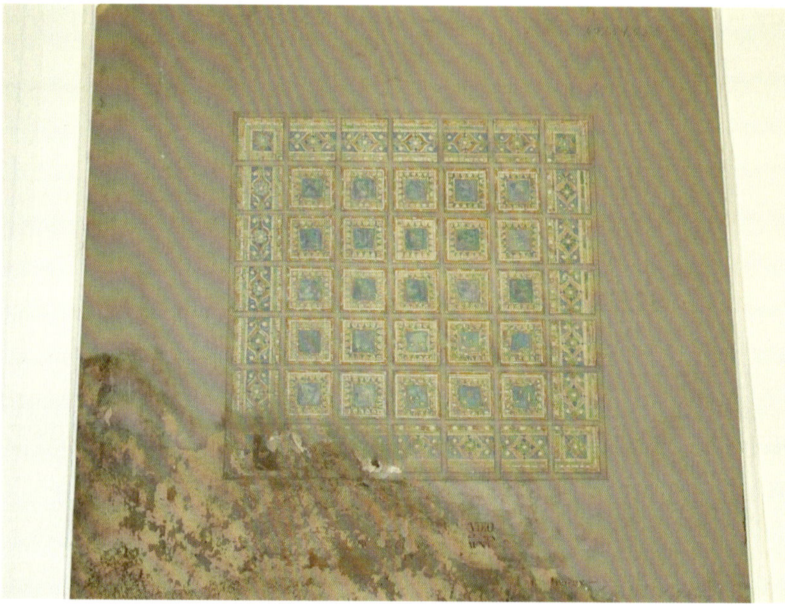

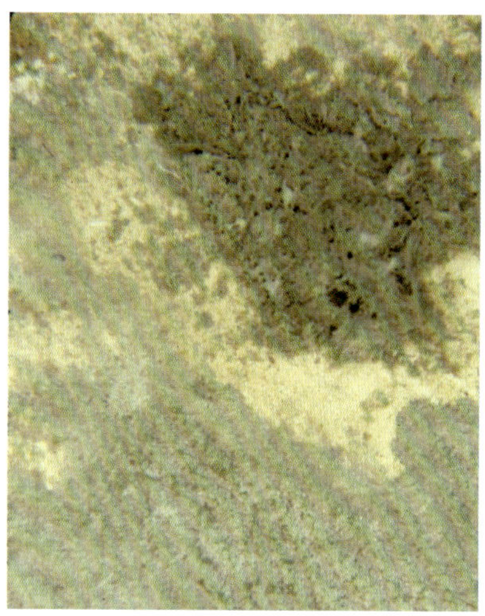

[3] Figure 3a shows deteriorating activity caused by mixed fungal contaminations on the bottom of the left side in *Mosaic Panel*. Figure 3b, mixed brown fungal stains scraped from the bottom of the left side in *Mosaic Panel*, 6.6×.

of Art. Based on analyses of the existing damage, it has been supposed that sometime between 1930 and 1967 the collection sustained water damage. And it is unclear how the handling of these 420 works, during the past decades, may have affected cross-contaminations among the drawings. Any records that could have been used to understand the original position of the works in relation to each other had not been recorded when the collection was first accessioned and later catalogued.

Thus, the investigations reported here indicate that culturing old fungal contamination may give partial results in terms of fungal species composition. With DNA methods it is possible to identify the primary colonizer fungi; probably those microorganisms are responsible for the staining and deterioration of the drawings. And at last, but not least important, this has implications in terms of restoration and conservation of artworks.

The fungal-induced stains on paper remain problematic. For instance, testing more selective and more effective procedures such as enzymatic treatments could be a valid alternative to those methods that have so far proven partially or totally unsuccessful. But, in order to remove the fungal hyphae and/or the fungal stains, there is the need to identify the fungal microorganism against which the enzymatic treatments are applied.

The implication for conservation of the Tiffany drawings is that the brown, and the gray to black spots are caused by the dematiaceous fungi *Chaetomium sp., Cladosporium sp.,* and possibly *Thielavia sp.,* and not by other species of fungi isolated by culture techniques from the surface of the drawings (e.g., *Aspergillus sp., Fusarium sp., Penicillium spp.*). Therefore treatment methods should be focused on removal of *Chaetomium sp., Cladosporium sp.,* and *Thielavia sp.* or their by-products that promote staining.

REFERENCES

ARAI, H., 2000. Foxing caused by fungi: Twenty-five years of study. International *Biodeterioration and Biodegradation*, 46, 181–188.

BALDWIN, A.M., 2003. An introduction to Tiffany studies. In: Koestler, R.J., Koestler, V.R., Charola, A.E., and Nieto-Fernandez, F.E., (Eds.), *Art, Biology, and Conservation: Biodeterioration of Works of Art*. The Metropolitan Museum of Art, New York, pp. 82–93.

GUO, L.D., HYDE, K.D., and LIEW, E.C.Y., 2001. Detection and taxonomic placement of endophytic fungi within from tissues of *Livistona chinensis* based on rDNA sequences. *Molecular Phylogenetics and Evolution*, 20, 1–13.

HERRMANN, B., AND HUMMEL, S., 1994. Ancient DNA. Springer-Verlag, New York.

KENNEDY, M.J., READER, S.L., and SWIERCZYNKI, L.M., 1994. Preservation records of micro-organisms: Evidence of the tenacity of life. *Microbiology*, 140, 2513–2529.

KORPI, A., PASANEN, A.L., PASANEN, P., and KALLIOKOSKI, P., 1997. Microbial growth and metabolism in house dust. *International Biodeterioration and Biodegradation*, 40, 19–27.

KRETZER, A., LI, Y., SZARO, T.M., and BRUNS, T.D., 1996. Internal transcribed spacer sequences from 38 recognized species of *Suillus sensu lato*: *Phylogenetic and taxonomic implications*. *Mycologia*, 88, 776–785.

LEVETIN, E.W., HORNER, E., CARPENTER, S., and MCGINNIS, M., 2002. A mycologist's guide to indoor mold investigations. In: *The Incubator* (Newsletter of the Mycological Society Supplement to *Mycologia*, 53, December, 2002, pp. 1–5.

PENNANEN, T., PAAVOLAINEN, L., and HANTULA, J., 2001. Rapid PCR-based method for the direct analysis of fungal communities in complex environmental samples. *Soil Biology and Biogeochemistry*, 33, 697–699.

RISTAINO, J.B., GROVES, C.T., and PARRA, G.R., 2001. PCR amplification of the Irish potato famine pathogen from historic specimens. *Nature*, 411, 695–697.

SZCZEPANOWSKA, H., and CAVALIERE A.R., 2000. Fungal deterioration of 18th and 19th century documents: A case study of the Tilghman Family Collection, Wye House, Easton, Maryland. *International Biodeterioration and Biodegradation*, 46, 245–249.

SZCZEPANOWSKA, H.A., and MOOMAW, W.R., 1994. Laser removal of fungus-induced stains from paper. *Journal of the American Institute of Conservation*, 33, 25–32.

SKOUBOE, P., FRISVAD, J.C., TAYLOR, J.W., LAURITSEN, D., BOYSEN, M., and ROSSEN, L., 1999. Phylogenetic analysis of nucleotide sequences from the ITS region of terverticillate *Penicillium* species. *Mycological Research*, 103, 873–881.

TRAVIS, H., IWEN, P.C., and HINRICHS, S.H., 2000. Identification of *Aspergillus* species using internal transcribed spacer regions 1 and 2. *Journal of Clinical Microbiology*, 38, 1510.

VAINIO, E., and HANTULA, J., 2000. Direct analysis of wood-inhabiting fungi using denaturing gradient electrophoresis of amplified ribosomal DNA. *Mycological Research*, 104, 927–936.

WHITE, T.J., BRUNS, S., LEE, and TAYLOR, J.W., 1990. Amplification and direct sequencing of fungal ribosomal RNA genes for phylogenetics. In: Innis, M.A., Gelfand, D.H., Sninsky, J.J., and White, T.J., (Eds.), *PCR Protocols: A Guide to Methods and Applications*, Academic Press, Inc., New York, pp. 315–322.

ZHOU, G., WHONG, W.Z., ONG, T., and CHEN, B., 2000. Development of a fungus-specific PCR assay for detecting low-level fungi in an indoor environment. *Molecular and Cellular Probes*, 14, 339–348.

ZYSKA, B., 1997. Fungi isolated from library materials: A review of the literature. International *Biodeterioration and Biodegradation*, 40, 43–51.

F.E. NIETO-FERNANDEZ[1]
S.A. CENTENO[2]
M.T. WYPYSKI[2]
M.P. DI BONAVENTURA[3]
A.M. BALDWIN[2]
R.J. KOESTLER[2]

ENZYMATIC APPROACH TO REMOVAL OF FUNGAL SPOTS FROM DRAWINGS ON PAPER

1 SUNY College at Old Westbury, Old Westbury, NY 2 The Metropolitan Museum of Art, New York

ABSTRACT A common and unwanted side effect of microbial growth on surfaces is the appearance of pigments that were not originally on the substratum. In the case of autotrophic organisms these pigments, i.e., chlorophylls and carotenoids, are associated with the photosynthetic processes. For heterotrophic organisms, such as fungi, pigments are synthesized as a response to environmental stress, and they appear to play a role in pathogenesis, in both plants and animals. The presence of one of these pigments, melanin, in the cell wall protects its constituents from enzymatic digestion. Melanin causes much of the black and brown staining on fungal-attacked paper surfaces, either attached to fungal structures or as "free" polymers attached to the paper. Removal of the melanin pigment without damaging the art is problematic. In this study an enzymatic approach to breakdown of melanin using lignin-degrading enzymes was tested. These enzymes are produced by white-rot fungi as part of their secondary metabolism and have been shown to successfully break down free melanin. The effectiveness of the treatment process was assessed using colorimetry, SEM (scanning electron microscopy), FTIR (Fourier transform infrared spectroscopy), and Raman spectroscopy.

INTRODUCTION The removal of fungal spots from cellulosic substrata is an ongoing problem in conservation science. Fungal spots can be caused by the secretion of metabolites that stain the paper or by the presence of pigmented fungal structures, especially melanins. Even after the fungus is dead the pigmented cell walls remain on the surface of the substratum, a phenomenon referred to as "melanin ghosts." These structures are particularly resistant to chemical and enzymatic degradation.

Melanins are high-molecular-weight polymers ranging in color from brown to black and composed of various types of phenolic or indolic monomers (Butler and Day, 1998; Bell and Wheeler, 1986). They are found in the cell wall of vegetative mycelia and reproductive structures such as sclerotia and conidia, and they may also be found as extracellular polymers formed in the medium around fungal cells (Bartnicki-Garcia and Reyes, 1964; Durrell, 1964; Bell and Wheeler, 1986; Butler and Day, 1998). They are recognized as macromolecules that are generally insoluble in water and aqueous acids (Bell and Wheeler, 1986). They protect the fungus from physicochemical degradation caused by exposure to UV radiation, high temperatures, oxidizing agents, enzymatic lysis, desiccation, and fungicides, and they also confer on the fungus metal binding properties, and virulence (Bloomfield and Alexander, 1967; Butler and Day, 1998; Jacobson, 2000).

Fungal melanins are derived from 1,8-dihydroxynaphtalene (DHN), found in Ascomycetes and Deuteromycetes, and γ-glutaminyl-4-hydroxybenzene (GHB) and cathecol found in Basidiomycetes (Bell and Wheeler, 1986). In the synthesis of DHN melanin, 1,3,6,8-tetrahydroxynaphtalene is initially formed by the head-to-tail joining and cyclation of acetate molecules. An alternating pair of reduction and dehydration reactions results in the formation of the immediate precursor 1,8-DHN. The mechanism of polymerization, however, is still unknown, and because of the difficulties involved in characterizing the intact polymer, its structure is poorly understood (Bell and Wheeler, 1986; Butler and Day, 1998; Jacobson, 2000).

Previous studies have tried proteases, chitinases, lipases, and glucanases to break down the fungal cell wall (Florian and Purinton, 1995; Bloomfield and Alexander, 1967). None of these treatments were successful in removing the pigmented fungal structures from the surface of the paper. The study reported herein gives preliminary results on the use of ligninases to digest melanized fungal spots.

Ligninase enzymes extracted from white-rot fungi have been shown to successfully cause depigmentation of conidia of the black yeast *Phaeococcomyces sp.* (Butler and Day, 1998). The present study used two different ligninases, a manganese peroxidase (MnP) and a laccase. MnP is a H_2O_2-dependent heme-containing enzyme that utilizes Mn^{+2} as a redox mediator. The enzyme has a manganese-binding site, and its principal function is to oxidize Mn^{+2} to Mn^{+3}. Mn^{+3} is an effective oxidant of phenolic substrates (Hofrichter, 2002; Martinez, 2002). Laccase is a copper-containing phenol oxidase that doesn't require H_2O_2 to oxidize substrates; instead, electrons are transferred to molecular oxygen to form water (Mayer and Staples, 2002).

The tests were carried out either on samples of unsalvageable matting from Tiffany drawings or on melanin extract from *Cladosporium sp.* The results of the treatments tested here were evaluated using colorimetry, SEM (scanning electron microscopy), FTIR (Fourier transform infrared spectroscopy) and RAMAN spectroscopy.

MATERIALS AND METHODS

CULTURES Two different species of fungi were used: DHN melanin-producing *Cladosporium sp.* (Butler and Day, 1998) isolated from the Tiffany drawings, and *Phanerochete chrysosporium*, a ligninase producing fungus, from the American Type Culture Collection (ATCC 24725). They were maintained in malt extract agar (MEA) and potato dextrose agar (PDA) plates. *P. chrysosporium* cultures were maintained under unstressed conditions to avoid production of melanin.

PAPER SAMPLE SELECTION AND TREATMENT REGIMEN Small strips of melanized fungal infested paper (0.5 × 2 cm) were cut from the unsalvageable matting surrounding one of the Tiffany drawings (MMA accession #67.654.318). The samples were subjected to a solvent treatment and an enzymatic treatment as follows:

Control 1: distilled water wash only

Control 2: enzymes and water wash

Control 3: solvent treatment and water wash

All of the experimentals (six) were subjected to the solvent and enzymatic treatments with lytic enzymes and chitinases. In addition, experimentals 1–3 were treated with laccase, and 4–6 with manganese peroxidase (see sections on solvent and enzymatic treatments below).

Solvent Treatment A sequence of solvents was used in order to provide a favorable hydrophilic environment for the subsequent enzymatic attack with chitinase and other lytic enzymes. The solvents chosen were: cyclohexane, diethyl-ether, trihydrofurane (THF), acetone, and ethyl alcohol. They were applied sequentially from least to most polar. The paper strips were immersed in each of the solvents for ten minutes, washed in distilled water, and then treated enzymatically. Three different controls were prepared: control 1 was not subjected to either the solvent or the enzymatic treatment, control 2 was subjected only to the enzymatic treatment, and control 3 was subjected only to the solvent treatment.

Enzymatic Treatment Two sets of enzymes were used. The first set was chosen to digest the main components of the fungal cell wall of Basidiomycetes: lysing enzymes containing glucanase (0.38 units), protease (0.85 units), and cell lytic activity (2153 units; SIGMA L8757); and chitinase (2.5 units; SIGMA C6137). The second set was chosen to digest nonspecifically the melanin in the cell wall of the fungus: manganese peroxidase (MnP, 600 units; Tienzyme, Inc.) or laccase (1500 units; Tienzyme, Inc.). The enzymes were applied to the paper strips sequentially by adding 1 ml of enzyme starting with the lysing enzymes followed by the chitinase and ending with either the MnP or the laccase. In the case of MnP the following reagents were also added as cofactors: 10 μM H_2O_2, 0.2 M sodium tatrate, and 0.1 M manganese sulfate. For each of the enzyme treatments the samples were incubated in 1.5-ml tubes for 24 hours at 32°C.

MELANIN EXTRACTION The mycelium of *Cladosporium sp.* was harvested from a MEA plate and homogenized using 0.2-mm glass beads in deionized water for 15 minutes. The mycelium was centrifuged and the supernatant containing the melanin kept for further processing. The extract was digested enzymatically, first with the lysing enzyme preparation, and then by the chitinase, at 32°C for 24 hours.

COLORIMETRY Evaluation of color changes was performed using a Minolta Chroma Meter CR-241 with a 0.3-mm aperture. The color values using CIE 1976 L*a*b* of control and experimental samples were measured before and after treatment. Measurements were taken from the same spot of each sample before and after treatment. A change of greater than 1.0 in any parameter was considered significant.

SCANNING ELECTRON MICROSCOPY The micrographs of the controls and experimentals [FIGS. 1–5] were taken using a LEO 1455VP SEM. Samples were gold-coated and examined under normal (high-vacuum) conditions at an accelerating voltage of 15 kV.

FTIR AND RAMAN MEASUREMENTS Fourier transform infrared spectroscopy (FTIR) and Raman spectroscopy were used to monitor chemical changes in the cell wall

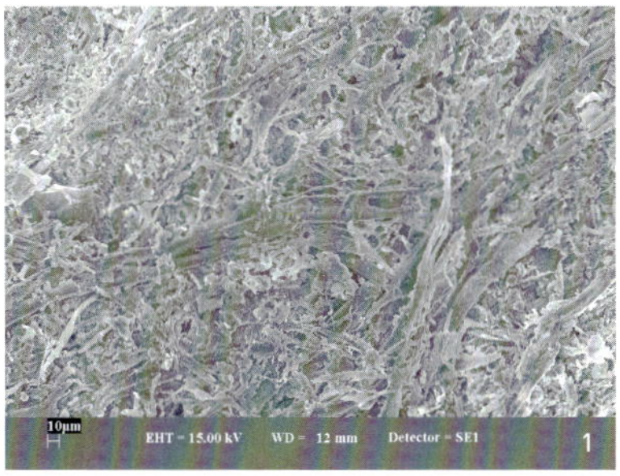

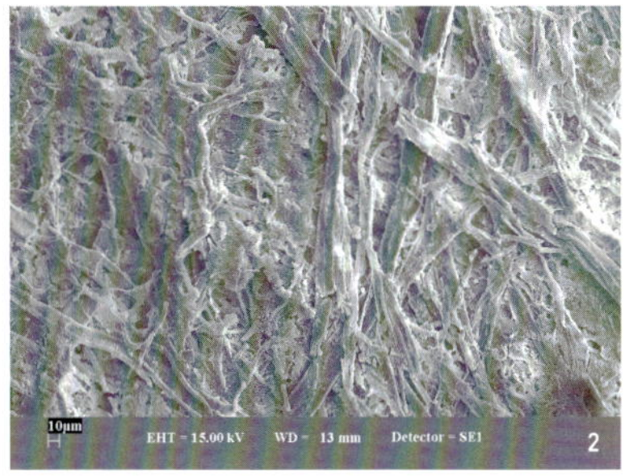

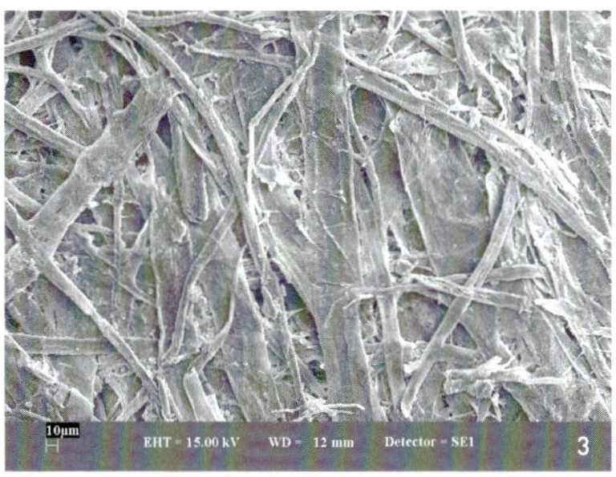

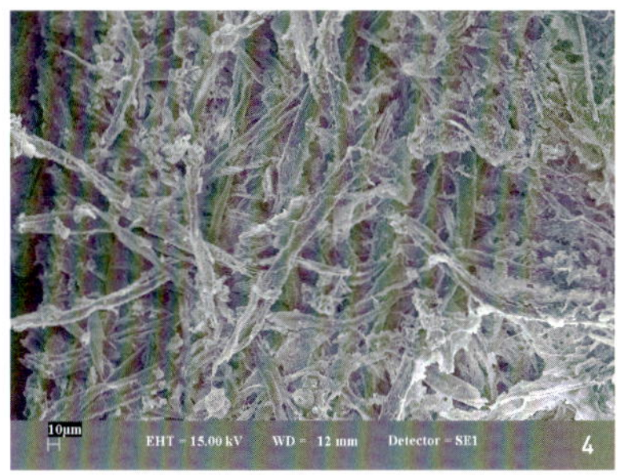

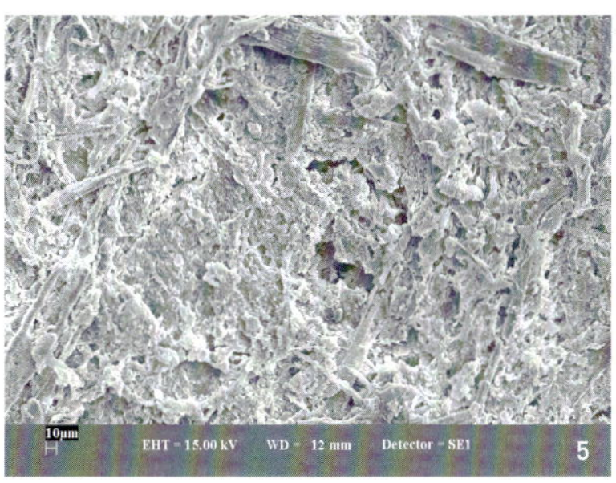

[1] SEM micrograph of fungal stained paper after washing in distilled water (Control 1). Original magnification = 500×. [2] SEM micrograph of fungal stained paper after enzyme treatment and distilled water wash (Control 2). Original magnification = 500×. [3] SEM micrograph of fungal stained paper after solvent treatment and distilled water wash (Control 3). Original magnification = 500×. [4] SEM micrograph of fungal stained paper after solvent and laccase treatment, and distilled water wash (Experimental 3). Original magnification = 500×. [5] SEM micrograph of fungal stained paper after solvent and MnP treatment, and distilled water wash (Experimental 6). Original magnification = 500×.

components with the enzymatic treatments. The FTIR was carried out on a Bio-Rad FTS 40 equipped with an UMA 500 microscope. Fungal hyphae and extracted and treated melanin samples were placed between the windows of a diamond anvil cell, and the spectra were recorded with a 4 cm^{-1} resolution.

Raman spectra of fungal material, present either on the paper strips cut from the mats surrounding the drawings or in the culture media of fungi isolated from the surface of the drawings (Di Bonaventura et al., this volume), were recorded using a Renishaw System 1000 spectrometer. A 514-nm laser beam was focused on clusters of fungal spores or hyphae using a Leica microscope. The laser spot size was approximately 2 microns in diameter, using a 50× lens. Laser powers up to 1 mW were employed, with accumulation times up to 120 seconds. Calibration was carried out using the 520.5 cm^{-1} line of a silicon wafer. It was also possible to record the Raman spectra of fungal spores or hyphae present on the drawings' paper surfaces without removing any sample material when a fiber optic probe attached to the spectrometer was used. Laser spot size in this case was 5 microns using a 20× lens.

As used in this study, FTIR and Raman are limited to providing qualitative data and are used for comparing spectra between treated and untreated samples within the same experimental design.

RESULTS AND DISCUSSION

COLORIMETRY Table 1 shows the results of the colorimetry test for L*. [TABLE 1] Differences of more than one unit between pre- and post-tests are considered significant. The results show that the treatment with the MnP enzyme yielded the most significant changes in the color of the sample measured as changes in L. The solvent control shows a significant difference in L* value, indicating that the solvent lightens up the substratum due to dissolution of materials other than the fungal melanins.

ELECTRON MICROSCOPY Scanning electron microscopic examination of the fiber microstructure of samples of gray to black-stained paper is shown in Figs. 1–5. The samples were examined over a range of magnifications from 50× to 10,000× (only the 500× are presented here). A comparison of the micrographs for the controls [FIGS. 1–3] with those of the enzymatic treatments [FIGS. 4 AND 5] shows evident

	L*		
	PRE	POST	DL
CONTROL 1	52.93	50.41	-2.52
CONTROL 2	47.05	51	3.95
CONTROL 3	46.82	57.77	10.95
LACCASE	51.96	53.38	1.42
LACCASE	48.16	48.93	0.77
LACCASE	30.81	43.44	12.63
MnP	51.58	63.19	11.61
MnP	37.73	52.68	14.95
MnP	41.62	53.43	11.81

[TABLE 1] Colorimetry data

distortion of the fiber structure of the paper in the treated samples. In figure 5, which corresponds to the treatment with MnP, the paper fibers appear to have become severely disrupted. [FIG. 5] In the solvent control [FIG. 3] adventitious material (nonpigmented fungal material) appears to be partially removed, which might explain the change in L* value with colorimetry seen in control 3 in comparison to controls 1 and 2.

FTIR AND RAMAN SPECTROSCOPY Figure 6 shows the FTIR spectra of melanized fungal hyphae from *Cladosporium sp.* (a) and nonmelanized hyphae from a *Phanerocahete chrysosporium* (b). [FIG. 6] The characteristic features of major fungal cell wall components dominate both spectra: proteins, polysaccharides, mainly glucans and chitin. [TABLE 2] The spectrum of a chitin reference sample (poly[1→4]-β-N-acetyl-d-glucosamine, Sigma) is shown in figure 6c.

Fourier transform infrared spectroscopy has been used as a standard method for the analysis of fungal polysaccharides (Gutierrez et al., 1996; Šandula et al., 1999; Mohaček-Grošev et al., 2001). Particularly, spectra of glucans, N-acetyl-α-d-glucosamine, and chitin have been reported by several authors (Šandula et al., 1999; Mohaček-Grošev et al., 2001; Kačuráková et. al., 2000; Kouach-Alix et al., 1994). Vibrational analysis of carbohydrates assigns the bands in the 1500–1000-cm^{-1} range to mixed modes (Kačuráková et. al., 2000; Vasko et al., 1971, 1972; Kouach-Alix et al., 1994; Sekkal et al., 1992); therefore only approximate assignments are possible in this range, particularly for such a complex system as fungal tissue.

The 1200–1000-cm^{-1} range is dominated by the pyranose ring vibrations overlapped with stretching vibrations of COH side groups and the COC glycosidic bond vibration. The 950–750-cm^{-1} anomeric region is also complicated by band overlapping, though the position of the C1H deformation bands in this range allows recognition of the main constituents in polymer mixtures (Kačuráková et. al., 2000). This range is particularly useful for differentiating α- and β-glucans (Šandula et al., 1999; Mohaček-Grošev et al., 2001). It should be taken into account that glucans form complexes with chitin and proteins and that the formation of these complexes changes the supramolecular structure of glucans, affecting the FTIR spectra in the 1150–750-cm^{-1} range (Šandula et al., 1999; Mohaček-Grošev et al., 2001). In the case of the fungal species analyzed, bands for both the α- and β-anomers were observed. [TABLE 2]

The broad peak at 1410 cm^{-1} in the spectrum of melanized *Cladosporium sp.* hyphae may have, in addition to the CH bending and OH bending modes of poly-

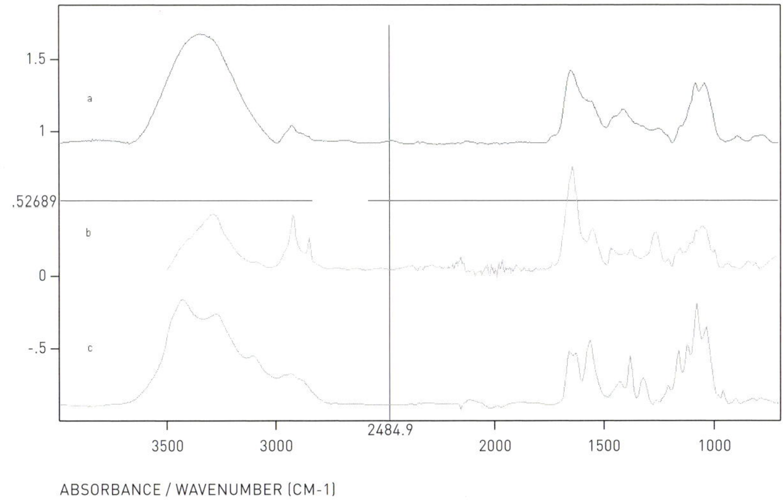

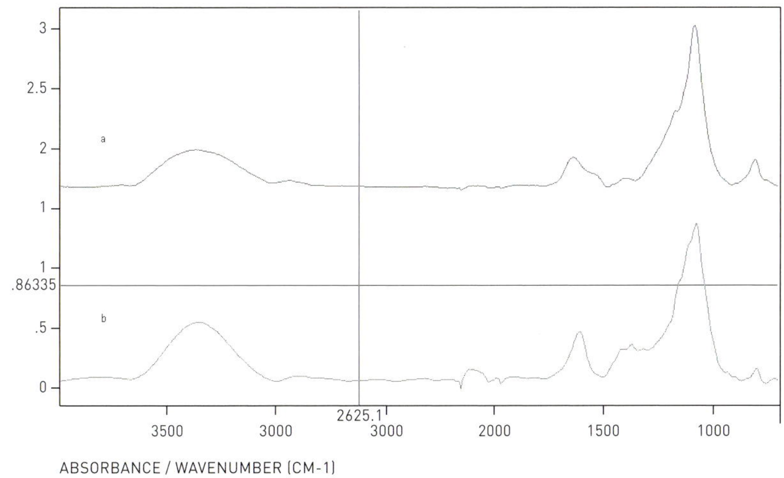

[6] FTIR spectra of melanized fungal hyphae from *Cladosporium sp.* (a), non-melanized hyphae from a *Phanerochaete chrysosporium* (b), and chitin reference sample (c). [7] FTIR spectra of *Cladosporium sp.* tissue digested using lytic and chitinase enzymatic preparations (a), and of the same tissue subsequently treated with MnP ligninase (b).

CLADOSPORIUM SP. hyphae FTIR BAND FREQUENCY (CM^{-1})	*P. CHRYSOSPORIUM* hyphae FTIR BAND FREQUENCY (CM^{-1})	PROPOSED VIBRATIONAL ASSIGNMENTS
3340 vs, vbr	3292 s, br	OH stretching
2927 m	2922 s	CH_2 stretching
2857 m	2851 s	CH_2 stretching
1736 sh	1738 w	C=O stretching
1650 s	1642 vs	Amide I (proteins, chitin)
1553 sh	1551 s	Amide II (proteins, chitin)
1459 sh	1467 m	CH_2 bending (glucan: 1461 cm^{-1})*
1410 m		CH bending, OH bending (saccarides) with COH bending (phenolic)
	1376 m	CH bending with OH bending contribution (saccarides, glucan: 1376 cm^{-1}, chitin: 1378 cm^{-1})*
1324 sh		
	1259 s	Glucan: 1260 cm^{-1} *
1250 m, br		
1201 sh	1204 w	Glucan: 1204 cm^{-1} *
1155 sh	1151 m	Contributions from COC stretching, CO stretching, CC stretching, CH deformation, COH deformation (chitin: 1158, 1075, 1031 cm^{-1}, glucan: 1160, 1078, 1042 cm^{-1})*
1080 s	1078 s	
	1110 sh	
	1050 s	
1041 s		
	995 m	
	932 w	CH deformation (α-glucan)*
890 w	894 sh	CH deformation (β-glucan)*
	842 w	CH deformation, saccarides
	808 w	CH deformation, saccarides
779 w		

s=strong, m=medium, w=weak, v=very, sh=shoulder

* Assignments of bands to glucan were done according to Mohaček-Grošev et al. (2001).

[TABLE 2] FTIR band frequencies (cm^{-1}) and assignments for melanized *Cladosporium sp.* hyphae and non-melanized *P. crysosporium* hyphae (spectra 6a and 6b).

saccharides, some contribution from a COH bending mode of phenolic groups from fungal melanin. Except for this contribution, melanin features are obscured by the more dominant ones due to the rest of the cell wall components.

The FTIR spectrum of *Cladosporium sp.* tissue digested using both lytic and chitinase enzymatic preparations is shown in FIG. 7A. Because of the complex heteropolymeric nature of DHN melanin, and the presumed presence of bound proteins, chitin, and glucans (Butler and Day, 1998) only a tentative assignment of the FTIR spectrum is possible. It should also be taken into account that insoluble degradation products of cell wall components are still present in the sample material.

Peaks at 1640 and 1539 cm^{-1} (amide I and II bands, respectively) in spectrum 7a indicate that the enzymatic digestion process did not succeed in removing all proteinaceous material or chitin. However, it does seem to have degraded glucan components: The peaks at 1459, 1201, 1041, and 890 cm^{-1} in spectrum 6a are no longer present in spectrum 7a. The strong peak at 1084 cm^{-1} in spectrum 7a may be due to vibrational modes of saccharides resulting from the decomposition of polymeric material during the enzymatic process. The shoulder at 1170 cm^{-1} together with the peak at 1401 cm^{-1} can be assigned to phenolic groups in fungal melanin (out of phase CCO deformation and C-OH bending, respectively).

When the fungal tissue, digested using the lytic enzymes and chitinase, was subsequently treated with MnP ligninase, it noticeably lightened. The corresponding spectrum is shown in FIG. 7B. The amide I and amide II bands are no longer present, suggesting that the enzymatic process removed proteins and chitin. The maximum at 1077 cm^{-1} and its shoulder in spectrum 7b are reflecting the presence of saccharides. The peak at 1609 cm^{-1} is attributed primarily to aromatic groups whose signal is enhanced by phenol functionalities (Pierce and Rast, 1995) resulting from the partial degradation of polymeric fungal melanin by the MnP ligninase. It has been found that when MnP reacts with phenolic or aminoaromatic subtrates, these are oxidized by hydrogen abstraction to form phenoxyl and amino-radicals, respectively (Hofrichter, 2002). Moreover, microbial melanin degradation has been found to cause a decrease in H-content and an increase in C-content, consistent with an increase in the aromaticity of the pigment (Zavgorodnyaya et al., 2002).

Raman spectroscopy has also been used to characterize fungal cell wall components and metabolites (Edwards et al., 1998; Spiro, 1987). Although it has proved very difficult to assign the vibrational bands from specific molecules in biological samples owing to the complex nature of their systems (Edwards et. al., 1995) it is often possible to infer the mechanism of biological processes from the behavior of specific molecular features.

Figure 8 shows the Raman spectra of melanized *Cladosporium sp.* hyphae (a) and spores (b) isolated from the surface of a Tiffany drawing in the 2000–100-cm^{-1} range. Tentative assignments are presented in TABLE 3. Spectrum c in figure 8 corresponds to a reference sample of ergosterol, a component of the cell membrane (Sigma). Strong scattering in the range 1700–1500 cm^{-1} is characteristic of aromatic and aliphatic C=C unsaturations (Lin-Vien et al., 1991). Bands in spectra a and b [FIG. 8] could be assigned to ergosterol, though the medium-weak peak at 1664 cm^{-1} in spectrum c is not evident in spectra a and b. Modes due to aromatic rings in fungal melanin may contribute to the absorption around 1600 cm^{-1}.

Characteristic bands of polysaccharides, such as COC ring modes and glycosidic links between sugar residues (800–900 cm^{-1}) and CC and CO stretchings of terminal skeletal modes (1100–1050 cm^{-1}) (Edwards et al., 1998), are not evident in spectra a and b in figure 8; neither are the characteristic bands of N-acetyl-glucosamine (Edwards et al., 1995), or the amide I (1680–1630 cm^{-1}) and amide III (1310–1250 cm^{-1}) bands of proteinaceous materials (Lin-Vien et al., 1991). Cell membrane fatty acids would be expected to give, in addition to features due to C=C unsaturations, C=O bands above 1700 cm^{-1}.

When *Cladosporium sp.* hyphae were digested using the lytic ezymes and chitinase, spectrum b in figure 9 was obtained. [FIG. 9] This spectrum presents features similar to those of the spectra of the intact spores and hyphae. When the sample was digested with MnP ligninase, bands in the range 2000–100 cm^{-1} gradually decreased as enzyme treatment time increased. This provided a means of evaluating the progress of the enzymatic treatment, even though melanin-characteristic functional groups could not be directly observed by Raman under the present experimental conditions.

CONCLUSIONS The applicability of the use of ligninases, in particular MnP and laccase, to break down pigmented fungal spots on paper surfaces is promising. Although the assessment of the enzymatic treatment proposed herein using colorimetry indicates that a significant lightening of the fungal spots takes place, SEM micrographs show that ligninases produce a distortion of the paper fiber structure as well. In addition, the visual appearance of the samples after treatment was not completely satisfactory from the conservator's point of view.

Cladosporium sp. HYPHAE RAMAN BAND WAVENUMBER (CM-1)	*Cladosporium sp.* SPORES RAMAN BAND WAVENUMBER (CM-1)	PROPOSED VIBRATIONAL ASSIGNMENTS
	1598s	C=C stretching aliphatic; CCH stretching, aromatic
1588s		
1455m	1453 m	CH_3,CH_2 deformation
1339 m	1322m	CH deformation

[TABLE 3] Raman band wavenumbers (cm-1) and assignments for Cladosporium sp. hyphae and spores. s = strong, m = medium. Spectra a and b in Fig. 8.

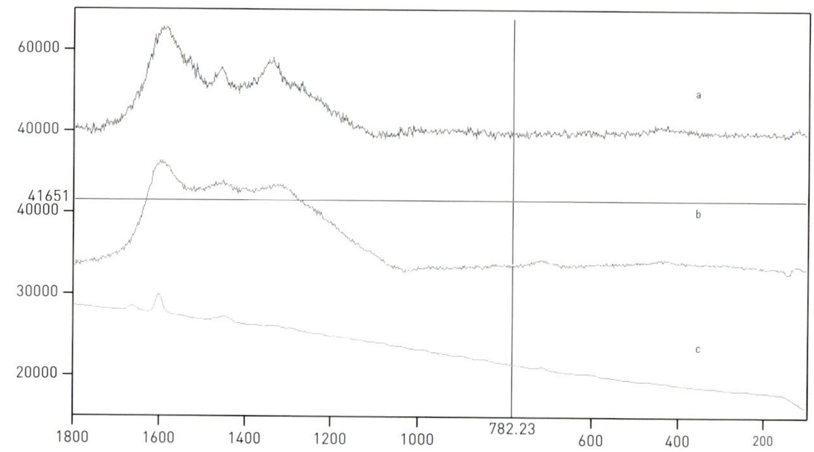

COUNTS / RAMAN SHIFT (CM-1)

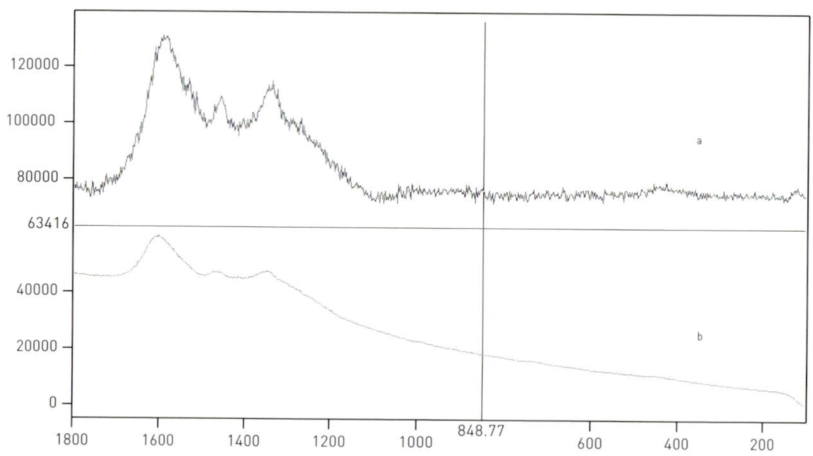

COUNTS / RAMAN SHIFT (CM-1)

[8] Raman spectra of melanized-*Cladosporium sp.* hyphae (a), melanized-*Cladosporium sp.* spores (b), and ergosterol reference sample (c). [9] Raman spectra of melanized-*Cladosporium sp.* hyphae (a), and *Cladosporium sp.* hyphae digested using the lytic enzymes and chitinase (b).

The use of a battery of solvents with increasing polarity appears to be effective in providing a favorable environment for the subsequent enzymatic attack with chitinase and other lytic enzymes.

Further research is needed to characterize the digestion process of melanin by MnP and laccase and its application to the bleaching of fungal spots on paper. It is necessary, for example, to determine the stoichiometry of the moieties in the catalytic process to optimize their concentrations and to avert any detrimental effects on the object. In addition, the use of pure enzyme preparations for MnP and laccase may prevent the unwanted digestion of the paper substratum.

Also of interest is the application of FTIR and Raman spectroscopy to monitor the progress of the enzymatic treatment. Although it is generally difficult to specifically assign vibrational bands in biological samples owing to the complex nature of their systems, it is often possible to infer the mechanism of biological processes from the behavior of specific molecular features. In this study, FTIR and Raman spectroscopy have proved valuable to follow up the breakdown process of *Cladosporium sp.* pigmented hyphae isolated from a Tiffany drawing.

REFERENCES

BARTNICKI-GARCIA, S., and REYES, E. 1964. Chemistry of spore wall differentiation in *Mucor rousii*. *Archives of Biochemistry and Biophysics*, 108, 125–133.

BELL, A.A., and WHEELER, M.H. 1986. Biosynthesis and Functions of Fungal Melanins. *Annual Review of Phytopathology*, 24, 411–51.

BLOOMFIELD, B.J., and ALEXANDER, M. 1967. Melanins and Resistance of Fungi to Lysis. *Journal of Bacteriology*, 93(4), 1276–1280.

BUTLER, M.J., and DAY, A.W. (1998). Fungal melanins: A review. *Canadian Journal of Microbiology*, 44, 1115–1136.

DI BONAVENTURA, M.P., DESALLE, R., EVELEIGH, D.E., BALDWIN, A.M., and KOESTLER, R.J. (this volume). Studies of fungal infestations of Tiffany's drawings: Limits and advantages of classical and molecular techniques. In: Koestler, R.J., Koestler, V.R., Charola, A.E., and Nieto-Fernandez, F.E. (Eds.), *Art, Biology, and Conservation: Biodeterioration of Cultural Property*. The Metropolitan Museum of Art, New York, pp. 94–109.

DURRELL, L.W. 1964. The composition and structure of walls of dark fungus spores. *Mycopath. Mycol. Appl.*, 23, 339–345.

EDWARDS, H.G.M., RUSSELL, N.C., WEINSTEIN, R., and WYNN-WILLIAMS, D.D. 1995. Fourier Transform Raman Spectroscopic Study of Fungi. *Journal of Raman Spectroscopy*, 26, 911–916.

EDWARDS, H.G. M., FALK, M.J., SIBLEY, M.G., ALVAREZ-BENEDI, J., and RULL F. 1998. FT-Raman spectroscopy of gums of technological significance, *Spectrochimica Acta* Part A: Molecular and Biomolecular Spectroscopy, 54 (7), 903–920.

FLORIAN, M.L., and PURINTON, N. 1995. Determination of location of stains in fungal spots and enzymatic removal of pigmented hyphae in paper. Preprints, *3rd International conference on biodeterioration of cultural property*, 4–7 July, 1995, Bangkok, Thailand, pp. 414–425.

GUTIERREZ, E., PRIETO, A., and MARTINEZ, A.T. 1996. Structural characterization of extracellular polysaccharides produced by fungi from the genus *Pleurotus*. *Carbohydrate Research*, 281, 143–154.

JACOBSON, E.S. 2000. Pathogenic Roles for Fungal Melanins. *Clinical Microbiology Reviews*, 13 (4), 708–717.

HOFRICHTER, M. 2002. Review: lignin conversion by manganese meroxidase (MnP). *Enzyme and Microbial Technology*, 30, 454–466.

KAČURÁKOVÁ, M., CAPEK, P., SASINKOVÁ, V., WELLNER, N., and EBRINGEROVÁ, A. 2000. FT-IR study of plant cell wall model compounds: pectic polysaccarides and hemicelluloses. *Carbohydrate Polymers*, 43, 195–203.

KOUACH-ALIX, I., and VERGOTTEN, G. 1994. A vibrational molecular force field of model compounds with biological interest. V. Harmonic dynamic of N-acetyl-α-D-glucosamine in the crystalline state. *Journal of Molecular Structure*, 323, 39–51.

LIN-VIEN, D., COLTHUP, N.B., FATELEY, W.G., and GRASSELL, J.G. 1991. *The Handbook of Infrared and Raman Characteristic Frequencies of Organic Molecules*. Academic Press, London.

MARTINEZ, A.T. 2002. Molecular biology and structure–function of lignin-degrading heme peroxidases. *Enzyme and Microbial Technology*, 30, 425–444.

MAYER, A.M., and STAPLES, R.C. 2002. Laccase: new functions for an old enzyme. *Phytochemistry*, 60, 551–565.

MOHAČEK-GROŠEV, V., BOŽAC, R., and PUPPELS, G.J. 2001. Vibrational spectrocopic characterization of wild growing mushrooms and toadstools. *Spectrochimica Acta* A, 57, 2815–2829.

PIERCE, J.A., and RAST, D.M. 1995. A Comparison of Native and Synthetic Mushroom Melanins by Fourier-Transform Infrared Spectroscopy. Phytochemistry, 39, 49–55.

ŠANDULA, J., KOGAN, G., KAČURÁKOVÁ, M., and MACHOVÁ, E. 1999. Microbial (1→3)-β-D-glucans, their preparation, physico-chemical characterization and immunomodulatory activity. *Carbohydrate Polymers*, 38, 247–253.

SEKKAL, M., LEGRAND, P., VERGOTEN, G., and DAUCHEZ, M. 1992. A vibrational molecular force field of model compounds with biological interest-III. Harmonic dynamics of α- and β-D-galactose in the crystalline state. *Spectrochimica Acta* A, 48 (7), 959–973.

SPIRO, T.G. 1987. *Biological Applications of Raman Spectroscopy*, Vol.1. Wiley, New York .

VASKO, P.D., BLACKWELL, J., and KOENIG, J.L. 1971. Infrared and Raman Spectroscopy of Carbohydrates. Part I: Identification of O-H and C-H related vibrational modes for D-glucode, maltose, cellobiose, and dextran by deuterium-substitution methods. *Carbohydrate Research*, 19, 297–310.

VASKO, P.D., BLACKWELL, J., and KOENIG, J.L. 1972. Infrared and Raman Spectroscopy of Carbohydrates. Part II: normal coordinate analysis of α-D-glucose. *Carbohydrate Research*, 23, 407–416.

ZAVGORODNYAYA ,YU.A., DEMIN, V.V., and KURAKOV, A.V. 2002. Biochemical Degradation of Soil Humic Acids and Fungal Melanins. *Organic Geochemistry*, 33, 347–355.

H.M. SZCZEPANOWSKA[1]
A.R. CAVALIERE[2]

ARTWORKS, DRAWINGS, PRINTS, AND DOCUMENTS—FUNGI EAT THEM ALL!

1 National Museum of Natural History, Washington, DC 2 Gettysburg College, Gettysburg, PA

ABSTRACT Our studies have focused on the examination of fungi infesting artworks and artifacts on paper. We have been concerned primarily with the mechanisms of deterioration by fungi and the role played by the composition of the substrates upon which the fungi grow. Several examples illustrate these relationships: (1) infestations that occurred after exposure to periodic floods (18th century Tilghman documents); (2) exposure to water spills (19th century watercolor and ink drawings); and (3) growth occurring in the microclimate of framed artworks exposed to fluctuating relative humidity (18th and 20th century pastel portraits).

Part of our investigation was aimed at the identification and numeration of the vast mycoflora that has been encountered on paper-based collections. Several groups of fungi appear to be common on the materials we examined. While Deuteromycetes (Fungi Imperfecti) dominate, we often find members of the Zygomycetes (bread molds), Ascomycetes (sac fungi) and Saccharomycetales (yeasts). Their growth is often linked to the materials upon which artwork and archival documents are rendered and the environment to which these works are exposed throughout their life. Moreover, fungus growth is often enhanced by the addition of artists' pigments and the great variety of materials, such as varnishes, glazes, and additives, applied later in the life of the artwork. The impact of the environment, specifically the time of the collections' exposure to moisture, greatly enhances both the number of species and the abundance of fungal growth. This report reviews the current status of our understanding of these fungus-substrate parameters, discusses their impact upon the preservation of collections, and concludes by offering a number of suggestions to consider for future practices when choosing methods of conservation.

INTRODUCTION Fungi are the primary organisms on earth responsible for the necessary process of biodegrading organic material and recycling of nutrients to the environment. However, deterioration of museum and archival collections by fungi is extremely detrimental because it lowers the quality and value of each artwork and artifact. Recognizing elements that contribute to the biodeterioration of any collection is the first step toward avoiding conditions that promote its occurrence. Understanding the causes and mechanisms of deterioration will provide a basis for developing strategies to prevent this damaging process.

Our studies have focused on the examination of fungi infesting artworks and artifacts on paper. The mechanism of deterioration by fungi may depend on the nature of the materials upon which they grow. While the chemical composition of the substrate and media at times determines the type of fungi that will invade, the structure of the media has impact on the aerial growth of the colonies. Moreover, the environment and the length of time of the collections' exposure to moisture greatly enhance the number of species and the abundance of their growth.

Samples of fungi encountered on paper-based collections were cultured and studied in order to identify and enumerate them. Moreover, we reviewed current conservation measures that aim to prevent fungal infestation and suggest methods to remedy the damage that they inflict upon artworks.

MATERIALS AND METHODS This investigation was conducted on artworks and documents delivered by owners of private collections to the Art Conservation Studio (studio of HMS) for the purpose of conservation. Pigments and supports were examined to the extent allowed by the owners. Examinations and analyses of artwork components were conducted in the Microbiological Laboratory and the Electron Microscopy Laboratory (ARC) of the Biology Department at Gettysburg College, Gettysburg, PA.

The following objects were investigated: pastel portraits dated 1860, 1886, and 1973; an ink wash by Lady Diana Beauclerk (1734–1808); a watercolor attributed to Thomas Birch, early 20th century; and an Audubon print, *Pelican* (1838), hand colored and glazed with gamboge. A collection of documents subjected to a series of floods, reported elsewhere (Szczepanowska and Cavaliere, 2000), supplied enormously rich material and is referenced here for comparative purposes.

Samples of fungi were collected either by lightly scraping the mold (hyphae) with a sterile lancet and transporting it to the laboratory in its sterile packet or by pressing a sterile cotton swab over the mold growth. When possible, samples of secondary supports were transferred to the laboratory wrapped in sterile paper. In some cases, clear cellophane tape was pressed over the sample and the hyphae adhering to the adhesive surface were examined directly (Butler and Mann, 1959).

Once in the laboratory, samples were examined directly and then cultured. In some cases, species producing distinctive conidiospores and conidia could be identified immediately *(Alternaria, Nigrospora, Cladosporium)*; however, in most cases, the fungi were allowed to grow in culture. All samples were plated on a series of traditionally employed fungal media—cornmeal dextrose agar (CMA), Czapek yeast extract agar (CYA), potato dextrose agar (PDA), and Sabouraud dextrose agar (SDA) (Difco, 1984). Cultures were allowed to incubate at room temperature (22–24°C). As soon as hyphal growth became apparent on the agar, the samples were subcultured to obtain unifungal isolates. In an attempt to keep our samples free from contamination, all transfers were performed in a Labconco sterile hood.

In our work we used a Nikon Optiphot with attached camera, FX-35 WA, for light microscopy; a Cambridge Stereoscan 100 and PTG Avalon EDS (for determination of pigment elemental composition as a possible source of nutrients for fungi); and a JEOL 5200 scanning electron microscope. The visual examination under light microscopy was recorded with traditional photography and digitally with a Fujifilm FinePix 4900 Zoom digital camera and Pixera digital microscope.

For light microscopy, specimens were mounted in either distilled water or in a 0.1% Acid Fuchsin solution. For electron microscopy all samples were processed utilizing standard methods of preparation (Dykstra, 1992; Bozzola and Russell, 1999). Small, 2-mm^2 samples of agar supporting fungal hyphae were removed from culture plates and submerged in Karnovsky's fixative for several hours. The samples were then subjected to several 15-minute washes in cacodylate buffer and dehydrated in a graded series of ethyl alcohol. The samples were then further dehydrated in a Tousimis Samdry critical point dryer, mounted on aluminum stubs, and layered with 10 nm of gold in a Denton Desk II gold coater. Under the scanning electron microscope specimens were examined, digitized, and photographed. Digital images were taken with a JEOL Digital Generator and photographed with a Polaroid 545 camera using Polaroid 55 P/N film.

CASE STUDIES

The types of objects that were selected for our study represent a large number of artworks and artifacts created on paper supports, such as pastels, drawings, prints, watercolors, and documents. Although an elevated humidity was the element that undoubtedly initiated fungal infestation in each case, we believe, based on our observations, that all, or certain elements within the artworks' structure either promoted or in some cases prohibited the growth of fungi. In an earlier observation, Strzelczyk (1981) reported that the pigments on polychrome wooden sculptures may actually have some inhibitory effects on fungal growth.

PASTELS

Media Pastel as a medium is a dry colored chalk combined with binder and filler. The pastel drawing is created by strokes of pastel stick on a support that is textured to ensure that the powder attaches to the surface. Natural chalks of various colors are derived from earth. Those occurring naturally are: gray, red, red-ochre, and white. Among pigments that were traditionally used for manufacture of dry-stick pastels are umber raw, natural and burnt sienna, and green earth (Flieder, 1982). Some of the natural pigments, such as those containing iron oxides and hematites, were found on brown and black of the pastel dated 1860.

The pigment powder is held together by a small amount of binding medium, 2–3% at most of gelatin solution, gum Arabic, or gum tragacanth. Historically, skim milk was also used as a weak binding material. Plasticizers such as honey or crystallized sugar were added to eliminate brittleness (Doerner, 1984). The composition of pastel itself and the method of its application, as powder lightly arranged on the surface, promotes aerial fungal growth, allowing enough air necessary for their development. Often the aerial growth on pastels is abundant and bushy. The observation that pastels are particularly prone to fungal infestation has been made elsewhere (James et al., 1997).

Support for Pastel Paintings Textured paper, pasteboards, canvas, and parchment, characterized by a distinct texture, provided the means to anchor the pastel powder and to prevent its dislodging. The pastel portraits that we investigated were rendered on pumice paper (1973), textured paper (1860), and paper stretched on canvas (1886).

The pastel portrait of a Gentleman (#99.20–1) created in the mid 19th century was drawn on paper and mounted to a board with brown, protein base glue. [FIG. 1]

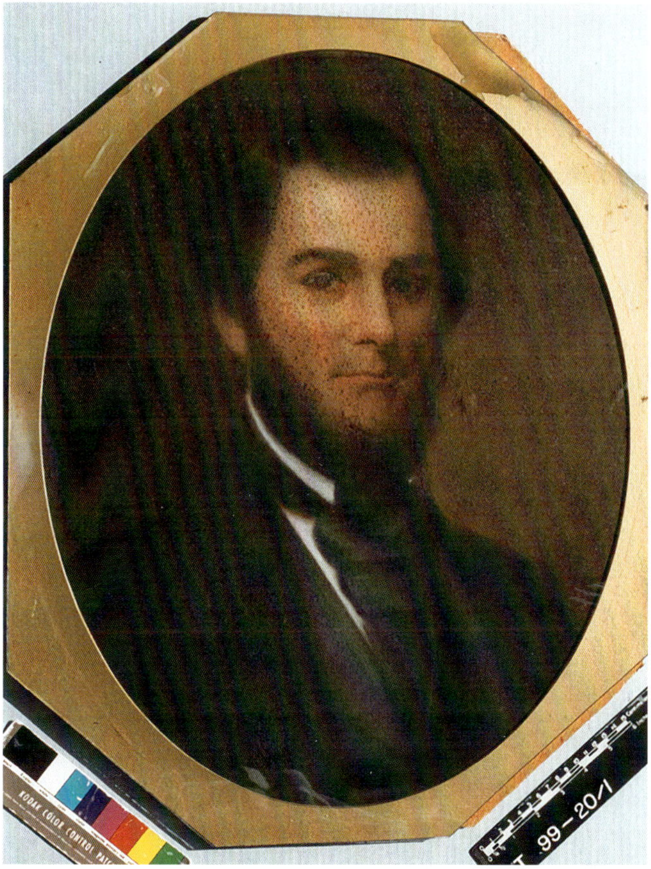

[1] *Gentleman*, pastel portrait, 1860 (#99.20-1). Overall view. Protein animal glue used for mounting the pastel to an acidic board supported heavy fungal infestation.

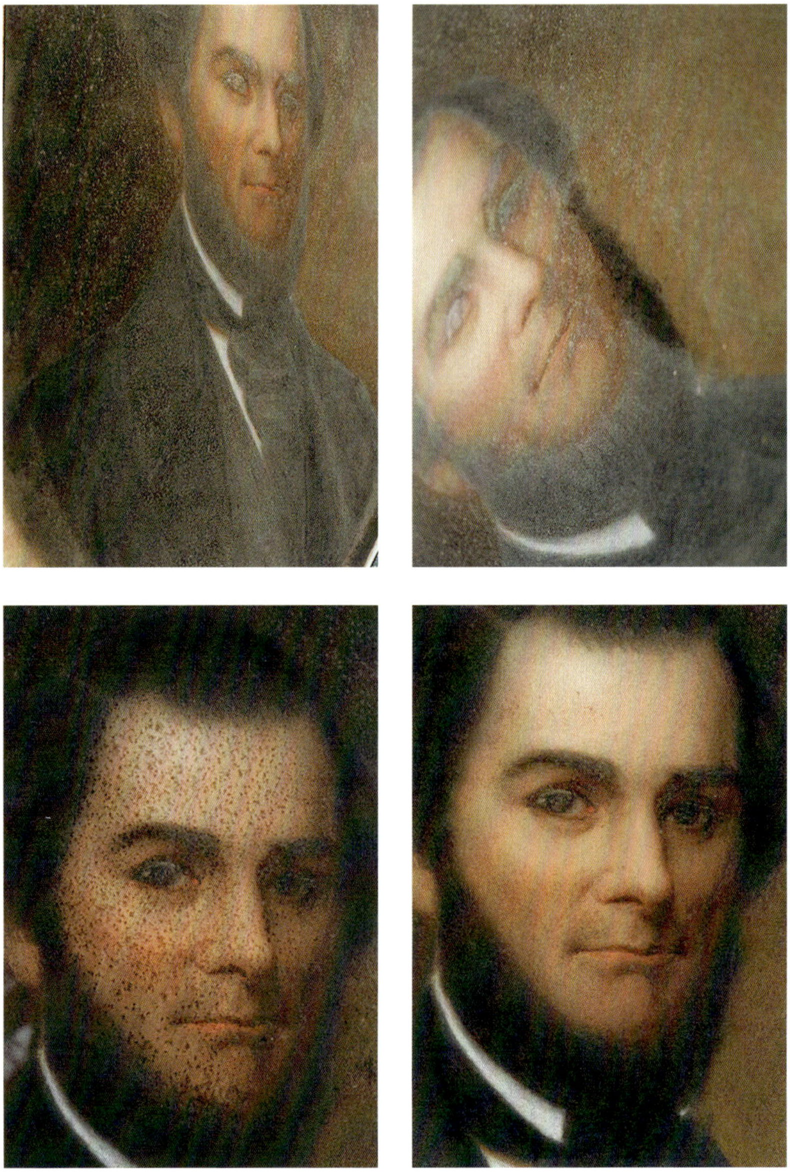

[2] *Gentleman*, pastel portrait, 1860 (#99.20-1). (a) Detail, angular view. Pattern of fungal infestation followed application of earth pigments. (b) Detail before and (c) after conservation treatment.

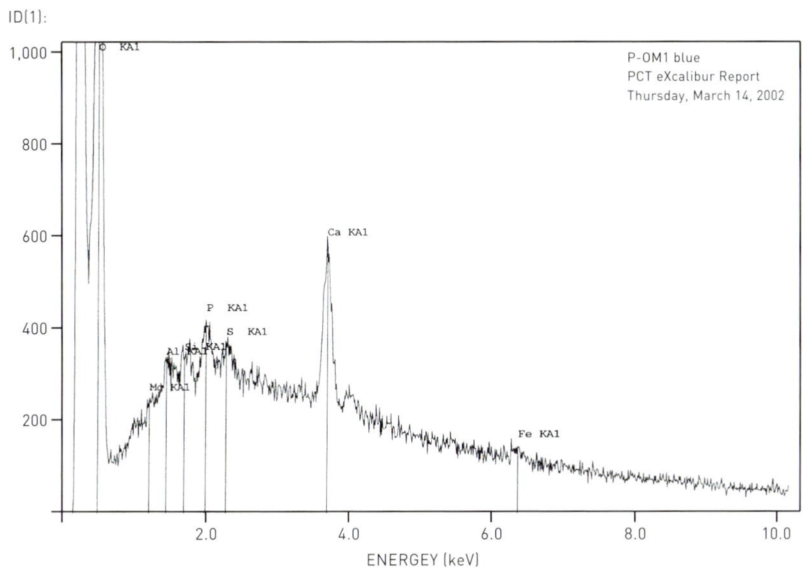

[3] *Gentleman*, pastel portrait, 1860 (#99.20-1). Spectrum produced by SEM Cambridge Stereoscan 100, EDS. Iron oxides in brown and black indicate earth pigments; dominating peak of Ca-base filler, calcium carbonate.

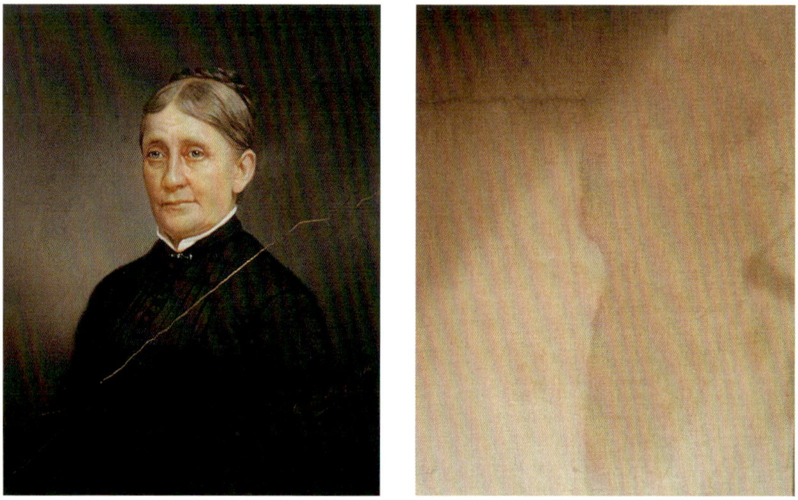

[4] *Hanna*, pastel portrait, 1886 (#02.20). (a) Recto and (b) verso. Pastel executed on paper mounted to canvas with protein, animal glue. Tideline on verso, indicating area exposed to water, corresponds to location of fungal colonies on recto.

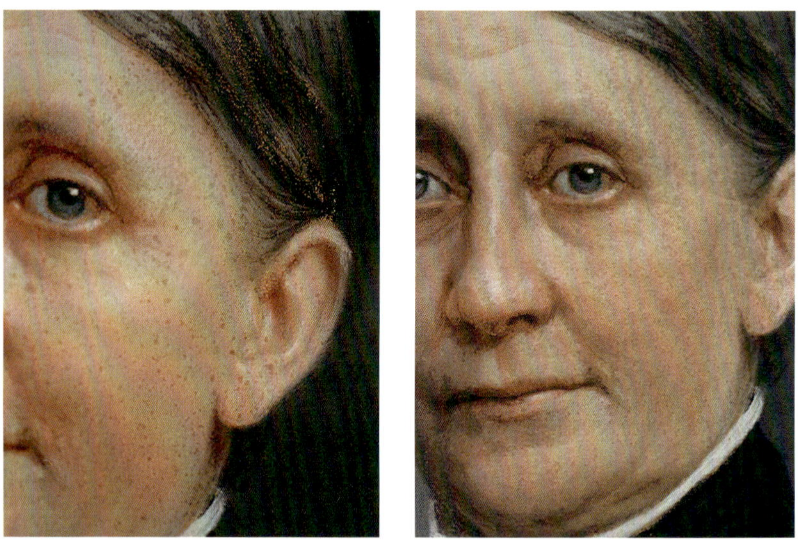

[5] *Hanna*, pastel portrait, 1886 (#02.20). (a) Detail of fungus colonies before treatment and (b) after conservation treatment.

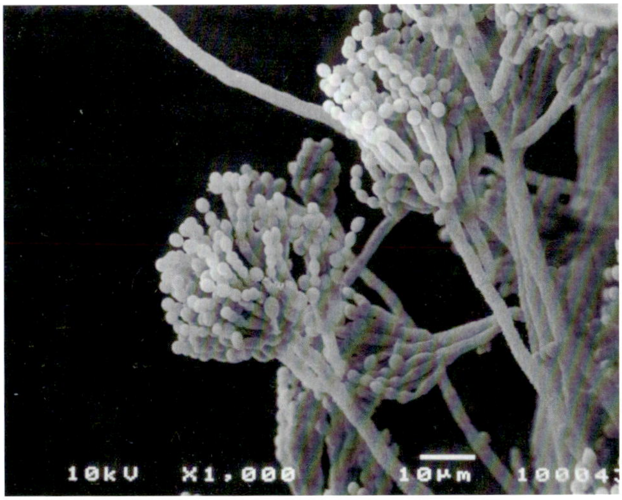

[6] *Hanna*, pastel portrait, 1886 (#02.20). *Penicillium sp.* (biverticillium group).

The pattern of extensive fungal growth is one of the best illustrations of the nutritional preferences of fungi in their choice of pigments. The aerial growth concentrated on the areas drawn with dark colors that were primarily earth pigments. [FIG. 2] Pigment analysis indicated the presence of iron oxides and hematites in brown areas of the gentleman's hair and black in the background. [FIG. 3]

The pastel of *Hanna* (1886; #02.20) was executed on textured paper lined with canvas. [FIGS. 4A–B] The brown adhesive used for lining was lightly applied and appeared to be a protein-based animal glue. Variations of gray palette dominated the portrait. Even the light pinkish tones of the face were combined with some amount of the gray pigment. Uniformity in pigment composition was reflected by uniformity of fungal colonies. [FIGS. 5A–B] Although there is no confirmation as to the type of dark pigment used in combination with gray, its presence in all colors is assumed to be the main nutrient source for the molds. [FIG. 6]

With the pastel portrait *Cecil Girl* (1973; #01.12–1), the sand-like paper was prepared traditionally by sprinkling pulverized pumice over the still wet surface of paper coated by starch paste. In addition, this portrait on the pumice-paper was attached to a board with a heavy layer of brown glue. [FIGS. 7C–D] The board was manufactured in Germany, under the name "Grumbacher, fine sanded buff pastel board." Removal of the board revealed yet another manufacturer—"Ersta made in

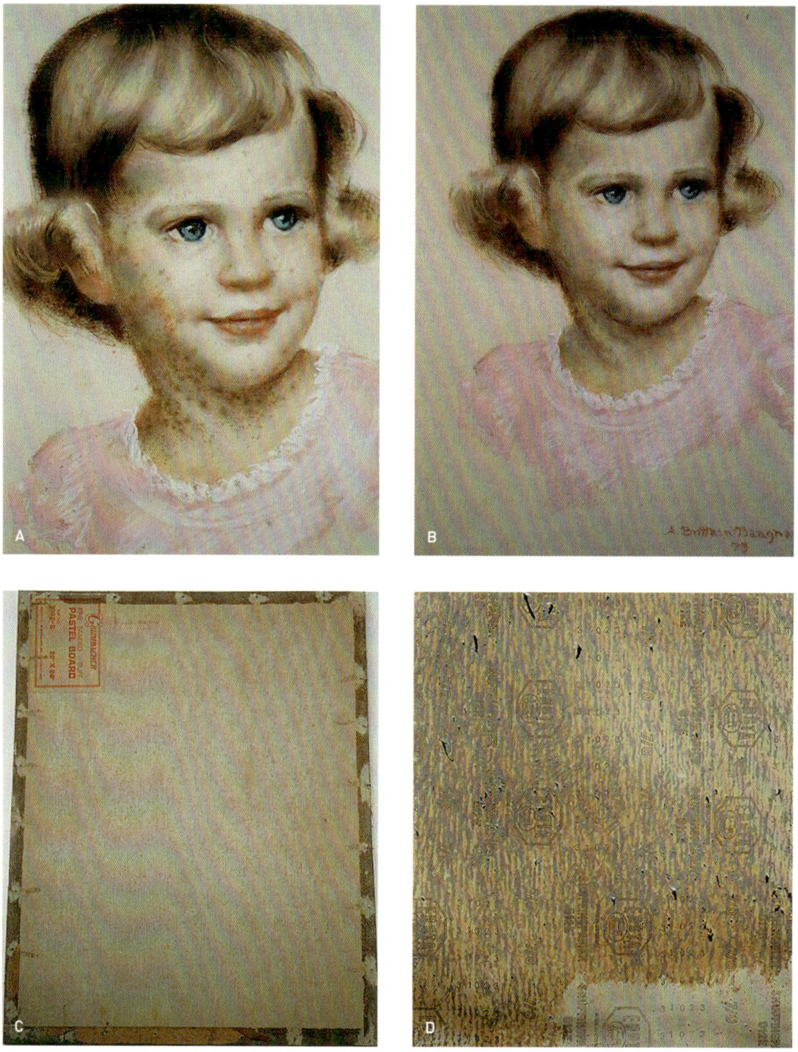

[7] *Cecil Girl*, pastel portrait, 1973 (#01.12-1). Before (a) and after (b) treatment. (c) Recto, portrait executed on pumice paper mounted to acidic board with heavily applied adhesive. Fungal infestation is shown as brown stains scattered on the girl's chin and neck. (d) Verso, board with insert showing adhesive layer. (e) After treatment close-up of face.

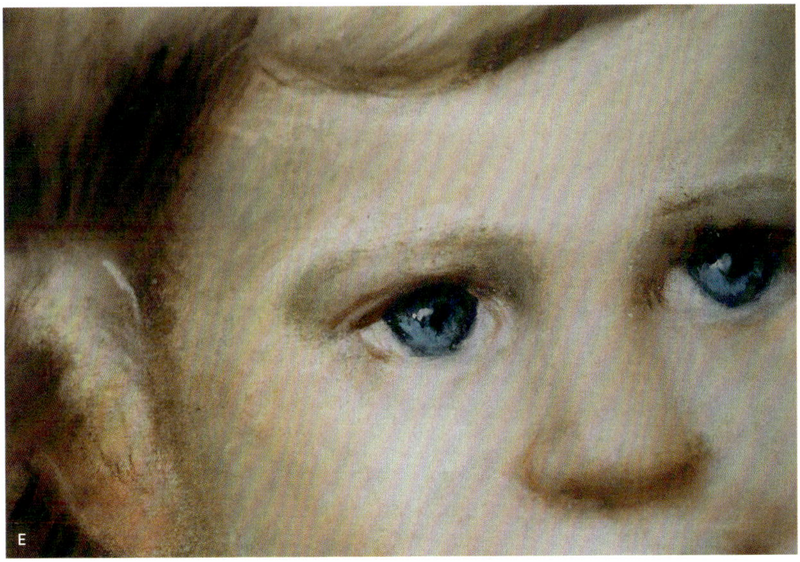
E

Germany, Snuffingpapier." A combination of all the elements, particularly the adhesive layers, promoted the fungal growth. [FIGS. 8A–G]

Conservation Treatment Pastels executed on textured paper form a loosely arranged layer of pigment that, when infested by fungi, promotes a light aerial (mycelial) growth. The colonies are often abundant and obscure the image, prompting the artwork's owner to seek an immediate treatment. Although the infestation often appears overwhelming, if treatment is applied in a timely fashion the removal of mold is usually successful. [FIGS. 2A–C, 5A–B] Fungal colonies growing on the surface of pastel powder are usually successfully picked up with a fine brush, 000, one colony after another. However, if the pastel layer is more compact and executed on a sand-paper (e.g., *Cecil Girl* portrait), the fungal growth is often more compact and often penetrates the layers underneath, discoloring the paper. The removal of these colonies as well as the discoloration is much more complex. One treatment that was successfully applied on the *Cecil Girl* portrait involved removal of stains from the background (non-image area) with a 0.2% aqueous solution of sodium borohydrite, a traditional bleaching technique, and local re-washing on a suction table. However, the discoloration formed on the image left a limited choice of treatment. In this case, the owner of the portrait was interested in the aesthetic improvement of the image and approved an inpainting of stains. Results of the treatment of *Cecil Girl* portrait can be seen in FIGS. 7B AND E.

WATERCOLOR

Media A watercolor painting is created with paint applied in very thin layers, allowing lightness of the ground to become an integral element of the art piece. To achieve transparency of colors, the paint requires a relatively small amount of pigment suspended in binding media—gum arabic, tragacanth or fish glue. In order to achieve an even flow of paint, plasticizers have often been used, such as glycerin, crystallized sugar or honey. The consistency of paint is a colloidal suspension of solid particles in a light watery solution of binding and plasticizer. The presence of gum in watercolor is essential to ensure the paints' capacity to stick to paper. Gum arabic also plays a crucial role in the maintenance of a stable dispersion of pigment particles in water until the film of wash has dried, and the colors are gummed in place.

Support To achieve the desired effects of the watercolor technique, and even distribution of pigment particles on the surface, the texture of that surface must be relatively even and uniform. Dry, rough paper, on the other hand, more appreciated in the 19th century, was used for quick handling of the drawing, capturing the paint only across the highest points of the sheet's surface. Historically, until the end of the 18th century, gelatin was used as surface sizing for western papers. According to 19th and early 20th century manuals and product sources (Cohn, 1977), the sizing that was used for preparation of watercolor followed Whatman's innovation that produced papers with hard-sized surface. However, any organic sizing was prone to fungal infestation by providing a source of nutrient.

The artist of the early-20th-century watercolor that we examined (work attributed to Thomas Birch) chose a medium textured paper. [FIG. 9A] The moderately applied sizing allowed detailing of the garden scene. Based on microscopic examination of the work, it became clear that the paint is thinly applied and partially absorbed by the support. The light sheen of the surface suggests use of gum arabic. However, this can be assumed only based on the history of watercolor media and information about this artist's technique and style of painting.

Areas of the watercolor that were exposed to water (as illustrated by the photograph of verso [FIG. 9B]) were infested by fungi that at one time had produced sexual reproductive perithecia. [FIG. 10] The fruiting structures were old, as evidenced by the fact that no spores were apparent and most of the bodies were broken. Those

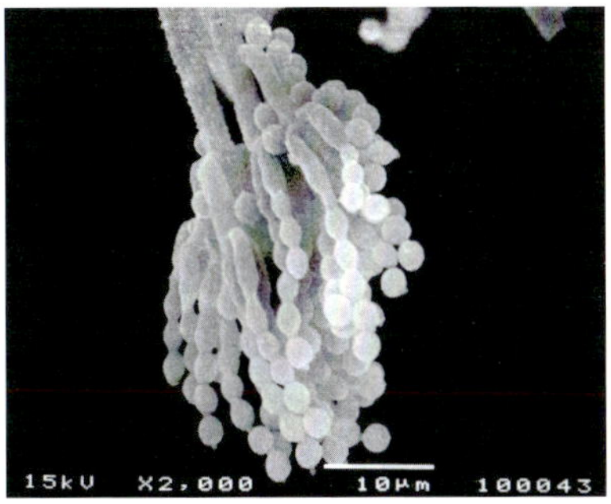

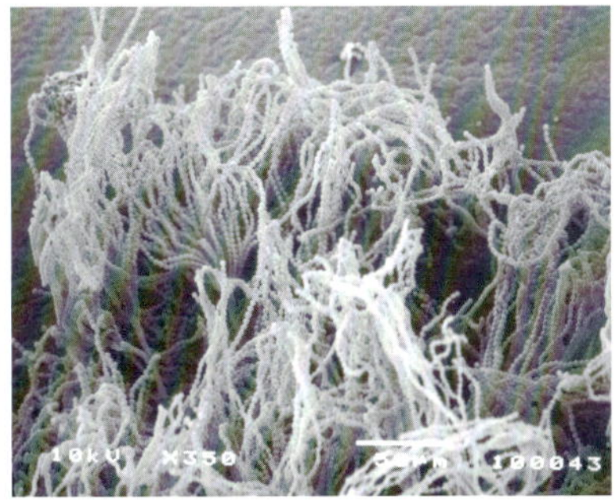

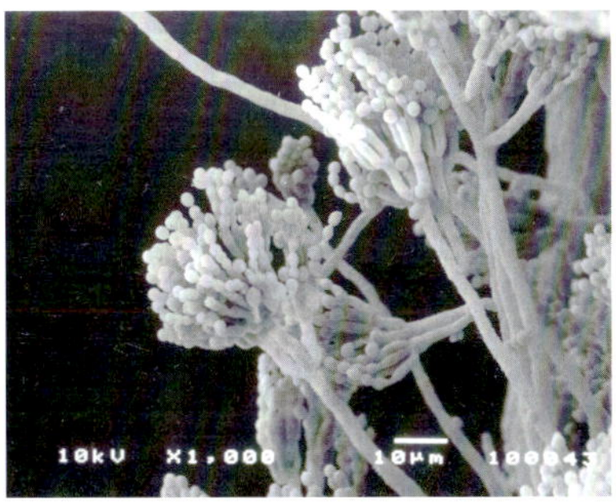

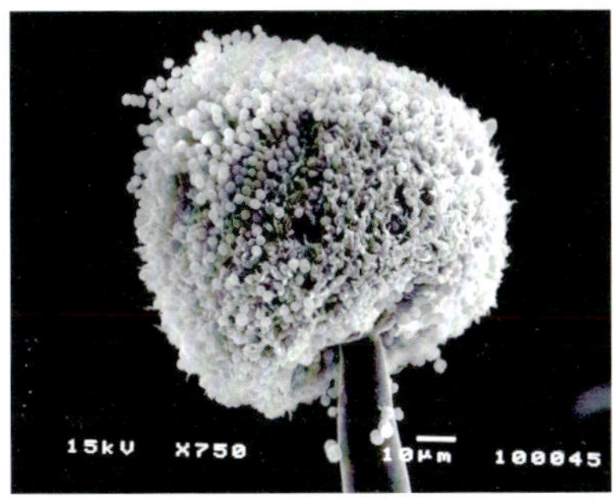

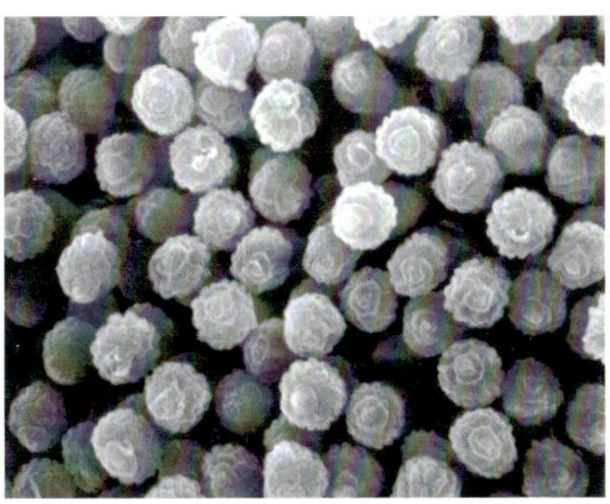

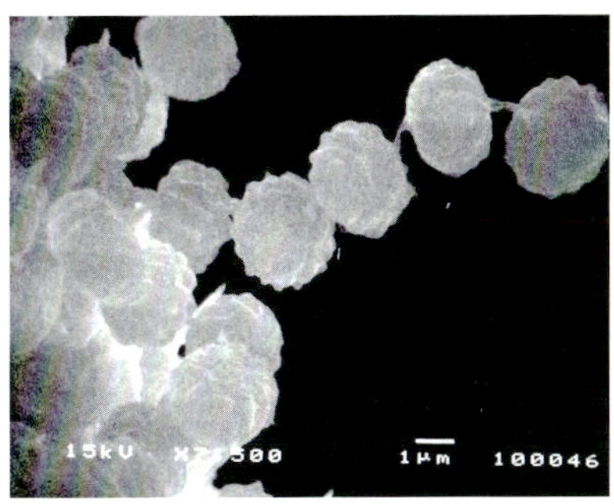

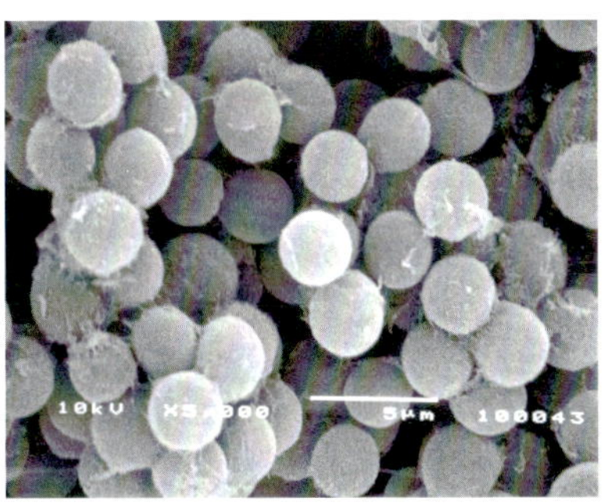

[8] *Cecil Girl*, pastel portrait, 1973 (#01.12-1):
(a) *Penicillium roqueforti* (2000×)
(b) *P. chrysogenum* (350×)
(c) *P. chrysogenum* (1000×)
(d) *Aspergillus niger* (750×)
(e) *A. niger* (5000×)
(f) *A. niger* (7500×)
(g) *Trichoderma viride* spores (5000×).

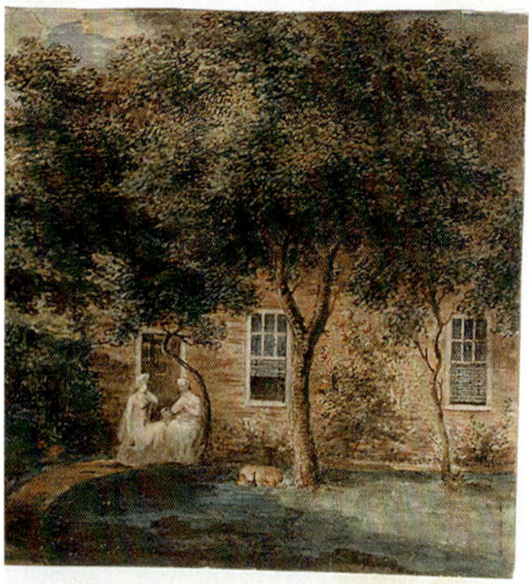

[9] Twentieth-century watercolor attributed to Thomas Birch (#4.29.2): (a) recto and (b) verso. Tideline on verso indicates a brief exposure to water.

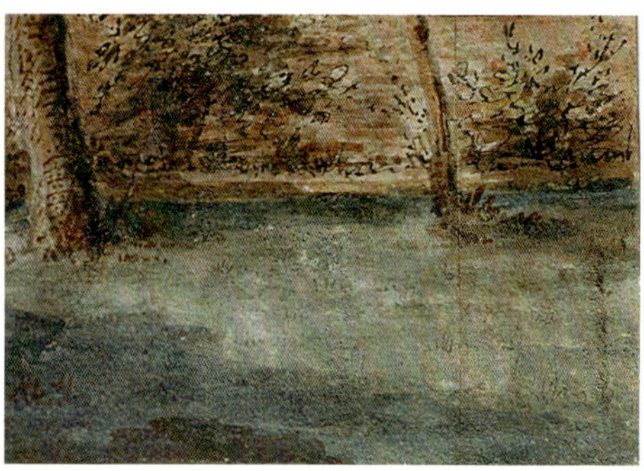

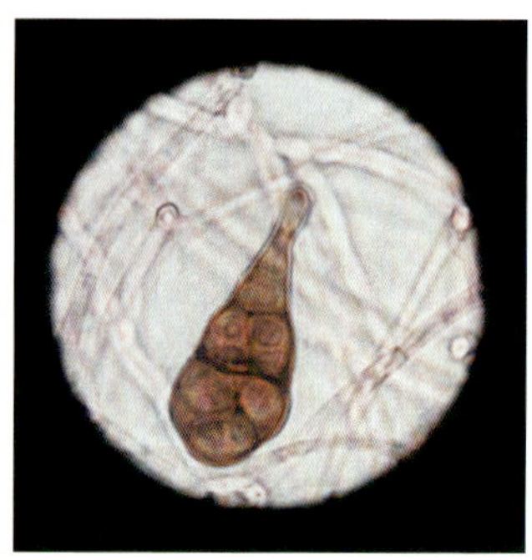

[10] Twentieth-century watercolor attributed to Thomas Birch (#4.29.2). Detail of Fig. 9, bottom right quadrant; insert showing perithecia embedded on recto in the area exposed to water. [11] Twentieth-century watercolor attributed to Thomas Birch (#4.29.2). *Alternaria alternata*; didymospore, showing longitudinal and transverse septa (1000×).

that were still intact were cultured but proved to be nonviable. However, other, active molds were found and cultured from the same watercolor. [FIG. 11]

Conservation Treatment Embedded perithecia could not be removed safely without adversely affecting the image. Laboratory testing proved that the perithecia did not contain active spores; therefore no treatment aiming to remove fungal residue was performed.

INK WASH

The technique of ink wash resembles watercolor, but is monochromatic. The composition of the medium is a mixture of a pigment (carbon black, brown bister, sepia, or other colors) with a vegetable gum. The distinction between colored ink and watercolor is not easy to make.

The ink wash medium in the investigated *Trees and Animals* (#01.53–4), created in the 18th century by Lady Diana Beauclerk (1734–1808), was very thinly and uniformly applied on paper. [FIGS. 12A–B] The fungal infestation that became visible on the surface was attributed to the composition of the artwork itself. The original drawing was mounted to a secondary support, tertiary paper and canvas. Each layer was interleaved with heavily applied paste. Once separated they revealed the full extent of the fungal infestation. [FIGS. 13A–C]

Conservation Treatment The drawing was separated from inferior secondary supports and washed to remove the primary source of nutrient-paste. Minor bleaching was carried out locally with a 0.2% aqueous solution of sodium borohydride. The artwork was re-washed after completion of the conservation treatment. [FIGS. 14A–B]

PRINT

We all live surrounded by prints created in a great variety of media, and volumes have been published about the artist-printing techniques. Therefore, our investigation focused not on the printing technique itself, but on a particular case of a print with gamboge glazing. The investigated print was an engraving on paper, *Brown Pelican* (#01.8), by John James Audubon, 1838. [FIGS. 15A–B] The gamboge glazing was frequently used by Audubon to enhance his green and brown colors.

Gamboge, a transparent yellow pigment, is a gum resin. It contains about 70% yellow resin, and 15–25% water-soluble carbohydrate gum. The major resin constituent, gambogic acid, contains a carboxylic acid group, and other elements such as phenolic compounds that should adversely affect the fungi and prohibit their

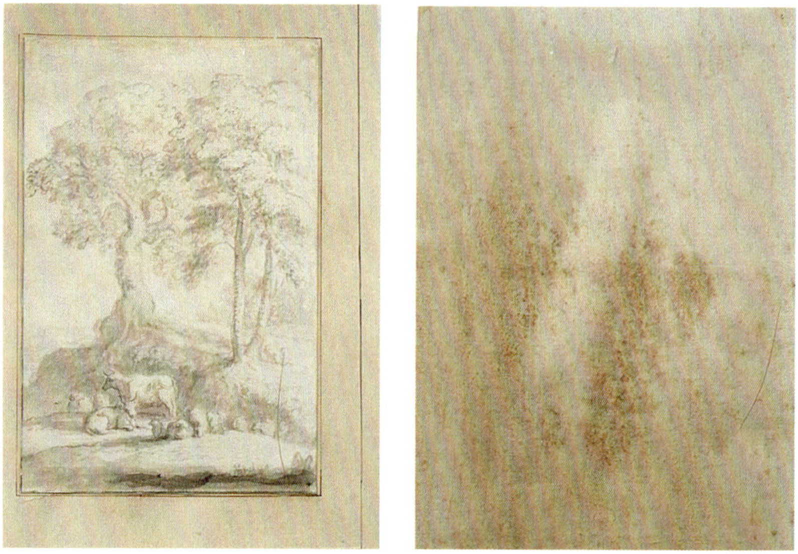

[12] *Trees and Animals,* ink wash by Lady Diana Beauclerk (1734–1808) (#01.53-4). (a) Recto and (b) verso before conservation treatment.

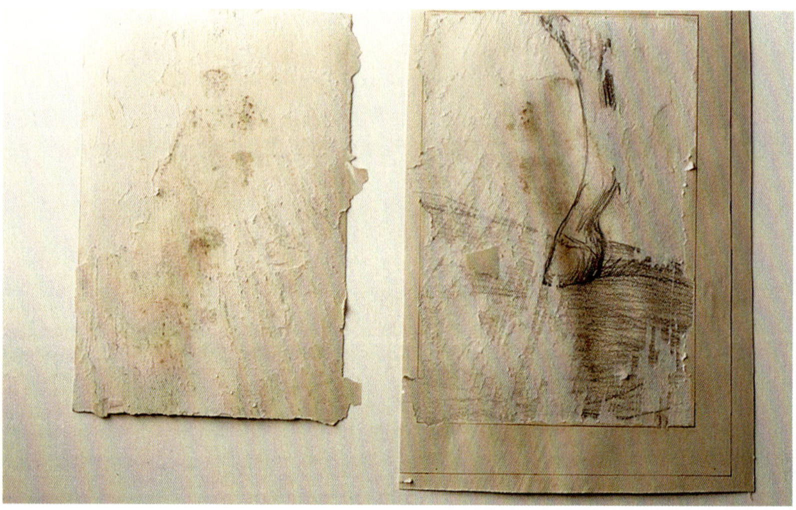

[13A] *Trees and Animals,* ink wash by Lady Diana Beauclerk (1734–1808) (#01.53-4). Paste used for mounting of several layers of secondary supports contributed to biodeterioration of the artwork.

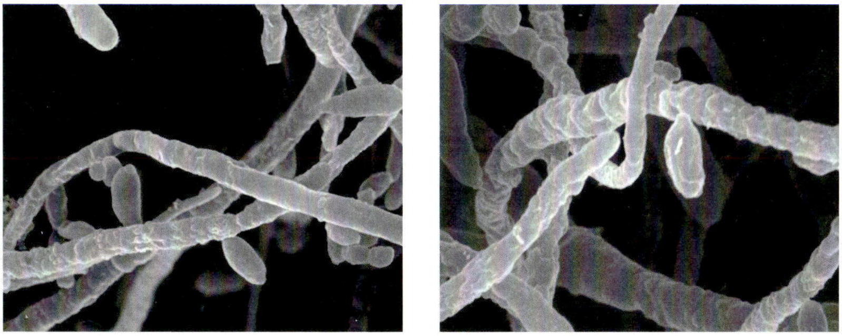

[13B,C] *Trees and Animals,* ink wash by Lady Diana Beauclerk (1734–1808) (#01.53-4): b) *Aureobasidium pullulans* (2000×); c) *Aureobasidium pullulans* (3500×).

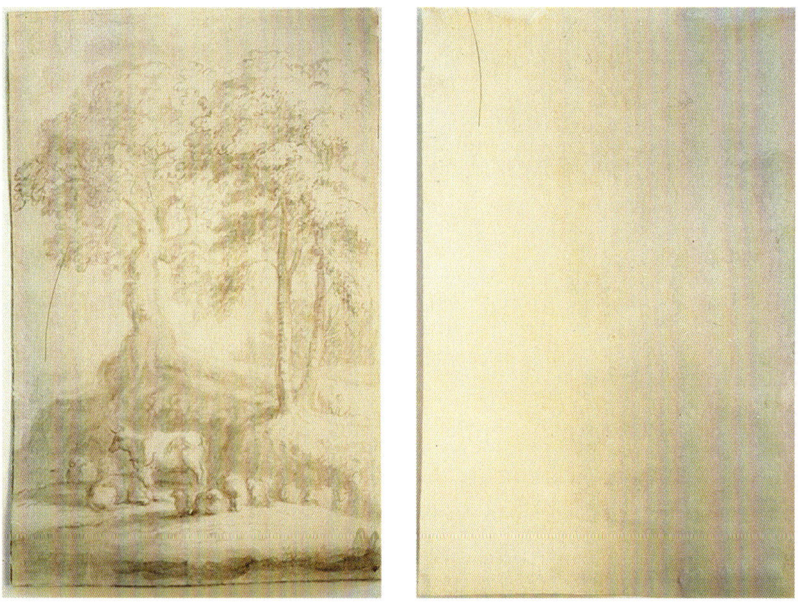

[14] *Trees and Animals,* ink wash by Lady Diana Beauclerk (1734–1808) (#01.53-4). (a) Recto and (b) verso after completion of conservation treatment. Fungal residue was removed mechanically and by washing and local bleaching with 0.03% sodium borohydrate.

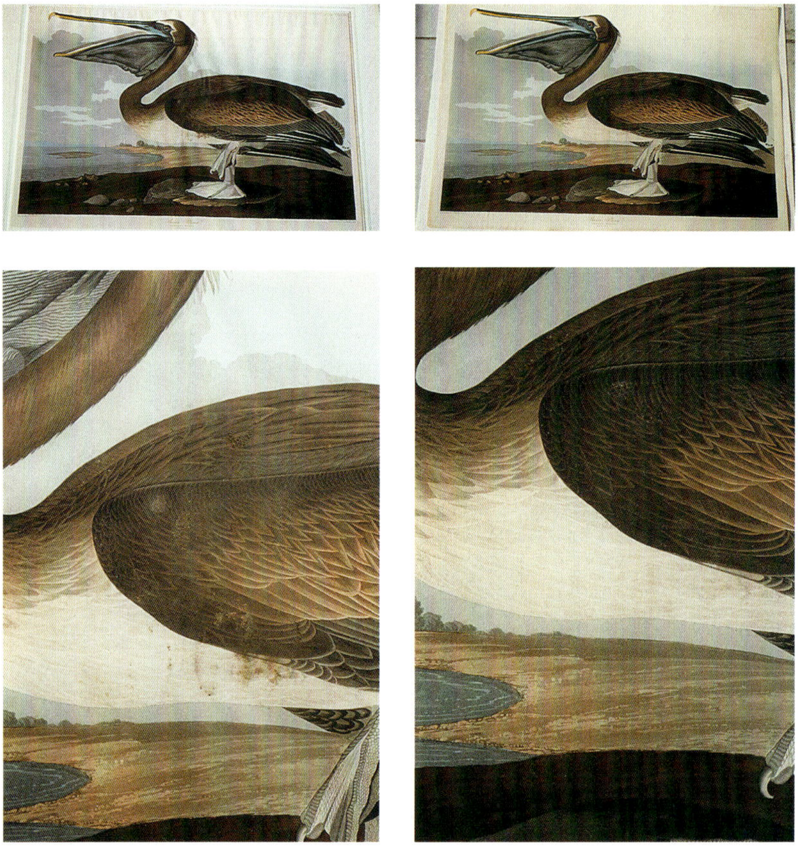

[15] *Brown Pelican,* 1838, John James Audubon (#01.8). Hand-colored engraving with gamboge glazing. (a) Recto before conservation (plus insert) and (b) verso after conservation (plus insert). Inserts illustrate details before and after conservation treatment.

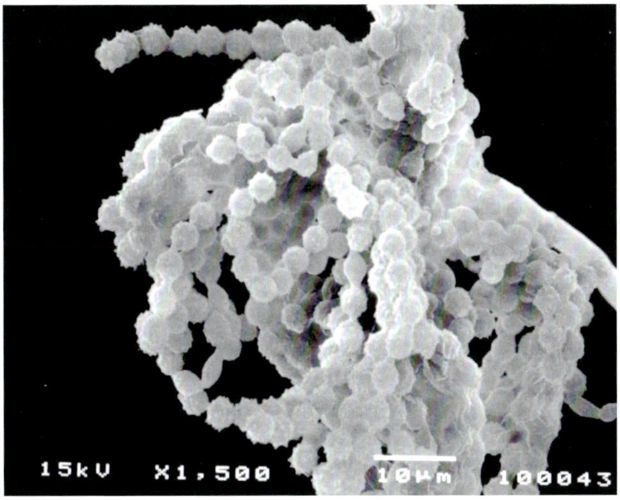

[16] *Brown Pelican*, 1838, John James Audubon (#01.8). *Aspergillus versicolor*, characterized by spherical conidial head (1500×).

development (Winter, 1977). In the case of the investigated Audubon print no such effect was noted. Fungi flourished on the areas glazed with gamboges. The laboratory testing established an abundance of mold. [FIG. 16]

Conservation Treatment The abundant growth of fungi was first removed mechanically with scalpels and brushes. Considering the presence of water-soluble yellow pigment gamboges, all subsequent aqueous treatment procedures were carried out locally on a suction table, avoiding the areas coated with gamboges. A 70% aqueous ethyl alcohol solution was introduced from verso for the purpose of disinfecting any remaining viable fungal elements.

BIODETERIORATION BY FUNGI

The results of our studies reveal the presence of a number of fungi infesting and growing on artworks, drawings, prints, and old documents. All major groups of the fungi are represented and very few are specific to a particular substrate. While the Deuteromycetes (Fungi Imperfecti) predominate, the mycoflora we encounter includes members of the Zygomycetes (so-called bread molds), Ascomycetes (sac fungi), and Saccharomycetales (yeasts) (Alexopoulos, Mims, and Blackwell, 1996). All of the species we encounter are ubiquitous in nature and are normally found growing on a large number of organic substrates, including soil particles, dung, living and dead plants and their parts, humus, and wood products. In some cases, they infest, infect, and parasitize wild or domesticated animals, as well as humans. In recent years, several of the species have been implicated as having a major role in causing allergies, respiratory infections, and the production of toxins responsible for a number of human illnesses in cases where the quality of indoor air has been compromised by water damage to the building (Morey et al., 1990; McNeel and Kreutzer, 1966; New York City Department of Health, 2000).

The species of fungi we encounter on works of art and documents have become old, familiar friends. Many of them, such as *Aspergillus* and *Penicillium* (Deuteromycetes) are ubiquitous. It is rare, indeed, if we do not find the growth of one or more species of *Aspergillus* appear or dominate water-damaged drawings, pastels, or prints. Common species that we almost always harvest are *A. nidulans, A. niger* [FIG. 8D], *A. oryzae, A. terreus,* and *A. versicolor.* [FIG. 15] While *A. flavus* and *A. parasiticus* have been noted on occasion, they do not appear often. We usually find a piece of artwork partially or substantially infested by a single species, generally covering a specific pigment on the work. However, it is not uncommon to find more than one species of *Aspergillus* claiming several territorial regions on a single piece of work. When several species are harvested from a single work, they are usually separated as distinct colonies. Although *Penicillium* is not as commonly found, it, too, is a sure bet among our cultures of damaged artworks. Two species that we encounter often are *P. chysogenum* and *P. roqueforti.* Only under rare conditions (*A. niger*), is it possible to distinguish between the two genera, *Penicillium* and *Aspergillus,* by the characteristics of their colonies. Positive identification separating the two genera as well as the species within each genus is only attainable through microscopic examination coupled with gross morphology of the colony. Several other conidial Deuteromycetes make up the familiar species line-up—*Alternaria, Aureobasidium, Cladosporium, Nigrospora,* and *Trichoderma.* All five of these produce distinctive conidiophores and/or conidia and are almost always harvested on our artworks and documents. The two genera of yeasts, *Rhodotorula* and *Saccharomyces* (Saccharomycetales), are often found at random growing with the conidial molds. *Rhizophus nigricans* (Zygomycetes) is not encountered with any frequency. The fungus grows very rapidly and sporulates quickly. It is considered opportunistic and may represent the first fungus in a successional series to establish on a substrate.

Chaetomium is a distinctive genus in the Ascomycetes and is readily identified by the characteristic hairs or setae surrounding the sexual reproductive structures (perithecia). Hundreds of species of *Chaetomium* occur in nature and several of them have been encountered in our work (Szczepanowska and Cavaliere, 2000).

CONSERVATION PRACTICES The conservation aspect of dealing with fungal infestation includes three major areas—environment, treatment of infestation that has occurred, and prevention.

The ambient environment in the storage or exhibit spaces is as important to the longevity of artworks as is the immediate microenvironment within the frame. Elevated humidity in the environment combined with stagnant air trapped within the frame will provide the most favorable conditions for mold growth. Often the colonies expand onto the glass surface. [FIG. 2B] The proper environmental condition for art collections is one of the most fundamental elements affecting them. Of equal significance is the impact of materials remaining in direct contact with the artwork itself, such as inferior backing boards, excessive animal glues, and cumbersome frames. Our studies indicate that a direct correlation exists between all elements of artworks and the fungi infesting them. Consequently, maintaining a proper environment, particularly with the RH under 65% and the use of inert materials that may come in contact with the artwork (framing, storage, exhibition, shipment) will reduce chances of fungal infestation.

Treatment Techniques The delicate nature of art surfaces poses limitations to the type of treatment that can be employed. A conservator often has to settle for minimal or no intervention in order to preserve the unity of the image. Surprisingly, although pastels are executed in one of the most vulnerable media, their airy and loose structure supports colonies loosely sitting on the surface, allowing for their gentle and successful removal. The severity of biodeterioration of prints, ink washes, or watercolors is often correlated with the length of time the artwork is exposed to water. The longer the period of exposure to moisture, the more severe the growth, staining, and formation of fungal structures, often embedded in the image. Secreted metabolic products create stains that cannot be removed by mechanical treatment; however, a local bleaching with sodium borohydride might be successful. The last resource in restoring esthetically disfiguring discoloration and stains caused by fungi could be inpainting of the affected areas.

CONCLUSIONS Our studies indicate that there are about a dozen ubiquitous fungi (or molds) that tend to infest artworks and documents on paper. Although these fungi, in most cases, do not appear to be substrate-specific, some appear to be promoted and supported by the medium used in creating the art piece, as well as by the nature of their support (acidic pH and excessive amount of starch glues and other organic adhesives). The growth of molds on any piece of artwork or document may be an indication of the fungi that were present and incorporated into the fabric of the artwork when it was rendered. Or they may represent airborne species that very recently landed on the work and found the pigments nutrient-rich and capable of sustaining their growth.

As with most substrates upon which fungi grow, each colony is distinct and represents a single species. Often colonies representing different species are in close proximity. However, even on pieces of art, fungi secrete chemicals that tend to prevent unwanted neighbors from getting too close. What is extremely interesting is that a successional pattern of fungal growth appears to occur on artworks and documents similar to that which occurs in nature. On paper-base substrates Zygomycetes (the bread molds) are followed by Deuteromycetes (mold), and only with longer exposures to moisture do the perithecial fungi of the Ascomycetes, such as *Chaetomium*, appear.

ACKNOWLEDGMENTS We wish to thank the art collectors for permission to publish their artworks in this investigation—Patricia Cecil (*Cecil Girl*, 1973); Robert Cockson (*Hanna* portrait, 1886); Martha Hopes (*Brown Pelican*, 1838); Mary McCabe (*Portrait of a Gentleman*, 1860); Pennsylvania Academy of Fine Arts (Thomas Birch, attr. watercolor); Mr. and Mrs. Peirce (*Trees and Animals* by Lady Diana Beauclerk). In addition, we acknowledge and thank Betty Siefert, Chief Conservator, Jefferson Patterson Park, Archeological Conservation Laboratory, for her assistance, and for access to a stereoscopic analysis microscope; Nancy Piatszyc, Supervisor at The Electron Microscopic Laboratory at Bronfman Science Center, Williams College, Williamstown, MA, for pigment analysis; and Rafal Szczepanowski, Applied Physics Laboratory, Johns Hopkins University, for his technical assistance in graphic enhancement.

REFERENCES

ALEXOPOULOS, C.J., MIMS, C.W., and BLACKWELL, M., 1996. *Introductory Mycology*. John Wiley and Sons, Inc. New York, NY.

BOZZOLA, J.J., and RUSSELL, L.D., 1999 *Electron Microscopy*, 2nd edition. Jones and Bartlett Publishers, Sudbury, MA.

BUTLER, E.E., and MANN, M.P., 1959. Use of cellophane tape for mounting and photographing phytopathogenic fungi. *Phytopathology*, 49, 231–232.

COHN, M.B., 1977. *Wash and Gouache, A study of the development of the materials of watercolors*. Published by the Center for Conservation and Technical Studies, Fogg Art Museum, Cambridge, MA.

Difco Manual of Dehydrated Culture Media and Reagents for Microbiology, 1984, 10th edition. Difco Laboratories, Inc. Detroit, MI.

DOERNER, M., 1984. *The materials of the artist and their use in painting, with notes on the techniques of the old masters*. Harcourt Brace & Company. San Diego, CA.

DYKSTRA, M., 1992. *Biological Electron Microscopy, Theory, Techniques, and Trouble-shooting*. Plenum Press, New York, NY.

FLIEDER, F., 1982. Study of the Conservation of Pastels. In: Brommelle, N.S., and Thomas, G., (Eds.), *Science and Technology in the Service of Conservation*. The International Institute for Conservation of Historic and Artistic Works. London, UK.

JAMES, C., CORRIGAN, C., and ENSHAIAN, M.C., 1997. *Old Masters Prints and Drawings*, Translated and edited by Marjorie B. Cohn, Amsterdam University Press, Colophon, Amsterdam.

MCNEEL, S.V., and KREUTZER, R.A., 1996. Fungi and indoor air quality. *Health and Environment Digest*, 10, 9–12.

MOREY, P.R., FEELEY, J.C. SR., and OTTEN, J.A., 1990. *Biological Contaminants in Indoor Environments*. AST STP 1071. American Society for Testing and Materials. Philadelphia, PA.

NEW YORK CITY DEPARTMENT OF HEALTH, 2000. *Guidelines for the assessment and remediation of fungi in indoor environments*. New York City Department of Health, Bureau of Environmental & Occupational Disease Epidemiology. New York, NY.

STRZELCZYK, A., 1981. Painting and Sculptures. In: Rose, A.H., (Ed.), *Economic Microbiology*, Vol. 6, Microbial Deterioration, Academic Press, New York, pp. 203–234.

SZCZEPANOWSKA, H., and CAVALIERE, A.R., 2000. Fungal deterioration of 18th and 19th century documents: a case study of the Tilghman Family Collection. Wye House, Easton, Maryland. *International Biodeterioration and Biodegradation*, 46, 245–249.

WINTER, J., 1997. *Gamboge, Artist Pigments, A Handbook of Their History and Characteristics*, Vol.3, Fitzhugh, E.W., (Ed.), National Gallery of Art, Washington, DC.

Y. VAN DYKE

PRACTICAL APPLICATIONS OF ENZYMES IN PAPER CONSERVATION

Winterthur Museum, Winterthur, DE, and The Metropolitan Museum of Art, New York

Paper conservators have long recognized the benefits of enzymes in the conservation treatment of works of art. Most commonly, hydrolase-type enzymes are employed in the conservation of works on paper to assist in the breakdown of adhesive residues from previous restorations or to facilitate the removal of secondary supports such as linings or mounts. The principal advantages of these enzymes are their specificity and efficiency in catalyzing hydrolytic cleavage of polymers such as proteins, polysaccharides, and lipids.

This work evaluates the effectiveness of two protease enzymes while taking into consideration cost effectiveness, optimal working conditions, and after-treatment effects. Efficacy of the enzyme treatments in the removal of proteinaceous adhesive from hot, cold, and rough-surfaced watercolor papers was evaluated. It was found that both enzymes performed very well in optimal conditions that ranged from a pH of 3.3 for pepsin to a pH of 7.5 for trypsin, both at room temperature. Color measurements of the sample substrates were recorded before and after enzyme treatments. After treatment L* values were very close to L* values for untreated control samples, confirming a panel assessment that rated visible reductions in adhesive residue. Half of the sample population was aged for one week in 75°c and 65% RH conditions. L* values showed very little differences between the before aged and after aged samples, indicating no signs of discoloration of either minor adhesive or enzyme residues. The fruition of this research was realized in the successful conservation treatment of an Indian miniature utilizing the protease enzyme trypsin that was incorporated into a gel; conservation treatment is discussed.

INTRODUCTION Enzymes are sometimes used in the conservation of works on paper to assist in the breakdown of adhesive residues from previous restorations or to facilitate the removal of secondary backings and mats. It is often necessary to remove these adhesives from paper artifacts because they may exacerbate the deterioration of the paper as they age or contribute to the unsightly and unacceptable physical and visual state of a work of art.

Enzymes make up the largest and most highly specialized class of protein molecules. These complex proteins, produced from living cells, are the primary instruments for the expression of gene action since they catalyze thousands of biochemical reactions. Within the enzyme collective is a group whose function is catalyzing hydrolytic cleavage of such biological polymers as proteins, polysaccharides, and lipids. These are known as hydrolases and are the type of enzyme most commonly used in paper conservation (Grattan et al., 1987).

Of all the types of enzyme-catalyzed reactions, hydrolytic transformations involving amide- and ester-bonds are the easiest to perform using proteases, esterases, or lipases. A large number of readily available enzymes possessing relaxed substrate specificities exist and are the main features that have made the hydrolases the favorite class of enzyme for organic chemists during the past decade. About two-thirds of the total research in the field of biotransformations has been performed using hydrolytic enzymes of this type, and new discoveries have been transpiring at rapid rates in the past decade (Faber, 1997).

Two types of protease enzymes have been chosen for comparison in this investigation—trypsin and pepsin. The principal advantage of enzymes is their specificity, which enables only one kind of material, such as starch or protein, to be acted on, and rapidly. Each enzyme has specific bond cleaving mechanisms and also differs in activity units. In order to evaluate the enzyme's effectiveness at removing a protein adhesive from a paper object, three types of papers were coated with a proteinaceous adhesive and subjected to enzyme immersion baths having the same concentrations of 200 units of activity per milliliter. The chosen papers are different in their surface characteristics because it is thought that the paper's topography may contribute to the retention of enzymes in the structure after treatment (Andrews, 1990). The enzymes' speed of action and their potential for use depend, however, on a number of factors which may not be well defined for specific use in paper conservation.

Paper conservators have long recognized the beneficial uses of enzymes in treatment of tenacious adhesives. A thorough review on the history of enzyme use in paper conservation up until 1983 is found in Pia De Santis's article (De Santis, 1983). In the past decade, a number of published experiments can be found with regard to the use of amylase enzyme (Andrews, 1990; Ericson, 1992), but none on the specific uses of proteases in paper conservation. A meaningful study was published in 1988, (Grattan et al., 1987) outlining the types of enzymes important to paper conservators and evaluating their performance in various assays. This paper characterized four protease enzymes by examining the effects of temperature, pH, concentration, and shelf life, as well as by evaluating their effectiveness as releasing agents for adhesives.

The published conservation literature illustrates that generally, experiments have been limited to only two or three protease enzyme structures. (Rickman, 1988; Grattan et al., 1987) These are: protease bovine, an enzyme from *Stryptomyces griseus*, and protease-pancreatic crude, all of which have substantially low activities ranging from 1.1 to 13.8 units/mg. These enzymes are also somewhat expensive and are commonly criticized as a high-cost conservation treatment.

Throughout the paper conservation literature on enzyme treatments, there is great concern about the ability to rinse or denature the enzymes after treatment is complete. Common recommendations for removing or denaturing enzymes include the use of hot water or alcohol baths; this advice appeared in the paper conservation literature as early as 1977 (Segal and Cooper, 1977). More recent studies (Ericson, 1992; Andrews, 1990) report on the use of various rinsing procedures after employing an alpha-amylase enzyme, and showed that these previously suggested denaturing steps were no more effective at removing enzymes than water rinses alone. This is an important discovery because, first, attempts to denature with hot water would be ineffective at the conventionally recommended temperatures (proteases reach their thermal optima in the range of 50–105°c), and second, sufficiently elevated temperatures could pose serious risks to the art object being treated. One of these studies was able to quantify the amount of residual enzymes, left after water rinsing, by using a radioisotope tagging system on the enzyme (Andrews, 1990). The research showed that these residues were on the magnitude of 0.5 $\mu g/cm^2$ of paper, or <2%.

The majority of the articles on enzymes found in the paper conservation literature describe treatment procedures as they relate to specific art objects (Hauser,

1987; Shelley, 1980; Wendelbo, 1974; Grattan et al., 1987; van der Reyden, 1988). One article addressed the effects of enzymes on a variety media of a work of art on paper (Burgess and Charrette, 1981). To date, only one partially published study has been found on the long-term effects of an enzyme treatment on a work of art on paper (Tse and Burgess, 1988), which shows the relative gap in the literature for research related to the after-treatment effects of enzymes.

The following investigation was developed primarily to gain insight into practical conservation treatment procedures followed when using enzymes, which included, among others: choice of enzyme, concentration of enzyme, and rinsing steps. In addition to evaluating the need for treatment protocols, it is necessary to approach the question of whether enzyme treatments have any adverse effects on the artifact itself. Issues involving physical changes of the surrogate objects will be examined, and include: efficiency of removing the unwanted proteinaceous adhesive, possible retention of enzymes within paper support, and aging of these objects with possible enzyme residues and subsequent color shifts, if any, produced by them.

The questions posed at the outset of this research were:

- Can a low-cost protease enzyme be found that is acceptable for use in paper conservation?
- Can a protease enzyme work at or close to room temperature, eliminating the literature recommendations for working temperatures of 40–50°C?
- Does the surface character of the paper being treated affect the enzyme's ability to remove adhesives or the degree of retention of the enzyme?
- Does the length of time of the rinse bath affect the overall appearance of the treated paper?
- Will there be a significant amount of enzyme residue left in the paper support if only water rinses are used and, if so, will these residues change color over time?

MATERIALS AND METHODS

DESIGN [FIG.1]

A comparison will be made as to the effectiveness of two protease enzymes, trypsin and pepsin. Three papers having different surface characteristics were brush-coated with two layers of an 8% solution of proteinaceous gelatin adhesive, which was stained with the protein-tagging chemical Rhodamine B isothiocyanate. The working procedure for the enzymes was by immersion, and enzymes with known activity rates at room temperature were selected. Each adhesive-coated sample was immersed into a 10-ml aqueous enzyme bath contained in a small, shallow plastic tray. All enzyme baths had a concentration of 200 activity units per ml. Samples remained in their respective baths for a total time of 30 minutes.

Half of the pepsin-treated samples and half of the trypsin-treated samples were rinsed in a half-hour de-ionized water bath, conditioned with the same buffering components as used in the enzyme immersion solution. The other half were rinsed in two half-hour baths; totaling 1-hour rinse time with rinse baths being refreshed after the first half hour. The samples were all dried in the same manner, being placed between polyester web and blotters and under glass and weights and allowed to dry for a period of 7 days.

Following treatment and drying, the enzymes' effectiveness at removing the adhesive from the samples was judged and evaluated by a panel assessment consisting of eight conservation specialists in addition to microscopic examination by the author. Both of these evaluation techniques assigned numerical codes to each sample, which were analyzed and averaged. Color measurements were also recorded before and after treatment with a Minolta chroma-meter. Variations on the effects of treatments relative to the surface characteristics of the paper were examined at this time and throughout the project. A portion of the sample population also underwent accelerated aging to track the changes that the retained residues exhibited. The aged samples were treated with the same evaluation procedures as noted above.

MATERIALS

Papers The choice of these specific surrogate materials for the experiment is rationalized as follows: In paper conservation, the nature of the paper support can vary widely from a thin sheet to a thick sheet, from a highly calendared paper to a rough-textured paper based on the sheet's porosity. Three types of papers have been chosen to represent a range of surface characteristics: (1) a hot pressed (smooth-textured) surface, (2) cold pressed (semi-textured), and (3) rough-textured watercolor paper, from T.H. Saunders, English papermakers. All were Waterford series, mould-made from 100% cotton, 90 lb/ream, neutral pH, four deckles, watermarked, white.

Gelatin Collagen is the structural protein of connective tissue in animals and fish, including the various parts such as skin, muscle tissue, bone, and hide. Gelatin is the result of the separation of the triple helical conformation of the collagen protein. Amino acids analysis of proteins used as adhesives shows the four major amino acids found in gelatin to be 24.7% glycine (GLY), 13.0% proline (PRO), 9.7%

glutamic acid (GLU), and 8.2% arginine (ARG) (Mills and White, 1994).

Gelatin is produced by breaking collagen's strong interactive bond sites between the hydroxyl of the hydroxyproline and the amino hydrogens of the adjacent glycine units through extensive boiling. This partial denaturing of the collagen form of the protein makes gelatin soluble in water. Solutions of leached material, such as gelatin, with a concentration of more than 2%, set on cooling to form very stiff gels, which have extremely powerful adhesive properties. Water-based solution adhesives become solidified through evaporation of the water and gain strength through secondary forces within the adhesive and between the adhesive and the adherend (paper support).

High densities of polar groups from the paper's surface interacting on a high molecular weight protein chain like gelatin yield some of the strongest cohesions. Surface deviations of some papers, with their varying degrees of amorphous or three-dimensional rough and open surface topography, make it possible for the protein adhesive to flow and penetrate into a relatively large surface area. In theory, rough surfaces have greater surface area and are better at interlocking than a smooth surface. It is therefore thought that stronger bonds will be formed with the "rough paper" as opposed to the smooth and slick "hot press." Additionally, it is believed that these cavities in the rough surface paper will also make it more challenging for the enzyme to get in and clip at these adhesive-bonding sites. The papers were brush coated twice with an 8% gelatin adhesive. In order to assist with visual assessment, colorimetry, and fluorescent densitometry evaluations, the gelatin was stained with the bright pink protein tagging chemical Rhodamine B isothiocyanate.

Enzymes As listed above, the most abundant amino acids found in gelatin are glycine, proline and hydroxyproline, glutamic acid, and arginine. The Sigma Chemical Co. catalog lists approximately 15 different protease type enzymes, including chrymotrypsin, collagenase, elastase, and papain. While one may be able to find enzymes that have specific bond cleavage sites extremely specific to that of gelatin, the cost of these enzymes can be prohibitive in choosing them for a practical treatment. For example, one vial of metalloendopeptidase, which contains approximately only 4 units of activity, costs $175.

Two of the three principal digestive protease enzymes were selected for review —trypsin and pepsin; trypsin is a pancreatic serine with substrate specificity based upon positively charged lysine and arginine side chains. Pepsin, an acidic protease, is the principal proteolytic enzyme of vertebrate gastric juice; its specificity is broad range and it demonstrates an esterase activity. The enzymes chosen belong to the group of hydrolases, which cleave the bonds of proteins, leading to shorter chain polypeptides and finally to amino acid molecules, which are water-soluble. These two particular enzymes were also chosen for their relatively high activity, purity, and affordability.

• Trypsin (EC 3.4.21.4)
Type IX from porcine pancreas
T 0134 crystallized, dialyzed, and lyophilized
Activity: 13,000–20,000 units per mg protein

• Pepsin A (EC 3.4.23.1)
P 6887 crystallized and lyophilized
Activity: 3200–4500 units per mg protein

The most commonly purchased form of enzyme is a water-soluble lyophilized powder. The number of grams of enzyme that will be weighed out, "M," when preparing "V" ml of an enzyme solution having a concentration of "C" units activity per ml solution is given by: $M = C \times V/A$, where "A" is the activity of the solid (Faber, 1997). Specific activity of each enzyme lot was provided by Sigma Chemical Co.: Typsin 0134, lot #076H1101, contains 15,200 units per mg protein, and pepsin 6887, lot #97H7465, contains 4500 units per mg protein. Therefore, it has been determined that the amount of trypsin and pepsin needed for 10 ml will be 0.00013 g and 0.00044 g per individual sample, respectively. However, a 750-ml volume of solution was made for ease of use, as well as to keep exposure of packaged enzyme to air to a minimum during weighing.

IMMERSION SOLUTIONS AND CONDITIONS In most cases, the pH of the enzyme solution must be controlled in order to maintain or accomplish the activity of the enzyme in use. In fact the use of a buffer to achieve optimal pH conditions is highly recommended. Some enzymes have optimum efficiency at a specific pH, while others exhibit a biphasic behavior in that they operate best at two distinct pH optima. An enzyme solution that is not buffered to maintain the optimum pH might be effective, but the action will be slower or may not occur at all.

The direct bearing of pH on an enzyme's effectiveness was, in this experiment, observed especially in the case of pepsin 6887. Initial trials were run with a solution that consisted of de-ionized water, enough pepsin for an activity of 200 units per ml, and 0.05% potassium phosphate that buffered the solution at pH 5.5. At this pH, pepsin 6887 was not active or was extremely sluggish.

Various alterations were made to the enzyme solution in hopes of resolving pepsin's lack of activity. Altering the concentration of the enzyme, adding a bile acid (thinking that a bile acid/surfactant was needed to combat the hydrophobic nature of the alkyl-ketene dimer sizing agent in the paper), increasing immersion times, and lowering the pH with citric acid were all attempted. In the end, pH proved to be the most influential element for induction of optimum activity for pepsin. The final working solution consisted of deionized water, 0.5% potassium phosphate, 0.5% citric acid, and 0.4316 g of pepsin 6887 (200 activity units/ml). This solution held a constant pH of 3.4 throughout the trial.

Final working solutions with trypsin enzyme consisted of 750 ml of de-ionized water, 0.01315 g trypsin 0134 (200 activity units/ml), and enough calcium hydroxide to maintain a pH of 7.5. All enzyme immersion treatments were conducted at room temperature, and continuously monitored for pH with a handheld pH meter and pH-indicating strips.

RINSING Influenced by the results of a recent study that examined the effectiveness of removing amylase enzymes with water rinses alone (Ericson, 1992), the following comparison of rinse baths was included. Two variations of rinsing procedures were followed in this study: each employing 10 ml of rinse water, but differing in time. Rinse 1 consisted of a half-hour immersion, and Rinse 2 consisted of two half-hour immersions, where a fresh bath was supplied for the second half hour. Composition of rinse water was dictated by the solution that each enzyme was active in. In theory, it is suspected that it is best to rinse initially with the same conditions minus the enzyme, so that denaturing of the enzyme will not take place while it possibly remains in the paper.

DRYING All samples were placed between polyester web and blotters, under glass and weights, and allowed to dry over a period of 7 days. This procedure is common practice within the paper conservation field and allows for equilibrium to be attained within the sheets, preventing planar distortions upon removal from the drying package.

MEASUREMENTS AND DATA

SAMPLING A total of 120 samples underwent enzyme treatments. [FIG. 1] Papers were cut into 3-cm squares. There were 20 samples per paper type with two en-

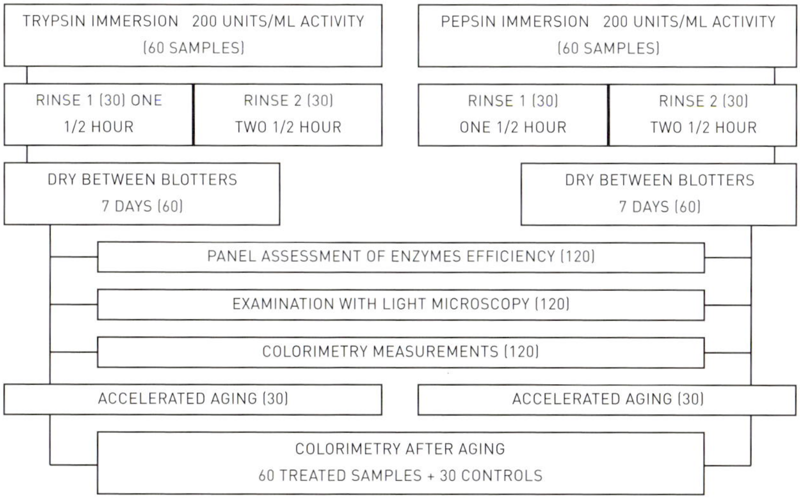

[1] Flowchart of experimental design including trail of 120 samples of hot press, cold press, and rough-surfaced papers.

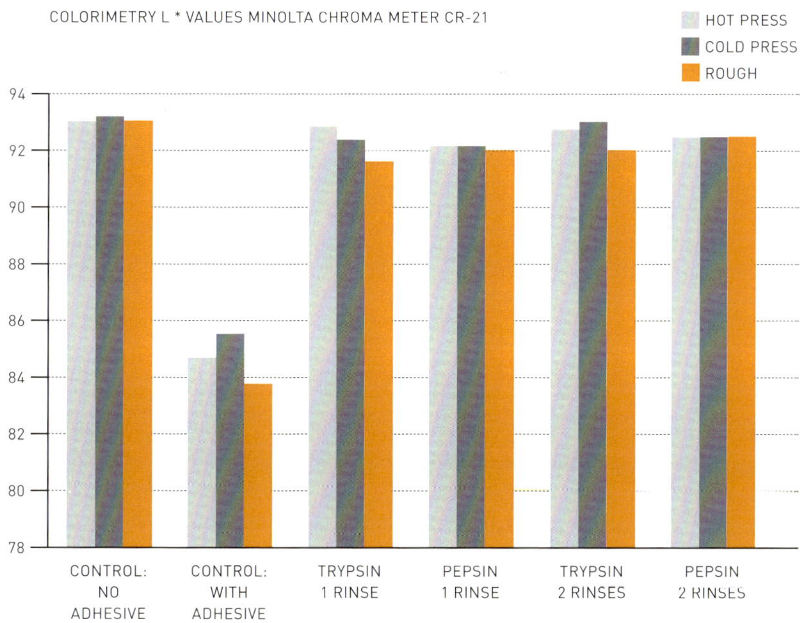

[2] L* value averages showing before and after treatment results.

zymes being compared, with several variations of control samples. An HB graphite pencil was used to label each sample on the verso to ensure that appropriate testing procedures are given to the correct sample. Throughout the evaluation processes, randomization of samples was conducted to avoid bias.

Paper samples were as homogeneous as possible. Sheets were large enough to accommodate the number of samples necessary (46) from each sheet. Conscientious selection of paper dictated that each sheet would be identical in all its components, including furnish, sizing, weight, and thickness. The only difference in the three sheets was in their surface character.

PANEL ASSESSMENT AND DETERMINATION OF EFFECTIVENESS OF TREATMENT The effectiveness of enzymes in removing protein adhesive was ranked according to original data based on human visual assessment. An eight-member panel consisting of paper conservators and conservation students judged the samples. While designing the experiment, the author believed that having the rare opportunity to utilize the eyes and minds of several trained professionals, with their extensive experience in the field of paper conservation, as well as those in training, could provide relevant information.

The samples were laid out randomly and a survey form was generated to make documentation easy and consistent. Lighting conditions in which the panel participants viewed the samples were made as consistent as possible by using a Scandles® light unit with a bulb type measuring 5300 Kelvin, in addition to natural light. All participants evaluated the samples from the same place in the paper conservation laboratory under the same conditions. Numerical codes were assigned to each sample to judge the effectiveness of the treatments: 4=excellent, 3=good, 2=fair, and 1=poor. Quantitative information was analyzed using averages and standard deviation based on the panel member's ranking system.

MICROSCOPIC EXAMINATION Viewed in normal raking light, under 6.5× magnification, each sample was evaluated in terms of the enzyme's effectiveness at removing the protein adhesive. Adhesive residue was easily discernible as a characteristic pink since the adhesive was tagged with the Rhodamine B isothiocyanate R-1755 (Sigma Chemical Co.), which has a bright pink color in normal light. A similar survey form was designed for the panel assessment, and the author judged the effectiveness of treatments by assigning each sample with the same numerical code. Samples were examined in a random order to avoid bias, and quantitative information was averaged and charted with standard deviation employed.

MEASUREMENT OF COLOR CHANGES A Minolta Chroma Meter CR-221 with a Xenon light source was employed, measuring an area 3 mm in diameter. Each sample was measured in four randomly chosen spots that were cut out of a Mylar template and used for each sample, and then averaged. The L*a*b* system was employed, taking into account perceived, not absolute, color differences. L measures light to dark, "a" measures red to green, and "b" measures yellow to blue. "L" values proved to be the most useful, and numerical data were generated and charted employing averages and standard deviation.

ACCELERATED AGING OVEN Half of the sample population underwent accelerated aging in a Blue M Electric AC Series Temperature/Humidity Chamber with MRS 7700 Control, and were compared to the unaged samples. At this time no standards exist for performing accelerated aging on paper artifacts. One of the most extensive investigations into the parameters one would use in accelerated thermal aging of paper with moisture recommends conditions of 70–80°C and 65% RH; these recommendations are endorsed by conservation scientist Robert Feller (Bansa and Hofer, 1984; Feller, 1994).

The MRS 7700 Controller is a circle chart recorder for measuring relative humidity. The instrument is capable of measuring, displaying, recording, and controlling relative humidity and/or temperature using dry bulb and wet bulb temperature. Conditions were set at 75°C and 65% RH and remained constant throughout the 7-day period.

A bookbinder's sewing needle was used to thread a piece of linen sewing thread through each of the 3-cm^2 paper samples so that they could be tied to the metal racks of the chamber without coming into contact with the metal.

RESULTS AND DISCUSSION

PANEL ASSESSMENT Overall, the results of the panel assessment show that both pepsin 6887 and trypsin 0134 worked well on all three paper types. Of the 120 samples averaged, 87% fell in the range of "good to excellent" and only 13% were given a "fair" rating, while none fell into the "poor" category. [FIG. 2]

An interesting trend can be noted when comparing the results of the various rinsing procedures. In all cases, the averages given to the samples that were rinsed twice are higher than those for the samples rinsed only once. However, this observation cannot necessarily be directly linked to the enzyme's activity. It may just be a case where prolonged exposure to water may have assisted in the release of whatever little water-soluble adhesive remained in the support. Nevertheless, rinsing in two half-hour baths resulted in brighter papers, and that fact alone holds practical significance. According to the results of the panel assessment, no correlation can be made between the different paper surfaces and the amount of adhesive removed.

MICROSCOPIC EXAMINATION Evaluating the three pepsin-treated papers, the hot press samples looked the best. Ninety-five percent of the hot press papers were given an excellent rating, while only 40% of the cold press and 25% of the rough-surface papers were rated excellent. These results could be interpreted as verification of the theory that the topographical surfaces of the papers contribute to the effectiveness of this particular enzyme. The larger cavities existing on rough-surface paper seem to make it more challenging for the enzyme to get in and clip at these adhesive bonding sites, or they may simply contain and hold larger amounts of adhesive, resulting in more adhesive residues. Examining the surfaces of the trypsin-0134-treated samples showed no differences between any of the paper types. All samples were rated excellent because no adhesive residues were visually discernible.

COLORIMETRY Focusing the evaluation of color measurements on the L values read with the Minolta chroma-meter, one can clearly see the excellent results of both enzymes employed in the study. Comparing the L values of the three paper types before treatment with no adhesive to those after treatment, L values returned to within 1 integer of the control. [FIG. 2]

L value differences between Rinse One and Rinse Two are relatively minute. One reason that little or no trend in the rinsing procedures can be seen from these results could be the limitations of the area evaluated by the chroma-meter. Since there were only four openings on the template, approximately 12% of the entire 3-cm^2 sample was being rated. During the course of the immersion treatments, it was observed that the samples had a tendency to float, especially from their edges, which resulted in some minor retention of adhesive on the edges of some of the samples.

ACCELERATED AGING OVEN According to the L values taken before and after aging, all samples were affected very little by the conditions of the temperature/humidity chamber. Differences in L values before and after aging were on the order of

0.5. Specimens that were rinsed twice still exhibit brighter values than those that were rinsed only once, but these differences in readings are very small as well. Statements cannot be made as to the degree of enzyme remaining in the papers and how they might change through accelerated temperatures and humidity. However, neither the minor amounts of remaining adhesive or possible enzyme residues left in the papers showed signs of rapid discoloration from the results obtained.

CONCLUSIONS OF EXPERIMENT

Clearly both trypsin and pepsin are capable of producing the desired effect of removing unwanted proteinaceous adhesives from all types of paper surfaces investigated. Both of these enzymes worked very well in room-temperature conditions, yet the pH conditions were quite different. Trypsin worked very well in the neutral pH range, while pepsin showed activity only in the acidic pH range of 3.3–3.4. It is very unlikely that the conditions in which pepsin 6887 operated could be used in paper conservation, but there are always exceptions to general rules, and an extraordinary circumstance may present itself. When employing enzymes for treatment, one should always test the pH of the paper and/or substrate being treated beforehand, as it is possible that the paper itself may act as a partial buffer or affect the enzyme's activity.

According to the panel assessment, there are significant visual benefits resulting in brighter papers from longer periods of rinsing. Results obtained from the aged samples showed that no discernible color shifts had taken place over the course of 1 week under accelerated temperature and humidity conditions. The conservator must also balance the risk to the object when considering the length of immersion time that may be needed with lower concentrations. Concentrations in the range of 50–250 units of activity per ml result in very little if any detectable residue, as long as in-situ denaturation is avoided.

Certainly, goals were met in finding an efficient and cost-effective enzyme; in addition, concerns regarding enzyme retention or clearance from the artwork are greatly diminished with such relatively small amounts of highly purified enzymes. Comparing the costs of two protease enzymes (trypsin 0134 and formerly recommended protease P5147) needed for a 1-liter bath at 200/units of activity illustrates clearly the more judicious choice.

Trypsin T0134 (15,200 units/mg protein)	0.013 g needed	$1.34
Protease P5147 (4 units/mg protein)	37.5 g needed	$1683.00

CONSERVATION CASE STUDY

BACKGROUND The challenging and specific conservation needs of one particular object presented themselves in the summer of 2000. An Indian miniature painting attributed to the artist Ebrahim, the son of Ruknuddin, one of the most famous painters in the Indian Rajput era, dating to ca. 1675, was under consideration for acquisition at the Metropolitan Museum of Art. [FIG. 3] Entitled *Khambhabati Ragini: A Page from a Dispersed Ragamala Series,* the work illustrates a ritual observance to the four-headed and four-armed Hindu god Brahma, who, according to Hindu mythology, was the creator and director of the universe. If acquired, the painting would be the first work by this artist to enter the MMA's collections and would join a small group of late-17th-century paintings from the Rajasthani state of Bikaner. The curator of Southeast Asian Art, Steven Kossak, was very positive about the prospect of acquiring the painting, but was understandably troubled by the damages it has sustained. With its acquisition hinging on whether or not improvements could be made to its overall physical state, initial examination and micro-tests prompted the use of enzymes in the design phase of conservation treatment. However, because of the painting's sensitivity to water, alternative methods of enzyme application would have to be found.

Indian miniatures are typically constructed of densely applied and burnished layers of transparent and opaque watercolor paint. This paint is applied over ink under-drawings on which a thin white watercolor ground or priming is laid down on thick multi-layered paper supports. The range of pigments and implements used to produce these pictures is relatively narrow; the extensive range of effects, including supremely delicate detail, results from a masterful handling of the materials, using brushes, pens, and burnishers. Inorganic pigments, typically bound with gums, are highly valued for their good covering power, relative physical and chemical stability, and resistance to fading. Damages incurred tend to be the result of external circumstances, such as poor housing, handling, neglect, or abuse, as opposed to any inherent vice in the materials.

CONDITION Measuring 16 × 13 cm, the major damage on this object relates to a former window mat attachment that was composed of poorly processed wood pulp. This window mat was insensitively applied with a loaded brush of gelatin, perhaps from the bottom of the glue pot, right on top of the painting itself. Later, the window mat brutally ripped off from the painting, taking with it a good deal of the left edge along with design material and primary support. All edges were left with

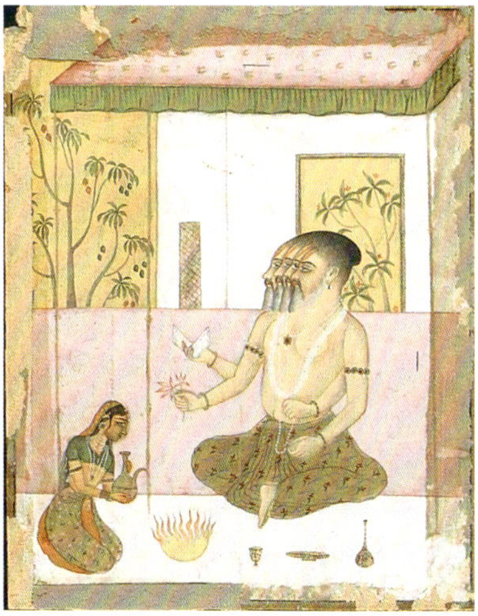

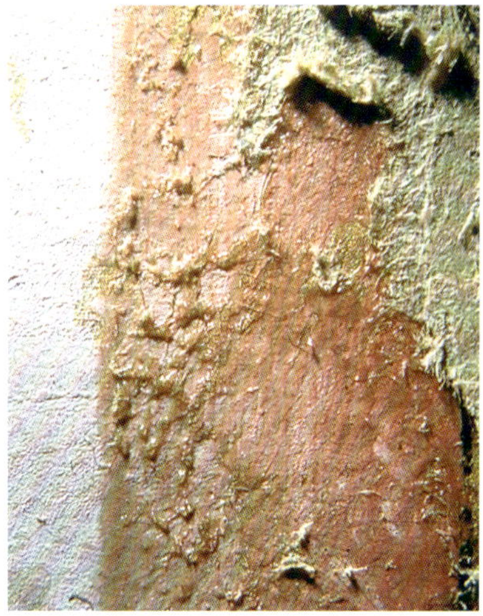

[3] Before treatment *Khambahavati Ragini*, MMA accession no. 2000.321. [4] Detail of adhesive with former mat window fibers seen along right side.

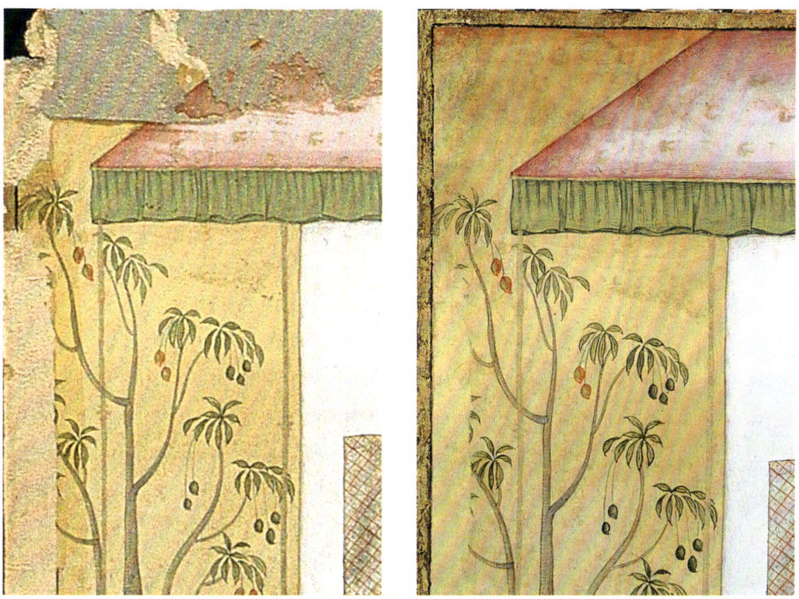

[5] Before (left) and after (right) treatment detail of upper left corner MMA accession no. 2000.321.

either complete loss (left edge) or a thick layer of adhesive residue and window mat fibers, resulting in a pigment layer that was covered and discolored with damages that were aesthetically unacceptable. A small sample of the adhesive was spot tested with an 0.1% solution of fluorescamine in acetone, which confirmed the adhesive to be a protein, fluorescing blue-green in long-wave ultraviolet light. [FIG. 4]

TREATMENT Keeping in mind the results of the initial tests and the vulnerabilities of an extremely water-sensitive paint layer, traditional methods of accretion removal like swabbing with cotton micro-swabs or swelling and scraping would simply not be possible. A conservation treatment strategy was developed that eliminated any mechanical action near the painting's surface and instead involved the use of an enzyme incorporated into an agarose gel. A homogeneous polysaccharide, agarose is derived from seaweed. Within the extracted polysaccharide, polymer chains vary in the number of helix-forming regions, and separations provide several types with varying gelling temperatures. A pure, low-gelling-temperature agarose was selected to be the carrier of the excellent performing protease enzyme trypsin T0134, the same one tested in the previous investigation.

Purchased in a powdered state, the agarose is added to water and heated to 80–100°C until polymer chains are solvated. The enzyme gel was successfully made by first preparing a 1.2% agarose gel. Once gel temperature reached 33°C, 500 units of activity/ml or 0.85 mg of the tyrpsin enzyme were incorporated. In order to avoid denaturation of the enzyme, one must wait until gel comes down to this temperature. Gel was then poured into a shallow petri dish and set within 2 minutes into a clear and ideal consistency for use.

There are numerous advantages for conservators when using an agarose gel, which include: (1) Gel is clear, allowing one to see substrate being acted upon; (2) minimal physical manipulation is necessary once in situ; (3) the concentration can be varied to alter wetness; (4) gel keeps well with refrigeration; and (5) gel can be tailored to specific shape (as compared with other poultices used, such as methylcellulose and hydroxy-propyl-methyl-cellulose). The 1.2% concentration of agarose was found to be a good working consistency in this particular case in that is was not too wet and not too rubbery. It is probably best to err on the side of excess wetness so that diffusion of the enzyme can take place and one can easily wick off excess moisture on a piece of chromatography paper.

Small gel blocks were cut with a bamboo spatula and placed on the painting's surface, and the real excitement began. Viewing through the binocular microscope and focusing through the gel block to the surface of the painting, one could see a

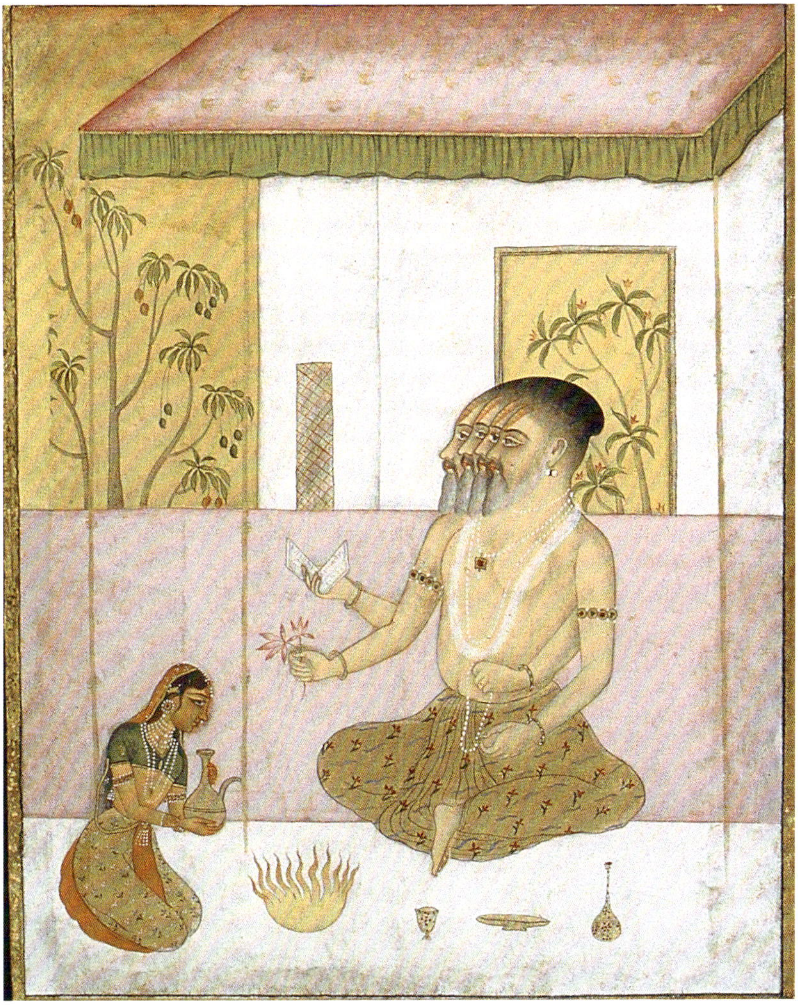

[6] After treatment with enzyme gel.

beautiful swirling pattern of amber colored adhesive moving within. The visuals were marvelous. The performance was ideal: Fast-moving ribbon-like patterns were swirling upwards, away from the object's surface into the gel block. Working in a systematic micro-layer progression, small gel blocks were selectively placed until all areas of adhesive accretions were successfully reduced.

A note of caution is in order, for although the enzyme gel eliminates the need for mechanical action, the 5 to 8 minutes of interaction resulted in a swelled, moderately soft and pliant paint surface and cellulose substrate. After gel blocks are removed, conservators will have to make empirical judgments in deciding when to place the exposed areas under polyester web, blotters, and weights in order to avoid unwanted planar distortions. [FIGS. 5 AND 6]

Final phases of the conservation treatment consisted of tying the picture together physically and visually. Losses in the primary support including delaminated layers were compensated for with paper similar in furnish and texture and toned to match the background colors so that the eye can focus on the all-important Supreme Brahma, central to the composition. [FIG. 6]

CONCLUSIONS Enzymes and their specificity have and will continue to be important and powerful tools for the conservator, allowing us to treat objects that are otherwise untreatable. With an understanding of the technical and physical nature of the objects, and as we learn and borrow from many disciplines, we continue to shape and refine treatment strategies to fit the needs of the unique and irreplaceable works of art in our care.

ACKNOWLEDGMENTS Thanks go to Richard C. Wolbers, Associate Professor, Winterthur/University of Delaware Program in Art Conservation (WUDPAC); Janice Carlson, Senior Scientist, Winterthur Analytical Laboratory; Harold M. Erickson, Conservation Department, Harry Ramsom Humanities Research Center, University of Texas at Austin; Fran Mayhew, Associate Professor, University of Delaware; John Krill, Head of Paper Conservation and Adjunct Professor, WUDPAC; Betty Fiske, Paper Conservator and Adjunct Professor, WUDPAC; Marjorie Shelley, Conservator in Charge, Sherman Fairchild Center for Works of Art on Paper and Photograph Conservation, The Metropolitan Museum of Art; and Steven Kossak, Curator of South and Southeast Asian Art, The Metropolitan Museum of Art.

REFERENCES

ANDREWS, T., 1990. An investigation into the removal of enzymes from paper following conservation treatment. *The Book and Paper Group Annual*, 9, 1–11

BANSA, H. and HOFER, H., 1984. Artificial aging as a predictor of paper's useful life. *Restaurator*, 6, 21–60.

BURGESS, H. and CHARETTE, C., 1981. Aspects of image safety in the use of enzymes in paper conservation. *International Council of Museums Committee for Conservation*, 6th Triennial Meeting, Ottawa, Preprints, 3, 10.1–10.13.

DE SANTIS, P., 1983. Some observations on the use of enzymes in paper conservation. *Journal of the American Institute for Conservation*, 23, 7–27.

ERICSON, H., 1992. Usage Recommendations for alpha-amylases: maximizing enzyme activity while minimizing artifact binding residues. *The Book and Paper Group Annual*, 11, 24–33.

FABER, K., 1997. *Biotransformations in Organic Chemistry*, Springer, London, pp. 50–67.

FELLER, R., 1994. *Accelerated Aging: Photochemical and Thermal Aspects*, The J. Paul Getty Trust, Los Angeles.

FLETCHER, S. and WALSH, J., 1979. The treatment of three prints by Whistler on fine Japanese paper. *Journal of the American Institute for Conservation*, 18 (2), 118–126.

GRATTAN, D., HILLAIRE, J., BURGESS, H., and MCCAWLEY, J.C., 1987. The characterization of enzymes for use in paper conservation. In: Petherbridge, G. (Ed.), *Conservation of Library and Archive materials and the Graphic Arts*, Butterworths, London, 15–24.

HAUSER, R., 1987. Whaling museum restores fifty logbooks. *The Abbey Newsletter*, 11 (5), 76–77.

MILLS, J., and WHITE, R., 1994. *The Organic Chemistry of Museum Objects*, Butterworths, London, p. 87.

RICKMAN, C., 1988. Wallpaper conservation, conservation today. In: Preprints of the UKIC 30th Anniversary Conference, 1988. London. United Kingdom Institute of Conservation, pp. 64–70.

SEGAL, J., and COOPER, D., 1977. The use of enzymes to release adhesives. *The Paper Conservator*, 2, 47–50.

SHELLEY, M., 1980. The conservation of the van Rensselaer wallpaper. *Journal of the American Institute for Conservation*, 20, 1–4.

TSE, S. and BURGESS, H., 1988. Effects of enzyme treatment on the degradation of paper. Symposium '88 Abstracts, Canadian Conservation Institute.

VAN DER REYDEN, D., 1988. Technology and treatment of a folding screen: comparison of oriental and western techniques. In: Mills, J., Smith, P., and Yamasaki, K.(Eds.) International Institute for Conservation of Historic and Artistic Works Preprints of the Contributions to the Kyoto Congress, The Conservation of Far Eastern Art, London, IIC, pp. 64–69.

WENDELBO, O., 1974. Extraction of papyri from gesso cartonnage: a new method based on enzymatic approach. *Proceedings from the 14th International Conference of Papyrology*, Oxford, pp. 41–52.

M.-L. E. FLORIAN

FUNGAL-PROBLEM MONITORING FOR HERITAGE COLLECTIONS: THE NEED FOR BASELINE REFERENCE LEVELS FOR FUNGAL STRUCTURES AND BETA-GLUCANS

Royal British Columbia Museum, Victoria, BC, Canada

INTRODUCTION The purpose of this paper is twofold—to show the need for suitable background or baseline standards for the interpretation of bioaerosol and surface sample changes, and to show the need for analyses of the fungal antigen beta-glucans, to determine the health hazard.

Fungal infestations can range from a few visually observable isolated colonies to an infested collection. In the visual absence of fungal structures, a moldy odor suggests fungal activity. A few isolated colonies on a heritage object are usually dealt with by isolating the object, mechanically cleaning it under a fume hood, and using a particulate mask and gloves for health protection. No sampling or monitoring is necessary, but fungal species identification may be pursued for documentation.

At the other extreme, the infested collection is decontaminated at the level of hazardous materials removal or asbestos abatement. Monitoring the bioaerosol of the air and surfaces in the area of the infestation and the recovery room used for cleaning the collection is done to determine the level of contamination. Monitoring of adjacent areas may also be done to determine if airborne fungal structures have spread.

Species identification is usually done as part of the documentation and also to determine if hyperallergenic or toxic fungal species are present. The cleanup usually does not wait for species identification, which can take several weeks using conventional culturing methods. Light and scanning electron microscopy can be used to identify genera and sometimes species origin of conidia; this takes less time than culturing. If polymerase chain reaction DNA templates are available for specific species, these can be used to identify suspect species within hours. Rarely—and then only in cases in which people have experienced allergic symptoms—are beta-glucans monitored.

In cases where there is just an odor, microbial volatile organic chemicals (MVOC) monitoring of the bioaerosol for fungal structures and chemical analysis of the volatile odors may be undertaken. The species origin of the MVOC can sometimes be determined by the presence of species-specific chemicals.

But are we doing a good job?

MONITORING

Can We Interpret the Results of Sampling? A surface sample may contain fungal structures, conidia, spores, hyphae, sclerotia, etc., that have settled from the air onto objects or are present as a result of an infestation. [FIG. 1]

An air sample contains the bioaerosol, airborne materials of biological origin. This is the component of air in which we are interested. It contains airborne fungal structures and volatile fungal metabolic chemicals, which we smell. Bacteria and their cell fragments may also be present and are a health concern.

We hope to find out the following from sampling or monitoring: what fungal structures are present, their relative amounts, whether they occur in numbers that present a hazard to collections or our health, sometimes the species, and sometimes their viability. We also take samples to determine if the numbers have decreased after cleanup of an infestation. The problem is that interpretation of all these analytical results is almost impossible. At best, all we know is what's in a specific volume of air or surface area, and in what relative amounts at that specific spot at that specific time. The main reasons we cannot interpret the results are that we do not have baseline references for comparison, and we do not have threshold level values (TLVs) indicating a health hazard or a threat to collections. Even samples taken at the time of an infestation and afterwards will show only a change in levels due to cleanup; we don't know the norm before the infestation because we sample after the fact. As yet we do not have scientifically established TLVs that indicate a health hazard. What's more, it is almost impossible to expect to ever have them because of the tremendous variability in human responses, the tremendous variability of environmental and fungal interactions, and the lack of knowledge of

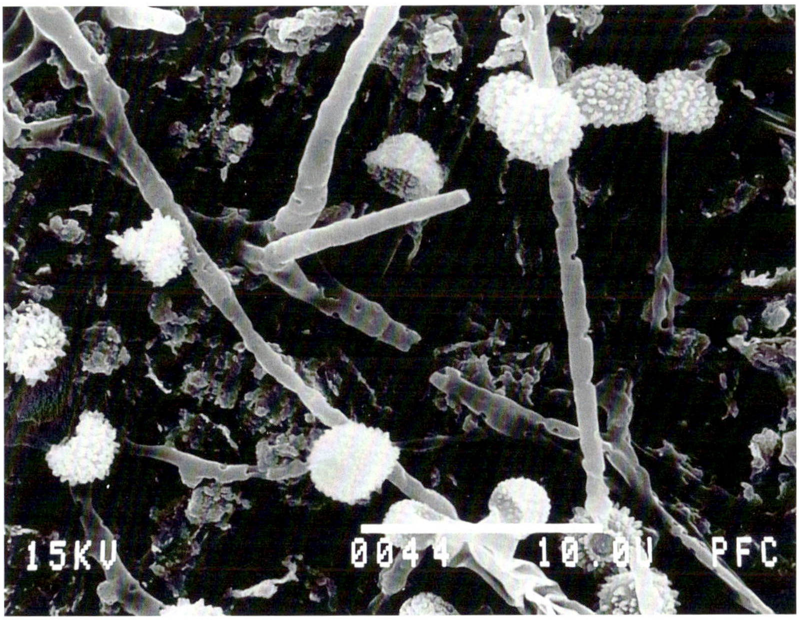

[1] Scanning electron micrograph of fungal structures—the ornamented conidia and threadlike hyphae—in a fungal fox spot (by Lesley Manning, Pacific Forest Center, BC, Canada).

the physiology of surface-growing heterotrophic fungi. Our knowledge comes from fungi grown as cultures on artificial media with a salubrious supply of water, not on austere surfaces of dead organic material.

This is not news. Today, toxicologists, industrial hygienists, and microbiologists are trying to cope with these weaknesses.

Sampling Method Biases Even the sampling methods have problems. There are many samplers and methods of sampling surfaces and the bioaerosol; the methods are reviewed by Willeke (1999) and Macher (1999). There are biases to all the methods (Macher, 1999; Health Canada, 1995; Florian, 2000, 2002). Biases are due to a variety of factors, including the aerodynamics of the conidia, air currents caused by air conditioning or people movement, room use, temperature stratification, and microenvironments. Figures 2 and 3 show the difference in sizes of the conidia of two common species. [FIGS. 2 AND 3] Comparing their sizes with their prevalence in air samples [TABLE 1] shows that the largest, heaviest, and easiest to settle are the commonest in surface samples. This certainly is an obvious bias.

Today the recommended monitoring or sampling method is collection by drawing a specific volume of air (air impaction) onto a collection slide along with surface sampling. Both should be analyzed by microscopic examination for total numbers of fungal structures and genera. These samples are combined with culturing to determine viable culturable species identification. The results of these methods show total numbers of fungal structures, some genera and some species names, and what's culturable. The culturable material is counted as colony forming units (CFU), and originates from viable conidia, spores, and hyphal fragments that can be grown on the specific media used. Species that need different culturing parameters but are viable are not shown in this count.

The Need for Background (Baseline) References Outdoor air is commonly used as a background reference, but is basically useless. Outdoor air and indoor air are normally quite different, the former being influenced by agriculture, industry, traffic, weather, and other factors (Bhandari et al., 1999; Khan et al., 1999; Ho, 1996; Halwagy and Taha, 1990) and the latter by air filters (El Mougith, 1993), and what use the building or rooms are being put to, e.g., office building space (Burge et al., 2000), bakery (Levy et al., 1999), hospital (Verma and Pandey, 1992), home. [TABLE 1] The air will be different in heritage collections, preparation, storage, and exhibit rooms.

Table 1, which compares outdoor and indoor air, shows that the two have fungi species in common. The usual comment is that the species are the same but there

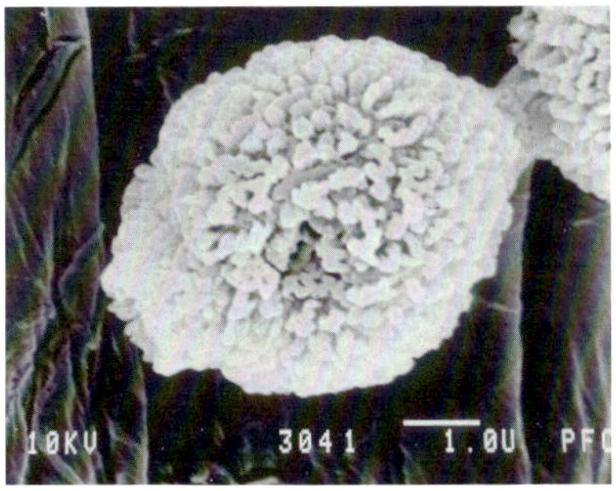

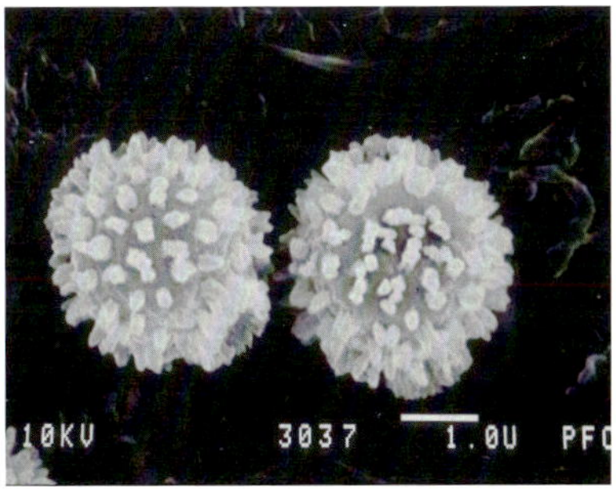

[2] Scanning electron micrograph of *Eurotium amstelodami*, an *Aspergillus* anamorph, in a fox spot on paper. Size is 5 × 3.5 μm. [3] Scanning electron micrograph of *Eurotium repens*, an *Aspergillus* anamorph in a fox spot on paper, 3 μm in diameter (by Lesley Manning, Pacific Forest Center, BC, Canada).

GENERA AND TYPES	OUTDOOR	INDOOR
Alternaria	821	541
Aspergillus/Penicillium	0	17
Ascospores	91	17
Basidiospores	30	17
Cladosporium	183	253
Smuts/Myxomycetes	304	68
hyalin fungi		34
brown fungi	213	101
hyphae fragments	426	710
algal spores	30	118

[TABLE 1] Data of outdoor air and indoor air of a home. The data show quite a different profile of air spores in each one. *Cladosporium* and *Alternaria* have large multicellular conidia, which is reflected in their large quantity, and the fact that they are the commonest airborne conidia worldwide. The high number of fungal fragments and the presence of *Aspergillus/Penicillium* indoors are significant. Data from John MacRae and Associates, 2000.

are fewer in indoor air and that there are often more *Penicillium* and *Aspergillus* species indoors because of amplifiers—e.g., food, vegetable storage, and humidifiers. It is obvious that samples of outdoor air and indoor air represent two quite different air profiles. Outdoor air has traditionally been used as a background reference because we do not have one for inside buildings. This is a strange cultural constraint; there is no reason why we should not have background references inside buildings for specific rooms. I am recommending to museums and archives that they include in their general environmental monitoring program four seasonal bioaerosol and surface samplings for fungal structures in specific indoor areas of concern, i.e., areas without adequate environmental controls or areas that contain a collection prone to fungal infestation, microenvironments, etc. The analysis of the samples should include total spore count, CFU, as well as enumeration of the beta-glucans. Monitoring for health hazard implications involves analysis of the presence of the fungal antigen beta-glucans in surface and bioaerosol samples (Burge and Ammann, 1999).

This background baseline for fungal structures and beta-glucans can be put in a computer databank that would be available for comparisons in the event of an infestation and can demonstrate a change and the degree of change from the norm. In the past, monitoring has always occurred when a problem arose; this is obviously too late.

BETA-GLUCANS: THE MAJOR FUNGAL ANTIGEN LOCATED IN ALL FUNGAL STRUCTURES, MYCOFIBRILS, AND BIOFILMS

Beta-Glucans Beta-glucans (Augustin, 1998) are considered as the causative agents of adverse effects in buildings with a history of water damage. They are considered to cause nonspecific inflammatory reactions, to increase the severity of asthma (Wan and Li, 1999), and to contribute to sick-building syndrome. They have been monitored in house dust (Chew et al., 2001) and air and surface samples in day care centers and homes (Wan and Li, 1999).

Once administered into the body of mice, they remain mainly in the liver and spleen, for a very long time. Beta-glucans are present in fungi as a structural component in the cell wall structure of all conidia (Wasylnka and Moore, 2000) and hyphae. Beta-glucan is present in several layers of the complex hyphal tip. They are also present in the noncellular, extracellular hyphal mycofibrils and in the biofilm. Beta-glucans may be in soluble or particulate forms. They are polymers that can have the form of a single or triple helix, or a random coil, or have cables or mycofibrils. The mycofibrils are less than 1 μm in diameter, so fragments can easily be airborne and inhaled. Extracellular alkaline soluble beta-glucans are phosphoglycoproteins, made up of glucose, although they may sometimes contain mannose and a protein similar to collagen. The extracellular material is probably the same as in the hyphal wall.

Beta-Glucans Analysis The antigenic characteristic of beta-glucans can be determined immunologically by the presence of antibodies IgE and IgG, which are formed by animals in response to beta-glucans (Burge and Ammann, 1999; Milton et al., 2001; Douwes et al., 1996). The chemical beta-glucans has been measured by fluorescent microscopy (Barrasa et al., 1998) and by proton nuclear magnetic resonance (Lowman and Williams, 2001). Beta-glucans were extracted from dust samples with hot detergent extractions (Douwes et al., 1996) and analyzed after the oxidation by sodium hypochlorite degradation (Miura et al., 1996). Florian and Puritan (1995) showed that the fungal fox spot discoloration, the biofilm, was removed by trisbuffer, while Florian (1996) showed that it was removed by a strong sodium hydroxide solution, but only the paper fibers in the center of the spot showed a phenolic-like discoloration. Thus the methodology for beta-glucans analysis is available. This analysis is more rapid and seems more meaningful than counting fungal structures. It is a more accurate count because it includes the mycofibrils that would not be included in the fungal structure count. Thus counting fungal structures does not give an accurate idea about the presence of beta-glucans, which are in fact a health hazard. Chew et al. (2001) reported that analysis of beta-glucans concentrations in floor dust is a surrogate for estimating airborne fungal exposure. This analysis could be undertaken seasonally as a baseline reference.

EXTRACELLULAR MYCOFIBRILS, FUNGAL BIOFILM, AND ADHESINS—ARE THEY ALL THE SAME?

Extracellular Mycofibrils Extracellular mycofibrils are on hyphae surfaces. They range in size from 10–50 nanometers to close to one micron in diameter and up to 25 microns in length. The thicker fibers are a result of cabling. They are composed of beta-glucans (Celerin and Day, 1998; Nicole et al., 1993; Larsen and Green, 1991) with a small component of protein that is similar to collagen. They have been observed by both transmission and scanning electron microscopy (Celerin and Day, 1998; Larsen and Green, 1992), as well as under light microscopy (Locci, 1972; Florian and Manning, 2000). Figures 4a–d show the mycofibrils on hyphae of *Aspergillus amstelodami* on a 145-year-old book paper. [FIGS. 4A–D] It has been suggested that the presence of mycofibrils is a response to an aerial growth

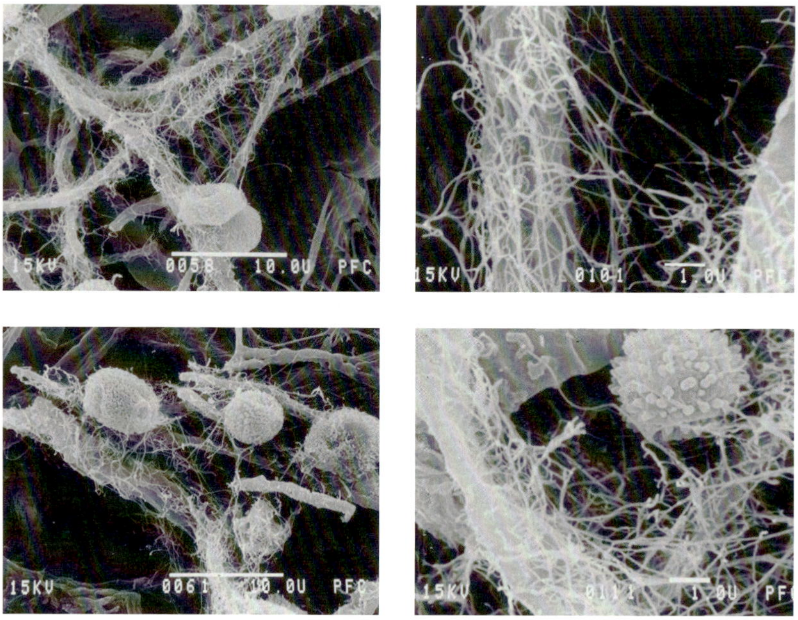

[4] Four different SEM micrographs showing the various amounts of mycofibrils on *Eurotium amstelodami* hyphae and associated conidia (by Lesley Manning, Pacific Forest Center, Victoria, BC, Canada).

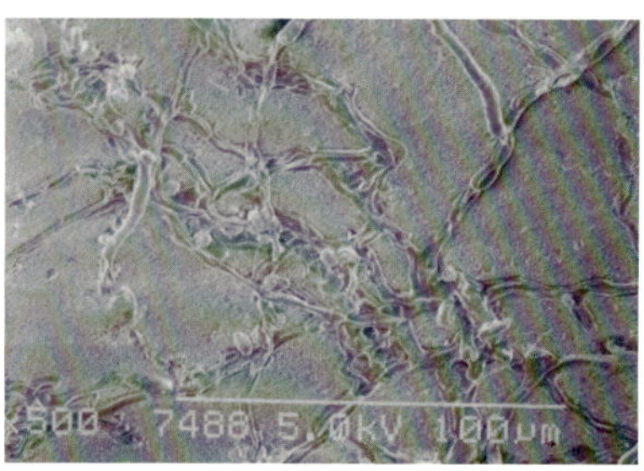

[5] Scanning electron micrograph of fungi growing on the surface of carded paper. The hyphae appear to be in troughs that suggest the presence of a dried beta-glucans slime (by Mary Mager, SEM Laboratory, Metallurgical Engineering, University of British Columbia, Victoria, BC, Canada).

or a genetic taxonomic feature. This may explain why some species are more allergenic than others. My own interpretation is that mycofibrils are a normal development on all hyphae but are not easily observed, or their amounts may vary due to physiological conditions, e.g., during water stress. The physiological state of the fungus—for instance, whether it is in the early growth stage or senescence, may result in different amounts of beta-glucans, or more may be produced during water stress—we do not know. Thus there is a need for more mycological research on heterotrophic fungi that form surface growth on dead organic dry materials.

Extracellular Biofilms A beta-glucans biofilm is laid down by an organism to make a microenvironment for its adhesion and survival on a surface. We are familiar with slimy biofilms produced by bacteria in water on various material surfaces, and most textbooks on biofilms (Lappin et al., 2001; van Losdrecht and Tijhuis, 1996; Doyal, 1999) discuss bacterial rather than fungal biofilms. But biofilms are also formed by fungi and are present on a substrate wherever there is or has been fungal growth. Chemically they are basically the same as those of bacteria (Robine, 1998), but they are formed on air-dried surfaces.

One can imagine the process and need for a biofilm formation. As soon as a conidium lands on a surface there is feedback from the surface, or substrate, to the conidium. If environmental parameters are conducive and conidium activation has occurred, a negatively charged polysaccharide (Wasylnka and Moore, 2000) or osmophilic layer (Tiedt, 1993) on the conidium wall will start to adsorb moisture and lay a biofilm to adhere itself to the surface. As the hyphae are formed, their outer surface contains water-soluble beta-glucans slime, [FIG. 5] which makes a microenvironment for the hyphae in which water is drawn and stored and enzymes are distributed.

Fungal beta-glucans biofilms have been shown to be present in both nonpathogenic yeast on glass (Masak et al., 2000) and the pathogenic yeast *Candida* (MacKenzie et al., 1997; Buurman et al., 1998), as well as in pathogenic *Aspergillus* species, and the skin fungus *Trichoderma pseudokoningii* (Matsuzawa et al., 1996). Its role with pathogens involves attaching itself to cells it invades.

Biofilms that include a mixture of microorganisms have been studied on inorganic heritage objects (Mitchell and Gu, 2000; De la Torre et al., 1993) and their relationship with small fauna (Gorbushina and Petersen, 2000) and Florian and Manning (2000) consider the fungal fox spot as a biofilm of surface fungi.

The red brown discoloration in old fungal fox spots is a biofilm (Florian, 1996; Florian and Manning, 2000). The biofilm stains with aniline blue, indicating the presence of beta-glucans. The aniline blue stains an area around and beyond the fungal structures.

The discoloration in the fungal fox spots (Florian and Manning, 2000) may penetrate several pages without containing any fungal structures at all (Florian, 1996). Amino acids have also been shown to be present in these spots (Florian and Manning, 2000). The large mycofibrils observed [FIG. 4] may be an artifact of the drying of the slime. It is a common characteristic of colloidal solutions that fibrous structures are formed on drying. It is not known if the antigenic characteristic of beta-glucans in a fungal fox spot is lost on aging. The chemical nature in the discoloration of the fungal fox spots is still unknown but must relate to the presence of biofilm aging products, i.e., browning of sugars, melanoidins, lipid sugar products, etc.

Understanding the details of mycofibrils and biofilms is important because these are major components in all fungal spots. They are a major source of the common allergenic beta-glucans; however, they are not monitored or considered in any monitoring methods. Also, after we remove the fungal structures by physical means, we do not consider the potential problem of the biofilm that remains. Will it be released into the air on aging? Will it cause discoloration in the future? It has been shown to be extracted by hot detergent, trisbuffer, and alkaline solutions. Could these be used by conservators according to conservation standards? Research is needed on this subject, not only to prevent discoloration on heritage objects, but also to prevent health hazards.

Extracellular Adhesin Jones (1994) discusses fungal adhesin and explains its role in colonizing of substrates and acquisition of domain. He makes a comparison between fungal adhesin and the biofilm of bacteria. The extracellular slimy sheath of the hyphae (Nicole et al., 1993) and sheaths or osmophilic layer of spores (conidia) (Tiedt, 1993; Wasylnka and Moore, 2000) are common features of fungi, and the slime may appear fibrous or nonfibrous. Their chemistry is not clear, but it is suggested that they are the same as exopolysaccharides secreted by fungi in liquid media and that the adhesive material may contain a polysaccharide plus a protein moiety, thus a beta-glucan. It is likely that Jones's (1994) fungal adhesin on materials may be the sheaths of hyphae and conidia, the mycofibrils described by Celerin and Day (1998) and Larsen and Green (1992) and the biofilms. All are beta-glucans and may be fibrous or nonfibrous. Their fibrous nature may not be seen because of lack of fine enough resolution in observation or because of a change in the physical state of a colloidal solution, i.e., its viscosity is altered by pH and water.

Mycotoxins and Microbial Volatile Organic Chemicals as Health Hazards Beta-glucan is an allergen or antigen, but there are also mycotoxins in the fungal structures of specific species, which are a health hazard if eaten or inhaled. Mycotoxins (Burge and Ammann, 1999) are fungal secondary metabolic products in the fungal structures. This is the concern with *Stachybotrys,* which grows on wet cellulosic building materials.

The odors and other microbial volatile organic chemicals, or MVOCs (Ammann, 1999) are secondary metabolic products; their monitoring and the study of their toxicity are still in the research stage. Fungal MVOCs emit the familiar moldy, mildewy, or musty odors. The odor of the secondary fungal metabolite 1-octen-3-ol in *Aspergillus niger,* and *Penicillium roqueforti,* varies depending on its concentration. In high concentrations the smell is resinous; in low concentrations like mushrooms, old paper, or mildew-like mustiness. *A. flavus* produces both 2-octen-1-ol and 1-octen-3-ol. *Chaetomium* produces both 2-methyl-isoborneol and geosmin, which cause its musty, earthy odor. *P. expansum* also produces geosmin. *P. casiecolum, P. roqueforti,* and *Botrytis cinerea* produce 2-methyl-isoborneol. All the above fungal species are common airborne and surface contaminates. The moldy odors in buildings have been shown to be caused by a combination of these secondary products, alcohols that are also terpenes. Terpenes are not unique to fungi; they are common products in the higher plants (examples are pyrethrin, clove oil, and lemon oil), and all are toxic at a specific dose. Florian (1998) has reviewed this problem in reference to the suggested use of natural products in insect and fungal prevention.

MVOC samples from sorbent sampling and concentration canisters are analyzed by gas and high performance liquid chromatography (HPLC) and mass spectroscopy. It is suggested that they are not at toxic levels but act only as an irritant or nuisance, but some infested closed chambers or small rooms certainly could have a hazardous buildup. In my view, all terpenes are suspect toxic chemicals. Again, it's a matter of dose; also, there is the possibility of accumulation from repeated exposures.

In assessing bioaerosol samples for the dose level, whether it is the numbers of fungal structures or the concentration of beta-glucans or MVOCs, the main question is, how much is too much? As of yet we do not know what the threshold level values are, that is, the levels at which a health hazard is present. The research has not been done mainly because of the many variables: the variable genetic makeup of individual people, the variability of the fungal species (for example, some 30% of *Stachybotrys* species do not produce the mycotoxin), and the variability of the environment in which the infestation grew. Thus the determination of health hazards is difficult.

CONCLUSIONS In conclusion, we must take a preventive role and establish background references to give us a baseline so that in the event of an infestation or odor we can logically interpret the monitoring results.

Monitoring must include total fungal structures as well as beta-glucans. If we are concerned about the health hazard we must undertake beta-glucans analysis and be aware of the unknown potential of MVOCs.

Because we still do not have all the answers we must proceed with caution. Thus when dealing with any fungal problem, no matter what the size of the infestation, all precautions must be taken regarding personal protection, and stringent aseptic techniques (Florian, 2000) should be used to prevent an increase of airborne fungal structures and cross-contamination of objects. Also, as a preventive measure, sterile materials must be used for conservation treatment (Florian, 2000, 2002).

We should be concerned about the presence of a biofilm left after the fungal structures in an infestation have been removed. It seems logical to remove it if possible.

REFERENCES

AMMANN, H.A., 1999. Microbial volatile organic compounds. Macher, J., (Ed), In: *Bioaerosol: Assessment and Control*, Cincinnati, OH, pp. 26–1 to 26–17.

AUGUSTIN, J., 1998. Glucans as modulating polysaccharides: Their characteristics and isolation from microbiological sources. *Biologia*, Bratislava, 53, 277–282.

BARRASA, J. M., GUTIERREZ, A., ESCASO, V., GUILLEN, F., MARTINEZ, M.J., and MARTINEZ, A., 1998. Electron and fluorescence microscopy of extracellular glucan and aryl-alcohol oxidase during wheat-straw degradation by *Pleurotus eryngii*. *Applied and Environmental Microbiology*, 64, 325–332.

BHANDARI, S., MISHRA, R.P., and OOMMACHAN, M., 1999. Seasonal variations in airspora at Jabalpur, India. *Geobios*, Jodhpur, 25, 131–136.

BURGE, H., and AMMANN, H.A., 1999. Fungal toxins and beta-(1–3)-D-glucans. In: Macher, J., (Ed), *Bioaerosols: Assessment and Control* ACGIH, Cincinnati, OH, pp. 24–1 to 24–13.

BURGE, H. A., PIERSON, D.L., GROVES, T.O., STRAWN, K.F., and MISHRA, S.K., 2000. Dynamics of airborne fungal populations in a large office building. *Current Microbiology*, 40, 10–16.

BUURMAN, E.T., WESTWATER, C., HUBE, B., BROWN, A.J.P., ODDS, F.C., and GOW, N.A.R., 1998. Molecular analysis of CaMnt1p, a mannosyl transferase important for adhesion and virulence of *Candida albicans*. *Proceedings of the National Academy of Sciences of the United States of America*, 95, 7670–7675.

CELERIN, M., and DAY, A.W., 1998. Sex, smut, and RNA: The complexity of fungal fimbriae. *International Journal of Plant Science*, 159, 75–184.

CHEW, G. L., DOUWES, J., DOEKES, G., HIGGINS, K.M., STRIEN, R.V., SPITHOVEN, J., and BRUNEKREEF, B., 2001. Fungal extracellular polysaccharides, beta (1–3) glucans and culturable fungi in repeated sampling of house dust. *Indoor Air*, 11, 171–178.

DE LA TORRE, M.A., GOMEZ-ALARCON, G., and PALACIOS, J.M., 1993. "In vitro" biofilm formation by *Penicillium frequentans* strains on sandstone, granite, and limestone. *Applied Microbiology and Biotechnology*, 40, 408–415.

DOUWES, J, DOEKES, G., MONTIJN, R., HEEDERIK, D., AND BRUNEKREEF, B., 1996. Measurement of beta (1–3) glucans in occupational and home environments with an inhibition enzyme immunoassay. *Applied and Environmental Microbiology*, 62, 3176–82.

DOYLE, R.J., 1999. *Biofilms*. Academic Press, San Diego, CA.

EL-MOUGITH, A.A., 1993. Mycoflora of dust particles accumulated on filters of air conditioning systems in United Arab Emirates. *Indian Journal of Mycology and Plant-Pathology*, 23, 260–266.

FLORIAN, M.-L.E., 1996. The role of the conidia of fungi in fox spots. *Studies in Conservation*, 41, 65–75.

FLORIAN, M.-L.E., 1998. Natural products—for insect and fungal control. *ICOM Ethnographic Newsletter* Oct. 18, 4–8.

FLORIAN, M.-L.E., 2000. Aseptic technique: a goal to strive for in collection recovery of mouldy archival materials and artifacts. *Journal of American Institute of Conservation*, 39, 107–115.

FLORIAN, M.-L.E., 2002. *Fungal Facts: Solving fungal problems on heritage collection.* Archetype Publishers, UK.

FLORIAN, M.-L.E., and MANNING, L., 2000. SEM analysis of irregular fungal fox spots in an 1854 book: Population dynamics and species identification. *International Biodeterioration and Biodegradation*, 46, 205–220.

FLORIAN, M.-L.E., and PURINTON, N., 1995. Determination and location of stains in fungal spots and enzymatic removal of pigmented hyphae in archival paper and textiles. With Editor-ICBCP-3, Bangkok, July 4–7,1995.

GORBUSHINA, A.A., and PETERSEN, K., 2000. Distribution of microorganisms on ancient wall paintings as related to associated faunal elements. *International Biodeterioration and Biodegradation*, 46, 277–284.

HALWAGY, M.H., and TAHA, M.A.H., 1994. Relation between airspora and weather conditions in Kuwait. *Egyptian Journal of Botany*, 33,235–255.

HEALTH CANADA, 1995. Appendix D: Sampling methods for fungal bioaerosols and amplifiers in cases of suspected indoor mould proliferation. In: *Fungal Contamination in Public Buildings: A Guide to Recognition and Management.* Environmental Health Directorate, Health Canada, Tunney's Pasture, Ottawa, Ont.K1A 0L2.

JONES, G.E.B., 1994. Fungal adhesion. *Mycological Research*, 98, 961–981.

KHAN, Z.U., KHAN, M.A.Y., CHANDY, R., and SHARMA, P.N., 1999. *Aspergillus* and other moulds in the air of Kuwait. *Mycopathologia*, 146, 25–32.

LAPPIN, S., HILARY, M., and BASS, C., 2001. Biofilm formation: Attachment, growth, and detachment of microbes from surfaces. *American Journal of Infection Control*, 29, 250.

LARSEN, M., and GREEN, F., 1992. Mycofibrillar cell wall extensions in hyphal sheath of *Postia placenta*. *Canadian Journal of Microbiology*, 38, 905–911.

LEVY, J.I, NISHIOKA, Y., GILBERT, K., CHENG, C., and BURGE, H., 1999. Variabilities in aerosolizing activities and airborne fungal concentrations in a bakery. *American Industrial Hygiene Association Journal*, 60, 317–325.

LOCCI, R., 1972. Scanning electron microscopy of ascosporic Aspergilli. *Rivista di Patologia Vegetale* Padova, 8, 1–172 Supplemento V.

LOWMAN, D.W., and WILLIAMS, D.L., 2001. A proton nuclear magnetic resonance method for the quantitative analysis on a dry weight basis of (1–3)-beta-D-glucans in a complex, solvent-wet matrix. *Journal of Agricultural and Food Chemistry*, 49, 4188–4191.

MACHER, J., 1999. Data analysis. In: Macher, J., (Ed), *Bioaerosols: Assessment and Control* ACGIH, Cincinnati, OH, pp. 13–1 to 13–16.

MACKENZIE, L.F, BROOKE, G.S., CUTRIELD, J.F., SULLIVAN, P.A., and WITHERS, S.G., 1997. Identification of Glu-330 as the catalytic nucleophile of *Candida albicans* Exo-b(1,3)-glucanase. *Journal of Biological Chemistry*, 272, 3161–3167.

MASAK, J., CEJKOVA, A., JIRKU, V., SIGLOVA, M., RYBARIKOVA, L., HRON, P., KOTRBA, D., and STRNAD, R., 2000. Controlled attachment of a yeast biodegrader to glass surfaces. Mededelingen Faculteit Landbouwkundige en Toegepaste Biologische Wetenschappen, Universiteit Gent, 65, 149–152.

MATSUZAWA, T., AMANO, Y., MITSURU, K., and KANDA, T., 1996. Mode of action of exo-beta-1,3-glucanase from *Trichoderma pseudokoningii* TM37 on various beta-D-glucans. *Seibutsu-Kogaku-Kaishi*, 74, 261–267.

MILTON, D. K., ALWIS, K.U., FISETTE, L., and MUILENBERG, M., 2001. Enzyme-linked immunosorbent assay specific for (1–6) branched, (1–3)-beta-D-glucan detection in environmental samples. *Applied and Environmental Microbiology*, 67, 5420–5424.

MITCHELL, R., and GU, J.D., 2000. Changes in the biofilm microflora of limestone caused by atmospheric pollutants. *International Biodeterioration and Biodegradation*, 46, 299–303.

MIURA, N.N., OHNO, N., ADACHI, Y., and YADOMAE, T., 1996. Characterization of sodium hypochlorite degradation of beta-glucan in relation to its metabolism in vivo. *Chemical and Pharmaceutical Bulletin* Tokyo, 44, 2137–2141.

NICOLE, M., CHAMBERLAND, H., RIOUX, D., LECOURS, N., RIO, B., GEIGER, J.P., and OUELLETTE, G.B., 1993. A cytological study of extracellular sheaths associated with *Rigidoporus lignosus* during wood decay. *Applied and Environmental Microbiology*, 59, 2578–2588.

ROBINE, E., 1998. Adhesion and survival of aerosolised bacteria on inert surfaces. *Comptes Rendus de l'Academie d'Agriculture de France*, 84, 101–110.

TIEDT, L.R., 1993. An electron microscope study of conidiogenesis and cell wall formation of conidia of *Aspergillus niger*. *Mycological Research*, 97, 1459–1462.

VAN LOOSDRECHT, M.C.M., and TIJHUIS, L., 1996. Biofilm structure, growth and dynamics: *Selected proceedings of the IAWQ International Conference and Workshop on Biofilm Structure, Growth and Dynamics. Noordwijkerhout*, The Netherlands, 30 August–1 September 1995. Elsevier Science, Tarrytown, New York.

VERMA, K.S., and PANDEY, M., 1992. Airborne fungal spores in a hospital ward at Jabalpur. *Proceedings of the National Academy of Sciences India*, Section B Biological Sciences, 62, 425–427.

WAN, G.H., and LI, C.S., 1999. Indoor endotoxin and glucan in association with airway inflammation and systemic symptoms. *Archives of Environmental Health*, 54, 172–179.

WASYLNKA, J.A., and MOORE, M., 2000. Adhesion of *Aspergillus* species to extracellular matrix proteins: Evidence for involvement of negatively charged carbohydrates on the conidial surface. *Infection and Immunity*, 68, 3377–3384.

WILLEKE, K., 1999. Air sampling. In: Macher, J., (Ed), *Bioaerosols: Assessment and Control* AGGIH, Cincinnati, OH, pp. 11–1 to 11–25.

TEXTILES

O. CIFERRI[1*]
M.E. BECKER[2]
E. DE ROSSI[1]

CHARACTERIZATION OF BACTERIA ISOLATED FROM NATURALLY AGED SILK FIBROIN

1 University of Pavia, Italy 2 Fukui University, Fukui, Japan
* Corresponding author

ABSTRACT Approximately 20 different bacterial strains were isolated from aged (3 to ca. 70 years old) silk samples from Italy and kimono linings from Japan. Six different isolates were found capable of colonizing fibroin fabric in the laboratory, and four of them were found to utilize the protein as a sole source of carbon and nitrogen for growth. Among these isolates a *Bacillus subtilis* and a *Micrococcus luteus* were selected for further studies, as they appeared to be endowed with a pronounced hydrolytic activity of purified silk fibroin. Both species produce more than one protease active on casein and fibroin and the proteolytic activity of *M. luteus* isolate is totally dependent on the presence of Ca^{++} ions. Work is underway to separate and characterize the different proteases produced by the two bacteria and to ascertain if fibroin is degraded by a specific hydrolase or rather by enzymes whose main substrates are other proteins.

INTRODUCTION Fibroin (degummed silk) is the most fragile of all materials used in artifacts of artistic or historic interest (tapestries, paintings, prints, garments, rugs, flags, etc.). To our knowledge, little has been reported on the microbial degradation of silk. Heyes and Holden (1926) investigated the ability of bacteria and fungi to grow on silk and to cause tendering and discoloration of the fiber as well as alterations in its dyeing properties. The tested fungi, mostly in the genus *Aspergillus,* seemed to grow on silk but without causing any tendering or discoloration. On the other hand, the bacterial species tested, all in the genus *Bacillus,* did not grow on silk. In another study, degummed silk held in soil appeared to be degraded, and bacteria of the genera *Bacillus, Pseudomonas, Serratia,* and *Streptomyces* were detected on the fabric and were probably responsible for the silk's degradation (Ishiguro and Miyashito, 1996). Recently, it has been reported that some of the bacteria isolated from fibroin buried in soil or from enrichment cultures of silk cocoons were capable of metabolizing fibroin in laboratory cultures and utilizing the protein as a sole source of carbon and nitrogen for growth (Seves et al., 1998). One such microorganism, *Variovorax paradoxus,* was shown to produce an extracellular acidic protease that hydrolyzed fibroin as well as other proteins (Forlani et al., 2000).

In this investigation we have isolated and characterized approximately 20 bacterial strains that were present on samples of commercial fibroin of different ages and provenance. Some of these strains were found to be capable of colonizing fibroin in the laboratory and of growing on chemically defined laboratory media in which the protein represented the only source of carbon and nitrogen. The capacity to grow at the expense of fibroin strongly suggests that these isolates colonize, and hence degrade, silk fabrics.

MATERIALS AND METHODS

Silk Samples An Italian twill silk fabric (82.9 g/m^2, pH 7.0) ca. 3 years old was used as a source of fibroin after degumming with a Marseille soap solution (1%, w/v) (sample I). Two Japanese kimono lining fabrics, approximately 20 and 70 years old, were utilized as the naturally aged fibroin samples (II and III). Both fabrics were plain weaves; however, the older sample was a pongee or hand-spun silk. Pulverized fibroin was purchased from Wako Pure Chemicals Industries.

Determination of Breaking Load and Elongation Samples were cut from both the warp and weft directions of the fibroin samples in accordance with ASTM D5035–95 test method, Breaking Force and Elongation of Textile Fabrics (ASTM, 1998). The samples were first cut into strips 35 mm wide and at least 150 mm long. They were then unraveled to a width of 25 mm. Ideally, one would test at least five specimens in the warp direction and eight in the weft. However, since these samples came from materials that had been in use and of limited size, this was not always possible.

Prior to testing, all samples were equilibrated to standard test conditions, 21±1°C and 65±2% RH. An automated Tensilon tensile tester at constant strain was used to evaluate the maximum load, breaking load, elongation at maximum load, and elongation at break.

Isolation of Microrganisms Present on Fibroin Samples (1 × 1 cm^2) of degummed silk were cut from the three fabrics, wetted with sterile physiological solution (SPS), and incubated under sterile conditions in a Petri plate containing a layer of filter paper soaked with sterile water as already reported (Seves et al., 1998). This procedure was followed in order to maximize the possibility of isolating all bacteria present on the fabric. This allowed, for instance, the germination of spores and the proliferation of cultures. At increasing intervals of time (up to 15–20 days), a silk sample was removed from each of the plates and re-suspended in 2 ml of SPS. After vigorous shaking (1 to 2 min in a Vortex apparatus), aliquots of the suspensions or their dilutions were plated onto solid Luria and Bertani (LB) medium (Sambrook et al., 1989) and then incubated at 27°c for up to 7 days. From the isolated colonies, pure bacterial cultures were prepared by streaking on fresh LB plates.

Identification of Bacterial Isolates The isolates were first examined for cell morphology, colony characteristics and appearance, presence of spores, and reaction to Gram stain. Gram staining and the presence of catalase and oxidase activities were tested as described by Smibert and Krieg (1994). Endospores were stained using the Schaeffer-Fulton staining method (Smibert and Krieg, 1994). The 20 apparently different bacterial isolates were classified by using standard API tests (BioMèrieux) for micrococci and staphylococci (API STAPH), and Gram-positive and Gram-negative bacilli (API 50CH and API 20NE, respectively). API strips were inoculated as recommended by the manufacturer and incubated at 37°c; results were read after 24–48 h. For the 16S rDNA sequences, extraction of DNA was carried out using a Qiagen Genomic DNA kit (Qiagen, Hilden, Germany). Amplification was performed in a MJ thermocycler using the *Escherichia coli* specific primers 27f (5'-AGAGTTTGATCATGGCTCAG-3') and 1489r (5'-TACCTTGTTAC GACTTCAC-3') (Brosius et al., 1978) and the micrococci-specific primers mc1 (5'-AGAGTTTGATCCTGGCTCAG-3') and mc2 (5'-GGCTACCTTGTTACGACTT-3'). All reactions were carried out in a total volume of 40 μl containing 20–40 ng of template DNA, 20 pmol of each primer, 200 μM of each dNTP, 4 μl of 10× PCR buffer (100 mM Tris-HCl, 15 mM MgCl2, 500 mM KCl, pH 8.3), and 0.5 U of DNA polymerase (Herculase® Enhanced DNA Polymerase, Stratagene, La Jolla, CA). The following thermocycling program was used: 3 min denaturation at 94°c, followed by 5 cycles consisting of 2 min at 94°c, 1 min primer annealing at 53°c, and

2 min extension at 72°C, and by 25 cycles consisting of 30 sec at 94°C, 30 sec at 53°C, and 1 min at 72°C, with a final extension step at 72°C for 10 min. All PCR products were analyzed by electrophoresis in 1.2% agarose gels before sequencing reactions, which were carried out by cycle sequencing using a Perkin-Elmer ABI 310 sequencer. All sequences were analyzed for homology using the NCBI BLAST program and the Ribosomal Database Project (Maidak et al., 1999).

Colonization of Fibroin by Pure Cultures of Bacterial Isolates To prevent growth of bacteria that could be inadvertently present on fibroin fabric, samples of the Italian degummed silk were sterilized by tyndallization as follows: The samples were suspended in distilled water and exposed to free-flowing steam at 100°C for 15 min on four consecutive days with intermittent incubation at room temperature.

Experimental inoculations by selected bacterial species isolated from different silk samples were performed as already reported (Seves et al., 1996, 1998). Briefly, bacterial cells were prepared and recovered from pure cultures in LB liquid medium by centrifugation. After washing with SPS, the cell sediment was resuspended in SPS and aliquots containing approximately 1×10^3 bacterial cells were pipetted onto 1×1 cm^2 squares of control fibroin (sample I). The squares were incubated under sterile conditions at 27°C and, at different time intervals, a square was removed from the incubation chamber, suspended in SPS, and shaken, and aliquots of the suspensions were then plated onto LB plates. After additional incubation at 27°C for four days, bacterial colonies were counted to assess the extent of fibroin colonization.

Utilization of Fibroin for Growth Cell suspensions, prepared as described in the previous section, were utilized to inoculate liquid Davis' minimal medium (Davis and Mingioli, 1950). When present, pulverized fibroin, having a final concentration of 2 mg/ml, was added to the medium devoid of all other carbon and nitrogen sources. The cultures were incubated at 27°C while being shaken (200 rpm) and, at different time intervals, aliquots of the suspensions or their dilutions were plated onto LB plates. The plates were incubated at 27°C and, after 2–4 days of incubation, bacterial growth was evaluated by counting the colonies appearing on the plates.

Other Procedures Proteolytic activities of the isolated microbial species were determined on LB plates containing either 2% (w/v) skim milk or 1% (w/v) fibroin. Hydrolysis halos were determined after 24 and 48 h of incubation at 37°C.

[TABLE 1] Summary of physical properties of silk fabrics

SAMPLE	MAX. LOAD (kgf)	EXTENSION AT MAX. LOAD (mm)	LOAD AT BREAK (kgf)	EXTENSION AT BREAK (mm)
Italian twill, 2–3 years (sample I)				
Warp Direction n=7	237.5	15.5	235.5	15.9
σ_{n-1}	± 2.8	± 0.7	± 3.0	± 0.8
Weft Direction n=6	244.4	16.6	207.6	17.9
σ_{n-1}	± 4.1	± 0.4	± 28.2	± 0.5
Japan, habutae ca. 20 years (sample II)				
Warp Direction n=9	123.9	11.5	122.6	11.5
σ_{n-1}	± 9.1	± 1.1	± 10.0	± 1.1
Weft Direction n=7	156.2	8.8	153.7	9.0
σ_{n-1}	± 11.9	± 1.2	± 15.9	± 1.0
Japan, pongee ca. 70 years (sample III)				
Warp Direction n=10	103.5	11.4	78.9	12.0
σ_{n-1}	± 8.1	± 0.9	± 20.0	± 1.0
Weft Direction n=3	170.1	9.3	151.2	9.7
σ_{n-1}	± 16.4	± 0.4	± 38.1	± 0.1

ASTM D5035-95 Test Method (1998)

RESULTS

Mechanical Properties of Naturally Aged Silk Fabrics The results of the tensile tests performed on the three samples are reported in TABLE 1. The 70-year-old silk pongee (sample III) was the weakest of all three samples. However, it is difficult to compare sample III with the other two, since it is woven from hand-spun yarns that are uneven in size. Because of this unevenness, the load is redistributed during the tensile tests to thicker yarns that can carry the load. This may account for the apparent strength of the weft yarns in sample III as well as their extension beyond the maximum load. For samples I and II, the load at break was very close to the maximum load. This suggests that the more uniform yarns distributed the stress more evenly throughout the testing and that there were no thicker yarns to carry the load further after reaching the maximum load. The modern Italian twill (sample I) is a thicker fabric than the habutae (sample II), which accounts for its higher overall strength.

All the samples are in fairly good condition; none of them tear easily or break with handling. Sample III appears to be the most fragile of the three, but whether this apparent weakness is due to age, deterioration, fabric construction, or some combination of these factors cannot be determined from these experiments. Further study, including comparison of fabrics having the same fiber content as well as fabric construction, would be required.

Bacterial Population from Silk Fibroin Samples As expected, the number of colonies isolated on solid laboratory media from the three fibroin samples showed a positive correlation with the age of the fabric, as did the number of morphotypes identified by Gram stain, cell morphology, and shape and color of the colonies. [TABLE 2] Indeed, only one morphotype was isolated from the control fibroin (sample I), four from the approximately 20-year-old fabric (sample II), and 15 from the oldest one (sample III). Interestingly, no fungi were isolated from any of the samples tested. Even though the medium utilized was not specific for molds, the LB medium is a rich, complex one that allows growth of fungi. This finding confirms previous reports indicating that soil fungi do not colonize silk (Seves et al., 1998) and that pure cultures of selected fungal strains do not attach to silk fabric (Seves et al., 2000). The bacteria isolated from the three samples were essentially Gram-positive bacilli. Only two out 20 were Gram-negatives and two were Gram-positive cocci.

Laboratory Colonization of Fibroin by Bacterial Cultures Isolation of a microorganism from a given substrate does not prove that the microorganism is metabolizing the substrate, as its presence may be incidental. To ascertain which of the strains

[TABLE 2] Morphotypes isolated from naturally aged silk samples

ORIGIN AND AGE	ISOLATE	COLONY CHARACTERISTICS	GRAM STAIN	CELL SHAPE
Italy, 2–3 years	A	Bright yellow; smooth and shiny; punctiform; convex; margin entire	+	Coccus
(sample I)	B	White; smooth and shiny; punctiform; flat; margin entire	+	Coccus
Japan, ca. 20 years	O	Whitish; smooth and shiny; punctiform; convex; margin entire	+	Bacillus
(sample II)	P	Yellowish; smooth and shiny; circular; slightly umbonate; margin entire	-	Bacillus
	U	Yellow; smooth and shiny; punctiform; slightly convex; margin entire	-	Bacillus
Japan, ca. 70 years	C	White; smooth and shiny; circular; flat; margin entire	+	Bacillus
(sample III)	D	White-yellow; shiny; circular; slightly umbonate; margin undulate	+	Bacillus
	E	White; shiny and mucoid; circular; slightly convex; margin filamentous	+	Bacillus
	F	White; transparent, smooth and shiny; punctiform; flat; margin erose	+	Bacillus
	G	Whitish; opaque; circular; flat; margin erose	+	Bacillus
	H	Whitish; opaque and rough; circular; flat; margin erose	+	Bacillus
	I	Whitish; shiny and rough; circular; flat; margin erose	+/–	Bacillus
	L	Whitish; smooth and shiny; punctiform; slightly convex; margin entire	+	Bacillus
	M	Whitish; shiny and rough; circular; flat; margin undulate	+	Bacillus
	N	White; shiny and mucoid; circular; concave; margin entire	+	Bacillus
	Q	White; smooth and shiny; irregular; concave; margin branched	+	Bacillus
	R	White; smooth and shiny; circular; convex; margin entire	+	Bacillus
	S	Salmon pink; shiny and rough; irregular; flat; margin branched	+	Bacillus
	T	Whitish; smooth and shiny; circular; flat; erose margin	+	Bacillus
	V	White; smooth and shiny; circular; flat; margin erose	+	Bacillus

isolated from fibroin were really growing on the protein, pure cultures of each isolate were prepared and used to inoculate sterile pieces of fibroin fabric. Of the 20 tested bacterial isolates, at least six were found capable of growing on sterile, contemporary fibroin. [FIG. 1] It is possible that failure of the other strains to infect degummed silk when tested in pure cultures may be due to requirements for some degradation products of fibroin brought about by one (or more) of the six other isolates. Thus, it is conceivable that a degradative community of microorganisms may colonize, and thus degrade, fibroin by hydrolyzing the protein (as is probably the case for the six isolates) or by utilizing the products derived from its hydrolysis. Indeed, after incubation for 14 days, the bacterial population that could be detached from the fabric increased by up to three orders of magnitude as compared to the number of cells used as the inoculum. It must be remembered that the assay only determines the culturable cells that are removed from the fabric when this is suspended in physiological solution and shaken. Since a bacterial biofilm was found on other silk fabrics (Seves et al., 1998), the cells that adhere tightly to the fabric will not be isolated and, as a consequence, the number of colony forming units reported in figure 1 underestimates the total microbial population colonizing the fabric. Of the six isolates that appeared to colonize fibroin in the laboratory, five were Gram-positive bacilli isolated from the oldest fabric (sample III). One was a Gram-positive coccus isolated from the control sample (sample I) but also found on the ca. 20-year-old fabric (sample II).

Classification of the Fibroin-Colonizing Bacteria Classification of the six strains that appeared capable of colonizing fibroin in the laboratory was first done utilizing standard tests for evaluating fermentation or utilization of selected carbohydrates (API tests). Morphologically, isolates E, M, N, I, and T appear to be Gram-positive endospore-forming bacilli, while isolate A is a Gram-positive coccus with cells arranged in tetrads. According to API tests, strains E, M, N, I, and T were allocated to the genus *Bacillus* and strain A to the genus *Micrococcus*. Whereas strains N, E, M, and A were identified as *B. subtilis, B. licheniformis, B. megaterium,* and *M. luteus,* respectively, strains T and I could not be identified at the species level. The classification of strains was confirmed by sequencing a portion of the DNA coding for the 16S ribosomal RNA. The sequences were compared with known sequences by using the Ribosomal Database Project (Maidak et al., 1999) and the classification is reported in TABLE 3.

With this approach, the identification of strains A, M, N, and E was confirmed. A large fragment (ca. 900 bp) of the 16S rDNA of isolate T was analyzed, showing

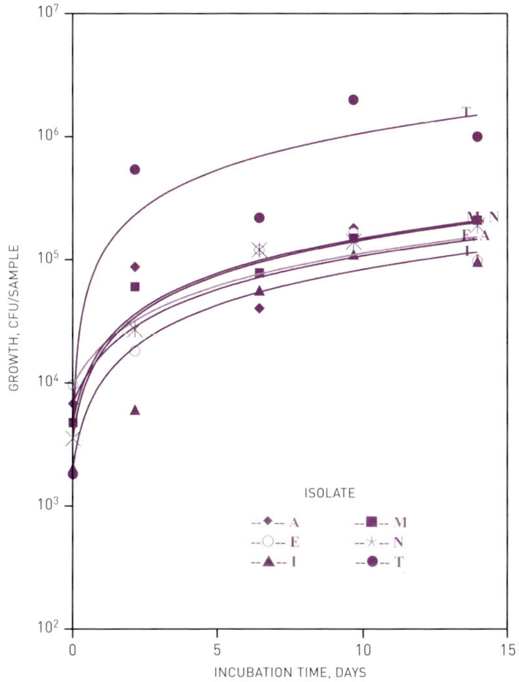

[1] Laboratory colonization of degummed silk by pure cultures of bacteria isolated from naturally aged silk fibroin.

ISOLATE	% SIMILARITY	WITH
A	99.5%	*Micrococcus luteus*
E	99.0%	*Bacillus licheniformis*
M	98.0%	*B. megaterium*
N	98.5%	*B. subtilis*
T	99.5%	*B. oleronius* and *B. sporothermodurans*
I	92.0%	*Bacillus sp.*

[TABLE 3] Identification of the bacterial strains isolated from degummed silk fabrics

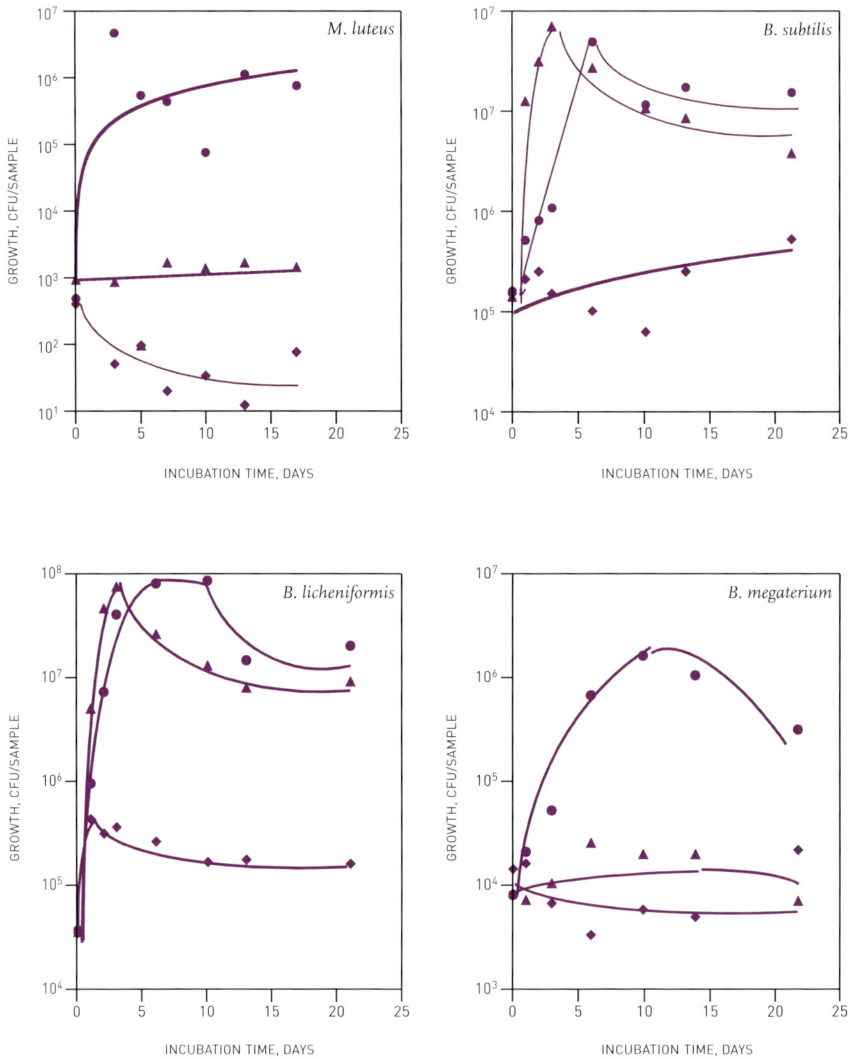

[2] Utilization of fibroin for growth by *M. luteus* and *B. subtilis, B. licheniformis,* and *B. megaterium*. The four strains were grown on complete Davis minimal medium (—▲—), or devoid of carbon and nitrogen sources (—◆—), or devoid of carbon and nitrogen sources but with fibroin (—●—).

the highest degree of similarity with sequences of *B. oleronius* and *B. sporothermodurans*. Finally, in the case of isolate I it was only possible to confirm it as a member of the genus *Bacillus*.

Utilization of Fibroin for Growth To confirm that colonization of fibroin in the laboratory was due to the metabolization of the protein, the six isolates were cultured in Davis complete minimal medium (Davis and Mingioli, 1950) as well as in the same medium in which the usual sources of carbon and nitrogen were omitted and substituted with pure fibroin. The results for isolates A, E, M, and N are reported in figure 2, since isolates I and T grew very little, if at all, on Davis' minimal medium, even in the presence of fibroin. The fact that pure cultures of isolates I and T did not grow or survive in this medium could be due to the presence on degummed silk of unknown compound(s) that stimulate bacterial growth and survival. Isolates A, E, M, and N grew quite well also when the fibroin powder was substituted by fragments of the fabric (data not reported). No growth was observed when the four strains were inoculated in minimal medium devoid of glucose, ammonium sulfate, and sodium citrate. [FIG. 2]

The results indicate that the four isolates may be subdivided into two groups: The first group, represented by *M. luteus* and *B. megaterium*, was unable to grow on Davis' minimal medium but grew when fibroin was used as the only source of carbon and nitrogen. These bacteria seem to have a growth requirement that presumably is satisfied by one or more of the amino acids present in fibroin. The other group, composed of *B. licheniformis* and *B. subtilis*, has no apparent nutritional requirements, as revealed by growth in complete minimal medium. Further, as depicted in figure 2, *B. subtilis* and *B. licheniformis* strains grew rapidly on fibroin, reaching a maximum (ca. 1×10^8 CFU/sample) after 6 days of incubation. A similar number of CFU was reached after 3 days when the two strains were grown in Davis' minimal medium. In any case, all four strains grew in a modified Davis medium in which fibroin represented the only source of nitrogen and carbon. These results confirm those reported in figure 1 that indicated that these isolates were able to colonize degummed silk.

Proteolytic Activity of M. luteus *and* B. subtilis The ability to utilize fibroin as a carbon and/or nitrogen source for growth by the four isolates suggests the presence of one or more proteases capable of hydrolyzing the protein and making the resulting amino acids available to the cells. To verify the presence of such enzyme(s), the *M. luteus, B. licheniformis, B. subtilis,* and *B. megaterium* isolates were grown on plates containing LB medium to which either skim milk proteins or fibroin were added. In the case of *B. licheniformis* and *B. megaterium*, colonies were surrounded by small halos of hydrolysis on the skim milk medium (data not shown). By contrast, as shown in figure 3, colonies of both *M. luteus* and *B. subtilis* were surrounded by large halos indicating hydrolysis of casein and other skim milk proteins. When the cell-free supernatant was assayed, a clearing zone was evident in the case of the filtrate from *B. subtilis* but not in that from *M. luteus*. [FIG. 3] The lack of activity of the *M. luteus* supernatant could be due to the production of endocellular proteases, to their instability, or to other reasons. Indeed, we have found that, contrary to the proteolytic activity of *B. subtilis*, that of *M. luteus* is totally dependent on the presence of Ca^{++} ions in the growth media. Indeed, if this strain was cultivated in the presence of calcium, a clear halo was evident on skim milk medium as well as when the cell-free supernatant was utilized (data not shown).

When fibroin was added to the solid medium, halos indicating hydrolysis of the protein were evident in the case of both bacteria (FIG. 4), thus confirming that fibroin is also degraded. It may be added that the halos were present also if the proteins (from skim milk or fibroin) were added to solid minimal media in which the proteins were the only source of carbon and nitrogen (data not shown).

DISCUSSION Naturally aged silk fibroin analyzed in this work appears to possess a microbial flora essentially composed of bacteria. Indeed, fungi were not isolated from new or old fibroin samples in agreement with previous results, indicating that even in soil burial experiments few fungi were isolated from silk kept in soil for up to 2 months (Seves et al., 1998). Further, laboratory attempts to induce colonization of fibroin by pure fungal cultures (some isolated by enrichment cultures of silk cocoons) were always unsuccessful (Seves et al., 1998) except if the fibroin samples were aged in the laboratory by heat treatment or by exposure to a Xenon lamp (Seves et al., 2000). Furthermore, both samples of the naturally aged silk fibroin used in the present investigation were refractory to colonization in the laboratory by pure cultures of the same species of *Aspergillus* and *Penicillium* that colonized artificially aged fibroin (data not reported).

These results, and those reported in previous investigations, confirm the susceptibility of silk fibroin to bacterial colonization and its substantial resistance to colonization by fungi. This conclusion contrasts with that reached by Sato, who reported that cells of *Aspergillus niger* or its extracts contain an acidic protease that had a hydrolytic activity of fibroin comparable to that of concentrated hydrochloric acid (Sato, 1976). However, other reports indicate the presence of different bacte-

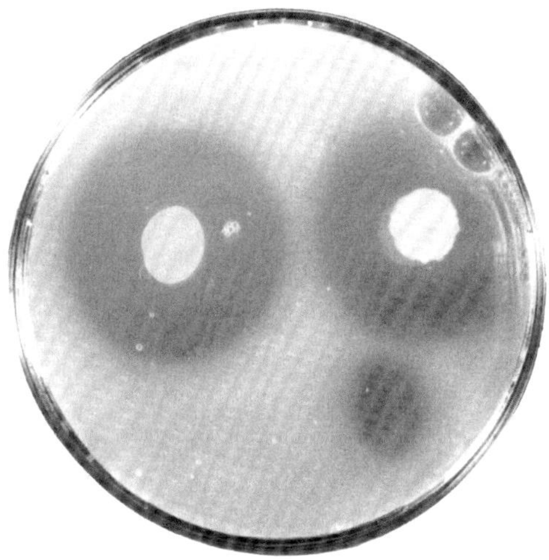

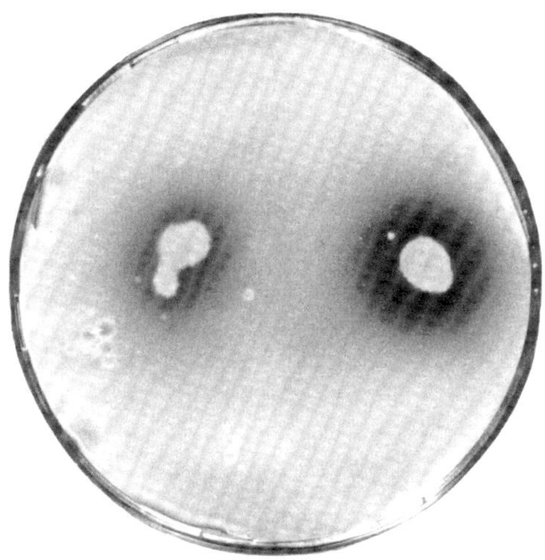

[3] Hydrolysis on solid media of casein by cells or cell-free supernatants of *M. luteus* and *B. subtilis* strains. Samples are (left to right and top to bottom): A and B, *M. luteus* and *B. subtilis* cells; C and D, *M. luteus* and *B. subtilis* cell-free supernatants. [4] Hydrolysis on solid media of fibroin by cells of *M. luteus* (A) and *B. subtilis* (B).

rial species and, to a lesser extent, protozoa, but not fungi in the water of workshops where silk is reeled off from cocoons (Ishiguro, 1989) and from silk buried in soil for up to two months (Ishiguro et al., 2000). Indeed, we have similarly reported that only bacteria were isolated from silk fibroin buried in soil and that fungi were isolated solely from enrichment cultures of cocoons, a material that contains, besides proteins, significant quantities of lipids and carbohydrates (Seves et al., 1998). Differences were observed in the types of bacteria isolated from aged fibroin and from fibroin buried in soil. Whereas in the present investigation the greatest majority (about 85%) of the bacteria isolated were Gram-positive, in the soil burial experiments and in the enrichment cultures of cocoons almost all bacteria isolated were Gram-negative (Seves et al., 1998). In this respect, the data from naturally aged silk are reminiscent of those we reported on the colonization of painted canvases ("mock paintings") by soil bacteria (Seves et al., 1996), where mostly Gram-positive bacteria of the genera *Bacillus* and *Micrococcus* were the main colonizers of "mock paintings." Only one *Bacillus, B. megaterium,* was isolated from the soil burial experiments, and the same species was present also on aged fibroin. In addition, the *B. megaterium* that came from the two sources was also capable of utilizing fibroin for growth.

Among the bacteria isolated from silk fabrics in this work and capable of utilizing fibroin as a sole source of nitrogen and carbon, *B. subtilis* and *M. luteus* seem to produce extracellular enzymes that hydrolyze fibroin. To our knowledge, few reports characterizing bacterial enzymatic activities degrading silk fibroin have been published. While two species of *Streptomyces* were shown to hydrolyze a variety of insoluble proteins, their activity on silk was very limited (Nakanishi and Yamamoto, 1974; De and Chandra, 1979). On the contrary, a protease isolated from *S. griseus* was reported to hydrolyze silk fibroin treated with a methanol solution, a treatment that changes the chemical and physical properties of the protein (Minora et al., 1990). Finally, a fibroin-degrading protease was isolated and partially characterized from a Gram-negative bacterium, *Variovorax paradoxus* (Forlani et al., 2000). Two Gram-positive bacteria, *M. luteus* and *B. subtilis,* isolated in the present investigation seem to be endowed with a high protease activity that acts also on fibroin. Many species of the genus *Bacillus* are well known as producers of a variety of proteases (Priest, 1977). *Micrococcus* species producing proteolytic enzymes have been used in dairy and meat fermentation; however, such proteases have not been extensively studied (Clark et al., 2000, and references cited therein). An alkaline protease from *M. luteus* that degrades elastin, casein, and β-lactoglobulin but not collagen and keratin has been described (Clark et al., 2000). Calcium was found to be essential for enzyme activity and stability, as was the case for the fibroin hydrolyzing activity of the *M. luteus* strain isolated in the present investigation.

Work is underway to separate and characterize the different proteases produced by the two bacteria and to ascertain if fibroin is degraded by a specific hydrolase (a fibroinase) or rather by enzymes whose main substrates are other proteins. Preliminary results revealed that both isolates produce more than one protease active on fibroin and other proteins such as milk casein. In the case of the *B. subtilis* strain, preliminary experiments seem to indicate that a fibroin-hydrolyzing enzyme is synthesized only when the strain is grown in a minimal medium in which this protein is the sole source of carbon and nitrogen. It is well known that specific proteases are produced only when bacteria, including the genus *Bacillus,* are grown in the presence of certain substrates (Rao et al., 1998).

In conclusion, the data presented here, as well as those previously published, indicate that bacteria rather than fungi appear to degrade silk fibroin and hence may present a significant threat to silk artifacts. Bacterial colonization, however, appears to be rather slow, and this may explain why "High quality kimonos have been passed down from several generations from grandmothers to granddaughters without bacterial damage of the silk fibroin, in spite of the very humid climate of Japan...." (Akai, 1997).

ACKNOWLEDGMENT This work was supported by grants from Progetto Finalizzato Beni Culturali of Consiglio Nazionale delle Ricerche (C.N.R.).

REFERENCES

AKAI, H., 1997. Anti-bacterial function of natural silk materials. *International Journal Wild Silkmoth and Silk*, 3, 79–81.

BROSIUS, J., PALMER, H.L., KENNEDY, J.P., and NOLLER, H.F., 1978. Complete nucleotide sequence of a 16S ribosomal RNA gene from *Escherichia coli*. *Proceedings of the National Academy of Sciences of the United States of America*, 75, 4801- 4805.

CLARK, D.J., HAWRYLIK, S.J., KAVANAGH, E., and OPHEIM, D.J., 2000. Purification and characterization of a unique alkaline elastase from *Micrococcus luteus*. Protein Expression and Purification, 18, 46–55.

DAVIS, B.D., and MINGIOLI, E.S., 1950. Mutants of *Escherichia coli* requiring methionine or vitamin B12. *Journal of Bacteriology* 60, 17–28.

DE, S., and CHANDRA, A.L., 1979. Degradation of organic nitrogenous wastes by a soil streptomycete. *Folia Microbiologica*, 24, 473–477.

FORLANI, G., SEVES, A.M., and CIFERRI, O., 2000. A bacterial extracellular proteinase degrading silk fibroin. *International Biodeterioration and Biodegradation*, 46, 271–275.

HEYES, T.F., and HOLDEN, H.S., 1926. The action of micro-organisms on silk. *Journal of the Society of Chemical Industry*, 45, T262–T265.

ISHIGURO, Y., 1989. Studies on cause and prevention of stains in the raw silk. In: "Research Reports of the Raw Silk Testing." *The Yokohama and Kobe Agricultural and Forestry Products Inspection Institutes* 42, 79–83.

ISHIGURO, Y., and MIYASHITA, M., 1996. Deterioration of silk caused by propagation of microbes. In: *The Third International Silk Conference*, Suzhou, China, The Organizing Committee, pp. 201–208.

ISHIGURO, Y., ZHOU, Y., and TAJIMA, F., 2000. Analysis of biodegradability of silk threads embedded in soil using image processing. *Proceedings of the 4th International Silk Conference*, State Bureau of Textile Industries, People's Republic of China, pp. 109–114.

MAIDAK, B.L., COLE, J.R., PARKER JR., C.T., GARRITY, G.M., LARSEN, N., LI, B., LILBURN, T.G., MCCAUGHEY, M.J., OLSEN, G.J., OVERBEEK, R., PRAMANIK, S., SCHMIDT, T.M., TIEDJE, J.M., and WOESE, C.R., 1999. A new version of the RDP (Ribosomal Database Project). *Nucleic Acids Research*, 27, 171–173.

MINORA, N., TSUKADA, M., and NATURA, M., 1990. Physico-chemical properties of silk fibroin membrane as a biomaterial. *Biomaterials*, 11, 430–434.

NAKANISHI, T., and YAMAMOTO, T., 1974. Action and specificity of a *Streptomyces* alkalophilic proteinase. *Agricultural and Biological Chemistry*, 38, 2391–2397.

PRIEST, F.G., 1977. Extracellular enzyme synthesis in the genus *Bacillus*. *Bacteriological Reviews*, 41, 711–753.

RAO, M.B., TANKSALE, A.M., GHATGE, M.S., and DESHPANDE, V.V., 1998. Molecular and biotechnological aspects of microbial proteases. *Microbiology and Molecular Biology Reviews*, 62, 597–635.

SAMBROOK, J., FRITSCH, E.F., and MANIATIS, T., 1989. In: "Molecular Cloning. A Laboratory Manual" Cold Spring Harbor Laboratory Press, vol. 3 Appendix A1.

SATO, M., 1976. The effects of molds on fibers and their products. VIII. Scanning electron microscopic study on the destruction of silk yarns damaged by molds. Kyoto-furitsu Daigaku Gakujutsu Hokoku: Rigaku, Seikatsu Kagaku vol. 27, pp. 59–64. (*Chemical Abstracts* vol. 87, 202944t, 1977).

SEVES, A.M., SORA, S., and CIFERRI, O., 1996. The microbial colonization of oil paintings. A laboratory investigation. *International Biodeterioration and Biodegradation*, 37, 215–224.

SEVES, A.M., ROMANÒ, M., MAIFRENI, T., SORA, S., and CIFERRI, O., 1998. The microbial degradation of silk: a laboratory investigation. *International Biodeterioration and Biodegradation*, 42, 203–211.

SEVES, A.M., ROMANÒ, M., MAIFRENI, T., SEVES, A., SCICOLONE, G., SORA, S., and CIFERRI O., 2000. A laboratory investigation of the microbial degradation of cultural heritage. In: Ciferri, O., Tiano, P., and Mastromei. G., (Eds.), *Of Microbes and Art. The Role of Microbial Communities in the Degradation and Protection of Cultural Heritage*, Plenum Press, pp. 121–133.

SMIBERT, R.M., and KRIEG, N.R., 1994. Phenotypic characterization. In: Gerhardt, P., Murray, R.G.E., Wood, W.A., and Krieg, N.R., (Eds.), *Methods for General and Molecular Bacteriology*, American Society for Microbiology, pp. 607–654.

STONE AND MURAL PAINTINGS

L. LAIZ
J.M. GONZALEZ
C. SAIZ-JIMENEZ*

MICROBIAL COMMUNITIES IN CAVES: ECOLOGY, PHYSIOLOGY, AND EFFECTS ON PALEOLITHIC PAINTINGS

Instituto de Recursos Naturales y Agrobiología, CSIC, Sevilla, Spain
* Corresponding author

ABSTRACT In this review we report on the culturable and nonculturable microbial communities present in caves. A high diversity was found using both methods; however, identification methods including PCR amplification of 16S rRNA genes (16S rDNA) and community fingerprinting by denaturing gradient gel electrophoresis (DGGE) resulted in a greater bacterial taxonomic diversity, and allowed the detection of unexpected and unknown bacteria. Growth of cave bacteria at different temperatures was investigated. The data indicate that most cave bacteria can grow at temperatures in the range 5 to 45°C. However, a higher diversity was obtained when incubation was carried out at 13°C, although incubation at this temperature drastically reduced the number of sporoactinomycete isolates. At 13°C the bacteria oxidized on average 87% of carbon sources provided, while incubation at 28°C reduced this to about 50% carbon source utilization. Selected actinobacteria isolates were tested for the formation of crystals. Strains from all tested genera except isolates of *Gordonia* and *Nocardia* produced vaterite and/or calcite. Production of Mg-calcite was restricted to strains of *Brachybacterium, Paenibacillus, Rhodococcus,* and *Streptomyces,* while struvite was only precipitated by an unidentified isolate. These findings indicate that actinomycetes may play a role in the formation of mineral deposits in caves.

INTRODUCTION In nature, chemical dissolution caused by naturally acidic waters transforms layers of permeable limestone into a dense network of tunnels, labyrinths, and large halls, most of which are adorned by stalactites and stalagmites, which form the karstic system. These subterranean karstic systems of horizontal caves and vertical abysses are distributed all around the world. Only a limited number of these caves are actually accessible as show caves, of which a few contain Paleolithic and Neolithic parietal art.

The Mediterranean Basin is characterized by the presence of a huge number of karstic caves. There are over 7000 karstic cave formations in Greece, the majority being in Crete. To date over 6500 such caves have been explored in Slovenia, and there are probably as many that are, as yet, unexplored. Many Mediterranean Basin countries have caves with Paleolithic paintings, such as the well-known Altamira, Tito Bustillo, Lascaux, Chauvet, and Niaux caves, located in Spain and France. Some others, such as Grotta dei Cervi, in Italy, contain valuable Neolithic paintings.

In spite of some available studies on geological, geochemical, and hydrological characteristics (Hoyos et al., 1998; Sanchez-Moral et al., 1999), there are very few studies on cave microbiology, most of them in connection with the conservation of paintings (Hoyos et al., 1998; Groth et al., 2001).

Certainly microorganisms play a role in cave ecosystems, as they are able to colonize bare rock surfaces, use the low organic matter present in dripping waters, and influence mineral crystallization processes. Cañaveras et al. (1999) suggested that bacteria present in caves may play a role in the formation of moonmilk deposits, as microbial communities predominantly composed of different species of the genus *Streptomyces* were found in association with hydromagnesite and needle-fiber aragonite deposits in the Altamira cave.

Just as caves are not uniform environments in terms of geological and geochemical characteristics, show cave managements exhibit striking differences as well. The Altamira and Lascaux caves have been heavily visited in the past, but had to be closed due to the deterioration observed in the paintings. Nowadays, visitors are limited to cave reproductions. In general, cave management today tends to reduce anthropogenic impacts by controlling visitors and microclimate (Hoyos et al., 1998).

In past years several show caves have been investigated in detail. These include Altamira, Tito Bustillo, La Garma, and Llonin caves in northern Spain, and Grotta dei Cervi in Porto Badisco, Italy (Groth and Saiz-Jimenez, 1999; Groth et al., 1999, 2001; Laiz et al., 1999, 2000). In this review we report on the ecology and physiology of cave bacteria.

ACTINOBACTERIA	GRAM – BACTERIA
Agromyces	*Acinetobacter*
Amycolatopsis	*Aeromonas*
Arthrobacter	*Chromobacterium*
Aureobacterium	*Chryseomonas*
Brachybacterium	*Enterobacter*
Brevibacterium	*Erwinia*
Cellulomonas	*Flavimonas*
Gordonia	*Janthinobacterium*
Kocuria	*Kingella*
Microbacterium	*Pseudomonas*
Micrococcus	*Serratia*
Micromonospora	*Variovorax*
Nocardia	*Xanthobacter*
Nocardioides	*Xanthomonas*
Nocardiopsis	
Rhodococcus	OTHER GRAM + BACTERIA
Rothia	*Bacillus*
Saccharothrix	*Paenibacillus*
Streptomyces	*Sporosarcina*
	Staphylococcus

[TABLE 1] Identified microbial genera

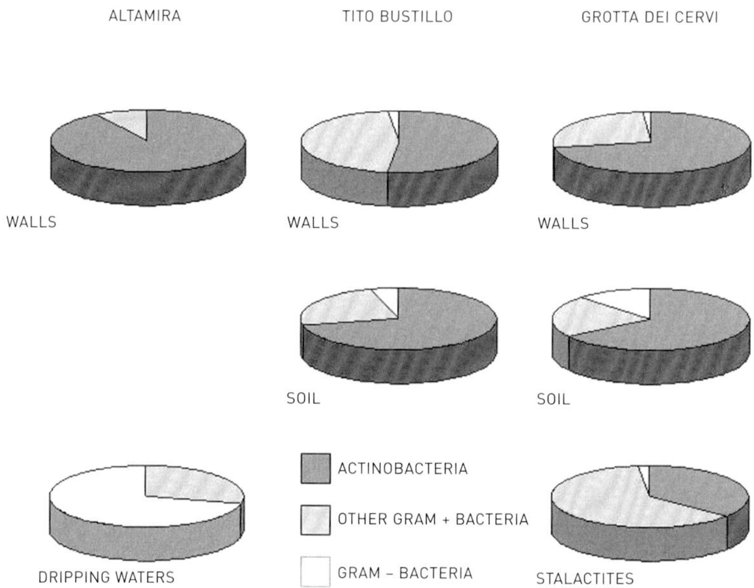

[1] Distribution of microbial isolates from cave ecosystems among different bacterial groups: Actinobacteria, other Gram positive bacteria, and Gram negative bacteria.

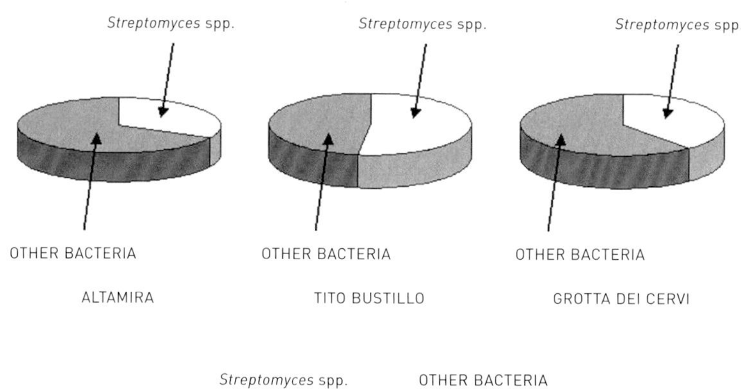

[2] Relative abundance of *Streptomyces* strains from Altamira Cave, Tito Bustillo Cave, and Grotta dei Cervi.

CAVE BIODIVERSITY Actinobacteria have been considered to be mainly confined to soils (Porter, 1971). However, it has been reported that they are found in nearly all natural habitats, including hypogean environments (Monte and Ferrari, 1993; Groth and Saiz-Jimenez, 1999).

Phototrophic microorganisms (cyanobacteria and algae) were absent in most studied caves due to the lack of light; actinobacteria growth was distributed all over the cave and could be observed on the active stalactites, on upper and lower parts of the rock walls, and in the cave soils. This seems to be a natural colonization, although apparently, at least in some cases, actinobacteria could be related to anthropogenic disturbances, as colonies were observed at the surface of the archaeologist's soil profile in Grotta dei Cervi.

In caves, samples of active stalactites, wall concretions, and rocks from the walls and ceiling of the galleries were investigated, and a high number of isolates was obtained. Thirty-seven genera were found among cave isolates. [TABLE 1] The most abundant genera were assigned to the actinobacteria division, particularly in rock walls and soils. [FIG. 1] In this division, *Streptomyces spp.* represented a large part of the isolated species. [FIG. 2] However, in stalactites, the most abundant species belonged to the low G+C Gram-positive bacteria of the genus *Bacillus,* although the most conspicuous and visible to the naked eye were actinobacteria.

Actinobacteria are well known for their ability to grow on very poor media, but also for their ability to use recalcitrant organic matter, e.g., lignocellulose residues and humic substances (Groth and Saiz-Jimenez, 1999). While guano input is appreciable in the ceiling and walls of the galleries and in the cave soils of Grotta dei Cervi, this is not the case in other caves, where organic matter was mainly present in dripping waters (Saiz-Jimenez and Hermosin, 1999).

The low or poor content of organic matter was noticed in the dripping water from stalactites, which increased during the spring season. However, considerable colonization was observed on the Grotta dei Cervi stalactites (Laiz et al., 2000). The pattern of the actinobacteria colonies originating from stalactites was similar to cooperative growth patterns reported for other bacteria, and described by Ben-Jacob et al. (1992). This colony morphology is often combined with unfavorable environmental conditions, and the bacteria have to develop sophisticated modes of cooperative behavior, which seems to be an adaptive response to poor nutrient flow. As Figure 3 shows, this type of growth was successfully reproduced in the laboratory with selected isolates. [FIG. 3] Groth and Saiz-Jimenez (1999) stressed that microbial communities of some caves usually rely on allochthonous input of dissolved and particulate organic matter that is transported from the surface.

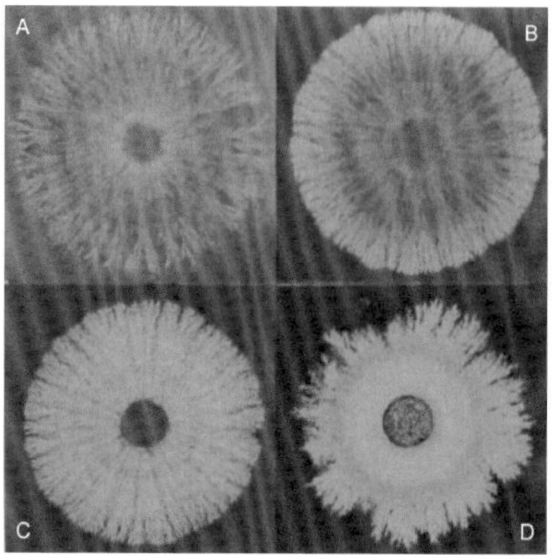

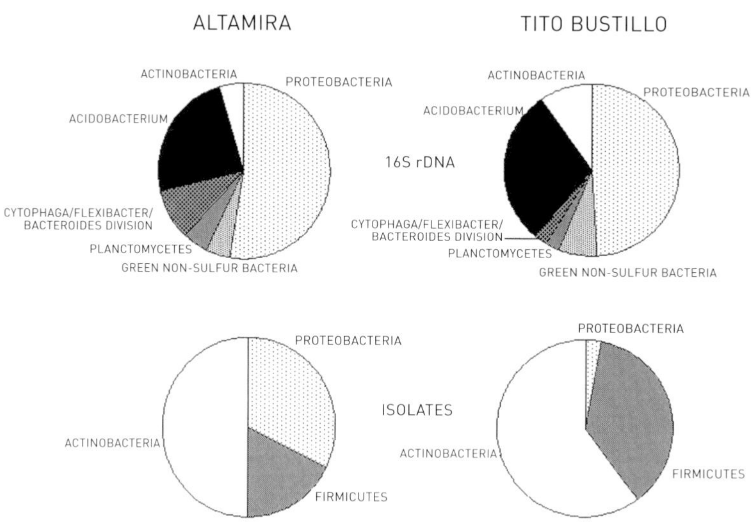

[3] Colonies of *Nocardiopsis* sp. grown on 4% agar and four different concentrations of peptone: 0.25, 2, 5, and 10%, from left to right, top to bottom. [4] Comparison of bacterial groups obtained by culture-dependent (isolation) and culture-independent (16S rRNA gene sequences) methods from Altamira and Tito Bustillo caves.

It has been reported that the actinobacteria isolated from the Altamira cave revealed taxonomic diversity, with the predominant isolates belonging to the genus *Streptomyces*. Members of the genera *Nocardia, Rhodococcus, Nocardioides, Amycolatopsis, Saccharothrix, Brevibacterium, Microbacterium,* and coccoid actinobacteria (family *Micrococcaceae*) were also found (Groth et al., 1999). The Grotta dei Cervi has a similar microbiological colonization but the biological diversity was somewhat higher. In spite of the facts that this cave was discovered in 1970, 91 years later than was Altamira, and that the total number of visitors since its discovery was very low (visitors came only for study purposes), its microbial colonization was surprisingly high. This is probably due to the high level of organic matter present in the cave, and to its nature (guano). A similar level of colonization was also noticed in the La Garma cave, discovered in 1995 and visited in 1998. In this period only 64 visitors to the cave were recorded, with a total of 45 hours of stay. Therefore, bacterial colonization seems to be independent of visits.

The studies on cultivated microorganisms in caves reveal information on only a minor portion of the cave population, and not a representative one. Many more microbial species are believed to be present, and these may have an important effect on cave walls. No information is available on uncultivated microorganisms in these caves, except some recently published papers (Schabereiter-Gurtner et al., 2002a, b). For analysis, samples were taken directly from the Paleolithic painting area in Altamira and Tito Bustillo. Without prior cultivation, nucleic acids were extracted from the sampled material and the phylogenetic relationship of PCR amplified and cloned 16S rRNA genes (16S rDNA) was analyzed. Genetic fingerprinting of the bacterial community was performed by denaturing gradient gel electrophoresis (DGGE), which allows the sequence-specific separation of partial 16S rDNA amplicons of the same length in polyacrylamide gels containing a denaturing gradient. To obtain phylogenetic information on individual bacterial members, 16S rDNA clone libraries were screened by DGGE. This approach gave us insight into a great bacterial taxonomic diversity, and allowed the detection of unexpected and unknown bacteria. Bacteria identified in the Altamira and Tito Bustillo caves were related to members of the cosmopolitan *Proteobacteria*, to members of the *Acidobacterium* division, *Cytophaga/Flexibacter/Bacteroides* phylum, green nonsulfur bacteria, *Planctomycetales* and *Actinobacteria*. The high number of clones most closely related to environmental 16S rDNA clones showed the broad spectrum of unknown and yet to be cultivated bacteria in these caves. [FIG. 4]

EFFECT OF TEMPERATURE ON BACTERIAL GROWTH The almost constant low temperature throughout the year in most of the studied caves suggested the possibility of an indigenous psychrotrophic microflora adapted to low temperatures that could be overlooked using standard microbiological procedures and incubation at higher temperatures (28°C). To investigate this possibility, soil samples from three different caves in northern Spain (Tito Bustillo, La Garma, and Llonin) were collected and studied at four different temperatures: 5, 13, 20, and 28°C. Enumeration, characterization of isolated species, and substrate utilization pattern analysis were carried out to investigate the influence of the different incubation temperatures on the isolates and their physiological versatility.

Incubation at the four different temperatures showed clear differences in viable cell counts (CFU). Cell counts at 28°c generally yielded two- to fourfold higher counts of CFU than at 5 or 13°c, while count differences between 20 and 28°c were comparatively lower. [FIG. 5]

A total of 101 strains were isolated—9 from Llonín, 21 from La Garma, and 71 from Tito Bustillo. Only 29 isolates corresponded to sporoactinomycetes, most of them isolated from Tito Bustillo at different temperatures. At 13°c only 6 sporoactinomycetes were isolated from Tito Bustillo, but at 28°c the figure was tripled, indicating that isolation of actinobacteria is temperature-dependent. This agreed with literature data indicating that optimum culturing temperature is 25–35°c, although some species grow at temperatures within the psychrophilic and thermophilic range (Holt et al., 1994; Groth et al., 1999). The relevance of actinobacteria in this cave was previously reported (Groth et al., 1999) and this high number of isolates is probably related to the input of organic matter and clays through periodic subterranean river flooding.

The most representative isolates obtained at different temperatures were selected for a study on the optimal growth temperature. Temperatures ranged from 5 to 45°c. The bacteria included two isolates from 5°c (*Kocuria sp.* and *Pseudomonas sp.*), four from 13°c (*A. viscosus, Bacillus sp., P. putida,* and *S. ureae*), one from 20°c (*Bacillus megaterium*), and one from 28°c (*B. halodurans*). Sporoactinomycetes were not included in the study due to their clear response to temperature isolation. The isolates showed different optimal growth temperatures. At 24 h, *B. halodurans* and *Bacillus sp.* had optimal growth at 28°c, while optimal growth for the rest was obtained at 38°c. At 5°c *B. halodurans, B. megaterium,* and *S. ureae* failed to grow, while the growth of *A. viscosus, Bacillus sp.,* and *P. putida* was very poor. At 45°c all isolates grew, some well (*A. viscosus, B. megaterium,* and *Kocuria sp.*). Independent of

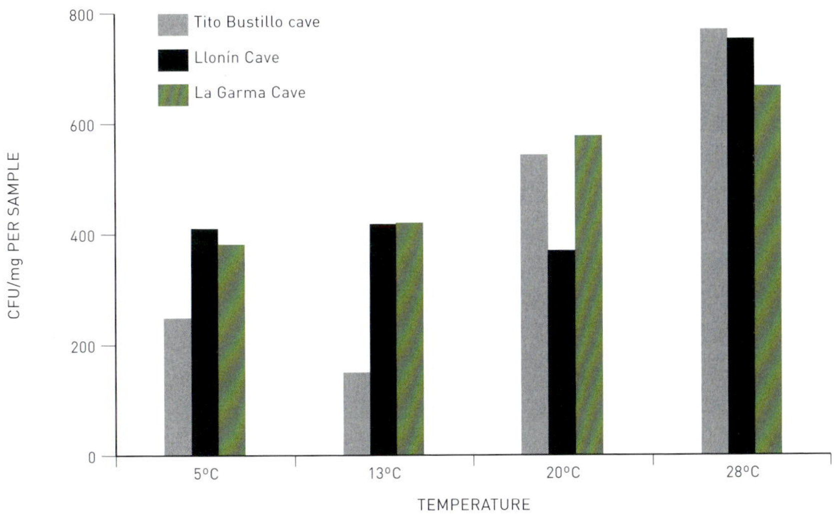

[5] Colony-forming units (CFU) per mg of sample at four incubating temperatures (5, 13 ,20, and 28°c).

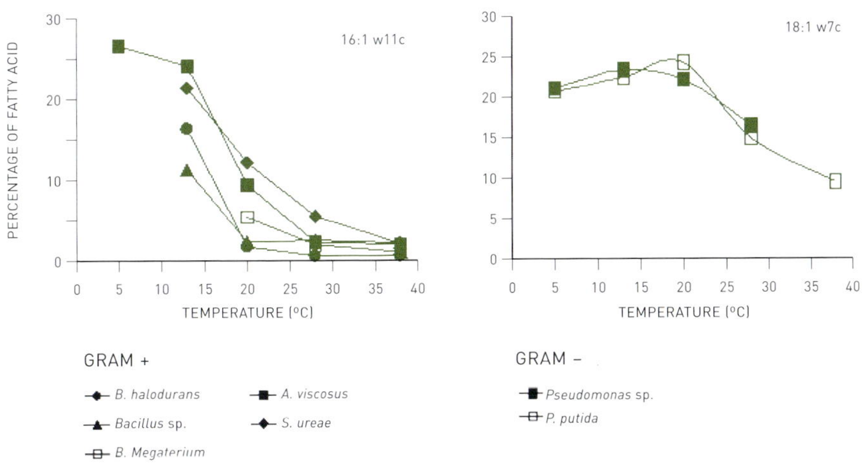

[6] Changes in relative composition of fatty acids (16:1w11c and 18:1w7c) at different temperatures.

the optimal growth temperature, the data show that these bacteria are able to grow comparatively well in a range of temperatures from 13 to 45°C.

Changes in fatty acid profile reflect an adaptive response to temperature. Variations with temperature were more pronounced for the fatty acid 16:1 w11c in Gram-positive and for the fatty acid 18:1 w7c in Gram-negative bacteria. Figure 6 shows the production of these fatty acids for selected isolates at different temperatures. [FIG. 6] Gram-positive bacteria fatty acid production of w11c was up to 16 times greater at 5°C than at 38°C (e.g., *A. viscosus*) while *S. ureae* had 12 times higher production at 13°C than at 38°C. For Gram-positive bacteria the increase in unsaturated fatty acid production appeared at 13°C, while for Gram-negative bacteria this increase (less apparent than that for Gram-positive bacteria) started at 20°C.

SUBSTRATE UTILIZATION A substrate diversity study was carried out with representative isolates covering three temperatures. [FIG. 7] The isolates were inoculated onto three Biolog GN microplates and incubated at 5, 13, or 28°C.

During incubation at their isolation temperature, the isolates from 5°C (*Kocuria sp.* and *Pseudomonas sp.*) oxidized numerous carbon sources. The highest number of carbon sources oxidized by these isolates occurred at 13°C. *Kocuria sp.* oxidized an intermediate number of carbon sources at 28°C, while *Pseudomonas sp.* oxidized a similar number of carbon sources either at 13 or 28°C. The rest of the isolates were not tested at this temperature, as no growth or poor growth were obtained in most of them.

The isolates from 13°C were the most active when incubated at their isolation temperature. In fact, *Bacillus sp., S. ureae, A. viscosus,* and *P. putida* metabolized, respectively, 91, 90, 89, and 62 out of 95 carbon sources, in comparison with 33, 3, 6, and 49 at 28°C.

The isolate at 20°C (*B. megaterium*) demonstrated a higher carbon source utilization at 13°C than at 28°C, while the isolate from 28°C (*B. halodurans*) was the only case in which a more efficient carbon source utilization at 28°C occurred.

It is remarkable that at 13°C a high number of carbon sources were utilized. In fact, on average, the tested bacteria oxidized 92% of all the carbohydrates, 82% of all the carboxylic acids, 88% of all the amino acids, and 84% of all other carbon sources. In contrast to these results, carbon sources utilized at 28°C showed lower efficiency (57, 42, 47, and 58%, respectively).

From the study on Biolog GN microplates, it appears that most bacteria isolated in the range 5–28°C oxidize carbon sources at incubation temperatures be-

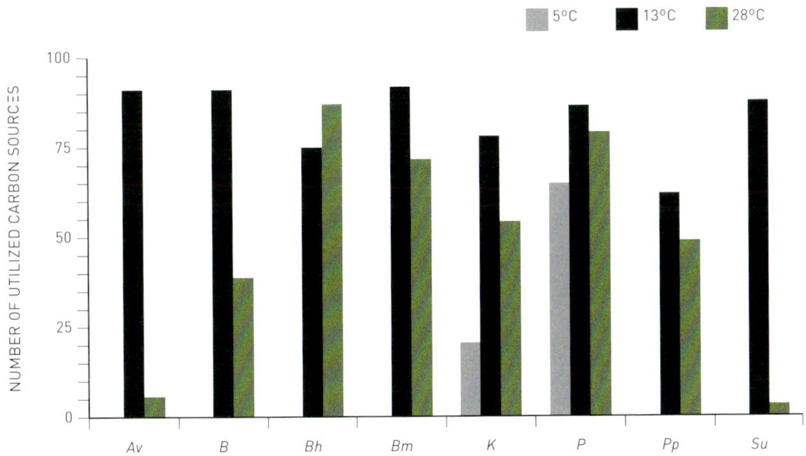

[7] Variations in the number of carbon sources utilized by different bacterial species at three incubation temperatures (5, 13, and 28°c). *Av*, *Arthrobacter viscosus*; *B*, *Bacillus* sp.; *Bh*, *Bacillus halodurans*; *Bm*, *Bacillus megaterium*; *K*, *Kocuria* sp.; *P*, *Pseudomonas* sp.; *Pp*, *Pseudomonas putida*; *Su*, *Sporosarcina ureae*.

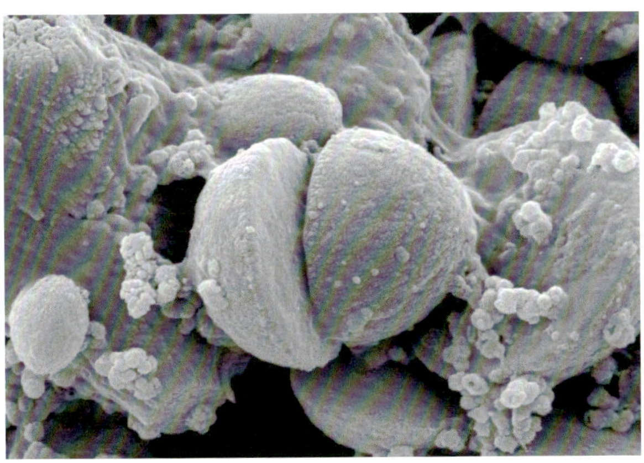

[8] Hemi-spherulite vaterite with a planar basement obtained from *Acinetobacter* sp.

tween 13 and 28°C, showing a wide metabolic versatility. The temperature of most caves in northern Spain and southern France is 13°C, the temperature at which a higher number of carbon sources is oxidized, thus indicating a high physiological adaptation. This suggests that cave bacteria can be very active in situ.

The oxidation of a higher number of carbon sources at 13°C could be explained by the fact that cave bacteria have the ability to alter membrane lipid composition for an efficient nutrient utilization at low temperatures (Russell, 1990, 1993). This can be observed in whole-cell fatty acid production at different temperatures, with an increase of unsaturated fatty acids in the lowest temperature range. This increase of unsaturated fatty acids is the most frequently reported response by Gram-negative bacteria to decreasing temperatures (Männistö and Puhakka, 2001).

Growth at low temperature seems to be correlated with a modulation of the outer membrane permeability. Dé et al. (1997) indicated that the response of *Pseudomonas fluorescens* to a temperature that falls under its optimum growth temperature is a decrease of outer membrane permeability to avoid the entry of toxic molecules. This effect is counterbalanced by the overproduction, at low temperature, of exocellular enzymes (lipases, proteases, etc.) for degrading macromolecules into small units, which provides *P. fluorescens* with carbon and energy sources.

CRYSTAL FORMATION A variety of different hygroscopic salts, including carbonates, chlorides, nitrates, and sulfates, can be found in caves (Cañaveras et al., 1999). Microbes are often harmful to Paleolithic paintings because they are related to constructive (mineral precipitation) and destructive (substrate dissolution) processes affecting different substrates (host-rock, speleothems, paintings, etc.) (Cañaveras et al., 1999, 2001). Therefore geomicrobiological studies were carried out to establish the role that microorganisms play in the microbial-mineral interactions that occur in hypogean environments.

This was one of the reasons for testing crystal production with a number of cave isolates. We tested 34 isolates in five different culture media: medium B-4, with three different carbon sources, medium M-2, and medium CT.

Medium B-4 (Boquet et al., 1973), containing calcium, magnesium, or ammonium acetate, resulted only in the production of crystals when calcium acetate was used as a carbon source. The production of calcite by a number of bacteria was previously reported in this medium by Boquet et al. (1973). Using the same medium, Laiz et al. (1999) found that *Acinetobacter spp.* produced vaterite (85%) and calcite (15%). [FIG. 8] It appears that the ability to form calcium carbonate polymorphs is widely distributed among environmental actinobacteria, as 19 out of 31 tested strains (61% of the actinobacteria) produced a considerable amount of crystals in both solid and liquid media. The tested microorganisms produced calcite alone (e.g., *Brachybacterium sp.*) or both vaterite and calcite (e.g., *Rhodococcus sp.*).

Medium M-2 containing both calcium and magnesium acetate was used for Mg-calcite formation (Gonzalez-Muñoz et al., 2000). Seven out of 34 strains produced crystals in this medium (20% of the tested strains). These seven strains also produced calcium carbonate polymorphs in the medium B-4. Crystalline aggregates composed of thin platelets of magnesium calcites (HMC, high magnesium calcite) in a rosette-like arrangement, resembling minute "desert-roses," were obtained from *Rhodococcus sp.* cultures.

Euhedral crystals of struvite, a magnesium ammonium phosphate ($MgNH_4PO_4 \cdot 6H_2O$), were found in the CT cultures (Perez-Garcia et al., 1989) of an unidentified strain that also produced calcite and Mg-calcite.

CONCLUSIONS Microbial diversity in selected habitats cannot be comprehensively studied by traditional approaches. One should be aware that only a minor fraction of the existing bacteria in an ecosystem is able to grow under laboratory conditions. However, high species diversity was observed in the studied caves. This diversity was even higher when identification methods included PCR amplification of 16S rRNA genes (16S rDNA) and community fingerprinting by denaturing gradient gel electrophoresis (DGGE). The approach used gave insight into a great bacterial taxonomic diversity, and allowed the detection of unexpected and unknown bacteria.

Cave bacteria have adapted to grow at low temperatures. The isolated bacteria, other than sporoactinomycetes, were psychrotolerants, as most of them can grow at 5°C and all grow between 13 and 45°C, although optimal growth temperatures occur between 28 and 38°C. No isolate had an optimum growth temperature below 20°C, and therefore none could be considered a true psychrophile. However, they were able to use different carbon sources at low temperatures, as shown by Biolog GN microplate tests.

Actinobacteria in caves were involved in biomineralization processes. This could open new insights into the role that these microorganisms can play in crystal formation and in cave geochemical cycles.

ACKNOWLEDGMENTS This research was funded by the European Commission, project EVK4-CT2000-00028.

REFERENCES

BEN-JACOB, E., SHMUELI, H., SHOCHET, O., and TENENBAUM, A., 1992. Adaptative self-organization during growth of bacterial colonies. *Physica A*, 187, 378–424.

BOQUET, E., BORDONAT, A., and RAMOS CORMENZANA, A., 1973. Production of calcite crystals by soil bacteria is a general phenomenon. *Nature*, 246, 527–528.

CAÑAVERAS, J.C., HOYOS, M., SANCHEZ-MORAL, S., SANZ-RUBIO, E., BEDOYA, J., SOLER, V., GROTH, I., SCHUMANN, P., LAIZ, L., GONZALEZ, I., and SAIZ-JIMENEZ, C., 1999. Microbial communities associated with hydromagnesite and needle-fiber aragonite deposits in a karstic cave (Altamira, Northern Spain). *Geomicrobiology Journal*, 16, 9–25.

CAÑAVERAS, J.C., SANCHEZ-MORAL, S., SOLER, V., and SAIZ-JIMENEZ, C., 2001. Microorganisms and microbially induced fabrics in cave walls. *Geomicrobiology Journal*, 18, 223–240.

DÉ, E., ORANGE, N., SAINT, N., GUÉRILLON, J., DE MOT, R., and MOLLE, G., 1997. Growth temperature dependence of channel size of the major outer-membrane protein (OprF) in psychrotrophic *Pseudomonas fluorescens* strains. *Microbiology*, 143, 1029–1035.

GONZALEZ-MUÑOZ, M.T., CHEKROUN, K.B., ABOUD, A.B., ARIAS, J.M., and RODRIGUEZ-GALLEGO, M., 2000. Bacterially induced Mg-calcite formation: role of Mg^{2+} in development of crystal morphology. *Journal of Sedimentary Research*, 70, 559–564.

GROTH, I., and SAIZ-JIMENEZ, C., 1999. Actinomycetes in hypogean environments. *Geomicrobiology Journal*, 16, 1–8.

GROTH, I., VETTERMANN, R., SCHUETZE, B., SCHUMANN, P., and SAIZ-JIMENEZ, C., 1999. Actinomycetes in karstic caves of Northern Spain (Altamira and Tito Bustillo). *Journal of Microbiological Methods*, 36, 115–122.

GROTH, I., SCHUMANN, P., LAIZ, L., SANCHEZ-MORAL, S., CAÑAVERAS, J.C., and SAIZ-JIMENEZ, C., 2001. Geomicrobiological study of the Grotta dei Cervi, Porto Badisco, Italy. *Geomicrobiology Journal*, 18, 241–258.

HOLT, J.G., KRIEG, N.R., SNEATH, P.H.A., STALEY, J.T., and WILLIAMS, S.T., 1994. *Bergey's Manual of Determinative Bacteriology*, Williams & Wilkins, Baltimore.

HOYOS, M., SOLER, V., CAÑAVERAS, J.C, SANCHEZ-MORAL, S., and SANZ-RUBIO, E., 1998. Microclimatic characterization of a karstic cave: human impact on microenvironmental parameters of a prehistoric rock art cave (Candamo Cave, northern Spain). *Environmental Geology*, 33, 231–242.

LAIZ, L., GROTH, I., GONZALEZ, I., and SAIZ-JIMENEZ, C., 1999. Microbiological study of the dripping waters in Altamira cave (Santillana del Mar, Spain). *Journal of Microbiological Methods*, 36, 129–138.

LAIZ, L., GROTH, I., SCHUMANN, P., ZEZZA, F., FELSKE, A., HERMOSIN, B., and SAIZ-JIMENEZ, C., 2000. Microbiology of the stalactites from Grotta dei Cervi, Porto Badisco, Italy. *International Microbiology*, 3, 25–30.

MÄNNISTÖ, M.K., and PUHAKKA, J.A., 2001. Temperature- and growth-phase-regulated changes in lipid fatty acid structures of psychrotolerant groundwater Proteobacteria. *Archives of Microbiology*, 177, 41–46.

MONTE, M., and FERRARI, R., 1993. Biodeterioration in subterranean environments. *Aerobiologia*, 9, 141–148.

PEREZ-GARCIA, I., RIVADENEYRA, M.A., and RAMOS-CORMEZANA, A., 1989. The influence of pH on struvite formation by bacteria. *Chemosphere*, 18, 1633–1638.

PORTER, J.N., 1971. Prevalence and distribution of antibiotic-producing actinomycetes. *Advances in Applied Microbiology*, 14, 73–92.

RUSSELL, N.J., 1990. Cold adaptation of microorganisms. *Philosophical Transactions of the Royal Society of London*, B 326, 595–611.

RUSSELL, N.J., 1993. Biochemical differences between psychrophilic and psychrotolerant microorganisms. In: Guerrero, R, and Pedrós-Alió, C., (Eds.), *Trends in Microbial Ecology*. Spanish Society for Microbiology, Madrid, pp. 29–32.

SAIZ-JIMENEZ, C., and HERMOSIN, B., 1999. The nature of the organic matter present in dripping waters from Altamira cave. *Journal of Analytical and Applied Pyrolosis*, 49, 337–347.

SANCHEZ-MORAL, S., SOLER, V., CAÑAVERAS, J.C., SANZ-RUBIO, E., VAN GRIEKEN, R., and GYSELS, K., 1999. Inorganic deterioration affecting the Altamira Cave, N Spain: quantitative approach to wall-corrosion (solutional etching) processes induced by visitors. *Science of the Total Environment*, 243/244, 67–84.

SCHABEREITER-GURTNER, C., SAIZ-JIMENEZ, C., PIÑAR, G., LUBITZ, W., and RÖLLEKE, S., 2002a. Cave Paleolithic paintings harbour complex and partly unknown microbial communities. *FEMS Microbiology Letters*, 211, 7–11.

SCHABEREITER-GURTNER, C., SAIZ-JIMENEZ, C., PIÑAR, G., LUBITZ, W., and RÖLLEKE, S., 2002b. Phylogenetic 16S rRNA analysis of Tito Bustillo Cave, Spain, and its Paleolithic paintings reveals the presence of complex and partly unknown microbial communities. *Environmental Microbiology* (in press).

E. ZANARDINI[1*]
P. ABBRUSCATO[1]
L. SCARAMELLI[1]
E. ONELLI[1]
M. REALINI[2]
G. PATRIGNANI[1]
C. SORLINI[1]

RED-STAINS ON CARRARA MARBLE: A CASE STUDY OF THE CERTOSA OF PAVIA, ITALY

1 University of Milan, Milan, Italy 2 Centre CNR "Gino Bozza," Milan, Italy
* Corresponding author

ABSTRACT There are numerous cases of red stains found on the Carrara and Condoglia marbles of Italian monuments. Examples can be found on the Certosa of Pavia façade and the Orvieto Cathedral, as well as the fountain statues of Villa Litta, near Milan. Although many studies have been performed in the past, there are still many questions that remain unresolved.

The aim of this paper is to provide a state-of-the-art report on the research conducted on the Certosa of Pavia, taking into consideration all the previous work performed by the authors, and to illustrate the results of the new campaign started in 2001 as part of a new conservation project. The microscopical analyses showed the presence of lead in the red stains that are distributed in correspondence to the intergranular microfissures of the calcite crystals. Microbiological and molecular approaches were adopted in order to investigate the microbial colonization of the red alterations, and to highlight the differences between the stained and unstained areas. Cultural microbiological techniques showed no significant differences in the microbial counts and a low biodiversity between stained and unstained samples; *Micrococcus sp., Bacillus sp.* and *Staphylococcus sp.* were shown to still be present, as we had found in previous work. In addition, *Bacillus sp.* and *Staphylococcus sp.* were shown to be lead-resistant up to 1500 mgl^{-1}.

Denaturing gradient gel electrophoresis analyses on red stained samples evidenced different profiles corresponding to a wider biodiversity in the unstained samples as opposed to the stained one.

INTRODUCTION Among the many alterations that occur to stone, of significant importance are the red stains that are observed on numerous Italian monuments, including the Certosa of Pavia, Orvieto and Siena Cathedrals, and the statues of the Galatea fountain in Villa Litta, near Milan.

With regard to the Certosa of Pavia [FIG. 1], red stains were first observed in 1844, exclusively on the Carrara marble and never on the other lithotypes (CNR Centre "Gino Bozza," 1988). These stains have progressively enlarged in the course of time.

While at the Certosa façade the red areas are localized only on its foundation stones [FIG. 2], on the fountain statues of the Villa Litta they are randomly distributed, and on the Orvieto façade these stains are mainly located in association with the gutters. In all these cases color variation of the stone is observed, but this does not affect the material properties of the stone.

The first studies on the Certosa façade, performed in 1986, demonstrated the presence of red-pigmented heterotrophic bacteria identified as *Micrococcus sp.* and *Halococcus sp.* and attributed the red coloration to organic pigments such as carotenoids (Bassi et al., 1986). Red-pigmented microorganisms were successively found in both the red stains present on the façade of Orvieto Cathedral and on the statues of the Galatea fountain in Villa Litta (Sorlini et al., 1994; Zanardini et al., 1994) and they were identified as *Flavobacterium sp.* and *Micrococcus sp.* together with a red yeast classified as *Rhodotorula minuta*. Photosynthetic microorganisms were found in all cases but not in association with the red stains; furthermore, no chemolithotrophic microorganisms were detected and there were a low number of mycetes.

The presence of red-pigmented microorganisms initially gave rise to the hypothesis that the red coloration could be attributed to organic pigments such as carotenoids. However, the low number of the red-pigmented microorganisms did not explain the strong marble coloration, and it seemed very unlikely that the carotenoids, being biodegradable, could have accumulated over time. [FIG. 3]

More recent studies by resonance-Raman spectroscopy have not shown the presence of organic pigments on altered marble samples, while normal micro-Raman spectroscopy has shown bands corresponding perfectly in both frequency and intensity ratio to those already observed for pure minium (red lead oxide, Pb_3O_4). Similar results were obtained on the red-stained Carrara marble samples

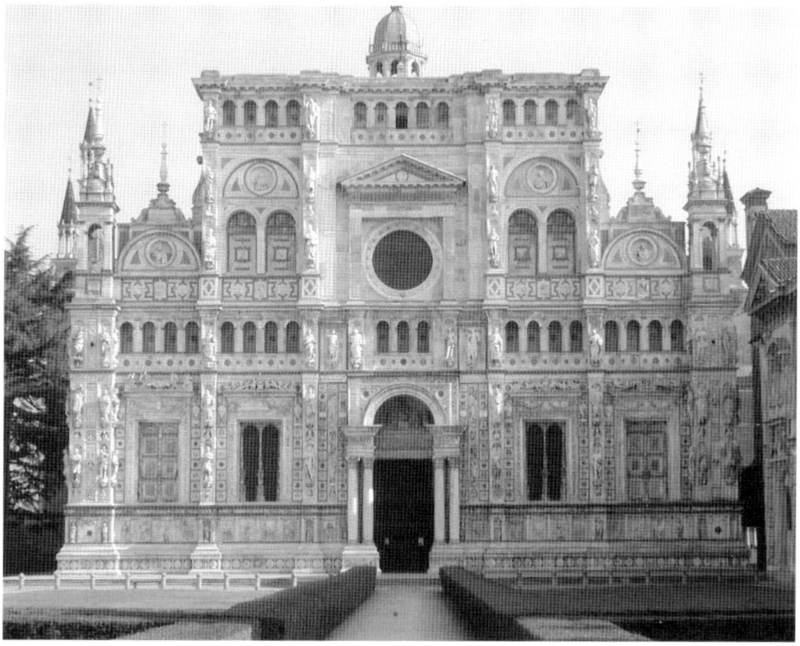

[1] Certosa of Pavia.

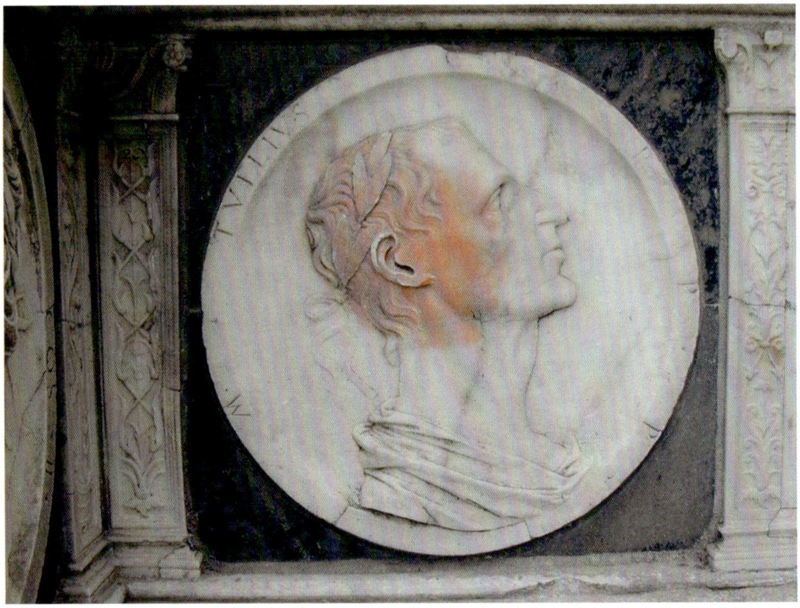

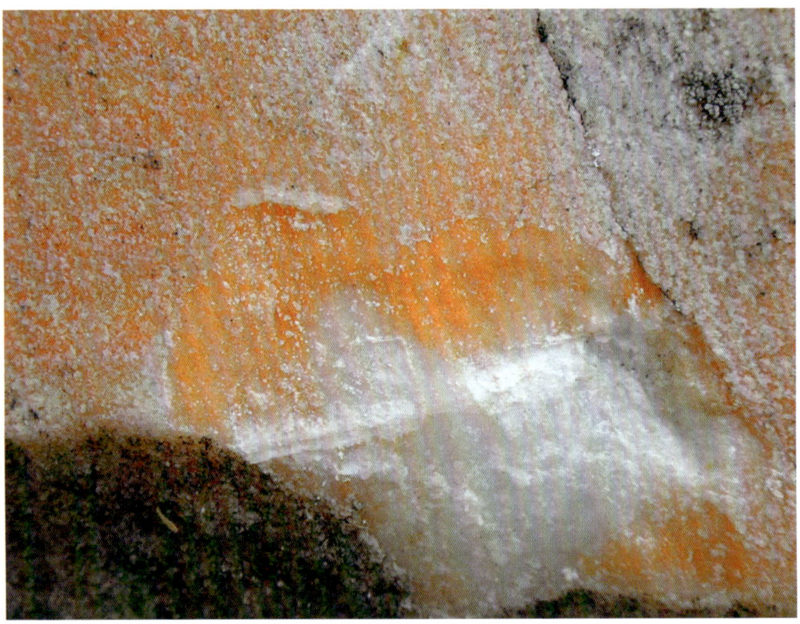

[2] Red stain on Carrara marble in the basement of the façade. [3] Detail of a sampling area.

from the façade of Orvieto Cathedral and from the Candoglia marble statues of the Galatea fountain in Villa Litta, near Milan (Sorlini et al., 1994; Zanardini et al., 1994 and 1997; Bruni et al., 1995), demonstrating that the red stains are of inorganic origin.

The presence of lead pigments in stone alterations has also been evidenced by other researchers, both on the Carrara marble of the Fonte del Labirinto in Florence (Matteini et al., 1986) and on wall paintings (Giovannoni et al., 1990).

In our work, different explanations as to the origin of the lead were suggested by the authors, although the mechanism of the lead oxidation was not clear. Since the chemical oxidation of lead seems improbable, some researchers have hypothesized a microbial role in the process (Petushkova and Lyalikova, 1986; Lyalikova and Petushkova, 1991).

With regard to the origin of the lead, different hypotheses were considered: For the statues of the Galatea fountain it could derive from the fountain pipes; for the Certosa façade, from atmospheric agents, from polishing materials, and/or from lead plates used between the marble blocks during the building's construction (Guidobaldi et al., 1984); and finally, for the façade of Orvieto Cathedral it could derive from the gutters.

As for the Certosa studies, in accordance with the request of the "Soprintendenza ai Beni Ambientali e Architettonici" of Milan, a new diagnostic campaign on the façade was begun in 2000 with the aim of providing new information for the conservation project. The objective is to augment the chemical and microbiological knowledge on such alterations and to study the microbial communities present in the stained and unstained areas using both microbiological and molecular methods. For the restoration, it was particularly important to confirm the presence of lead in the red areas, and to investigate the chemical-physical decay of the marble and the possible relationship among microbial colonization, red colorations, and the presence of lead.

Moreover, the new marble sampling would allow more detailed investigations on the lead-resistant microflora already evidenced in our previous works (Zanardini et al., 1997).

MATERIALS AND METHODS

SAMPLES Samples of the Certosa Carrara marble, from both stained and unstained areas, were taken from the façade, transferred in sterile conditions to the laboratory, and analyzed. The powder for culture analyses was obtained by scraping the superficial layer of every sample using a sterile scalpel, to a depth of about 1 mm.

MICROSCOPICAL OBSERVATIONS

Optical Analysis The morphology of red areas on the surfaces was directly observed with a Wild Makroskop M420. Cross-polished sections of the stained marble have been observed by a Leitz Ortholux in order to evaluate the depth of the coloration.

SEM Analysis Scanning electron microscopic observations of the surface were carried out on gold-coated samples, in order to investigate the presence of bacteria and fungi on the decayed surface, and on carbon-coated samples in order to evaluate the presence of lead and/or other elements in relation to the red stains. Observations were carried out using a Jeol 5910 LV microscope equipped with an IXRF 2000 spectrometer (for energy-dispersive X-ray spectrometry).

TEM Analysis Transmission electron microscopic (TEM) observations were carried out to study the lead-resistant bacterial cytoplasmatic ultrastructure and to investigate if the lead had accumulated in the cytoplasm. Cells were encapsulated in sodium alginate 5% (solidified in 0.2 M $CaCl_2$) and fixed in 3% glutaraldehyde (in 0.1 M cacodylate buffer, pH 6.9). After a few days, the samples were repeatedly rinsed in the same cacodylate buffer and post-fixed in 1% osmium tetroxide (in cacodylate buffer) for 3 h at 4°c. Subsequently, samples were dehydrated in ethanol in ascending grades up to absolute and embedded in Araldite epoxy. Ultrathin sections, obtained using a Reichert Jung Ultracut E microtome (Reichert-Jung, Vienna, Austria), were stained with 3% uranyl-acetate and lead citrate (Reynolds 1963). Ultrathin sections were observed with a Jeol 100 SX transmission electron microscope.

MICROBIOLOGICAL ANALYSES

Culture Media The following culture media were employed for the detection of different microbial groups: Plate Count Agar (PCA, Oxoid) and Malt Agar (Oxoid) for aerobic heterotrophic bacteria and molds and yeasts, respectively; CHU and Detmer mineral media for cyanobacteria for chlorophyta; Stanier medium for nitrifying bacteria; and Pochon and Tardieux medium for sulfur-oxidizing bacteria (Normal Committee, 1988).

Identification of Bacterial Isolates For the identification of some isolates, after standard microbiological procedures (Gram staining, oxidase, catalase, etc.), biochemical characterization using API Staph, and API 20 NE, API 50 CH, and ID 32 C was performed.

Growth Test in Presence of Lead—Enrichment Cultures For the growth tests in the presence of lead, microbial consortia from the red marble areas and isolated strains were inoculated in Plate Count Broth and lead acetate ($(CH_3COO)_2Pb$) was added at different concentrations (250, 500, 750, 1000, 1300, and 1500 mgl^{-1}). Microbial growth was measured after 6 days of incubation by spectrophotometric measurements of the optical density at 600 nm of the broth-cultures and by ATP total content determination using the luminometer Biocounter Lumac M 1500 P (Hawronskyi and Holah, 1997).

MOLECULAR ANALYSIS

Amplification with Primers Specific for Bacillus cereus *Group* Total genomic DNA was extracted from the pure cultures following the procedure described by Ausubel et al. (1991). A PCR protocol specific for the *Bacillus cereus* group (Daffonchio et al., 1999) was used to confirm the affiliation of one of our lead resistant isolates to this group. DNA extracted and amplified were run (100 mV, Power Pac 300, Biorad) on a variable-percentage agarose gel (0.8–2%), depending on DNA fragment sizes, in 0.5× Tris Borate EDTA (TBE) buffer, stained in ethidium bromide, visualized and photographed under UV light.

DGGE Analyses Denaturing gradient gel electrophoresis analyses were employed for the study of the microbial communities, both on the stained and unstained samples, in order to ascertain the differences between the two communities. DNA extraction from marble samples was performed following the protocol reported by Schabereiter-Gurtner et. al. (2001). For DGGE analyses, molecular protocols reported by Gurtner et al. (2000) and Piñar et al. (2001) were adopted. The DGGE profiles were visualized on acrilamide gel by silver staining.

RESULTS AND DISCUSSION

MICROSCOPICAL OBSERVATIONS

Optical Microscopy To study the diffusion and the penetration of the red-stained areas on the marble, observations were made using a stereomicroscope, which revealed that the red coloration is present on both the surface and the inner layer of the samples (up to a depth of 1.5 mm) [FIG. 4]; furthermore, observing at higher magnitudes on cross-polished sections, it was found that the red coloration corresponds to the intergranular micro-fracture of the calcite crystals.

It is also important to note that the stains present different red color tonality, ranging from orange to dark red. In addition to the red stains that were the object of our study, other alterations were also evidenced, although they were less pronounced and more diffused; an example is the green patinas on the marble, suggesting the presence of photosynthetic microorganisms.

The optical analysis also permitted observation of the poor state of conservation of the material, both in the stained and unstained areas, showing a granular de-cohesion and loss.

Scanning Electron Microscopy SEM observations confirmed an evident corrosion of the material both on stained samples and unstained ones [FIG. 5], suggesting that the red coloration may not be related to the deterioration of the material but rather that this is dependent on chemical processes.

The presence of microflora both on stained and unstained surfaces appears to be very low: Fungal hyphae were only occasionally observed, and then mainly on the unstained marble [FIG. 6], while no bacteria were visible, and this is probably because they are located in the micro-fissures of the stone. On the stained surface the presence of a lamellar-like layer was observed [FIG. 7]; EDS analysis showed the lead to be the most relevant element [FIG. 8], in contrast to the unstained areas, where lead was not detected. Back-scattered electron (BSE) observations showed that the presence of lead corresponds to the intergranular micro-fractures of the marble [FIGS. 9 AND 10], as had also been observed for the distribution of the red coloration. In the micro-fissures, a sort of patina was observed, suggesting a mixture of bacterial cells and lead [FIGS. 11 AND 12], although further investigations into this matter are necessary.

MICROBIOLOGICAL ANALYSES

The Presence of Microorganisms Chemolithotrophic microorganisms (nitrifying and sulfur-oxidizing bacteria) and photosynthetic microorganisms were not found in either the stained or unstained samples. No significant differences between the microbial counts of the stained and unstained samples were observed: the heterotrophic aerobic bacterial and fungi counts were 10^4 UFC g^{-1} and 10^2–10^3 UFC g^{-1}, respectively. [FIG. 13] Furthermore a low phenotypic biodiversity of bacterial colonies was observed both in the stained and unstained areas; the only exception was

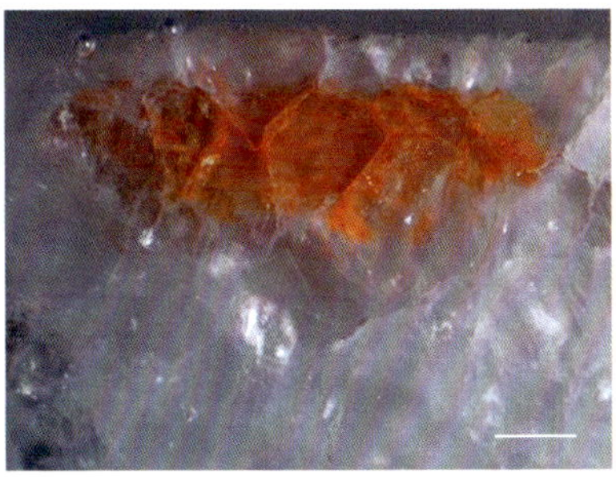

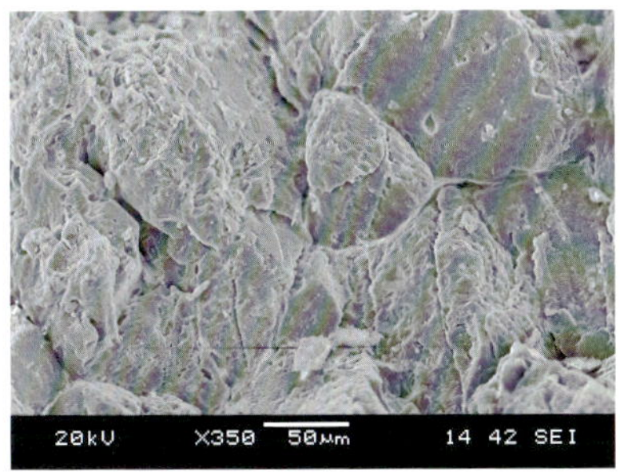

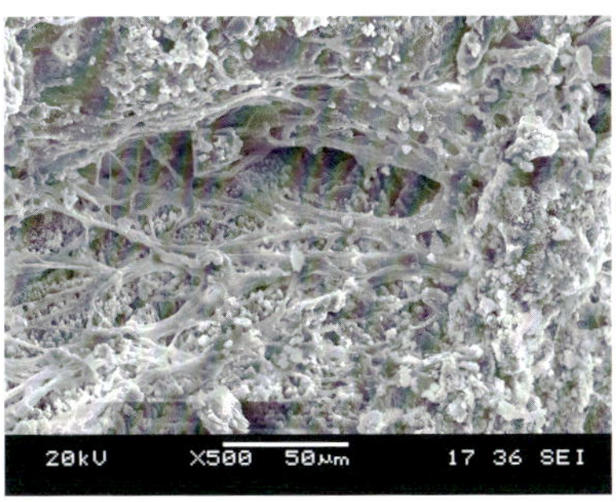

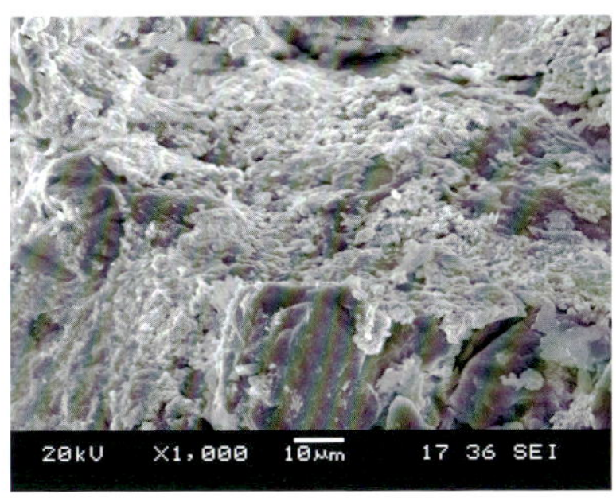

[4] Cross polished section of stained Carrara marble (bar = 250 μm). [5] Stained Carrara marble surface with evident decay processes (SEM). [6] Unstained marble surface with fungal presence (SEM). [7] Lead layer present on red stained surface (SEM).

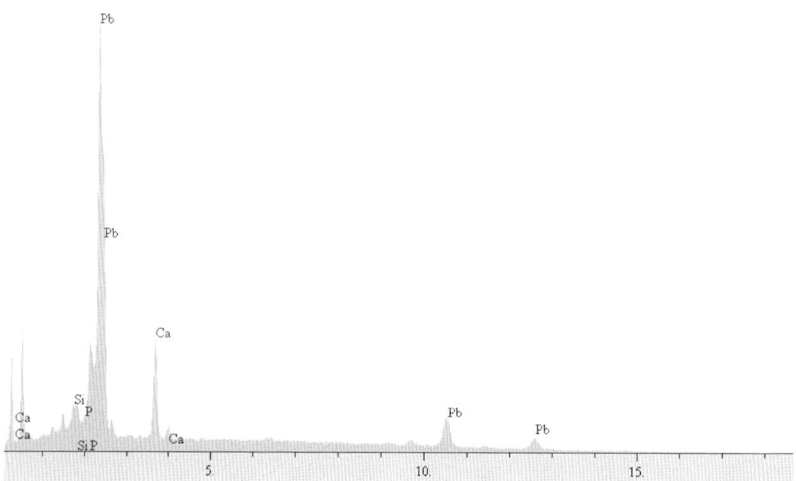

[8] EDS analysis of the layer shown in Fig. 7.

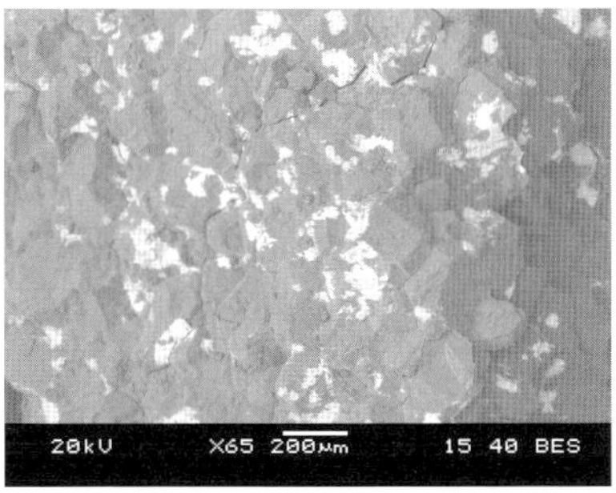

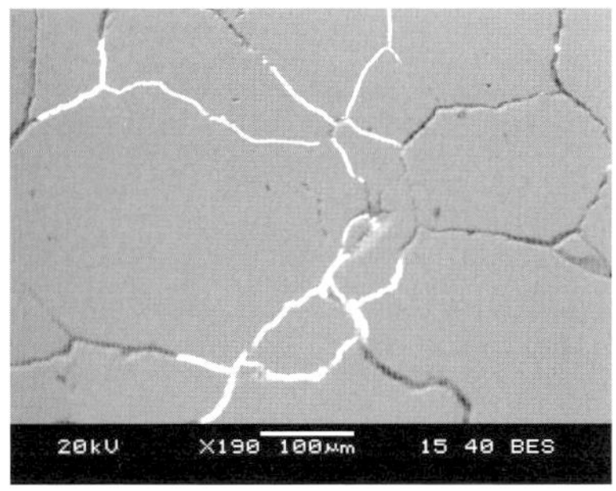

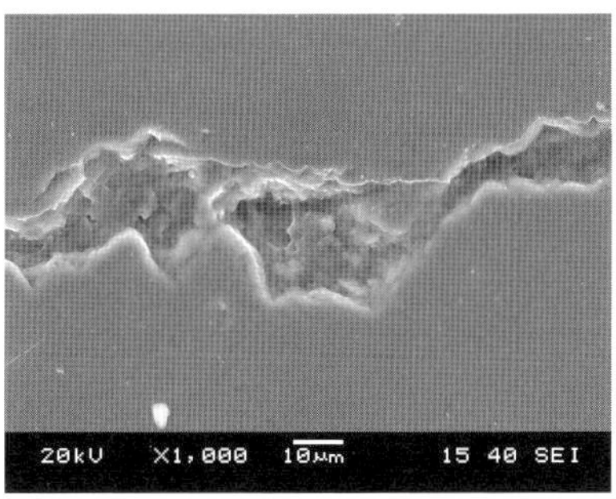

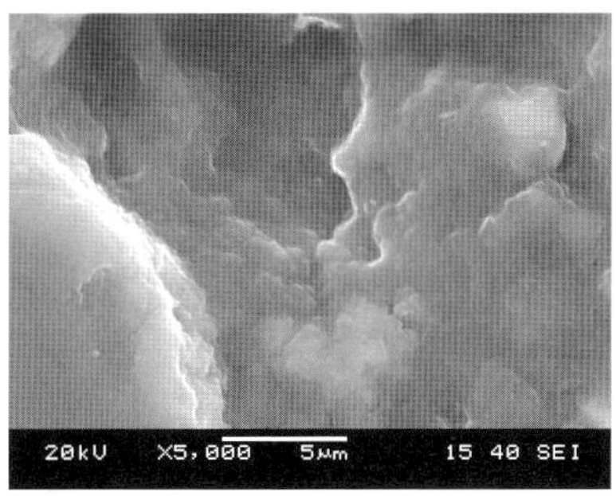

[9] Stained Carrara marble surface. The back scattered image shows the lead distribution (SEM). [10] Back scattered image of a cross polished section of a stained marble. The lead is located in the intergranular micro-fractures (SEM). [11] Cross polished section of a stained marble. Micro-fracture in which lead has been detected (SEM). [12] Detail of Fig. 11, in which microrganisms seem to be associated with lead (SEM).

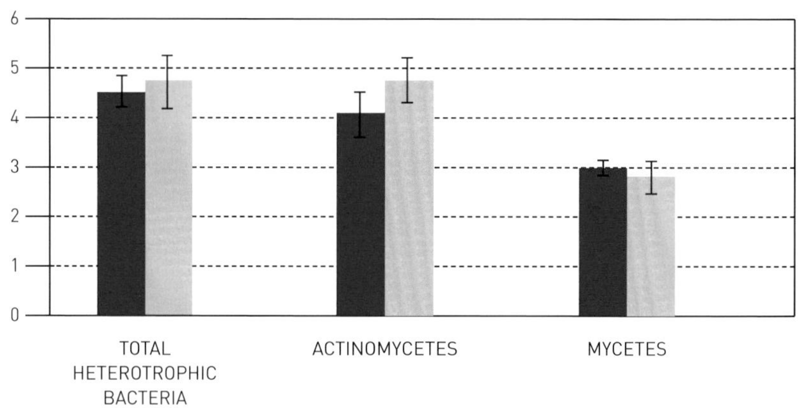

[13] Microbial counts of stained (green) and unstained (gray) areas.

the presence of red-pink-pigmented bacteria identified as *Micrococcus roseus* and *Micrococcus sp.*, which were isolated only from the stained samples. Among the other isolates identified were *Bacillus sp., Chriseomonas luteola, Staphylococcus sp.*, and a yeast, *Rhodotorula minuta;* these microorganisms had previously also been found in our other work (Salvadori et al., 1994; Sorlini et al., 1994; Zanardini et al., 1994 and 1997; Bruni et al., 1995). Other authors have reported the presence of pigmented bacteria on altered stones and wall paintings (Eppard et al., 1996; Oltmanns et al.,1999; Urzì et al., 2001).

Lead Enrichment Cultures The microbial growth of the consortia in the presence of different lead concentrations was monitored at time zero, and at 1 and 6 days by ATP content determinations, as this gives a good indication of the total viable biomass present in the enrichment broth-cultures. The microbial consortium in the stained areas grew up to a maximum lead concentration of 1500 mgl^{-1} and showed no growth at a lead concentration of 1700 mgl^{-1}. [FIG. 14] The comparison between the stained and unstained areas showed that, at higher lead concentrations (750, 1000, 1300, and 1500 mgl^{-1}), the consortium from the stained samples was significantly more resistant than that from the unstained samples. [FIG. 15]

Lead-Resistant Strains The growth of bacterial isolates in the presence of different lead concentrations was monitored at time zero, 1 day, and 6 days by ATP content determinations. Among the bacterial isolates, *Bacillus sp.* and *Staphylococcus sp.*, identified by API System, were resistant to the highest lead concentrations (1300 and 1500 mgl^{-1}). We had observed this previously in Zanardini et al., 1997, where none of the red-pigmented bacteria were shown to be lead-resistant.

Bacillus species are widely spread in the environments and were found in numerous artistic stoneworks (Salvadori et al., 1994; Zanardini et al., 1997; Daffonchio et al., 2000). Because they are spore-forming they can resist a wide range of temperatures, salinity, and pH and they survive dehydration and exposure to toxic compounds and UV radiation.

With regard to the *Staphylococcus* genus, the normal habitat of these microorganisms is human and animal skin, so it is difficult to comprehend its presence on the stone, but many microbiological tests, such as those for physiological, morphological, and growth characteristics, confirmed the API identification. Molecular analyses are in progress in order to confirm the classification of this bacterial strain. The heavy-metal resistance of *Staphylococcus sp.* strains has also been reported by other authors (Levinson et al., 1996; Roane, 1999).

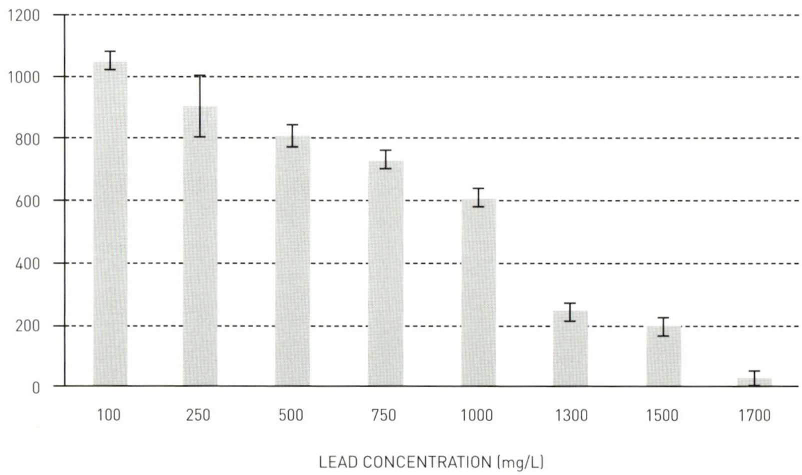

[14] Lead resistance of the microbial consortium present on stained samples.

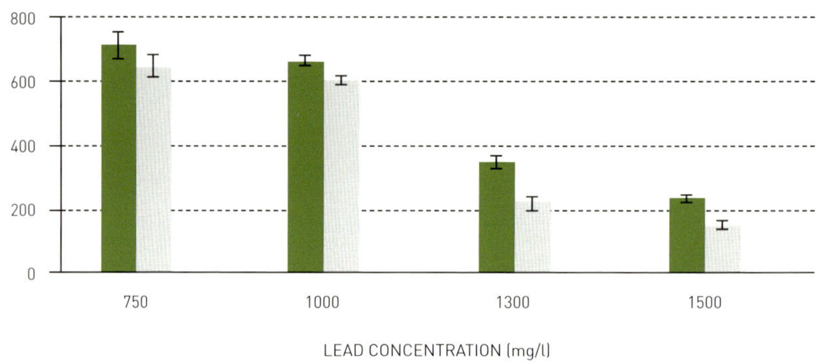

[15] Microbial growth in lead enrichment cultures of stained (green) and unstained (gray) samples.

Transmission electron microscopic observations performed on *Bacillus sp.* and *Staphylococcus sp.* grown at the higher lead concentrations showed that in both cases the lead was not accumulated in the cytoplasm [FIGS. 16 AND 17]; moreover, at the lead concentration of 1500 mgl^{-1} numerous *Staphylococcus* cells undergoing division were present, while for *Bacillus sp.* cellular lysis was observed.

MOLECULAR ANALYSES Molecular analyses were performed in order to confirm the identification of the *Bacillus sp.* lead-resistant strain and to study the microbial communities present in the stained and unstained areas.

Amplification with Primers Specific for Bacillus cereus *Group* The PCR procedure with primers specific for the Bacillus cereus group was performed on the bacterial isolate and on the enrichment cultures of the stained marble at lead concentrations of 750, 1000, 1300, and 1500 mgl^{-1}. The expected band of 749 bp was found, confirming the affiliation of this strain to the *Bacillus cereus* group. Moreover, the signal was also present at the various concentrations, confirming the lead resistance of this microorganism. [FIG. 18] The band at the lead concentration of 1500 mgl^{-1} is weaker compared to the others at the lower concentration, confirming the toxicity of the lead.

Our previous studies have already demonstrated the presence of *B. cereus* strains in marble samples from the Certosa of Pavia (Zanardini et al., 1997) and in other deteriorated stoneworks such as those from the Ca' d'Oro palace in Venice, where this isolate also showed a good halo-tolerance (Salvadori et al., 1994). Recently the occurrence of halo-tolerant/halophilic bacterial communities in altered stones has been reported (Saiz-Jimenez and Laiz, 2000).

DGGE Analysis Differential gradient gel electrophoresis analyses were performed both on stained and unstained marble samples and on the enrichment cultures at different lead concentrations. Our preliminary DGGE results showed differences among the profiles of the microbial communities in stained and unstained areas that had not been found using the traditional culture techniques. In fact, the microbial community of the stained samples showed a lower biodiversity that was probably due to the selective pressure of the lead present in the red areas. [FIG. 19]

In order to evaluate the biodiversity in microbial consortia in the presence of lead, DGGE analysis was performed on the enrichment cultures of both stained and unstained samples at lead concentrations of 750 and 1500 mgl^{-1}. [FIG. 20] The DGGE profiles obtained show that in the stained sample there are more bands as compared to those in the unstained sample at the highest lead concentrations, thus

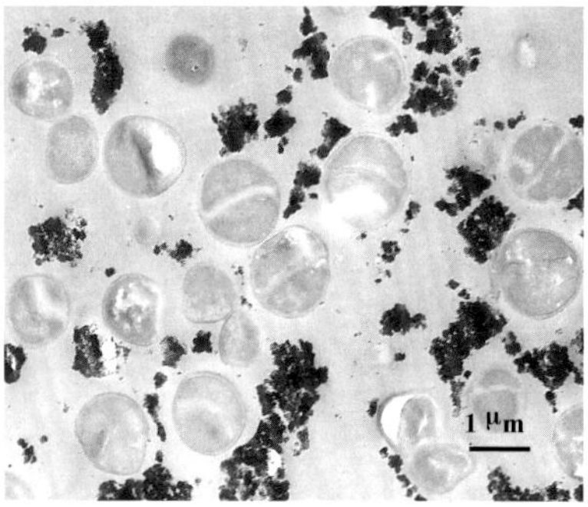

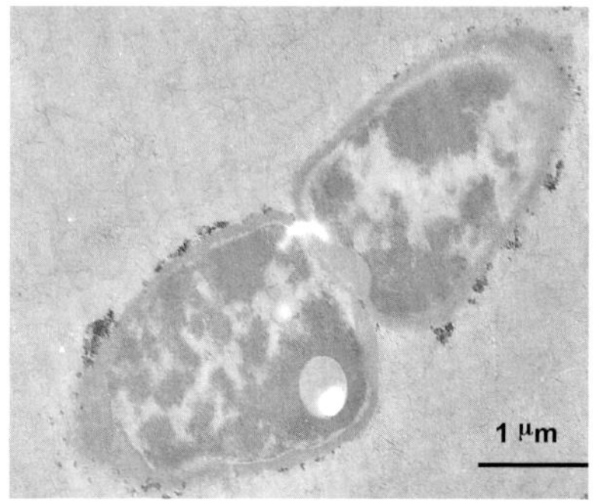

[16] Extracellular lead presence in cocci grown at lead concentration of 1500 mg/L (TEM). [17] Extracellular lead presence in rods grown at lead concentration of 1500 mg/L (TEM).

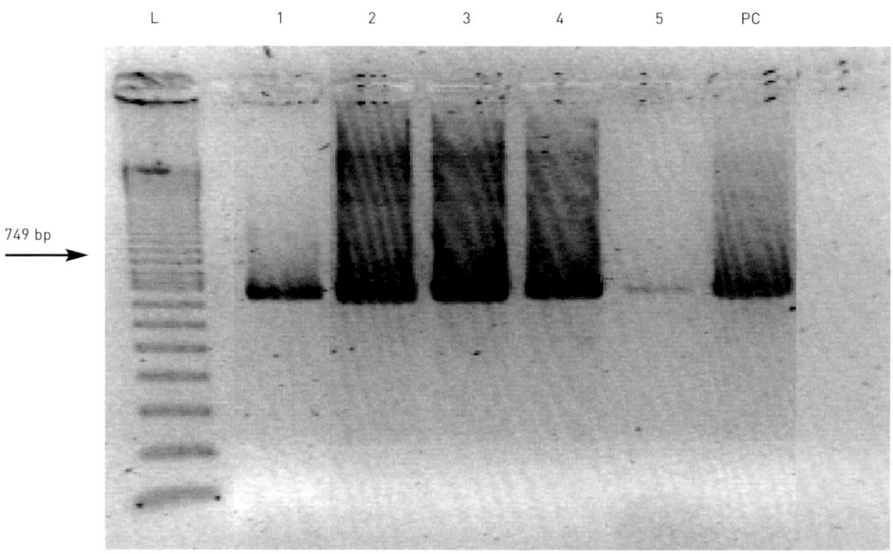

[18] Amplification with SG-749f and SG-749r primers. Lanes: L – ladder 100 bp, 1 – *Bacillus sp.* isolate, 2–5—750, 1000, 1300, 1500 mg/l; PC – Positive control; NC – Negative control.

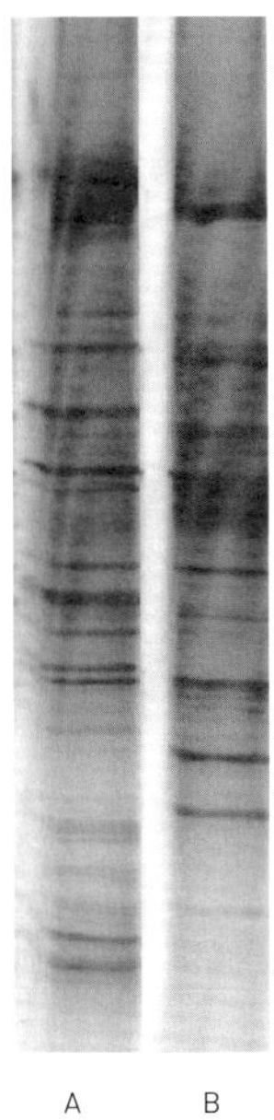

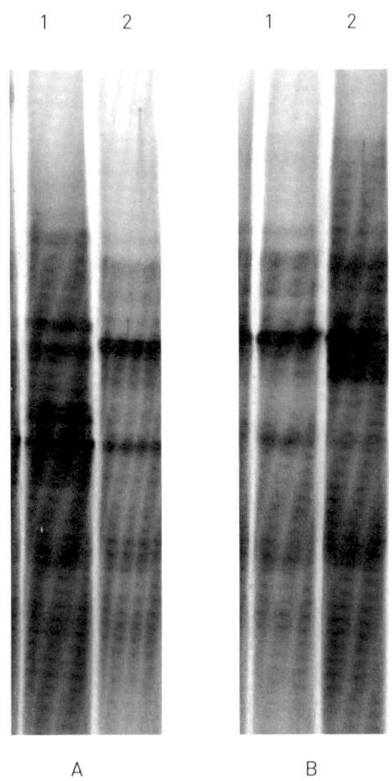

[19] DGGE profiles of marble samples: Lanes: A – unstained sample, B – stained sample. **[20]** DGGE profiles of enrichment cultures at different lead concentration. A – unstained sample, B – stained sample. Lanes: 1 – 750 mg/l; 2 – 1500 mg/l.

showing an adaptation to lead by the microbial community present in the red areas. [FIG. 20] More detailed analyses are still in progress.

With regard to the study of microbial diversity in stoneworks, some authors have recently applied DGGE analysis showing the importance of molecular tools in the detection of uncultivable microorganisms (Rölleke et al., 1996, 1998 and 1999; Gurtner et al., 2000).

CONCLUSIONS This research on the red stains from the Certosa of Pavia was performed four years after our previous study, permitting a more detailed analysis of the diffusion and penetration of the red coloration in the marble. The research showed that the stains are present both on the surface and up to a depth of about 1.5 mm, and are concentrated in the intergranular micro-fractures of the calcite crystals. No difference in the state of conservation of the material was observed between the stained and unstained areas.

In the red-stained marble areas, lead distribution was associated with the intergranular micro-fissures of the marble, confirming a correlation between the red coloration and the lead. In the red areas, no chemolithotrophic or photosynthetic microorganisms were found, nor were significant differences seen in the cultivable microbial counts between the stained and unstained samples.

Although red-pigmented microorganisms were found in the stains, their presence does not appear to be related either to the red coloration or to the presence of lead; indeed, none of the red-pigmented bacteria proved to be lead-resistant.

In stone deterioration pigmented microorganisms are frequently found, and the protective role of pigments such as carotenoids vis-à-vis UV radiation is well known. Lead-resistant isolates (*Bacillus sp.* and *Staphylococcus sp.*) were also found, and TEM observations revealed that these microorganisms do not accumulate this element in the cytoplasm.

With regard to our preliminary DGGE analysis, we observed the effect that the presence of lead had on the microbial communities on marble samples: The profiles showed a lower diversity in the stained areas as compared to the unstained areas. Further investigations by sequence analysis of DGGE-separated DNA fragments could probably help us characterize the microbial community present in the red areas.

In conclusion, even if the red stains seem to be strictly related to the presence of lead, the role of the lead-resistant microorganisms is as yet unclear, because it was not possible until now to determine whether they are directly involved in the lead oxidation or if they only tolerate its presence. Our future research will focus on this aspect of the problem.

REFERENCES

AUSUBEL, F.M., BRENT, R., KINGSTON R.E., MORE D.D., SEIDMAN, J.G., SMITH, J.A., and STRUHL, K. 1991. *Current protocols in molecular biology.* John Wiley and Sons, New York.

BASSI, M., FERRARI, A., REALINI, M., and SORLINI, C. 1986. Red stains on the Certosa of Pavia: A case of biodeterioration. *International Biodeterioration,* 22, 201–205.

BRUNI, S., CARIATI, F., BIANCHI, C.L., ZANARDINI, E., and SORLINI, C. 1995. Spectroscopic investigation of red stains affecting the Carrara marble façade of the Certosa of Pavia. *Archaeometry,* 37, 249–255.

C.N.R. CENTRE "G. BOZZA" 1988. *La Certosa di Pavia: passato e presente nella facciata della chiesa.* Consiglio Nazionale delle Ricerche, Roma.

DAFFONCHIO, D., BORIN, S., FROVA, G., GALLO, R., MORI, E., FANI, R., and SORLINI, C. 1999. A randomly amplified polymorphic DNA marker specific for the *B.cereus* group is diagnostic for *B. anthracis. Applied Environmental Microbiology,* 65, 1298.

DAFFONCHIO, D., BORIN, S., ZANARDINI, E., ABBRUSCATO, P., REALINI, M., URZÌ, C., and SORLINI, C. 2000. Molecular tools for the detection and the analysis of the diversity of bacteria on deteriorated stonework. In: Ciferri, O., Tiano, P., and Mastromei, G. (Eds.), *Of Microbes and Art: the role of microbial communities in the degradation and protection of cultural heritage.* Kluwer Academic/ Plenum Pub. Co., New York, pp.21–38.

EPPARD, M., KRUMBEIN, W.E., KOCH, C., RHIEL, E., STALEY, J.T., AND STACKEBRANDT, E. 1996. Morphological, physiological, and molecular characterisation of actinomycetes isolated from dry soil, rocks, and monument surfaces. *Archives of Microbiology,* 166, 12–22.

GIOVANNONI, S., MATTEINI, M., and MOLES, A. 1990. Studies and developments concerning the problem of altered lead pigments in wall painting. *Studies in Conservation,* 35, 21–25.

GUIDOBALDI, F., LAURENZI-TABASSO, M., and MEUCCI, C. 1984. Monumenti in marmo di epoca imperiale a Roma: indagini sui residui di trattamenti superficiali. *Bollettino d'Arte,* 24, 121–134.

GURTNER, C., HEYRMAN, J., PIÑAR, G., LUBITZ, W., SWINGS, J., and RÖLLEKE, S. 2000. Comparative analyses of the bacterial diversity on two different biodeteriorated wall paintings by dgge and 16s rRNA sequence analysis. *International Biodeterioration and Biodegradation,* 46, 229–239.

HAWRONSKYI, J.M., and HOLAH, J. 1997. ATP-a universal hygiene monitor. Trends in *Food Science and Technology,* 8, 79–84.

LEVINSON H.S., I. MAHLER, P. BLACKWELDER and HOOD T. 1996. Lead resistance and sensitivity of *Stahylococcus aureus. FEMS Microbiological Letters,* 145, 421–425.

LYALIKOVA, N.N., and PETUSHKOVA, Y.P. 1991. Role of microorganisms in the weathering of minerals in building stone of historical buildings. *Geomicrobiology Journal,* 9, 91–101.

MATTEINI, M., MOLES, A., MANGANELLI, C., FRATINI F., LAZZARONI L., and PECORARO, M. 1986. Schede di restauro: Fonte del Labirinto o di Fiorenza. In: *ODP Restauro,* Opus Libri, Florence: pp. 172–186.

NORMAL COMMITTEE 1988. Raccomandazioni Normal B 9/88: Microflora autotrofa ed eterotrofa: tecniche di isolamento in coltura. C.N.R.-I.C.R. Publishers, Rome.

OLTMANNS, T., FÖLSTER, H., and PETERSEN, K. 1999. Characterization of red-pigmented bacteria from wall paintings. In: *Extended Abstracts: Of Microbes and Art, an International Conference on Microbiology and Conservation*, Florence, pp. 86–90.

PETUSHKOVA, Y.P. and LYALIKOVA, N.N. 1986. Microbiological degradation of lead-containing pigments in mural paintings. *Studies in Conservation*, 31, 65–69.

PIÑAR, G., SAIZ-JIMENEZ, C., SCHABEREITER-GURTNER, C., BLANCO-VARELA, M.T., LUBITZ, W., and RÖLLEKE, S. 2001. Archaeal communities in two disparate deteriorated ancient wall paintings: detection, identification and temporal monitoring by denaturing gradient gel electrophoresis. *FEMS Microbiology Ecology*, 37 (1), 45–54.

REYNOLDS E.S. 1963. The use of lead citrate at high pH as an electron opaque stain in electron microscopy. *Journal of Cell Biology*, 17, 208.

ROANE T.M. 1999. Lead-resistance in two bacterial isolates from heavy metal—Contaminated soils. *Microbial Ecology*, 37, 218–224.

RÖLLEKE, S., MUYZER, G., WAWER, C., WANNER, G., and LUBITZ, W. 1996. Identification of bacteria in a biodegraded wall painting by denaturing gradient gel electrophoresis of PCR-amplified gene fragments coding for 16SrRNA. *Applied and Environmental Microbiology*, 62 (6), 2059–2065.

RÖLLEKE, S., WITTE, A., WANNER, G., and LUBITZ, W. 1998. Medieval wall paintings—a habitat for Archea: identification of archea by denaturing gradient gel electrophoresis (DGGE) of PCR-amplified gene fragments coding for 16S rRNA in a medieval wall painting. *International Biodeterioration and Biodegradation*, 41, 85–92.

RÖLLEKE, S., GURTNER, C., DREWELLO, U., DREWELL, R., LUBITZ, W., and WEISSMANN, R. 1999. Analysis of bacterial communities on historical glass by denaturing gradient gel electrophoresis of PCR-amplified gene fragments coding for 16S rRNA. *Journal of Microbiological Methods*, 36, 107–114.

SAIZ-JIMENEZ, C. and LAIZ, L. 2000. Occurrence of halotolerant/halophilic bacterial communities in deteriorated monuments. *International Biodeterioration and Biodegradation*, 46, 319–326.

SALVADORI, O., SORLINI, C., and ZANARDINI, E. 1994. Microbiological and biochemical investigations on stones of the Cà d'Oro façade (Venice). *Proceedings of Third International Symposium on the conservation of Monuments in the Mediterranean Basin*, Venice, June: pp. 343–347.

SCHABEREITER-GURTNER, C., PIÑAR, G., LUBITZ, W., and RÖLLEKE, S. 2001. An advanced molecular strategy to identify bacterial communities on art objects. *Journal Microbiological Methods*, 45, 77–87.

SORLINI, C., ZANARDINI, E., ALBO, S., PRADERIO, G., CARIATI, F., and BRUNI, S. 1994. Research on chromatic alterations of marbles from the fountain of Villa Litta (Lainate, Milan). *International Biodeterioration and Biodegradation*, 33, 153–164.

URZÌ, C., BRUSETTI, L., SALAMONE, P., SORLINI C., STACKEBRANDT E., and DAFFONCHIO D. 2001. Biodiversity of Geodermatophilaceae isolated from altered stones and monuments in the Mediterranean basin. *Environmental Microbiology*, 3 (7), 471–479.

ZANARDINI, E., BRUNI, S., CARIATI, F., RANALLI, G., and SORLINI, C. 1994. Investigation on the red spots present on the Carrara marble façade of Orvieto Cathedral. *Proceedings of Third International Symposium on the Conservation of Monuments in the Mediterranean Basin*, Venice, June: pp. 349–352.

ZANARDINI, E., ANDREONI, V., BORIN, S., CAPPITELLI, F., DAFFONCHIO, D., TALOTTA, P., SORLINI, C., RANALLI, G., BRUNI, S., AND CARIATI, F. 1997. Lead-resistant microorganisms from red stains of marble of the Certosa of Pavia, Italy and use of nucleic acid-based techniques for their detection. *International Biodeterioration and Biodegradation*, 40, 171–182.

C. MCNAMARA[1]

T.D. PERRY[2]

M. ZINN[2]

M. BREUKER[3]

R. MÜLLER[4]

G. HERNANDEZ-DUQUE[5]

R. MITCHELL[1*]

MICROBIAL PROCESSES IN THE DETERIORATION OF MAYA ARCHAEOLOGICAL BUILDINGS IN SOUTHERN MEXICO

1 Harvard University, Cambridge, MA 2 Federal Institute for Environmental Science and Technology (EAWAG), Duebendorf, Switzerland 3 Northeast Cultural Resources Center, Lowell, MA 4 Beth Israel Deaconess Medical Center and Harvard Medical School, Boston, MA 5 Universidad del Mayab, Merida, Yucatan, Mexico

* Corresponding author

ABSTRACT Carbonate stone has been used extensively in the construction of Maya archaeological buildings in southern Mexico. A diverse microbial community is normally found on historic limestone, and may contribute significantly to deterioration of the stone. A variety of methods have been used to investigate microbial deterioration of stone, including culturing, molecular analysis, SEM, and NMR. In this study, we evaluated the use of culture-based methods for assessment of ethanol as a biocide and began evaluation of a new method, micro-computed x-ray tomography (μCT), to quantify stone biodeterioration. Swab samples were collected from a Maya pyramid in Ek Balam, Yucatan, Mexico. The number of colony forming units (CFU) following growth on culture media and measurement of total bacteria, determined by DAPI-staining, were compared between two rooms, one that was cleaned with 50% ethanol 15 days prior to sample collection, and one that had not been cleaned for approximately one year. No significant differences in either CFU or total bacteria estimated by DAPI-staining were detected between the two rooms. These data suggest that use of culture-based methods may be appropriate for studies comparing the effects of biocidal treatments. Over 700 isolates were plated on media containing $CaCO_3$ to determine which isolates had the ability to dissolve $CaCO_3$ through production of acidic metabolites. Approximately 5% of our isolates had the capacity to dissolve $CaCO_3$. Aggressive $CaCO_3$ dissolving isolates were identified through analysis of the sequence of a portion of the 16S rDNA. Most isolates were bacteria normally found in soil and water. Analysis of stone samples by μCT was performed after treatment with the bacterial metabolites citric and oxalic acid. Deterioration of the stone treated with citric acid was evident, while there was minimal change in the stone treated with oxalic acid. The μCT analysis provided sensitive and high-resolution analysis of stone deterioration. We plan to expand use of this technique in our continuing investigation of the microbial processes involved in the deterioration of limestone in Maya sites in southern Mexico.

INTRODUCTION Carbonate stone has been used extensively in the construction of historic and culturally important structures (Gauri and Bandyopadhay, 1999). However, it is highly susceptible to physical, chemical, and biological deterioration. Evidence is accumulating that microbial metabolites frequently contribute to the deterioration of limestone structures (Koestler et al., 1985). The microbial community found on limestone includes fungi, algae, cyanobacteria, and heterotrophic and chemolithotrophic bacteria, both in surface biofilms (Mitchell and Gu, 2000) and within endolithic communities (Nienow and Friedman, 1993).

Microbial growth on stone is commonly analyzed using methods that rely on the ability of microorganisms to grow on culture media. In recent years, molecular techniques that exploit natural variation in DNA sequences have been developed to enumerate and identify microorganisms. Each of these methods has both advantages and disadvantages. For example, despite the simplicity and ease of use of culture-based techniques, they routinely detect less than 1% of microorganisms present in environmental samples (Jannasch and Jones, 1959). Molecular techniques avoid the selective bias of culturing, but require significant expertise, and may introduce other sources of error, such as formation of chimeric sequences during PCR (Wang and Wang, 1997).

The inability to detect the majority of organisms limits the usefulness of culture-based techniques for studies in which the goal is an accurate description of the complete microbial community. However, in studies where the goal is a comparison between treatments, analysis of a fraction of the community may be sufficient (Lemke et al., 1997). For the practicing conservator, culture-based methods may often be appropriate.

For the conservation scientist, evaluation of the effects of microorganisms on stone objects depends not only upon determination of the size and composition of the microbial community, but equally on its function (i.e., impact of the microorganisms on the stone). It is likely that community function will vary both with species composition and environmental conditions (von Canstein et al., 2002), making analysis of community function essential.

The impact of microorganisms on stone has been examined using a variety of techniques. Scanning electron microscopy (SEM), which is used to observe surface deterioration, fails to detect changes below the surface. Nuclear magnetic resonance measures changes in pore size distribution within stone, but gives no information about the microorganisms. Acoustic wave velocity provides information about subsurface discontinuities in stone, but results generally must be correlated with other measurements, such as change in stone mass.

X-ray computed tomography (CT) has been used extensively for the nondestructive visualization of objects in medical research (Berland, 1987) and paleontology (Conroy and Vannier, 1984), as well as in studies of art objects (Lal, 1992), soil (Phogat and Aylmore, 1989) and sediments (Holler and Kögler, 1990). Recently, CT has been used to examine stone samples from building materials (Bentz et al., 1995; Jacobs et al., 1995). We have adapted high-resolution three-dimensional micro-computed tomography (μCT) (Rüegsegger et al., 1996; Müller and Rüegsegger, 1997) to analyze the deterioration of carbonate stone in our study. The high resolution of this instrument (approximately 17 μm) provided a useful technique to quantitatively determine the effects of microbial metabolites.

The purpose of our investigation was to evaluate microbial deterioration of limestone used in Maya archaeological sites in southern Mexico. We carried out a microbiological analysis of two rooms in a pyramid in Ek Balam, and initiated an investigation of the application of micro-computed tomography to determine the processes involved in stone deterioration.

METHODS

Microbiological Analysis Samples were collected from two rooms at the top of the main pyramid in Ek Balam, Yucatan, Mexico. Room 43 was cleaned with 50% ethanol approximately 15 days prior to sample collection. Replicates (n = 3) were collected every 50 cm along a horizontal transect 10 cm from the base of the western wall. Room 44 was cleaned with 50% ethanol approximately one year prior to sample collection. Replicates (n = 3) were collected in a vertical transect every 50 cm from the base of the western wall to a height of 2 m. Two sets of samples were collected from biofilms on the stone surface using sterile cotton swabs. One set of swabs contained 0.03% NaCl, 15 % glycerol, and 0.01% Triton-X and were frozen for later analysis by culturing. A second set of swabs contained 1% formaldehyde and were stored at 4°c for epifluorescence microscopy.

Bacteria were cultured on modified nutrient agar (mNA) (Leff and Meyer, 1991), oligotrophic peptone-yeast extract-glucose-vitamin (PYGV) medium (Swings and Descheemaeker, 1995), alkaliphilic medium (pH 10.5) (Horikoshi, 1998), tryptic soy agar (TSA) with 10% $MgSO_4 \cdot 7H_2O$ (Saiz-Jimenez, 2000), and chemolithotrophic agar (Mitchell and Gu, 2000). Plates were incubated for three weeks at room temperature and colony forming units (CFU) were enumerated. Isolates were collected and transferred to a medium containing $CaCO_3$ (Di Bonaventura et al., 1999) to determine which organisms had the potential to degrade carbonate stone through the production of acidic metabolites.

Isolates capable of dissolving $CaCO_3$ were cultured in nutrient broth and DNA was extracted from cells using the DNeasy Tissue Kit (Qiagen Inc., Valencia, CA). A portion of the 16S rDNA gene was amplified using the primers 341f (5'-CCTACGGGAGGCAGCAG-3') (Muyzer et al., 1993) and 907r (5'-CCCCGT-CAATTCATTTGAGTTT-3') (Muyzer et al., 1995). PCR reactions were conducted in 50-µl volumes and contained 25 pmol of each primer, 0.2 mM of each dNTP, 5.0 µl of 10x PCR buffer (200 mM Tris-HCl pH 8.4, 500 mM KCl), 2 mM $MgCl_2$, 2 U of Taq DNA polymerase, 4 µl of template DNA from the extractions, and deionized water. PCR was performed in a Robocycler Gradient 96 thermocycler (Stratagene, La Jolla, CA) using a touchdown PCR program (Schabereiter-Gurtner, 2001). Products were cleaned using the QIAquick PCR Purification Kit (Qiagen Inc., Valencia, CA) and eluted in the final step with deionized water. PCR products were sequenced at the DNA Core Facility, Department of Microbiology and Molecular Genetics, Harvard Medical School, using the BigDye Terminators v2.0 Cycle Sequencing Ready Reaction kit and a Model 3100 Capillary DNA Sequencer (Applied Biosystems, Foster City, CA). Sequences were compared to aligned sequences contained in the Ribosomal Database Project using the Sequence Match function (Maidak et al., 2001).

Direct counts of bacteria from the biofilms were made on samples preserved with formaldehyde. Bacteria were stained for 5 min with 1.0 $\mu g\ ml^{-1}$ 4',6-diamidino-2-phenylindole (DAPI), concentrated by filtration (15 kPa vacuum) onto a 0.22-µm pore size black polycarbonate membrane (Poretics, Livermore, CA), and rinsed with 1.0 ml deionized water (Porter & Feig, 1980). Bacteria were enumerated using epifluorescence microscopy.

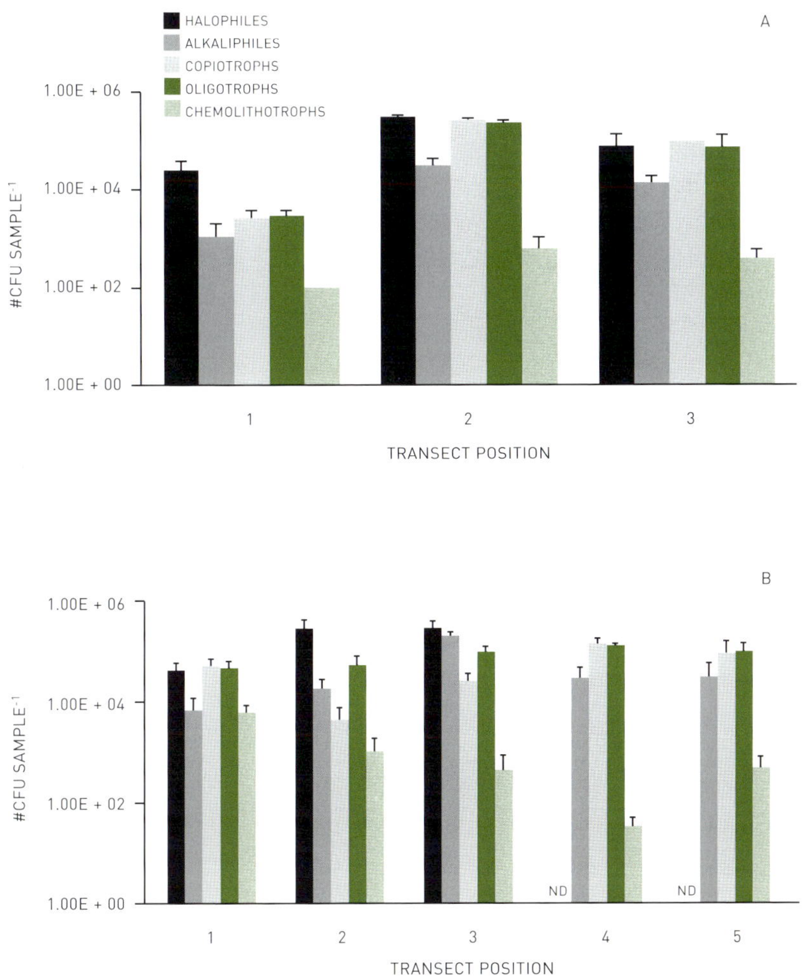

[11] Number of colony forming units (CFU) per sample from room 43 (A) and room 44 (B). Mean±se, n = 3. ND = No data.

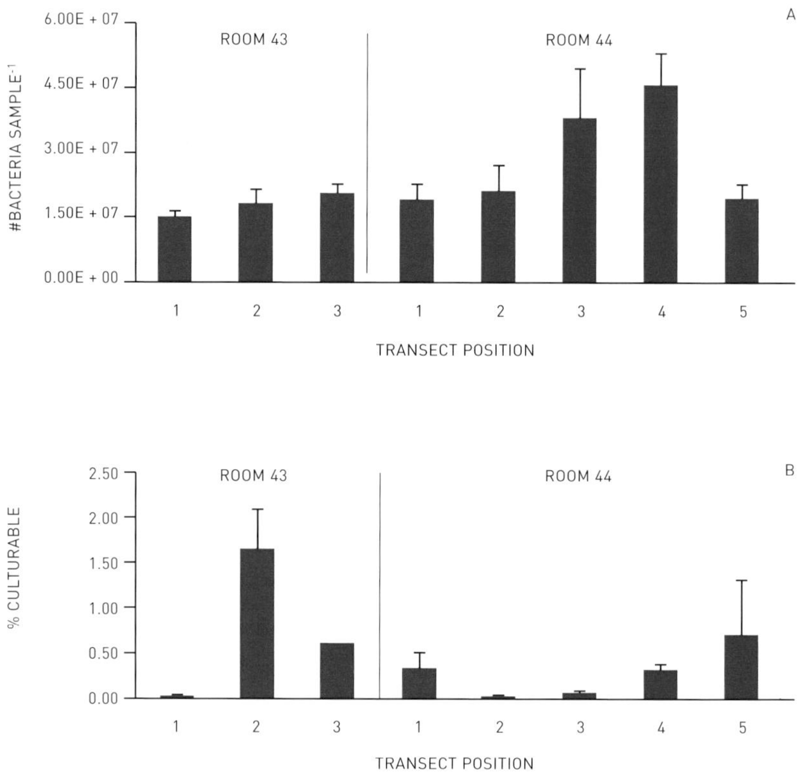

[2] Number of bacteria per sample determined by staining with DAPI (A). Percentage of bacteria that were culturable, calculated using number of CFU on the copiotrophic medium and the number of DAPI stained cells (B). Mean±se, n = 3 for both.

Statistical analysis was performed using SPSS 10.0 (SPSS Inc., Chicago, IL). The number of CFU on each medium, the number of cells stained with DAPI, and the percentage of bacteria that were culturable were compared between rooms using independent sample t-tests ($\alpha = 0.05$). The percentage of bacteria that were culturable was calculated using the number of DAPI stained cells and the number of CFU enumerated on the copiotrophic medium, and then arcsine-square root transformed prior to analysis.

Micro-Computed Tomography Analysis Prior to use, the composition of stone samples was analyzed using a Cameca MBX electron microprobe (average ± standard deviation composition of stone was: MgO 0.45±0.03, CaO 55.24±0.28, MnO 0.01±0.01, FeO 0.01±0.01, SrO 0.03±0.02, BaO 0.00±0.00, CO_2 43.86±0.23). The stone was cut into blocks (average weight: 18 g, average size: 4 × 4 × 1 mm), which were washed with distilled water, dried at 100°C for 15 hours, and cooled to room temperature in a desiccator. Samples were then glued with silicone adhesive (DAP Inc., Baltimore, MD) onto a plastic μCT sample holder, and cured for 24 h at 70°C.

Stone samples were scanned using the μCT as described below. Samples were then transferred to microcentrifuge tubes containing 1.0 ml 250 mM citric or oxalic acid. After a 4-day treatment samples were dried as described above, and scanned a second time.

Analysis was performed using a desktop μCT imaging system (μCT 20, Scanco Medical AG, Bassersdorf, Switzerland). The spatial resolution of the system was defined by the 10% contrast level in the modulation transfer function resulting in a spatial resolution of 28 μm (Rüegsegger et al., 1996). For each sample, a total of 51 to 102 micro-tomographic slices, using a slice increment of 17 μm, were acquired depending on the height of each sample (0.9–1.7 mm). Stone samples were segmented from background using a global thresholding procedure (Müller and Rüegsegger, 1997). Morphometric indices such as total stone volume and stone surface area were determined using direct three-dimensional morphometry (Hildebrand et al., 1999).

RESULTS Samples for microbiological analysis were collected from the limestone surface in two rooms in the main pyramid at Ek Balam, Yucatan, Mexico, and grown on five different media. Room 43 was cleaned with 50% ethanol 15 days prior to sample collection, while Room 44 had not been cleaned for approximately one year. Halophiles, copiotrophs, and oligotrophs were the most numerous organisms found in both rooms. [FIG. 1] Halophiles made up the largest group, with as many as 3×10^5 CFU per sample in some locations.

Alkaliphiles and chemolithotrophs were the least numerous bacteria in our samples. Alkaliphiles were always present in low numbers in room 43 (recently cleaned), but reached 2×10^5 CFU per sample at 100 cm in room 44 (cleaned one year earlier). The number of chemolithotrophs cultured from the limestone was smaller than any of the other groups of bacteria.

The total number of bacteria in our samples, determined by staining with DAPI, was at least two orders of magnitude higher than the number of CFU. [FIG. 2A] Consequently, the percentage culturable was < 1% in most samples. [FIG. 2B]

T-tests were used to compare the bacterial community between rooms. There were no significant differences in the number of CFU between rooms on any of the media (halophiles: $t = -0.621$, $p = 0.544$; copiotrophs: $t = 1.649$, $p = 0.116$; oligotrophs: $t = 0.557$, $p = 0.583$; alkaliphiles: $t = -1.449$, $p = 0.161$; chemolithotrophs: $t = -1.242$, $p = 0.227$). There were also no significant differences between rooms in the number of total bacteria ($t = -2.021$, $p = 0.056$), or in the percentage of bacteria that were culturable ($t = 1.529$, $p = 0.143$).

We determined the ability of the bacteria living on the limestone at the Ek Balam site to dissolve $CaCO_3$. Over 700 isolates were collected and plated on a medium containing $CaCO_3$. Of these organisms, 8.5% of isolates from room 43 and 2.9% of isolates from room 44 were capable of dissolving the $CaCO_3$. Ten isolates were selected for identification using analysis of a portion of the 16S rDNA. [TABLE 1] Molecular analysis showed that a wide range of bacteria capable of limestone dissolution are present in the Ek Balam limestone. Most of the isolates were closely related to gram-negative bacteria commonly found in soil and water. However, *Salmonella bongori* is normally found in lizard feces (Le Minor, 1984). Two uncommon gram-positive bacteria were present on the limestone: *Staphyloccocus sciuri*, which is part of the normal flora of rodents (Schleifer, 1986), and *Intrasporangium calvum*, which is closely related to the actinomycetes (Lechevalier, 1986). Isolates 86AA and 86AC were most similar to clone G21, an uncultured groundwater organism located within the *Acinetobacter baumanii* subgroup of the γ-proteobacteria.

Organic acids, such as citric and oxalic acids, are commonly produced by microorganisms on limestone (Di Bonaventura et al., 1999). We determined the potential effectiveness of micro-computed tomography (μCT) to detect biodeterioration of stone samples after treatment with the bacterial metabolites citric and oxalic acid. Deterioration occurred when the stone was exposed to citric acid. Dissolution

ISOLATE	ROOM	TAXA WITH MOST SIMILAR SEQUENCE	% SIMILARITY
8	43	*Intrasporangium calvum*	96.2
194	43	*Staphylococcus sciuri*	98.0
86AA	43	clone G21 within *Acinetobacter baumanii* subgroup	97.9
86AC	43	clone G21 within *Acinetobacter baumanii* subgroup	97.9
101A	43	*Chryseobacterium gleum*	84.3
153	43	*Pseudomonas fulva*	99.3
236B	43	*Pseudomonas putida*	98.5
342B2	44	*Pseudomonas putida*	99.3
539	44	*Salmonella bongori*	94.4
561	44	*Salmonella bongori*	94.0

[TABLE 1] Bacteria capable of dissolving $CaCO_3$ in biofilms from Maya limestone. Identification was based on similarity of 16S rDNA sequences to those in the Ribosomal Database Project II (Maidak et al., 2001).

was clearly visible in the two-dimensional slices. [FIG. 3A] Three-dimensional composite images showed alteration of the surface morphology. [FIG. 3B]

In contrast to the stone treated with citric acid, no dissolution was detected following exposure to oxalic acid. [FIG. 4A] However, there was a slight difference in appearance of the stone surface [FIG. 4B], probably the result of formation of a calcium oxalate surface layer.

In our study, we tested the capability of the μCT to quantitate deterioration of stone through extraction of morphometric data from scan images. Morphological changes in the limestone caused by citric acid resulted in a large increase in surface area due to increased roughness. [FIG. 5] However, dissolution of the stone caused an overall decrease in volume. There was little change in surface area or volume of the marble treated with oxalic acid.

DISCUSSION The high temperature and humidity of the climate in southern Mexico provide ideal conditions for formation of microbial biofilms on limestone structures. In order to protect the stone from biodeterioration, the ruins are periodically cleaned using ethanol as part of the normal maintenance program. An aqueous solution of ethanol (50%) is used for several reasons. It is inexpensive and easy to obtain. It evaporates readily and presents minimal danger to the stone, conservators, or local environment. Because biological growth is a recurrent and persistent problem in tropical environments, the goal of the maintenance program is not complete eradication or long-term prevention of growth. The treatment was designed to remove visible signs of microbial growth, such as spotting or discoloration (A. Alonso, personal communication).

We examined the effect of 50% ethanol on the bacterial community of the limestone by comparing a surface that was recently cleaned to one that had not been cleaned for approximately one year. Differences between the culturable bacterial communities of the two rooms could not be detected. Similarly, analysis of the total bacterial community using DAPI staining indicated that there was no difference between the size of the bacterial communities in the two rooms.

The popularity of molecular techniques has been accompanied by a general disparagement of conventional culturing techniques in environmental microbiology (e.g., Brock, 1987), primarily because culture methods frequently detect less than 1% of the community. In our study, conventional culture on growth media yielded less than 1% of the bacterial community. Yet, our data suggest that cultural analysis may be an appropriate method to compare treatments, since it yielded similar conclusions to those reached through analysis of the size of the total community. For

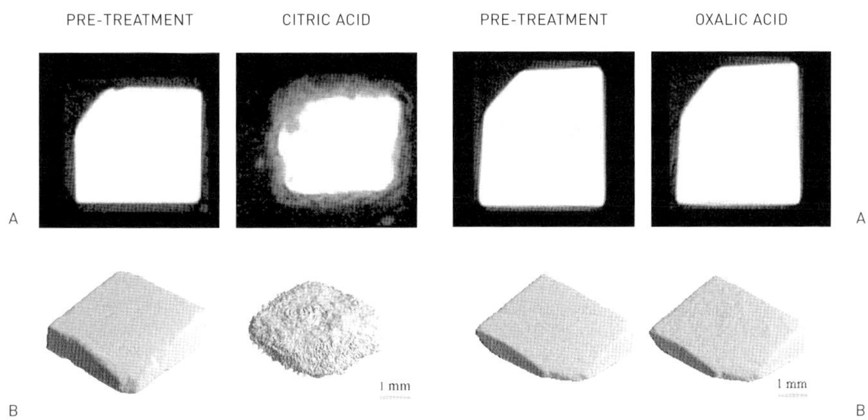

[3] Frontal sections of the stone sample before and after a 4-day treatment with 250 mM citric acid (A). Three-dimensional composite images of the stone sample before and after a 4-day treatment with 250 mM citric acid (B). [4] Frontal sections of the stone sample before and after a 4-day treatment with 250 mM oxalic acid (A). Three-dimensional composite images of the stone sample before and after a 4-day treatment with 250 mM oxalic acid (B).

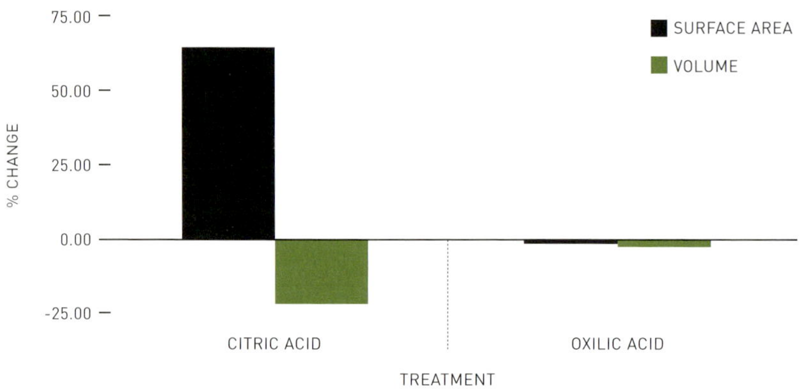

[5] Percentage change in surface area and volume of stone samples treated for 4 days with 250 mM citric or oxalic acid.

conservators interested in the effectiveness of cleaning methods, knowledge of the impact of the treatment on a portion of the community (i.e., the culturable portion) may be sufficient to evaluate its effectiveness.

More than 5% of isolates cultured from limestone in Maya archaeological buildings were capable of dissolving $CaCO_3$ when grown on either carbon- or salt-rich media. Molecular analysis of some of these bacteria indicated that the $CaCO_3$-dissolving community is comprised of both common soil and aquatic bacteria, and bacteria normally present in the local fauna. Our investigation gave no indication of their activity under natural conditions. However, our research suggests that a significant portion of the biofilm community has the potential to damage limestone structures, and therefore warrants further investigation.

Because the phenotype of microorganisms may change in response to environmental conditions, simple identification of the organisms may supply minimal information about detrimental effects on the stone. While a variety of methods have been used to evaluate microbial deterioration of stone, each has significant disadvantages. We have begun to test the application of a new technique, micro-computed tomography (μCT), to provide highly accurate quantitative data describing deterioration of limestone.

While x-ray computerized tomography has been used in a number of fields, few studies have attempted to adapt the technique to stone building materials (Bentz et al., 1995; Jacobs et al., 1995). The methodology used in our study provides significant advantages over previous studies. The μCT instrument used in our investigation yielded resolution of 28 μm with a slice increment of 17 μm (compared to 80 μm and 1 mm, respectively, for the instrument used by Jacobs et al., 1995). In addition to enhanced resolution, the smaller slice increments enabled the use of much smaller samples. This is an important consideration in the study of cultural heritage materials, where there are frequently limits on the size of samples that can be obtained. Bentz et al. (1995) were able to obtain excellent resolution (< 10 μm) using the X-2B x-ray microtomography beamline of the National Synchrotron Light Source at Brookhaven National Laboratory in New York. However, this is a large facility and similar instruments are not readily available to small research laboratories. The instrument used in our study was able to achieve comparable resolution using a widely available bench-top unit.

Investigations of limestone deterioration could benefit significantly from data obtained using μCT analysis. Citric acid caused severe deterioration of limestone. However, the effects of microbial metabolites on carbonate stone are complex, in that not all acids cause immediate dissolution of the stone. Oxalate has a high affinity toward calcium, and the acid, while attacking the crystalline structure of $CaCO_3$, forms a more insoluble salt (calcium oxalate) with the released calcium from the stone. Oxalic acid produced by microorganisms forms whewellite or weddellite, calcium oxalate mono or dihydrate minerals, respectively, when calcium ions are available (del Monte et al., 1987). These calcium oxalate minerals are relatively insoluble (when compared to calcium carbonate), and it has been suggested that a calcium oxalate layer may protect stone from further deterioration (Di Bonaventura et al., 1999). Micro-CT analysis has the potential to provide information about both the surface layer formation and events occurring beneath the surface leading to deterioration.

Our preliminary analysis indicates that μCT is able to detect changes in stone samples caused by microbial metabolites. We have also identified a number of isolates that are capable of producing sufficient amounts of acid metabolites to dissolve $CaCO_3$ in culture, indicating that they have the potential to cause deterioration of carbonate stone in situ. The combination of molecular techniques and quantitative measurement of changes in stone using μCT may lead to valuable insights into the biodeterioration of carbonate stone, both in Maya archaeological building materials and in limestone used in other historic sites.

ACKNOWLEDGMENTS We thank M. Helms for comments and suggestions during this project. A. Alonso assisted in sample collection at Ek Balam and provided valuable information about the site. This research was supported by grants to Harvard University from the National Science Foundation (BES-9906337) and from the Samuel H. Kress Foundation. Additionally, this work was supported by a grant from the Consejo Nacional de Ciencia y Tecnologia (CONACYT) to the Universidad del Mayab.

REFERENCES

ALESIANI, M., CAPUANI, S., CURZI, F., MARAVIGLIA, B. 2000. Evaluation of stone pore size distribution by means of NMR. In: Fassina, V. (Ed.), *9th International Congress on Deterioration and Conservation of Stone*, Venice, Italy, Elsevier, New York, pp. 579–585.

BENTZ, D.P., MARTYS, N.S., STUTZMAN, P., LEVENSON, M.S., GARBOCZI, E.J., DUNSMUIR, J., SCHWARTZ, L.M. 1995. X-ray microtomography of an ASTM C109 mortar exposed to sulfate attack. In: Diamond, S., Mindess, S., Glasser, F.P., Roberts, L.W., Skalny, J.P., Wakeley, L.D. (Eds.), *Microstructure of Cement-based Systems/bonding and Interfaces in Cementitious Materials.* Materials Research Society, Pittsburgh, PA, pp. 77–82.

BERLAND, L.L. 1987. *Practical CT: Technology and Techniques.* Raven Press, New York.

BROCK, T.D. 1987. The study of microorganisms in situ: Progress and problems. *Symposia of the Society for General Microbiology* 41, 1–17.

CONROY, G.C., VANNIER, M.W. 1984. Noninvasive three-dimensional computer imaging of matrix-filled fossil skulls by high resolution computed tomography. *Science*, 226, 456–458.

DEL MONTE, M., SABBIONI, C., ZAPPIA, G. 1987. The origin of calcium oxalates on historical buildings, monuments and natural outcrops. *The Science of the Total Environment*, 67, 17–39.

DI BONAVENTURA, M.P., DEL GALLO, M., CACCHIO, P., ERCOLE, C., LEPIDI, A. 1999. Microbial formation of oxalate films on monument surfaces: bioprotection or biodeterioration? *Geomicrobiology Journal*, 16, 55–64.

GAURI, K.L., BANDYOPADHYAY, J.K. 1999. Carbonate stone: Chemical Behavior, Durability, and Conservation. John Wiley and Sons Inc., New York.

HILDEBRAND, T., MÜLLER, R., LAIB, A., DEQUEKER, J., RÜEGSEGGER, P. 1999. Direct 3-D morphometric analysis of human cancellous bone: Microstructural data from spine, femur, iliac crest, and calcaneus. *Journal of Bone and Mineral Research*, 14, 1167–1174.

HOLLER, P., KÖGLER, F.-C. 1990. Computer tomography: a nondestructive, high resolution technique for investigation of sedimentary structures. *Marine Geology*, 91, 263–266.

HORIKOSHI, K. 1998. Alkaliphiles. In: Horikoshi, K., Grant, W.D. (Eds.), *Extremophiles: Microbial Life in Extreme Environments*, Wiley, New York, pp. 155–180.

JACOBS, P., SEVENS, E., KUNNEN, M. 1995. Principles of computerized tomography and applications to building materials. *The Science of the Total Environment*, 167, 161–170.

JANNASCH, H.W., JONES, G.E. 1959. Bacterial populations in seawater as determined by different methods of enumeration. *Limnology and Oceanography*, 4, 128–139.

KOESTLER, R.J., CHAROLA, A.E., WYPYSKI, M., LEE, J.J. 1985. Microbiologically induced deterioration of dolomitic and calcitic stone as viewed by scanning electron microscopy. In: Felix, G. (Ed.), *Vth International Congress on Deterioration and Conservation of Stone.* Presses Polytechniques Romandes, Lausanne, pp. 617–626.

LAL, U.S. 1992. Computer assisted tomography (CAT) for art objects. *Studies in Museology*, 26, 1–4.

LE MINOR, L. 1984. Section 5: Facultatively anaerobic gram-negative rods. Genus III: Salmonella. In: Krieg, N.R., Holt, J.G. (Eds.), *Bergey's Manual of Systematic Bacteriology*, Vol. 1. Williams and Wilkins, Baltimore, pp. 427–458.

LECHEVALIER, H.A. 1986. Section 17: Nocardioforms. In: Sneath, P.H.A., Mair, N.S., Sharpe, M.E., and Holt, J.G. (Eds.), *Bergey's Manual of Systematic Bacteriology* Volume 2. Williams and Wilkins, Baltimore, pp. 1458–1579.

LEFF, L.G., MEYER, J.L. 1991. Biological availability of dissolved organic carbon along the Ogeechee river. *Limnology and Oceanography*, 36, 315–323.

LEMKE, M.J., BROWN, B.J., LEFF, L.G. 1997. The response of three bacterial populations to pollution in a stream. *Microbial Ecology*, 34, 224–231.

MAIDAK, B.L., COLE, J.R., LILBURN, T.G., PARKER, C.T. JR., SAXMAN, P.R., FARRIS, G.M., OLSEN, G.J., SCHMIDT, T.M., TIEDJE, J.M. 2001. The RDP-II (Ribosomal Database Project). *Nucleic Acids Research*, 29, 173–174.

MITCHELL, R., GU, J.-D. 2000. Changes in the biofilm microflora of limestone caused by atmospheric pollutants. *International Biodeterioration and Biodegradation*, 46, 299–303.

MÜLLER, R., RÜEGSEGGER, P. 1997. Microtomographic imaging for the nondestructive evaluation of trabecular bone architecture. *Studies in Health Technology and Information*, 40, 61–79.

MUYZER, G., DE WAAL, E.C., UITTERLINDEN, A.G. 1993. Profiling of complex microbial populations by denaturing gradient gel electrophoresis analysis of polymerase chain reaction-amplified genes coding for 16S rRNA. *Applied and Environmental Microbiology*, 59, 695–700.

MUYZER, G., TESKE, A., WIRSEN, C.O., JANNASCH, H.W. 1995. Phylogenetic relationships of Thiomicrospira species and their identification in deep-sea hydrothermal vent samples by denaturing gradient gel electrophoresis of 16S rDNA fragments. *Archive fur Microbiology*, 164, 165–172.

NIENOW, J.A., FRIEDMANN, E.I., 1993. Terrestrial lithophytic (rock) communities. In: Friedmann, E.I. (Ed.), *Antarctic Microbiology*, Wiley-Liss, Inc., New York, pp. 343–412.

PAPIDA, S., MURPHY, W., MAY, E. 2000. The use of sound velocity determination for the non-destructive determination of physical and microbial weathering of limestones and dolomites. In: Fassina, V. (Ed.), *9th International Congress on Deterioration and Conservation of Stone*, Venice, Italy, Elsevier, New York, pp. 609–617.

PHOGAT, V.K., AYLMORE, L.A.G. 1989. Evaluation of soil structure by using computer assisted tomography. *Australian Journal of Soil Research*, 27, 313–323.

PORTER, K.G., FEIG, Y.S. 1980. The use of DAPI for identifying and counting aquatic microflora. *Limnology and Oceanography*, 25, 943–948.

RÜEGSEGGER, P., KOLLER, B., MÜLLER, R. 1996. A microtomographic system for the nondestructive evaluation of bone architecture. *Calcified Tissue International*, 58, 24–29.

SAIZ-JIMENEZ, C., LAIZ, L., 2000. Occurrence of halotolerant/halophilic bacterial communities in deteriorated monuments. *International Biodeterioration and Biodegradation*, 46, 319–326.

SCHABEREITER-GURTNER, C., PINAR, G., LUBITZ, W., ROLLEKE, S. 2001. An advanced molecular strategy to identify bacterial communities on art objects. *Journal of Microbiological Methods*, 45, 77–87.

SCHLEIFER, K.H. 1986. Section 12: Gram-positive cocci. In: Sneath, P.H.A., Mair, N.S., Sharpe, M.E., and Holt, J.G. (Eds.), *Bergey's Manual of Systematic Bacteriology*, Volume 2. Williams and Wilkins, Baltimore.

SWINGS, J., DESCHEEMAEKER, P. 1995. The role of heterotrophic bacteria in the degradation of stones. In: Interactive physical weathering and bioreceptivity study on building stones, monitored by computerized x-ray tomography as a potential non-destructive research tool, *Protection and Conservation of the European Cultural Heritage Research Report* No. 2, 1995.

VON CANSTEIN, H., KELLY, S., LI, Y., WAGNER-DÖBLER, I. 2002. Species diversity improves the efficiency of mercury-reducing biofilms under changing environmental conditions. *Applied and Environmental Microbiology*, 68, 2829–2837.

WANG, G.C.-Y., WANG, Y. 1997. Frequency of formation of chimeric molecules as a consequence of PCR coamplificationof 16S rRNA genes from mixed bacterial genomes. *Applied and Environmental Microbiology*, 63, 4645–4650.

J. KARBOWSKA-BERENT

MICROBIODETERIORATION OF MURAL PAINTINGS: A REVIEW

Nicolas Copernicus University, Toruń, Poland

ABSTRACT Mural paintings are frequently subject to changes: bleaching, colored stains and deposits associated with peeling and flaking off of the paint layer, and changes in color of susceptible pigments. These alarming phenomena are observed worldwide. The results of a considerable number of studies mostly implicate bacteria, hyphal fungi, and algae in these developments. Among bacteria both heterotrophic and autotrophic organisms, including thiobacilli and nitrifiers, are observed. In addition, halophilic bacteria, capable of growing on salt-affected mural paintings; actinomycetes; and cyanobacteria are often isolated. Among the hyphal fungi the most frequently mentioned species are *Acremonium spp.*, *Aspergillus versicolor*, *A. niger*, *Chaetomium globosum*, *Cladosporium herbarum*, *C. sphaerospermum*, *Penicillium brevi-compactum*, and *Phoma glomerata*. The algae occurring on mural paintings are mostly unicellular green algae of the family *Chlorophyceae*.

Specific ecological conditions prevail on mural paintings; these are characterized by a low number of carbon and energy sources, varying moisture levels, and mostly low temperatures. The microorganisms occurring on the surface of the mural painting form a specific and complex ecosystem, called biofilm, consisting of populations of many species. The composition of microflora on a mural painting depends, first, on the climatic conditions of its environment, as well as on the materials used in the painting itself and for its conservation. For oligotrophic microorganisms, the mineral and organic components in the dust and soot settling on mural paintings over centuries may provide sufficient nutrition. Possible sources of carbon on mural paintings are binders, such as casein and wheat paste, some pigments, as well as synthetic polymers used for conservation. The consumption of the binder by microorganisms leads to its impairment and, consequently, to the flaking off of pigment particles devoid of any binding factor. Another important factor leading to the destruction of mural paintings is organic acids excreted by microorganisms growing on them. These acids dissolve the painting's mineral components, especially calcium carbonate. Some of them can form complexes with metal ions taken from the mineral components of the painting, changing its composition.

Over the course of time quantitative and qualitative changes in the microflora on mural paintings occur under the influence of abiotic factors, e.g., cyclic diversification depending on the season of the year, or changes caused by the microorganisms themselves. The pioneer microorganisms, which are first to settle on the mural painting, change the properties of the habitat, making it more suitable for populations of more and more species. The pioneer species effect these changes through their development, i.e., the secretion of metabolites, slimes, the decomposition of the particular components, and hindering water evaporation. The destructive activity of the microorganisms in the biofilm on the painting is not limited to its surface; they penetrate deep inside the paint layer and the plaster, and their destructive effect can be even stronger in the deeper layers than on the surface.

INTRODUCTION Mural paintings are frequently subject to changes: vast whitenings, colored stains and deposits, associated with peeling and flaking off of the paint layer, and, not seldom, change in pigment color. These alarming phenomena, observed worldwide, have induced art conservators to join with microbiologists in a common effort to find the causes. The results of a considerable number of studies point to the participation of various groups of microorganisms in these developments (Strzelczyk, 1981a, b; Agrawal et al., 1989; Rebrikova, 1993; Garg et al., 1995; Ciferri, 1999). The authors mention both autotrophic (photosynthesizing and chemosynthesizing) and heterotrophic bacteria, as well as hyphal fungi, and algae. Lichens, and even insects, are also reported to live on mural paintings. Observations and studies suggest that the occurrence and the composition of microflora on a mural painting depend, in the first place, on the climatic conditions of its environment, as well as on the materials used in the painting itself or for its conservation.

MURAL PAINTINGS—MATERIALS AND PAINTING TECHNIQUES The term "mural painting" includes all painting techniques used in decorating walls. Generally, a mural painting consists of three layers, each with a different function, thickness, and technique, as well as technology of production. These are:

• The construction support—most frequently a brick wall, less so a wall made of stone (e.g., limestone or marble), wood, and nowadays also concrete.

• Undercoat—mural paintings most frequently are laid on mineral mortar, which consists of lime and sand mixed with water, sometimes with an addition of gypsum and some fillers, such as ground brick, pumice, or marble powder. There are generally two layers in the mortar: the inner one, more rough and thick *arriciato*, and the outer one, thinner and smoother, on which the painting is done—*intonaco*. In Byzantine-Russian paintings, organic fillers, mainly vegetable fibers, were used instead of sand in order to extend plaster drying time. Those were: cut-up straw, flax, sawdust, jute, or hemp fibers. Mural paintings on wooden support were laid on glue undercoat (mostly animal glue) with an addition of chalk.

• The paint layer—made in various techniques, using both mineral binders (e.g., *fresco buono* and *fresco secco*) and organic binders (mainly in the casein and glue techniques, less frequently in the tempera or oil techniques). The original painting layer can later be covered with numerous layers of overpainting and lime wash. The durability and state of preservation of the painting depend to a large extent on the technique and the materials used in it. Among the most durable are paintings made in the *fresco buono* technique, which has been used ever since antiquity.

In that technique pigments mixed with water were used to paint on fresh, damp mortar consisting of slaked lime (calcium hydroxide) and pure, dry sand, mixed in varying proportions. The binding of pigments was effected through the penetration of the dissolved calcium hydroxide from the mortar through the capillaries between the pigment grains onto the surface, and its transformation was effected by carbon dioxide from the air into microcrystalline, transparent calcium carbonate, according to the reaction:

$$Ca(OH)_2 + CO_2 = CaCO_3 + H_2O.$$

The calcium carbonate resulting from the carbonization bound the pigment particles with one another and with the surface while at the same time covering them with a glassy, transparent coat. Thanks to the chemical binding of the pigments the paintings are closely united with the mortar, forming a hard, waterproof layer.

The fresco support was not only damp but also strongly alkaline, which is why only alkali-resistant pigments could be used. The most adequate were considered to be mineral pigments, such as calcium white ($CaCO_3$), ochres ($Fe_2O_3 \cdot 3H_2O$) of various colors, oxide red (Fe_2O_3), green earth ($FeSiO_3$), azurite ($2CuCO_3 \cdot Cu(OH)_2$), and smalt ($CoO \times nK_2SiO_3$), as well as organic blacks, viz. charcoal and vine black and soot. This kind of mural painting surface was then nearly completely devoid of organic compounds, which constitute a potential medium for microorganisms.

Besides the *fresco buono*, several other techniques are found, which are derived from it. One of them is the *fresco secco* technique, i.e., painting on dry mortar. Pigments were mixed in milk of lime instead of water. The principle of pigment consolidation was the same as in *fresco buono*, but because of inferior penetration of the binder the paintings created with this technique are less durable.

In the casein technique the pigment was mixed in a weak solution of lime casein. Of all organic binders casein best tolerates an alkaline environment. Lime casein, which was obtained from skim curds mixed with slaked lime, reacts with excess calcium to produce calcium caseinate, a very durable compound that is relatively resistant to microbial decay. In this technique too, it was possible to use only specific fresco pigments, complemented with zink white (ZnO).

Lime mortar after complete binding (ca. one year) is pH neutral on the surface, and it was possible to paint on it in the glue and tempera techniques. In the glue technique, the binder used was protein glue obtained by boiling animal hides; in the tempera technique, egg emulsions, egg and linseed oil emulsions, or emulsions made of egg, varnish, and carbonic lotion can be used as binders (Ślesiński, 1983a, b). All those binders are susceptible to microbiodeterioration under conditions favorable to the development of microflora.

Even this brief review shows that the susceptibility of mural paintings to microbiodeterioration depends on the technique with which they are made. Most components of mural paintings are inorganic compounds. On the other hand, the development of microflora is favored by the presence of organic compounds, such as casein, egg binders, and some substances used in conservation, as well as dust and soot accumulating on wall surfaces for centuries. Those slender sources of energy, carbon, and nitrogen are sufficient for many oligotrophic microorganisms to develop. Microbial attacks are intensified by the presence of chemical pollutants in the air, mainly SO_2, ammonia, and nitrogen oxides, a presence that has increased with the progress of civilization. A correlation has been observed between the increase in air pollution and the occurrence of chemolithotrophic bacteria, particularly thiobacilli and nitrifiers. Aggressive chemical compounds cause microchinks to develop in the plaster, where microorganisms find better development conditions than on its smooth surface.

A review of publications on the subject leads to the conclusion that on a damp wall every representation will be deteriorated, irrespective of the technique with which it was created. Every ecological niche of the wide range that develop meets the requirements of at least one species of microorganism. These requirements involve adequate moisture of the substratum, temperature, and pH.

CONDITIONS FAVORABLE TO THE DEVELOPMENT OF MICROFLORA ON MURAL PAINTINGS

Specific ecological conditions prevail on mural paintings; these are typically a low number of carbon and energy sources, mostly low temperatures, and varying moisture levels. Petersen et al. (1995) found that the distribution of microorganisms depended more on the degree of moistness of the painting layers, rather than on the presence of organic binders, i.e., a potential source of carbon and energy. The occurrence of moisture on mural paintings can be caused by:

- The penetration of precipitation onto the painting or its being exposed to rainfall;
- Ascent of ground water up the walls; and
- Condensation of water vapor caused by changes in temperature and relative humidity of the air surrounding the painting (Grzesikowa, 1965; Mazur, 1990; Roznerska and Mikołajczyk, 1995).

Water activity making possible the development of microorganisms comes within the range 0.60–0.99, which corresponds to about 60–100% relative air humidity. Optimum water activity for the development of most fungi is ca. 0.85–0.95; for that of most bacteria, 0.96–0.99. With moistness of the material corresponding to about 70% relative air humidity most fungal spores are unable to germinate, but the mycelia do grow. Weirich (1989) reports that spores of all *Aspergillus* strains and those of half of *Penicillium* and *Cladosporium* strains isolated from mural paintings germinated at 70–90% relative air humidity, while for the germination of *Alternaria* spores even 85–90% humidity was too low.

Many authors studying microbiodeterioration of mural paintings provide some information on the microclimate prevailing in the buildings containing the paintings. Those data, however, are usually limited to general statements of wall dampness, and thus cannot be compared. Only some authors give more detailed descriptions of the source of the moisture, take into account the effect of the seasons, and report the values of the temperature and relative air humidity in the buildings. Even less information can be found concerning the methods of improving the microclimate in historic buildings. Moisture content in painting samples has been studied in Romanian monasteries; it ranged from 0.6% (on dry wall) to 4% (on damp, infested walls) (Savulescu and Ionita, 1971; Ionita, 1973). The dampness of walls in that case was caused by water vapor condensation on the surface of cold walls. In a baroque church in Einsiedeln, Switzerland, the occurrence of fungi on ceiling paintings was made possible by the influx of cold air through cracks in the ceiling, which caused water vapor to condense, and poor ventilation, made even worse by double-glazing of the windows (Raschle, 1983).

Particularly exposed to water vapor condensation are mural paintings done in an oil technique, due to the hydrophobic properties of the painting layer. In the cathedral in Alexandrov, Russia, Rebricova (1989) observed dark-brown and black fungal deposits covering mural oil paintings on the northern, cooler wall. In the window niches, where ventilation was poor, the deposits were reminiscent of soot.

Other authors report data concerning the air in buildings containing wall paintings deteriorated by microorganisms. Berner et al. (1997) report that the RH in St. Georgen parish church in Styria, Austria, ranged from 60 to 95%. In St. Claire's refectory in St. Damian monastery in Assisi that value in January was 70% at a temperature of 12°C (Guglielminetti et al., 1994). In St. Eusebio crypt in Pavia the temperature reached 2–4°C in winter, 14–15°C in spring and autumn, and 19°C in summer, while the RH ranged from 70% in summer to 95% in the other seasons (Bianchi et al., 1980). Excessive relative air humidity (90–100%) and uncontrolled climate (6–7°C in winter and 18°C in summer) resulted in the development of microorganisms, mainly actinomycetes, on paintings in many crypts, tombs, and vaults in Italy. After restoration but without any improvement in microclimate, several years later the paintings were again covered with a powdery deposit, while the introduction of microclimate control stopped the development of actinomycetes (Giacobini et al., 1987). Improvement of climate by installing ventilation resulted in a 10^2–10^3-fold reduction in the number of microorganisms on the walls of the cathedral in Ferapontovo, Russia (Rebricova, 1989).

On the other hand, no traces of microorganisms were found in Japanese tumuli decorated with mural paintings in spite of high RH (95–100%). However, microorganisms appeared shortly after the tumuli had been excavated and opened. The factor that inhibited their growth for hundreds of years was most probably low-molecular amines, which had been found in the air of a tumulus before it was opened. These came from the disintegration of the bodies buried in the tumulus and were found to show disinfecting properties even at extremely low concentrations. As they occurred in gas form, they escaped the moment the tumulus was opened, and microorganisms started to attack the paintings (Arai, 1983, 1984).

Water content in a brick wall exceeding 3–4% or more may be evidence of capillary ascent of moisture up the wall. Old moisture barriers may be inoperative or absent altogether; moreover, the water conditions around the buildings may have changed for the worse. In many monuments of Old Russian architecture capillary ascended moisture has filled up the pores of building materials to a height of 2–2.5 m above ground level (toward the end of a dry summer this was reduced by about 1 m). The zone between 1.5 and 2.5 m above ground level is subject to the greatest seasonal thermodynamic changes, which results in geochemical and microbiological processes of the highest intensity (Rebricova, 1993).

The composition of the sets of microorganisms occurring on mural paintings is also affected by the amount of light, by the temperature, and by the pH of the substratum. Heterotropic microorganisms can develop in darkness, although they need a small amount of light to develop their reproductive organs. Sunlight, particularly UV radiation, is harmful to them, with the exception of Deuteromycetes of the *Dematiaceae* family because of the melanin in their spores and/or hyphae. On paintings lit by sunlight or artificial light, autotrophic microflora can develop: These include cyanobacteria, green algae, and, less frequently, diatoms. Sometimes, a small amount of light coming in through a window is sufficient for these to develop. At night they switch over to heterotrophic feeding.

The effect of the remaining factors, temperature and pH, on the development of microorganisms on mural paintings has only rarely been the subject of studies. At temperatures below 0°c the activity of microorganisms is inhibited, while at higher temperatures it is the amount of humidity that is the more important factor. It seems obvious that warm and humid climate is conductive to attacks of microflora on all kinds of cultural heritage objects. On the other hand, microorganisms developing on mural paintings in temperate climates are rather psychrophilic and develop better in the 4 to 10°c temperature range than at room temperature (Daca, 1979; Rebricova, 1993). Likewise, Raschle (1983) noticed that *Aspergillus glaucus*, isolated from a vault painting in Einsiedeln, developed better at 18°c than at 20 or 25°c. Weirich's (1989) results, on the other hand, demonstrate that optimum temperature for the development of most microorganisms found on exterior paintings in Switzerland was somewhat higher, i.e., 20–25°c for fungi and 20–30°c for bacteria.

MICROORGANISMS DETERIORATING MURAL PAINTINGS

Bacteria Nowadays, bacteria are considered by many authors to be pioneer microorganisms, that is, the first organisms to attack mural paintings, which then, through their presence and the secretion of metabolites, make them accessible to other microorganisms (Ciferri, 1999). One of the earliest mentions of the importance of bacteria on mural paintings can be found in Tonolo and Giacobini (1961), who observed, on a number of paintings, powdering and separation of the paint layer as well as granular deposits on their surface, caused by the presence of huge numbers of coccus bacteria of the species *Sarcina lutea*.

The role of bacteria in the deterioration of mural paintings was also discussed by Lazar (1971), who studied mural paintings in the Cozia monastery in Romania. Large areas of the painting were covered with whitish, closely adhering powder, making considerable fragments of the representations illegible. From the areas attacked, 90 strains of bacteria, mainly heterotrophic, were isolated. Among them sporulating forms of the genus *Bacillus* predominated. Besides those, species belonging to the genera *Micrococcus, Sarcina, Arthrobacter, Mycobacterium, Nocardia, Pseudomonas,* and, probably, *Thiobacillus* were identified. Those bacteria represented types widely spread in nature, with great potential for adaptation. The presence of some of them was certainly accidental, but a few could have been associated with the development of white patches. In order to carry out observations on deteriorating reproductions, 40 bacterial strains were secondarily inoculated onto uninfested fragments of a painting. About one half of them, and in particular *Arthrobacter sp., Bacillus pumilus, Micrococcus roseus, M. sp.,* and probably *Thiobacillus sp.* and, to a lesser extent, *Sarcina lutea* and *Bacillus subtilis*, produced symptoms similar to natural ones. Although this experiment is in line with Koch's postulates, it is dangerous for the original painting, and therefore controversial.

A number of studies devoted to the deterioration of mural paintings by bacteria have been carried out in Germany. In some paintings, dating from various times between the 10th and 20th centuries, reddish stains have been observed, spreading over areas up to several square meters. It has been found that the factor responsible for these stains are G+ (Gram positive) true bacteria containing red pigment and belonging to the genera *Micrococcus* and *Nocardia*. Examinging 13th-century paintings in St. Peter's church in Eilsum, north Germany, investigators isolated *Micro-cosccus agilis* (reclassified as *Arthrobacter agilis*) and *Nocardia corynebacteroides*. The carotenoids contained in the cells of those bacteria were responsible for the pigmentation (Oltmanns et al., 1999).

Very frequently, parallel with microbiological deterioration of mural paintings, salt efflorescences occur as a result of water migrating in the wall and dissolving various salts (sulfates, nitrates, carbonates, and chlorides) contained in brick, mortar, or the ground. The water with the dissolved salts moves in the wall in accordance with the laws of diffusion. As the solution dries up on the wall surface, the salts crystallize, loosening the adhesion of the paint layers to the support, or even detaching its particles. Salt crystallization was the main cause of dramatic losses in the paint layer of the Romanesque mural painting in Nonnberg Abbey in Salzburg (Petersen and Hammer, 1992). It can be expected that microorganisms colonizing such niches are adapted to surviving under extreme conditions. And so they are: Many strains growing on salt-affected paintings show various degrees of halotolerance, or even halophily. Such bacteria were found on the mural painting in Nonnberg, as well as on 13th-century paintings in the chapel in the castle of Herberstein, Austria. The latter were studied at the same time as the rich mural paintings in the tombs of the Roman necropolis of the 1st–2nd century A.D. in Carmona, Spain. In both places large numbers of heterotrophic bacteria were found: Samples taken from the paintings contained up to 2.1×10^9 cfu/g. Unlike those from Carmona, the microorganisms from Herberstein grew in largest numbers on media with 10% NaCl, which is evidence of the presence of high numbers of halophilic bacteria. A total of 210 strains were isolated from the Carmona samples, 151 from the Herberstein samples. Identification by means of gas chromatography of methylated fatty acids revealed the presence of 90% G+ bacteria among them. The most frequently found species were *Bacillus licheniformis, B. pumilis, B. spp., Micrococcus spp., M. luteus, Kocuria varians, Paenibacillus polymyxa*, and *Bacillus megaterium* (Heyrman et al., 1999).

There is now growing criticism of the traditional techniques of isolation and culturing of strains. The prevailing opinion is that, using them, it is possible to detect only 1–5% of the species occurring in the material under study, while most remain undetected. That is why new methods taken from the field of molecular biology are being introduced into the identification of bacteria occurring on mural paintings. These techniques, based on electrophoresis of PCR-amplified gene fragments encoding 16S rRNA, followed by sequencing them, made it possible to identify strains of bacteria that were extremely difficult or impossible to detect by traditional methods (Rölleke et al., 1996). Studies carried out on samples from the Herberstein painting resulted in the detection of eubacteria belonging to the genera *Clostridium* and *Frankia* or to closely related genera, as well as of extremely salt-tolerant *Halomonas*, which can grow on substrata containing up to 32% salt. The detection of halophilic eubacteria induced the authors to search the painting for *Archebacteria*, which are known for their ability to develop under extreme conditions, such as a highly saline environment. Using archaeal specific primers for 16S rDNA amplification, the presence of representatives of *Archebacteria*, viz., a species closely related to *Halobacterium halobium*, was actually found (Rölleke et al., 1998). The authors presume that it is those extremely halophilic bacteria that are producing carotenoids that yield the red pigment that may be responsible for the reddish stains on the painting. In other samples from the Herberstein chapel painting other *Archebacteria* species were detected, some of which are closely related to the extremely halophilic genus *Halococcus;* for others, no related species are known. Strains related to *Halococcus* and representative of the alkalophilic genus *Natronobacterium* were also detected on the aforementioned paintings in the Roman necropolis in Carmona, Spain (Piñar et al., 1999). Those are very important discoveries, since up until now *Archebacteria* have not been taken into account as organisms participating in biodestruction of works of art. However, the methods of molecular biology fall short in that they do not make it possible to distinguish whether the DNA under study comes from living or dead organisms, and whether the identified species actually take part in the destruction of the painting or their occurrence on it is accidental (Ciferri, 1999).

Other instances of adaptation to living under extreme conditions were found while studying mural paintings in the Krummhorn-Campen parish church in north Germany. The paintings there were covered in places with 20 layers of lime wash and overpainting. From these paintings researchers isolated 16 strains of bacteria showing various degrees of alkaliphily; all of these strains were capable of growth at pH 10.5 (Petersen et al., 1995). Their presence was probably due to the local occurrence of strongly alkaline saline waters, which results in the pH of the walls in Campen being higher than it usually is in old plasters. Moreover, it was found that all the strains studied were able to grow under microaerophilic conditions. That ability made it possible for them to grow in deeper layers of the painting, where the oxygen supply was reduced.

Heterotrophic bacteria are mentioned in a number of other studies; thus *Staphylococcus conhi* and *Pseudomonas maltophyla* were the dominant species on two mural paintings in St. Damian monastery in Assisi (Carruba et al., 1999). From external Swiss paintings Weirich (1989) isolated a number of bacterial strains, most of which belonged to the genera *Bacillus* (e.g., *B. subtilis, B. polymyxa*), *Micrococcus* (*M. roseus*), and *Arthrobacter*. The occurrence of heterotrophic bacteria is also reported by Agrawal et al. (1988), Karpovich-Tate and Rebrikova (1990), and Petersen and Hammer (1992).

Using traditional isolation methods, five other bacterial strains, besides the halophilic bacteria, were isolated from the Herberstein chapel paintings. After their 16S rDNA fragments were sequenced, they were identified as microorganisms belonging to actinomycetes or related to them (*Arthrobacter globiformis, Streptomyces griseus*, and an unidentified species of *Pseudonocardiaceae*), as well as microorganisms belonging to *Proteobacteria* (*Acinetobacter lwofii*) (Rölleke et al., 1996).

The importance of actinomycetes has also been pointed out by other authors. Tonolo and Giacobini (1961) discussed their gradually increasing development on the vault of Chiesa di S. Clemente in Rome. There were noticeable changes on the painting in the form of an intensive, compact, and widespread whitish deposit, accompanied by powdering of the painting layer. A microscopic study of samples revealed large numbers of tangled filaments of streptomycete pseudomycelium.

Streptomycete strains were also found by Bassi and Giacobini (1973) on mural paintings covered with a white "patina" very similar to salt efflorescences. Giacobini et al. (1987) found them on paintings from numerous excavations, crypts, grottos, and tombs in Italy. Streptomycete growth on these paintings appeared in the form of a whitish powder or a colorless, shiny layer, sometimes with fine, light-gray grains. The colors in the places attacked were faded and "dusty." The number of streptomycete cells ranged from 5×10^3 to 1×10^6/g of sample, depending on the sampling place. About 200 streptomycete strains were isolated from 13 paintings, 60 of which were identified to the species level. Most commonly represented were *Streptomyces rectus flexibilis, S. albus, S. cinereoruber, S. griseolus, S. vinaceus*, and species of the genus *Nocardia*.

Actinomycetes of the genera *Streptomyces, Micromonospora*, and *Nocardia* were also isolated from paintings in St. Panteleimon church in Boyan, Bulgaria (Hadjiulcheva and Gesheva, 1982), and representatives of the genera *Streptomyces* and *Nocardia* were taken from paintings in Santa Maria de la Rabida monastery in Huelva, Spain (Saiz-Jimenez and Samson, 1981).

White deposits, sometimes with yellowish or greenish tints, were also observed on mural paintings in monuments of Old Russian architecture, such as the Orthodox churches in Meletovo, Ferapontovo, and Jaroslavl. Most Old Russian mural paintings were made in the fresco technique, complemented with tempera. Later overpaintings were made mostly in the tempera or oil technique. The binders used were chicken egg yolk, wheat glue, or animal glue. In the course of drying, and later natural aging of the painting layer on lime plaster, reactions were going on, which provided the painting with a certain durability and resistance to microorganisms. In some plasters, preserved remnants of organic fillers have survived until the present day; in other, damp ones they have become centers of microorganism development. In the whitish deposits occurring on the surface of the paintings under study, large amounts of salts, mainly sulfates, have been found, as well as very numerous microbial cells, the most dominant being actinomycetes, *Arthrobacter sp.*, *Bacillus spp.*, and nitryfying bacteria, among others. Some fungi have also appeared in the cultures, but their numbers were 10–1000-fold lower than those of bacteria and actinomycetes (Rebricova, 1989, 1993).

On the mural paintings in St. Fiodor's chapel in the cathedral in the Pafnutii-Borovski monastery, too, the dominant microorganisms were streptomycetes and bacteria of the genera *Arthrobacter* and *Bacillus* (Karpovich-Tate and Rebrikova, 1990). The paintings, dating from the mid-17th century, were painted in mixed technique with egg-bound tempera paints. Restorers fixed them with egg-yolk emulsion. Within several years after the restoration, a compact, yellowish-white deposit appeared on the surface of the paintings, as well as on the plaster and brick up to a height of 1.5 m above the floor. At the same time the stone and brick showed symptoms of intense weathering, but no destruction or alterations of color in the painting could be seen under the deposit (see also below). The number of heterotrophic microorganisms ranged from 10^4 to 10^6 cells/g of sample.

Beside numerous heterotrophic bacteria, other organisms found fairly frequently on mural paintings are autotrophic bacteria, known to grow on historic stone and walls (Strzelczyk, 1981a; Leźnicka, 1991; Griffin et al., 1991; Bock and Sand, 1993). Sorlini et al. (1982), in their work on the walls of the 12th-century Palazzo della Ragione in Milan, found, in addition to heterotrophic microorganisms, very large numbers of thiobacilli, particularly *Thiobacillus thiooxidans* and *T. thioparus*. The authors presume that the presence of large numbers of thiobacilli was due to the high SO_2 concentration (mean, 0.07 ppm; max., 0.244 ppm) in the air of central Milan. Sulfur dioxide pollution of the atmosphere was also the cause of thiobacilli's appearing on the mural paintings in Santa Maria de la Rabida monastery in Huelva, Spain (Saiz-Jimenez and Samson, 1981). On the other hand, the mural paintings in the cathedral in Pafnutii-Borovski Monastery supported, in addition to numerous heterotrophic microorganisms, nitrifying bacteria, which, according to researchers, were the main biological agent causing the destruction of paintings, plaster, brick, and stone (Karpovich-Tate and Rebrikova, 1990). Nitrifiers were also detected on paintings in St. Panteleimon church in Boyan, Bulgaria (Hadjivulcheva and Markova, 1983).

Another group of autotrophic bacteria destroying mural paintings, namely photosynthesizing cyanobacteria (earlier blue-green algae) have been cited several times, mainly in Italy, as one of the factors causing deterioration of mural paintings. Cyanobacteria (beside green algae) were responsible for the development of a red-violet deposit on paintings in Villa Adriana in Tivoli (Andreolli and Rascio, 1982).

They were also found in the basement of Domus Aurea in green deposits that appeared on surfaces lit by lamps for visitors. Their development was so rapid that in 1981 the lighting was discontinued (Grilli Caiola et al., 1987). The dominant microorganisms in those deposits were cyanobacteria of the family *Oscillatoriaceae* —*Lyngbya pseudoramosa* Hoffman and *Lyngbya parva* Hoffman and green algae (Albertano et al., 1991). On a number of Italian mural paintings yet another cyanobacterium was detected—*Nostoc sp.*—capable of binding atmospheric nitrogen (Tomaselli et al., 1979).

FUNGI A large number of mold species are reported to cause deterioration of paintings. Many authors consider them to be very important microorganisms responsible for the deterioration of mural paintings (Garg et al., 1995). At times antibacterial components (chloramphenicol, streptomycin, penicillin) have been added to the media in order to isolate fungi from paintings (Jeffries, 1986; Guglielminetti et al., 1994). However, other studies show that fungi can only be secondary colonizers, getting their nutrition from dead cells of other microorganisms or their metabolites (Saiz-Jimenez and Samson, 1981; Sorlini et al., 1982; Karpovich-Tate and Rebrikova, 1990).

One of the earliest mentions of the significance of fungi can be found in the publications of Tonolo and Giacobini (1961). Among the alterations caused by molds, these researchers distinguish those on the picture surface and those in its structure. The alterations to the surface are aesthetic in nature: The fungi deteriorate color vividness, and cause blemishes or black stains. The alterations inside the painting are the result of the hyphae growing deep inside it through the painting layer and the support. The evidence of their activity consists of raising and curling of the paint layer or crater-like cracks. Also, decreased cohesion of the paint layer is observed, as well as its extreme friability, which cannot be brought about only by alterations of a physicochemical nature. Tonolo and Giacobini point to the activity of fungi, in particular those of the genera *Penicillium, Aspergillus, Cephalosporium*, and some of the family *Dematiaceae*, as the cause of those alterations. Some of the fungi produce perithecia, pycnidia, or stromata.

The occurrence of fungi on mural paintings is reported by Romanian workers. They studied 14th–17th-century mural paintings in the Cozia monastery as well as in monasteries in northern Moldova and in a church in Curtea de Argeş, both inside and outside the buildings. They were created by local Moldovian painters, who developed their own technique by combining Byzantine technique with Central European ones. To the *intonaco* layer they added organic components: hemp and flax fibers, tow, cut-up straw, and barley husks. As binders they used calcium caseinate, and as pigments mineral compounds, as well as vegetable pigments, pigmented clays, calcium carbonate, and charcoal.

These paintings stimulated the growth of heterotrophic microorganisms. The pH of the painting layer ranged from 6.5 to 7 (7.5–8 for plaster water solution). It was noted that, of all the pigments used, it was the pigmented clays that were the most deteriorated. About 30 fungal species were isolated, with more of them from the inside of monasteries than from outdoor paintings. They caused all kinds of deterioration, from small spots on the surface to deep changes in the wall structure. Particular attention was paid to *Phoma glomerata*, which is characteristic of the painting layer, and whose appearance signals its deterioration. [TABLE 1] The Romanian authors also presented a long list of fungal species with descriptions of the damaged spots from which they were isolated [TABLE 2] (Savulescu and Ionita, 1971; Ionita, 1973). However, the authors emphasize that much damage develops as a result of the cooperation of a number of microbial species, and that particular species can be accountable for the development of various symptoms of deterioration.

Greenish-brown and black stains occasioned the undertaking of microbiological studies of two 1480 mural paintings ("St. Augustine" by Botticelli and "St. Jerome" by Ghirlandaio) in the Ognissanti church in Florence. The paintings were restored in 1969 after being damaged in the flood of 1966, but several years later the stains appeared. From Botticelli's painting researchers isolated 15 strains, most of which were strains of *Penicillium brevi-compactum* Dierckx and *P. purpurogenum* Stoll, accompanied by *Cladosporium cladosporoides* (Fres.) de Vries. Among the 13 strains isolated from Ghirlandaio's fresco the majority were strains of *Aspergillus versicolor* (Vuill.) Tiraboshi, and beside them there were strains of *Cladosporium sphaerospermum*. Moreover, a number of other species previously found on mural paintings were isolated (*Botrytis cinerea, Penicillium notatum, Trichoderma harzianum, Aspergillus flavus, P. rugulosum, Engyodondium sp.*), as well as some whose occurrence on mural paintings had not been reported (*A. sydowii, Coniothyrium cerealis, Oidiodenron cerealis, P. meleagrinum, Mycelia sterilia, Rhodotorula sp.*, and *Tilletiopsis sp.*) (Sampo and Luppi Mosca, 1989). The differences in composition between the microflora found on two very similar paintings in the same building probably resulted from the different locations of the paintings in the church, and, consequently, the slightly differing microclimates (Ciferri, 1999).

Other authors, too, present long lists of mold species isolated from mural paintings. A number of fungal species [TABLE 1] were isolated from very ancient (the oldest come from 2nd–1st-century B.C.) mural paintings in Buddhist rock-cut shrines at Ajanta, India (Agrawal et al., 1988). The paintings were made on still damp, very thin and smooth mortar consisting of mud with vegetable fibers, paddy husks, grass, and other fibrous organic materials. Glue or gum was used as binder for pigments (mainly mineral). White, brown, or black superficial encrustations on the paintings in some caves were caused by fungi. Additional nutrition for them was the bat-excreta found in nearly all caves.

Weirich (1989) studied the microflora of ten external mural paintings in Switzerland, in both urban and rural environments. He isolated about 20 species of fungi [TABLE 1] and some bacteria. He found no differences in species composition between the microflora isolated from the two environments.

About 20 fungal species [TABLE 1] were isolated from two medieval paintings in Austria (in the Herberstein chapel and in St. Georgen parish church in Styria). The Herberstein paintings were covered with numerous ochre, brown, black, and pink stains. The greater numbers and diversity in the fungal colonies in the Herberstein chapel were probably due to the close vicinity of the wine cellar and the consequent presence of organic acids, which could be food for the fungi (Berner et al., 1997).

More than 40 fungal species [TABLE 1] were isolated from two 17th-century mural paintings in the St. Chiara refectory of the St. Damian monastery in Assisi ("Crocifisso con Francesco Giovane" and "Mensa di S. Chiara") (Guglielminetti et al., 1994).

Careful perusal of the lists of isolated microorganisms makes it possible to distinguish the most frequently occurring species. Particularly noteworthy is the genus *Cladosporium*, whose representatives have been reported by many authors. Bassi and Giacobini (1973) blame *Cladosporium sp.* for causing dark-brown spots penetrating through the entire thickness of the painting to the very plaster. Sometimes it was isolated even from places free from dark spots. The fungus is characterized by its considerable tolerance toward light and changes in external conditions (temperature, humidity), which permits it to grow on damp walls and stone objects. Its frequent occurrence in those places is also accounted for by the presence of large numbers of its spores in the air, its relative resistance to lysis by actinomycetes, and its co occurrence with mycophilic fungi (Rebrikova, 1993). Bianchi et al. (1980) isolated *Cladosporium cladosporoides* (Fres.) de Vries and *Penicillium lanosum* var. *coeruleum* Westling from the pustules on the paintings on the vaults in S. Eusebio crypt in Pavia (northern Italy). Saiz-Jimenez and Samson (1981) isolated fungi associated with thiobacilli and heterotrophic bacteria from the mural

SPECIES	LOCATION OF MURAL PAINTINGS					
	MONASTERIES IN ROMANIA	SHRINES IN AJANTA, INDIA	RUSSIA	SWITZERLAND (OUTSIDE)	ST. DAMIAN'S MONASTERY IN ASSISI, ITALY	HERBERSTEIN CHAPEL AND ST. GEORGE CHURCH IN STYRIA, AUSTRIA
Acremonium charticola (Lindau) Gams			+			+
Acremonium spp.			+		+	+
Alternaria alternata (Fr.) Keissler		+		+	+	
Aspergillus flavus Link.	+		+			
Aspergillus niger Van Tieghem	+	+		+	+	
Aspergillus sydowii (Thom et Church)		+	+			+
Aspergillus versicolor (Vuill.) Tiraboshi	+	+	+	+	+	+
Aspergillus candidus		+				+
Aureobasidium pullulans (de Bary) Arnaud			+		+	
Botrytis cinerea Pers ex Fr.			+	+		
Chaetomium globosum Kunze	+	+	+		+	
Chaetomium murorum Corda	+				+	
Chaetomium spp.					+	+

Cladosporium sphaerospermum Penz.		+	+	+	+	+
Cunninghamella echinulata Thaxter	+		+			
Engyodontium album (Limber) de Hoog (synonymy *Tritirachium album* (Limber) Sacc., *Beauveria alba* (Limber) Sacc.	+		+			+
Epicoccum nigrum Link		+	+			
Fusarium oxysporum Schlecht.		+		+	+	
Fusarium solani (Mart.) Sacc.				+		
Mycelia sterilia		+	+		+	
Penicillium brevi-compactum Dierckx				+	+	+
Penicillium chrysogenum Thom	+		+	+	+	
Penicillium expansum Link. ex Gray					+	
Penicillium verrucosum Dierck			+		+	+
Phoma glomerata (Corda) Wollen. & Hochapf.						
Trichoderma viride Pers. Ex Grey	+		+	+		
Trichthecium roseum (Pers.) Link	+			+		
Verticillium lecanii (Zimm.) Viegas			+			+
Yeasts					+	+

[TABLE 1] Fungi most commonly found on selected mural paintings (Savulescu and Ionita, 1971; Agrawal et al., 1988; Rebricova, 1989; Weirich, 1989; Guglielminetti et al., 1994; Berner et al., 1997)

SYMPTOMS OF DETERIORATION OF PAINTING IN SAMPLING PLACE	FUNGI FOUND IN SAMPLE
Pinkish colored stains	*Penicillium lilacinum* Thom. *Acremonium roseum* (Oudem.) W. Gams
Light-brown powders	*Scopulariopsis brevicaulis* (Sacc.) Bainier
Brown and black stains, cracks, chicks	*Claosporium herbarum* (Pers.) link. *Stemphylium piriforme* Bonorden *Chaetomium globosum* Kunze *Ulocladium chartarum* (Preuss) Simmons
Whitish stains after drying up of condensed water drops	*Penicillium chrysogenum* Thom. *Aspergillus flavus* Link.

[TABLE 2] Fungi from Romanian mural paintings (according to Ionita, 1973, selected instances)

paintings in Santa Maria de la Rabida monastery in Huelva, Spain; most of the organisms were *Cladosporium sphaerospermum* (75–88% of the isolates), *Engyodontium album* (a little more than 10% of the isolates) and *Aspergillus versicolor*. Among fungi isolated from the mural paintings in the Nativity of the Virgin Cathedral in Pafnutii-Borovski Monastery, beside nitrifiers and actinomycetes, 65% were colonies of *Cladosporium sphaerospermum*, accompanied by colonies of mycophilic fungi: *Beauveria bassiana, Acremonium sp., Hyalodendron sp., Verticillium lamellicola*, and *Geomyces pannorum* (Karpovich-Tate and Rebrikova, 1990).

Other fungi most commonly found on mural paintings are: *Aspergillus versicolor, A. niger, Penicillium brevi-compactum*, and species of the genera *Acremonium, Fusarium*, and *Chaetomium*. [TABLE 1] Fungi of the genus *Aspergillus* are xerophilic, and therefore able to grow under the variable humidity conditions prevailing on many mural paintings.

In addition to fungi deteriorating mural paintings, some authors report yeasts (among others, *Candida sp.*) (Curri, 1979; Rebricova, 1989).

Most of the paintings studied have yielded from several to several dozen species of microorganisms. However, from the mixed-technique paintings in St. Andrew's Chapel in Canterbury Cathedral only one species was isolated: the white mold *Beauveria alba* (Limber) Sacc. (Jeffries, 1986) [synonym: *Engyodontium album* (Limber) de Hoog]. The fungus appeared in abundance after restoration over about a 7-m^2 area of the painting, causing pigment particles to fall off, exfoliation of the painting surface, and powdering of the weakened painting layer.

Only one species, *Aspergillus penicilloides* Spegazzi, was isolated from the brown spots occurring on the ancient Egyptian paintings in Tutankhamen's tomb. It is an absolutely xerophilic fungus, one of those held responsible for the rusty spots, called foxing, frequently found on historic paper (Arai, 1993). Another species of the genus *Aspergillus* was mainly accountable for the infestation of the painting in Einsiedeln in Switzerland (Raschle, 1983).

In Poland, too, a number of valuable mural paintings have been subjected to microbiological studies. On the mural paintings in the palaces in Wilanów and Nieborów, in the castle in Lublin, and in the old church in Czerwińsk, the following fungi were found: *Mucor mucedo* (L.), *Rhizopus nigricans* Ehrenb., *Aspergillus sp., Penicillium sp., Pyronema domesticum* (Sow.) Rehm, *Plicaria muralis* Sow., *Cephalosporium acremonium* Corda, *Sporotrichum roseum* Link., *Torula murorum* Corda, and *Pullularia pullulans* Berkh and *Stemphylium alternariae* (Sacc.) (Ważny, 1965).

As a result of studying a gray-white deposit on the surface of mural paintings in St. Joseph's baroque church in the post-Cystersian complex in Krzeszów, Daca

(1979) isolated a set of 14 species of fungi, the dominant ones being *Cladosporium cladosporoides* (Fres.) de Vries, *Aureobasidium pullulans* (de Bary) Arnaud, *Candida humicola* (Daszewska) Diddent at Ladder, *Papulaspora sporotrichoides*, and *Sclerofomina sp*. The isolated strains were subject to biochemical and physiological studies (see below).

Strzelczyk and Modrzejewska (1999), studying mural paintings in St. Jacob's gothic church in Toruń, found streptomycetes to be dominant. The paintings were covered with a widespread, delicate, powdery, whitish deposit. From the samples taken, 16 streptomycete strains were isolated and subjected to biochemical studies (see below). A number of studies have been carried out on the microflora occurring on mural paintings in churches of the Toruń neighborhood: the cloister church in Skępe and the parish churches in Górzno and Okonin. Fungi of the genera *Cladosporium* and *Penicillium* have been found, as well as numerous bacteria, including streptomycetes (Strzelczyk et al., unpublished).

Algae Algae, being autotrophic, occur in very damp places exposed to natural and artificial light. The most susceptible to their attack are therefore paintings decorating the façades of buildings and underground paintings illuminated for visitors. Symptoms of algal attack are stains, streaks, and deposits of various hues of green. The occurrence of algae on mural paintings, particularly unicellular green algae of the family *Chlorophyceae*, has been discussed by Tonolo and Giacobini (1963).

A drastic example of the deteriorating effect of algae on drawings on rock was found in the Lascaux Cave, where there are about 600 prehistoric presentations dating back to 15,000–13,000 B.C. and representing mainly animals (steppe horses, bisons, ibexes, etc.). Lascaux Cave, discovered in 1940, was opened to visitors in the years 1948 to 1963. The reason this attraction was then closed was the appearance on the paintings of abundant green deposits due to the growth of the unicellular green alga *Bracteacoccus minor* (order *Chlorococcales*). It was found that the algae appeared as a result of the installation of lights, which were on almost nonstop, and of the influx of visitors, which resulted in considerable increase in CO_2 concentration and humidity level (Lefevre, 1974). The algae continued to reproduce even after the closure of the cave, in complete darkness, changing over to heterotrophic feeding.

The opening to visitors in 1951 of the basement of Domus Aurea in Rome resulted in the development of green deposits on the paintings. Besides cyanobacteria, which were dominant (see above), green algae were found to grow there.

These were *Chlorella sp., Pseudococcomyxa simplex* (Mainx) Fott., *Pseudopleurococcus printzii* Vischer, and *Chlorococcum sp.* (Albertano et al., 1991).

Abundant green deposits were found in St. Peter's Church in Eilsum, Germany (Krumbein and Petersen, 1990). The deposits occurred in places where moisture was coming up from the ground, in the apse and in the window niches. The deposits were found to contain green algae and diatoms, as well as cyanobacteria, actinomycetes, and fungi. After the flood in 1988 black-green bubbles the size of pinheads appeared; they were agglomerations of spherical algae associated with fungal hyphae.

If access to light is not completely cut off, algae can live even under lime washes. That phenomenon was observed by Petersen and Krumbein (1995) on mural paintings in Immingen, Germany.

Lichens Reports of damage to mural paintings caused by lichens are rather scarce and concern externally exposed paintings. Tonolo and Giacobini (1961) pointed out the special danger posed by crustose lichens, whose apothecia sometimes penetrated through the painting layer and settled in the mortar. In Palazzo Farnese at Caprarola in Italy, 16th-century mural paintings decorating the galleries were found to be damaged by the crustose lichens *Dirina massiliensis* forma *sorediata.* They attacked mainly parts covered with yellow and brown pigments, while rather avoiding red pigment, which probably contained metal ions inhibiting their growth (Seaward and Giacobini, 1988). Genus *Dirina* is known to include lichens typical of the Mediterranean region, growing on rocks. *Dirina massiliensis* is one of the lichens inhabiting continuously moist habitats and avoiding direct insolation (Bystrek, 1977).

Mechanical removal of lichens from the paintings in Palazzo Farnese proved extremely troublesome, as, together with the thalli, fragments of the painting layers were detached from the support, so that it was necessary to reconstruct several square meters of the paintings (Seaward and Giacobini, 1988). It seems therefore clear that removing live lichen thalli grown with their rhizoids several millimeters deep into the support is absolutely not recommended. The only way to get rid of lichens is to apply chemical disinfectants, which would cause their death and desiccation, after which their removal from the painting poses no problem.

Other Organisms The presence of numerous microbial cells, particularly tangles of fungal hyphae, as well as dust, vegetable, and animal remnants create, especially in warm climates, the right conditions for the development of insects. On the mural paintings in the Buddhist rock-cut shrine at Ajanta, India, numerous silverfish (*Lepisma saccharina*) were found, as well as beetles of the family *Tenebrionidae*, known for attacking dry leaves, cow dung, etc. Their activity in the paintings probably consists of their settling in the plaster, which they loosen by feeding on the organic components it contains. Moreover, the presence of butterfly larvae and pupae and of cockroaches was detected on them (Agrawal et al., 1988).

Studies of samples taken from Italian paintings carried out by Guglielminetti et al. (1994) revealed, through scanning electron microscopy, the presence of various forms of mites, probably feeding on fungi, and particularly numerous near or inside micropits covered with mycelia.

ECOLOGICAL DYNAMICS As can be seen from the above review, the microflora of mural paintings is diversified, and which species are found depends upon the environmental factors affecting the living conditions in the specific ecosystem, i.e., the mural painting. Most commonly it is heterogeneous and consists of populations of a number of microbial species, linked together with all kinds of interrelationships. A lot of information concerning the role of microorganisms on the painting can be obtained by studying their abilities to decompose the particular components of the painting as well as their antagonistic properties, e.g., their lysing the cell walls of other microorganisms.

Some authors describe quantitative and qualitative alterations in the microflora on mural paintings, over the course of time. These can be cyclic alterations, taking place under the influence of abiotic factors, such as alterations in microclimate (allogenic succession), or slow alterations caused by the microorganisms themselves (autogenic succession). Thus, Lazar (1971) observed cyclic diversification of the microfloral composition depending on the season and on the related changes in temperature and relative air humidity—in the summer the growth of bacteria was markedly more intense than in spring and autumn. Rebricova (1993) found maximum numbers of microorganisms in late spring and early summer due to increased moisture condensation on the walls, still cold after winter, while minimum numbers were found in autumn. That regularity seems common in churches in climates with a cyclic recurrence of seasons of the year. Church caretakers should be aware of the danger of condensation and more intense growth of microorganisms in late spring and summer, and should prevent the influx of warm spring air into the cold churches.

Several types of ecological succession of microbial communities have been observed, which slowly and gradually transform the painting as they grow on it. The pioneer microorganisms, those that first settle on the mural painting, change the properties of the habitat, making it more suitable for populations of more and more species. They accomplish this through their development—the secretion of metabolites, the decomposition of particular components of the painting, and the hindering of water evaporation, The latest opinion is that these pioneer organisms are often bacteria, although up until recently most studies have focused on fungi (O'Neill, 1988; Ciferri, 1999).

If the painting abounds in organic compounds, the succession can take a different course. First, fast-growing microorganisms develop; these use up the easily assimilable components of the paintings. Then, other microorganisms follow, ones that grow more slowly and are able to decompose more complex organic compounds (Strzelczyk, 1981a).

On mural paintings with adequate access to light, such as exterior ones, the pioneer role can be taken by cyanobacteria or algae. The metabolites secreted by algae in times of drought (amino acids, monosaccharides, growth substances, vitamins) and their dead cells can be fed on by heterotrophic microorganisms: bacteria and fungi.

On the mural painting in Santa Maria de la Rabida Monastery at Huelva, Spain, heterotrophic microorganisms were observed to succeed the pioneer chemoautotrophic ones (Saiz-Jimenez and Samson, 1981). According to the authors, the pioneer microorganisms on that painting were thiobacilli oxidizing sulfur compounds (about 65,000 cfu/g). Their occurrence was due to the location near the monastery of industrial works emitting large amounts of pollutants into the atmosphere, SO_2 being one of them. The sulfuric acid forming in the reaction of SO_2 with oxygen and water dissolved the $CaCO_3$ in the painting, which increased the porosity of the surface. The gypsum forming in that reaction ($CaSO_4 \times 2\ H_2O$) as white crystals caused the appearance of whitenings on the surface of the painting. Besides those, there were green-black spots on the paintings. In addition to thiobacilli, secondary colonizers—fungi and heterotrophic bacteria—were also isolated from the green-black spots and the whitenings. Their source of organic compounds was dead thiobacilli cells, and the effect of their development was further deterioration of the painting, such as the detachment of fragments of the painting layer. Similar relationships between thiobacilli and heterotrophic microorganisms were reported by Sorlini et al. (1982), as they studied the microflora in Palazzo della Ragione in Milan.

Another instance of succession of microorganisms was observed on the paintings in St. Fiodor's Chapel in the Cathedral of the Nativity of the Virgin in Pafnutii-Borovski Monastery. According to Karpovich-Tate and Rebrikova (1990) the pioneer microorganisms were nitrifying bacteria. Secondarily there was a development of heterotrophic becteria and fungi, mainly *Cladosporium sphaerospermum*, using the extracellular polysaccharides secreted by the nitrifiers and their dead cells. The spores and hyphae of the fungus were probably one of the food sources for actinomycetes and the so-called mycophilic fungi *Beauveria bassiana, Acremonium sp., Hyalodendron sp., Verticillium lamellicola,* and *Geomyces pannorum.* The authors think that those nutritional interrelationships among populations of microorganisms were most important, although heterotrophic microorganisms could also use organic substances contained in dust, the binders, and the egg-yolk emulsion used for restoration of the paintings.

Actinomycetes and mycophilic fungi were found to be capable of lysing the cell walls of pioneer bacteria and yeasts, as well as of decomposing fungal cell components: glucans, mannose, chitin, and proteins. They contained dextranase, and some, levanase (Karpovich-Tate and Rebrikova, 1990). Levan and dextran, polysaccharide components of bacterial capsules, can serve as sources of nutrition for fungi on painting layers (O'Neill, 1986). According to Rebricova (1989, 1993), the presence of mycophilic fungi and actinomycetes on the painting is evidence of long-lasting succession and stabilization of relationships among microbial populations on the painting.

MECHANISMS OF DETERIORATION OF MURAL PAINTINGS BY MICROORGANISMS There are two kinds of damage to mural paintings caused by microorganisms: aesthetic (stains, whitenings masking the drawing, alterations in pigment color) and structural (cracking and exfoliation of painting layers due to decomposition of binders). These two types of damage usually occur jointly. Often, in the initial phase of development, only scattered fungal spots can be seen, and only later, consequent to the penetration of hyphae deep into the painting and the decomposition of binders, the painting layer starts to flake off. Studies on the mechanisms leading to the microbiodeterioration of mural paintings are still incomplete, but as of now it appears that the most important mechanisms are (Krumbein and Petersen, 1990):

• The development of a biofilm on the surface of the painting;
• The enzymatic decomposition of the organic components of paintings, mainly binders;

- The secretion of aggressive acids; and
- Oxidation and reduction reactions.

Not all of the microorganisms isolated from samples taken from mural paintings are responsible for damage. Some of them, e.g., those contained in dust, were isolated accidentally and are of no significance for the painting. In order to establish which microorganisms are accountable for the alterations on the painting, experiments are carried out testing the biochemical capacities of the isolated microorganisms for decomposing the binders used and the conservation materials, as well as for producing chemical alterations in pigments.

The microorganisms occurring on the surface of the mural painting form a complex and still not fully understood ecosystem, called a biofilm. Biofilms occurring in home and industrial water distribution systems have been studied fairly thoroughly (Morton and Surman, 1994). It should be noted that the occurrence of biofilms is not limited to the solid/liquid contact area; they can also be found at the solid/air interface, e.g., on the surface of buildings and monuments, as well as on painting covers. A biofilm consists of microorganisms (bacterial cells, fungal hyphae, algae, myxophyta, etc.) occurring on the surface and the substances they secrete. Colonization of microorganisms in a new site is facilitated by extracellular polysaccharide substances, EPS), i.e., glycocalyx. The glycocalyx facilitates adhesion of the cell to the surface, increases its enzymatic activity, and constitutes a protective physical barrier against biocides.

The prevailing biofilm models present it as a number of islands consisting of microcolonies, separated by vast void areas, which with time become overgrown. These phenomena on the surface of mural paintings were observed by Rebricova (1993). Palmer and White (1997) found that the initially single-species microcolonies transformed into multispecies associations as the biofilm developed into maturity. Part of the biofilm's harmful effect is that it keeps the moisture in deeper parts of the painting by impeding evaporation. The swelling and shrinking of a biofilm with climatic changes result in tensions in the biofilm/paint layer contact area and progressive cracking and flaking off of the paint layer (Petersen, 1996).

The destructive activity of biofilm microorganisms on a painting is not limited to its surface; they penetrate deep inside the painting layer and the plaster. All the layers of the painting can be infested by microorganisms, and their destructive effect can be even stronger in the deeper layers than on the surface. In the deeper layers they can find better development conditions, as water evaporation there is slower. Tonolo and Giacobini (1961) found that fungi can grow between the paint layer and the undercoat, which results in progressive detachment of the paint layer and decay of the mortar beneath it. Petersen (1995) observed vigorous growth of microorganisms on the underside of flakes of the desquamating paint layer. On some paintings microorganisms could be found underneath many layers of lime wash and overpaintings. Such deep penetration was due to the use for conservation of the consolidants Klucel and Paraloid B72. On the other hand, consolidation of the painting with injection of lime and casein solution resulted in growth of microorganisms in injection sites, even under plaster (Petersen, 1995).

Many microorganisms produce thick slime capsules, particularly under climatic conditions of varying humidity. Slime-producing fungi have been found in books (Rosa and Strzelczyk, 1979). Slime capsules help retain moisture, and thus make it possible for microorganisms to grow under drought conditions. Prolonged drought causes slimes to dry up and shrink, and the microorganisms then form spores and sclerotia, while vegetative forms die. In situations of frequent changes in moisture content the mechanical tensions, associated with the alternately swelling and shrinking slimes, are conducive to loosening the structure of the painting (Krumbein and Petersen, 1990).

Heterotrophic bacteria and fungi need a source of organically bound carbon for their growth and development. For oligotrophic microorganisms, the mineral and organic components in the dust and soot settling on mural paintings over centuries may prove sufficient. The possible sources of carbon on mural paintings are binders, such as casein, wheat paste, animal glue or, less frequently, egg binder or oil; some pigments; as well as slimes produced by microrganisms and their dead cells. The consumption of the binder by microorganisms leads to its impairment and, consequently, to the falling off of pigment particles devoid of any binding factor.

A number of authors tested the capability of isolated microorganisms to decompose the particular components of mural paintings. Understanding the enzymatic spectrum and physiological properties of microorganisms helps estimate the amount of risk posed by the isolated strains to the painting. Thus, the strains of microorganisms isolated by Weirich (1989) from external mural paintings produced enzymes capable of decomposing casein and, to a lesser degree, gelatin, fat, and starch. Streptomycete strains isolated by Strzelczyk and Modrezejewska (1999) vigorously decomposed starch, the sources of which were the remainders of binders and of kneaded bread, which used to be utilized for removing the dirt from painting surfaces.

Among pigments, a scanty source of carbon for microorganisms can be organic pigments, such as vegetable black and earth pigments: umber and ochre (Petersen and Hammer, 1992; Rebricova, 1993).

Sampo and Luppi Mosca (1989) pointed out another nutrient source for fungi can be materials used for the restoration of mural paintings: calcium caseinate, animal glue, and, to a minor extent, vinyl polyacetate. Even Masonite® plates are attacked by fungi. Petersen and Hammer (1992) point out that hydroxypropylocellulose and ammonium carbonate solution used for cleaning paintings is a potential nutrient source for microorganisms. Fungi also decompose cellulose fibers from the remainder of desalting compresses (Berner et al., 1997). Strzelczyk et al. (1999) found that microorganisms from paintings were capable of decomposing vinyl polyalcohol films and, to a lesser degree, vinyl polyacetate films.

Synthetic polymers have begun to be used to reinforce mural paintings because of their greater resistance to microorganic attack. Heyn et al. (1996) analyzed in detail Mowiol 4–98 (consolidant based on vinyl polyalcohol), Mowilith DM 5 (vinyl acetate/butyl acrylate), Primal AC 33 (terpolymer of ethyl acrylate, methyl metacrylate, and ethyl metacrylate), and Paraloid B-72 (copolymer of methyl acrylate and ethyl metacrylate). They tested their susceptibility to the attacks of 13 strains of bacteria, fungi, and yeasts isolated from mural paintings. These researchers carried out SEM observations of polymer samples overgrown by microorganisms, studied changes in pH in the course of the culture, identified the organic acids that developed, and measured respiration of the microorganisms growing on the polymers. They found out that all the polymers studied, often used for restoration of mural paintings, were liable to microbiological decomposition. The most active of the microorganisms in decomposing synthetic polymers proved to be yeasts and bacteria, while fungi showed markedly less capacity to do so. The strains used in the experiment preferentially decomposed Mowiol 4–98 and Mowilith DM 5. As the microorganisms developed, organic acids, mainly acetic acid, developed on those consolidants. The most resistant among the polymers studied proved to be Primal AC 33 and Paraloid B-72, which were affected by microbial attack only to a low degree. Similar differences in the susceptibility of synthetic polymers were also observed by other authors (Zyska, 1977).

It has also been found in situ that unless the microclimate inside the building was improved, the attack of microorganisms on paintings consolidated with synthetic polymers not only was not stopped, but in some cases was even enhanced. It has been found that those accelerated deterioration processes were induced by contamination of the consolidant with bacteria before its application onto the painting. Also, the consolidants changed the microclimate of the painting by obstructing water evaporation from it. Paraloid B-72 decomposition was observed after it had been applied to a 20th-century painting in Salvador da Bahia, Brazil. Microorganisms isolated from a Romanesque painting in Nonnberg, Austria, consolidated with Klucel E (hydroxypropylocellulose), were able to use it as nutrient source. Likewise, the microorganisms isolated from a 12th-century painting in Idensen, consolidated with vinyl polyacetate, were able to feed on it. In the latter case, it was found that the destruction of polyacetate films was started by microorganisms living in the deeper layers. A drastic increase in microorganism activity was found with the use of methyl cellulose for sticking Japanese tissue-paper on the surface of mural paintings and other works of art (Petersen et al., 1993; Petersen, 1995). Furthermore, Rebricova (1993) observed fungal growth on vinyl polyacetate under laboratory conditions as well as on mural paintings previously consolidated with it.

Organic acids, which may be secreted by the microorganisms growing on paintings, dissolve the mineral components of mural paintings, particularly $CaCO_3$. Secretion of acids by microorganisms deteriorating mural paintings has been detected by many authors (Petushkova and Lyalikova, 1986; Weirich, 1989; Petersen and Hammer, 1992; Petersen et al., 1995). The most frequently detected were oxalic, formic, citric, gluconic, and acetic acids. Some of those acids are capable of forming complex compounds with bi- and trivalent metal ions coming from the mineral components of the substratum, thus changing its qualitative composition. The acidity of the substratum is increased by CO2 forming as a result of microbial respiration and reacting with water to produce carbonic acid. Acids, both organic and mineral, reduce the pH at the place on the painting attacked by microorganisms and dissolve the $CaCO_3$. Carbonic acid reacts with the insoluble $CaCO_3$, which results in the formation of soluble acidic calcium carbonate:

$$CaCO_3 + H_2CO_3 = Ca(HCO_3)_2$$

That process leads to a deterioration of the plaster's binding properties, to its dissolution, and to partial weathering.

Other processes affecting paintings are oxidation, reduction, and transfer of metal ions, which may result in changes in color in metal-containing pigments. An example can be the change of blue azurite ($2\ CuCO_3 \cdot Cu(OH)_2$) into green copper acetate effected by acetic acid according to the reaction:

$$2\ CuCO_3 \cdot Cu(OH)_2 + CH_3COOH = 3\ Cu(CH_3COO)_2 + 2\ CO_2 + 4\ H_2O$$

Umber has been observed to fade due to the influence of fungi, which under laboratory conditions showed extreme tolerance toward iron, and even mercury (Petersen and Hammer, 1992). Sorlini et al. (1987) demonstrated that some isolated fungal strains were capable of growth on gelatin media with an addition of malachite, and they caused the pigment to change its color.

Petushkova and Lyalikova (1986) found that the microorganisms living on the mural paintings in St. John Divine Church in Rostov the Great (Yaroslav region of Russia) were responsible for the oxidation of white lead ($2\ PbCO_3 \cdot Pb(OH)_2$) to lead dioxide (PbO_2), the presence of which was detected in the brown and black spots on the surface of the painting. It has been found that those microorganisms are also able to oxidize other lead pigments used in mural paintings: yellow massicot (PbO) and red lead (Pb_3O_4). Microscopic studies of media in which lead was transformed revealed large numbers of bacteria of the *Mycoplasma* group. The capability for oxidizing white lead and massicot was found in 13 out of 18 bacterial strains isolated from stains on paintings, even if there were no lead-containing pigments there. Some of those bacteria were identified as belonging to the genus *Arthrobacter*. The authors suggest two possible mechanisms of transformation of lead-containing pigments by the bacteria. One is oxidation in a reaction with hydrogen peroxide, which is a harmful metabolite of many microorganisms:

$$2\ PbCO_3 \cdot Pb(OH)_2 + 3\ H_2O_2 = 3\ PbO_2 + 2\ H_2CO_3 + H_2O$$

The products of this reaction, carbonic acid and lead dioxide, which is a strong oxidant, increase the deterioration of the paintings.

Other bacteria studied by Petushkova and Lyalikova were able to transform the lead pigments mentioned above to black lead sulfide. That compound formed as a result of a reaction with hydrogen sulfide, a byproduct of decomposition of sulfur containing amino acids. On the paintings studied by the authors the bacteria could use animal glue as source of sulfur.

Alterations of colors on paintings can occur as a result not only of chemical reactions of metabolites secreted by microorganisms with components of the painting, but also as a result of the production by microorganisms of metabolites which themselves are color compounds. Those compounds can occur in the spores and, sometimes, in mycelial hyphae, or they can be excreted into the painting. *Cladosporium* strains, like other representatives of the *Dematiaceae*, have melanin in their spores and/or hyphae, which is why their growth can be seen in the form of the commonly occurring black or brown spots or deposits. The greenish stains on the paintings' surface can be accounted for by the presence of fungi, e.g., *Penicillium spp.* and *Aspergillus spp.*, many of which produce green spores, or by that of algae if the place is sufficiently damp and lighted. The microorganisms responsible for white, powdery deposits or scarcely visible veils can be streptomycete strains (Bassi and Giacobini, 1973; Giacobini et al., 1987; Strzelczyk et al., 1999). Bacteria causing red stains were found to contain pigments of the carotenoid group (Rölleke et al., 1998; Oltmanns et al., 1999). The ability to secrete color compounds into the substratum has been found in vitro in strains of *Aspergillus sydowii* and *Chrysosporium pannorum* isolated from paintings (Berner et. al., 1997), as well as in many streptomycetes and strains of *Chaetomium* and *Fusarium*. Some of those compounds can be antibiotiotic in character or can change their colors depending on the pH of the substratum.

Because of the diversity of components and of surrounding conditions, there are no two identical paintings, under identical conditions, and with identical microbiological problems. Not all mechanisms of mural painting deterioration are well understood. In fact, all the microorganisms that can develop on paintings have not even been detected yet. However, with current knowledge on the subject, it is possible to work out the right means to stop the deterioration processes. The susceptibility to microbiological attack of even the most advanced synthetic conservation materials leads to the conclusion that the only effective way of preventing the attack is ensuring adequate microclimate, in which microorganisms are unable to develop.

ACKNOWLEDGMENTS The author wishes to acknowledge her gratitude to Prof. Alicja B. Strzelczyk, Bogumiła Rouba, and Maria Poksińska for valuable remarks in the course of her work on the present review.

REFERENCES

AGRAWAL, O.P., DHAWAN, S., GARG, K.L., SHAHEEN, F., PATHAK, N., and MISRA, A. 1988. Study of Biodeterioration of the Ajanta Wall Paintings. *International Biodeterioration* 24, 121–129.

AGRAWAL, O.P., DHAWAN S., AND GARG, K.L. 1989. Microbial deterioration of paintings—a review. *Intach Conservation Centre, Lucknow*, India.

ALBERTANO, P., LUONGO, L., and GRILLI CAIOLA, M. 1991. Influence of different lights on mixed cultures of microalgae from ancient frescoes. *International Biodeterioration* 27, 27–38.

ANDREOLI, C., and RASCIO, N. 1982. Preliminary report on two prokaryotic forms from a pink powder growing on Villa Adriana's (Tivoli) frescoes. *Caryologia* 35, 350–351.

ARAI, H. 1983. Microbiological studies on the conservation of mural paintings in tumuli. In: Emoto, Y. and Miura, S. (Eds.), *International Symposium on the Conservation and Restoration of Cultural Property. Conservation and Restoration of Mural Paintings.* Tokyo National Research Institute for Cultural Property, Tokyo, pp. 117–124.

ARAI, H. 1984. Antimicrobial factors found in virgin tumuli. In: Barry, S., Houghton, D.R., Llewellyn, G.C., and O'Rear, C.E. (Eds.), *Biodeterioration 6, Proceedings of the 6th International Biodeterioration Symposium*, C.A.B. International, Slough, UK, pp. 363–368.

ARAI, H. 1993. Relationship between fungi and brown spots found in various materials. *Biodeterioration of Cultural Property: Proceedings of the 2nd Internatinal Conference*, Oct. 5–8, 1992, Yokohama, Japan. International Communication Specialists, pp. 320–336.

BASSI, M., and GIACOBINI C. 1973. Scanning electron microscopy: A new technique in the study of the microbiology of works of art. *International Biodeterioration Bulletin* 9, 57–68.

BERNER, M., WANNER, G., and LUBITZ W. 1997. A comparative study of the fungal flora present in medieval wall paintings in the chapel of the Castle Herberstein and in the parish church of St Georgen in Styria, Austria. *International Biodeterioration and Biodegradation* 40, 53–61.

BIANCHI, A., FAVALI, M.A., BARBIERI, N., and BASSI, M. 1980. The use of fungicides on mold-covered frescoes in S. Eusebio in Pavia. *International Biodeterioration Bulletin* 16, 45–51.

BOCK, E., and SAND, W. 1993. The microbiology of masonry biodeterioration. *Journal of Applied Bacteriology* 74, 503–514.

BYSTREK, J. 1997. *Podstawy lichenologii*, Wyd. UMCS, Lublin.

CARRUBA, G., SPERA, G., CARETTA, G., D'ALÓ F., RADAELLI, A., GUGLIELMINETTI, M., PONTIERI, E., SPOSINI, T., BISONI, F., and DE GIULI MORGHEN, C. 1999. Identification, molecular biotyping and utrastructural sudies of bacteria found in two frescoes of the monastery of St. Damian Assisi. In: *Extended Abstracts: Of Microbes and Art, an International Conference on Microbiology and Conservation*, Florence, pp. 29–31.

CIFERRI, O. 1999. Microbial degradation of paintings. *Applied and Environmental Microbiology* 65, 879–885.

CURRI, S.B. 1979. Biocide testing and enzymological study on damaged stone and frescoes surfaces: Preparation of antibiograms. *Biochemistry and Experimental Biology* 15, 97–104.

GARG, K.L., KAMAL K.J., and MISHRA, A.K. 1995. Role of fungi in the deterioration of wall paintings. *The Science of the Total Environment* 167, 255–271.

GIACOBINI, C., DE CICCO, M.A., TIGLIE, I., and ACCARDO, G. 1987. *Actinomycetes* and biodeterioration in the field of fine art. In: Houghton, D.R., Smith, R.N., and Eggins, H.O.W. (Eds.), *Biodeterioration* 7, Elsevier Applied Science, London & New York, pp. 418–423.

GRIFFIN, P.S., INDICTOR, N., and KOESTLER R.J. (1991). The biodeterioration of stone: A review of deterioration mechanisms, conservation case histories and treatment. *International Biodeterioration* 28, 189–202.

GRILLI CAIOLA, M., FORNI C., and ALBERTANO P. 1987. Characterization of the algal flora growing on ancient roman frescoes. *Phycologia* 26, 387–390.

GRZESIKOWA, H. 1965. *Nadmierne zawilgocenie budowli jako czynnik niszczący malowidła ścienne*, Bibl. Muzealnictwa i Ochrony Zabytków, Seria B, Tom XI, 156–179.

GUGLIELMINETTI, M., DE GIULI MORGHEN, C., RADAELLI, A., BISTONI, F., CARRUBA, G., SPERA, G., and CARETTA, G. 1994. Mycological and ultrastructural studies to evaluate biodeterioration of mural paintings. Detection of fungi and mites in frescoes of the monastery of St Damian in Assisi. *International Biodeterioration and Biodegradation* 33, 269–283.

HADJIVULCHEVA, E.N., and GESHEVA, R. 1982. Actinomycetes isolated from the Boyana Church mural paintings. *Comptes Rendus de l'Academie Bulgare Des Sciences* 35, 71–74.

HADJIVULCHEVA, E.N., and MARKOVA, T.H. 1983. Model study of the intensification of atmosperic ammonia in samples of mural paintings from the Boyana Church. *Comptes Rendus de l'Academie Bulgare Des Sciences* 36, 1431–1434.

HEYN, CH., PETERSEN, K., and KRUMBEIN, W.E. 1996. Untersuchungen zum mikrobiellen Abbau in der Denkmalpflege eingesetzter synthetischer Polymere, *Zeitschrift für Kunsttechnologie und Konservierung* 1/10, 87–106.

HEYRMAN, J., MERGAERT, J., and SWINGS, J. 1999. Diversity of bacterial heterotrophs present in biofilms on roman and medieval mural paintings. In: *Extended Abstracts: Of Microbes and Art, an International Conference on Microbiology and Conservation*, Florence, pp. 96–101.

IONITA, I. 1973. Contributions to the study of the biodeterioration of the works of art and historical monuments. IV. Fungi involved in the deterioration of mural painting from the Monasteries of Moldavia. *Revue Roumaine de Biologie – Botanique* 18, 179–189.

JEFFRIES, P. 1986. Growth of *Beauvaria alba* on mural paintings in Canterbury Cathedral, *International Biodeterioration* 22, 11–13.

KARPOVICH-TATE, N., and REBRIKOVA, N.L. 1990. Microbial communities on damaged frescoes and building materials in the Cathedral of the Nativity of the Virgin in the Pafnutii-Borowskii Monastery, Russia. *International Biodeterioration* 27, 281–296.

KRUMBEIN, W.E., and PETERSEN, K. 1990. Mikroorganismen beschleunigen den Zerfall mittelalterlicher Wandgemälde, Arbeitsh. Denkmalpfl. *Niedersachsen* 8, 115–120.

LAZAR, I. 1971. Investigations on the presence and role of bacteria in deteriorated zones of Cozia Monastery painting. *Revue Roumaine de Biologie – Botanique* 16, 437–444.

LEFEVRE, M. 1974. La "maladie verte" de Lascaux. *Studies in Conservation* 19, 126–156.

LEŹNICKA, S. 1991. Rola drobnoustrojów w niszczeniu zabytków z kamienia. *Postępy Mikrobiologii* 30, 375–393.

MAZUR, A. 1990. *Przyczyny niszczenia malowideł ściennych*, ASP, Warsaw.

MORTON, L.H.G., and SURMAN, S.B. 1994. Biofilms in Biodeterioration—a Review, *International Biodeterioration and Biodegradation* 34, 203–221.

O'NEILL, T.B. 1988. Succession and interrelationships of microorganisms on painted surfaces. *International Biodeterioration* 24, 373–379.

OLTMANNS, T., FÖLSTER, H., and PETERSEN, K. 1999. Characterization of red-pigmented bacteria from wall paintings. In: *Extended Abstracts: Of Microbes and Art, an International Conference on Microbiology and Conservation*, Florence, pp. 86–90.

PALMER, R.J., JR, and WHITE, D.C. 1997. Developmental biology of biofilms: implications for treatment and control. *Trends in Microbiology* 5, 435–440.

PETERSEN, K. 1996. Aspects of microbial "subsurface colonization" of mural paintings. *Restauratorenblätter* 16, 103–110.

PETERSEN, K., and HAMMER, I. 1992. Biodeterioration of romanesque wall paintings under salt stress in the Nonnberg Abbey, Salzburg, Austria. In: Emoto, Y. and Miura, S. (Eds.), *International Symposium on the Conservation and Restoration of Cultural Property. Conservation and Restoration of Mural Paintings*. Tokyo National Research Institute for Cultural Property, Tokyo, pp. 263–277.

PETERSEN, K., HEYN, CH., and KRUMBEIN, W.E. 1993. Degradation of synthetic consolidants used in mural paintings restoration by microorganisms. In: *Proc. of Journées d'Études de la S.F.I.I.C.*, Dijon, pp. 47–58.

PETERSEN K., YUN, Y., and KRUMBEIN, W.E. 1995. On the occurence of alkalitolerant and alkaliphilic microorganisms on wall paintings and their interaction in restoration/consolidation. In: *Proc. of the 3rd International Conference on Biodeterioration of Cultural Property*, 1995, Bangkok, pp. 371–385.

PETUSHKOVA, J.P., and LYALIKOVA, N.N. 1986. Microbiological degradation of lead-containing pigments in mural paintings. *Studies in Conservation* 31, 65–69.

PIÑAR, G., LUBITZ, W., RÖLLEKE, S., and GURTNER, C. 1999. Identification of Archaea in deteriorated ancient wall paintings by molecular means. In: *Extended Abstracts: Of Microbes and Art, an International Conference on Microbiology and Conservation*, Florence, pp. 46–50.

RASCHLE, P. 1983. Experience of combating moulds during restoration of ceiling paintings in a Swiss baroque monastery church. In: Oxley T.A., and Barry, S. (Eds.), *Biodeterioration* 5, John Wiley & Sons Ltd, New York, pp. 427–433.

REBRICOVA, N.L. 1989. Some ecological aspects of protection of old russian wall paintings from microbiological deterioration. In: *Proceedings of the International Conference on Biodeterioration of Cultural Property*, Lucknow, India, pp. 20–25.

REBRICOVA, N.L. 1993. Micromycetes taking part in deterioration of old Russian wall paintings. In: Garg, K.L., Garg, N., and Mukerji, K.G. (Eds.), *Recent Advances in Biodeterioration and Biodegradation*, Vol. I, pp. 205–232.

RÖLLEKE, S., MUYZER, G., WAWER, C., WANNER, G., and LUBITZ, W. 1996. Identification of bacteria in a biodegraded wall painting by denaturing gradient gel electrophoresis of PCR-amplified gene fragments coding for 16S r RNA. *Applied Environmental Microbiology* 62, 2059–2065.

RÖLLEKE, S., WITTE, A., WANNER, G., and LUBITZ, W. 1998. Medieval wall paintings—a habitat for archaea: Identification of archaea by denaturing gradient gel electrophoresis (DGGE) of PCR-amplified gene fragments coding for 16S r RNA in a medieval wall painting. *International Biodeterioration and Biodegradation* 41, 85–92.

ROSA, H., and STRZELCZYK, A. 1979. Badania nad zdolnością tworzenia śluzów przez grzyby wyodrębnione ze skamieniałych książek. *Acta Universitatis N. Copernici*, Nauki Humanist.-Społ. 91, 119–128.

ROZNERSKA, M., and MIKOŁAJCZYK, P. 1995. Malarstwo ścienne. *Przyczyny powstawania zniszczeń*. Wydawnictwo UMK, Toruń.

SAIZ-JIMENEZ, C., and SAMSON, R.A. 1981. Microorganisms and environmental pollution as deteriorationg agents of the frescoes of the monastery "Santa Maria de la Rabida", Huelva, Spain. 6th Triennial Meeting of the International Council of Museums Committee for Conservation, Ottawa, Canada.

SAMPÔ, S., and LUPPI MOSCA, A.M. 1989. A study of the fungi occuring on 15th century frescoes in Florence, Italy. *International Biodeterioration* 25, 343–353.

SAVULESCU, A., and IONITA, I. 1971. Contributions to the study of the biodeterioration of the works of art and historic monuments. I. Species of fungi isolated from frescoes. *Revue Roumaine de Biologie – Botanique* 16, 201–206.

SEAWARD, M.R.D., and GIACOBINI, C. 1988. Lichen-induced biodeterioration of Italian monuments, frescoes and other archaeological materials. *Studia Geobotanica* 8, 3–11.

ŚLESIŃSKI, W. 1983a. Techniki malarskie. *Spoiwa mineralne*. Arkady, Warsaw.

ŚLESIŃSKI, W. 1983b. Techniki malarskie. *Spoiwa organiczne*. Arkady, Warsaw.

SORLINI, C., SACCHI, M., AND FERRARI, A. 1987. Microbiological deteroration of Gambara's frescoes exposed to open air in Brescia, Italy. *International Biodeterioration Bulletin* 23, 167–179.

SORLINI, C., ALLIEVI, L., SACCHI, M., and FERRARI, A. 1982. Microorganisms present in deteriorated materials of the "Palazzo della Ragione" in Milan. *International Biodeterioration Bulletin* 18, 105–110.

STRZELCZYK, A.B. 1981a. Paintings and sculptures. In: Rose, A.H. (Ed.), *Economic Microbiology, Microbial Biodeterioration*, Vol. 6. Academic Press, London, pp. 203–234.

STRZELCZYK, A.B. 1981b. Stone. In: Rose, A.H. (Ed.), *Economic Microbiology, Microbial Biodeterioration*, Vol. 6., Academic Press, London, 61–79.

STRZELCZYK, A.B. , and MODRZEJEWSKA, J. 1999. In: *Extended Abstracts: Of Microbes and Art, an International Conference on Microbiology and Conservation*, Florence.

TOMASELLI, L., MARGHERI, M.C., and FLORENZANO, G., 1979. Indagine sperimentale sul ruolo dei cianobatteri e dlle microalghe nel deterioramento di monumenti ed affreschi. In: *3° Congresso Internazionale sul Deterioramento e la Conservazione della Pietra*, Venice, Padua, Italy.

TONOLO, A., and GIACOBINI, C. 1961. Microbiological changes on frescoes. *Contributions to the IIC Rome Conference*, pp. 62–64.

TONOLO, A., and GIACOBINI, C. 1963. Microbiological changes on frescoes. In: Thomson, G. (Ed.), *Recent Advances in Conservation*, Butterworths, London, 62–64.

WAŻNY, J. 1965. Mikroorganizmy występujące na malowidłach ściennych, *Bibl. Muzealnictwa i Ochrony Zabytków*, Seria B, XI, pp. 151–156.

WEIRICH G., 1989. Untersuchungen bei Mikroorganismen von Wandmalereien, *Material und Organismen* 24, 139–159.

ZYSKA, B. 1977. *Mikrobiologiczna korozja materiałów*, Wyd. Nauk.-Techn., Warsaw.

P. ALBERTANO

METHODOLOGICAL APPROACHES TO THE STUDY OF STONE ALTERATION CAUSED BY CYANOBACTERIAL BIOFILMS IN HYPOGEAN ENVIRONMENTS

University of Rome "Tor Vergata," Rome, Italy

ABSTRACT The synergistic action of physicochemical and biological agents causes severe deterioration problems on stone surfaces in hypogean archaeological sites. Methods presently applied to investigate the structural and ecophysiological features of phototrophic microbial communities that develop in proximity to lamps will allow us to understand the damage mechanism and to set up control and prevention strategies to limit the growth of cyanobacteria-dominated biofilms.

INTRODUCTION Methods to control, prevent, and monitor phototrophic biofilms that cause damage to rock surfaces are presently very scarce, although the colonization of stone monuments and archaeological remains by this type of microbial community is widespread all around the world and particularly abundant in humid areas. Exposed lithic faces are the natural substratum to which microorganisms adhere and grow according to the bioreceptivity of rocks and the environmental selection. Microbial diversity and biodeteriorative activity are dependent on the carbon and energy sources of microorganisms, whose metabolic activity produces irreversible changes and biomineralization of substrata. The presence of bacteria, fungi, and phototrophic organisms can influence mineral formation and dissolution, and increase physical and chemical damage of stones exposed to natural and anthropogenic alteration (Erlich, 1996; Warscheid and Braams, 2000). In this type of terrestrial environment, most of the organisms settle at the surface and form more or less thick biofilms that originate from the development of airborne cells and spores. However, microbial colonization may produce different patterns and appear as a patchy distribution of cells that accumulate in fissures, cracks, or subsurface and deep layers, depending on the porosity and state of conservation of the material, as well as on the ecological requirements of individual species (Dormieden et al., 2000; Tomaselli et al., 2000; Videla et al., 2000; Ascaso et al., 2002).

CYANOBACTERIA-DOMINATED BIOFILMS IN HYPOGEA In underground archaeological sites, colonizing microorganisms are usually distributed on the mineral surface layer, while few develop beneath it or actively bore into the calcareous substrata. These sites are usually characterized by high relative humidity, constant temperature throughout the year, and extremely low photon flux densities provided by sunlight penetrating through holes in the ceilings or artificial light sources. However, the light available for photosynthesis is usually provided by the illumination systems used to allow viewing of paintings, frescoes, stuccoes, and marbles of historic and artistic value during visitors' hours. Close to the entrances and throughout the photic zone, phototrophic microorganisms form more or less thick, differently pigmented biofilms, which cause aesthetic, physical, and chemical damage to the stone surfaces. Terrestrial cyanobacteria occur as the dominant phototrophic microorganisms, since few species of eukaryotes are able to develop under the low irradiance conditions available for photosynthesis, and only terrestrial species of sciaphilous coccal and filamentous green algae, diatoms, and mosses have been reported. The species composition of these epilithic microbial communities is, therefore, mostly determined by the available light (Albertano, 1993).

However, bacteria, and to a lesser extent fungi, are always present in the same microhabitats colonized by cyanobacteria and microalgae, and populations of heterotrophic bacteria, mostly actinobacteria, have been found in close association with phototrophs (Albertano and Urzì, 1999). Generally, a synergistic biodeteriorative effect on stone surfaces is possibly achieved by the concomitant growth of phototrophic and heterotrophic populations, as has been shown for other microbial communities in confined environments. Bacteria and fungi, in fact, are able to use the organic matter produced by phototrophs to release acidic organic compounds that dissolve the minerals of the substratum.

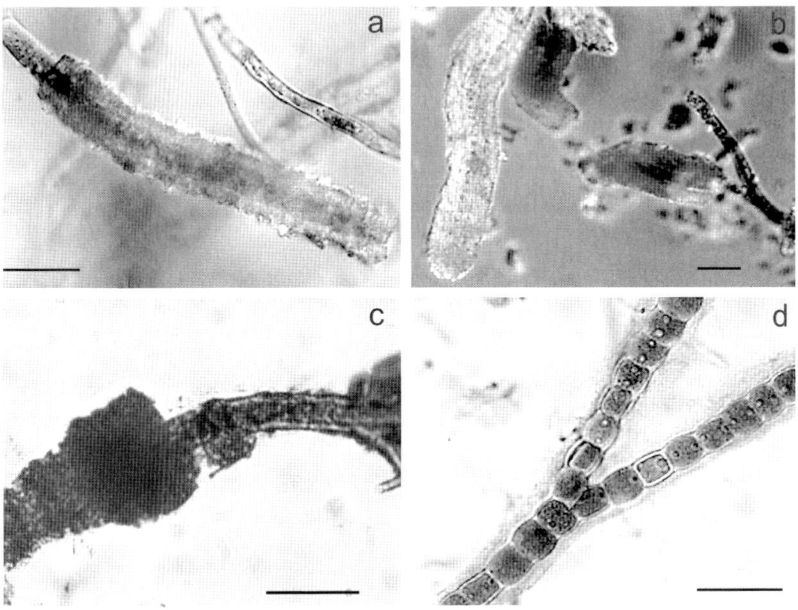

[1] Light micrographs of *Scytonema julianum* (a, c) and *Loriella osteophila* (b, d), two calcifying cyanobacteria present in the Christian catacombs of Rome, Italy. Both these species show heavy carbonate precipitates on the polysaccharide sheaths that surround their trichomes. Cytochemical staining with Ruthenium Red (c) and Alcian blue pH 2.5 (d) evidenced the abundance of acidic polysaccharides. Bars = 10 μm.

CURRENT METHODOLOGICAL APPROACHES Most studies of microbial deterioration have dealt with the quantification and identification of the different types of microorganisms within the stone community. Chroococcal and filamentous cyanobacteria have been identified as the most important photosynthetic organisms in Roman hypogea (Albertano and Urzì, 1999). To date, the list of cyanobacterial taxa found in catacombs and necropolises includes about 20 taxa. Among these, the calcifying species *Scytonema julianum* and *Loriella osteophila* [FIGS. 1–4] might be considered the most deleterious because of their ability to mobilize calcium ions from calcareous substrata, as do the other cave-inhabiting cyanobacteria *Geitleria calcarea* and *Hepyzonema pulverulentum* (Hernandéz-Mariné and Canals, 1994; Arino et al., 1997; Hernandéz-Mariné et al., 1999). However, species belonging to the genus *Leptolyngbya* and *Fischerella* [FIG. 4B, D] are also able to bond significant amounts of calcium to their sheaths (Albertano, 1997).

The application of light and electron microscopy to the study of biofilm-forming microbial communities offers valuable possibilities in the visualization of cyanobacterial biofilms, providing the means to characterize structural interactions between species, and in resolving ultrastructural details at a micro- and nanometer scale. A variety of microscopy techniques have been used for the examination of microorganisms in the hypogean environment to assess their taxonomic affiliation and nutritional status (Albertano et al., 1991b, c, 1994, 1995; Albertano, 1997). Light microscopy [FIG. 1], epifluorescence, confocal laser microscopy [FIG. 2], scanning [FIG. 3], and transmission electron microscopy [FIG. 4] allow us to understand the organism relationships within the polymicrobial population and the interactions between mineral substrata and microorganisms (Urzì and Albertano, 2001). It should be noted that many microorganisms that develop in the hypogean environment are unknown at the genus or species level, and some of them have been reported to possess adaptive features that are interesting from an evolutionary point of view (Albertano et al., 2000a).

Man-made hypogea, like natural caves, can be considered environments with extreme light conditions, and, as with other extreme habitats, they can be thought to harbor low-diversity communities (Albertano, 1993). About 200 cave cyanobacteria and 150 eukaryotic microalgae have been reported from low-light environments all over the world, but many of them are probably not true cavernicolous species (Hoffmann, 2002). Identification based on morphological characteristics is sometimes difficult. In addition, studies of the 16S rRNA of other cyanobacterial mats showed the presence of many genotypically different cyanobacterial populations with phenotypes morphologically and cytologically similar or essentially

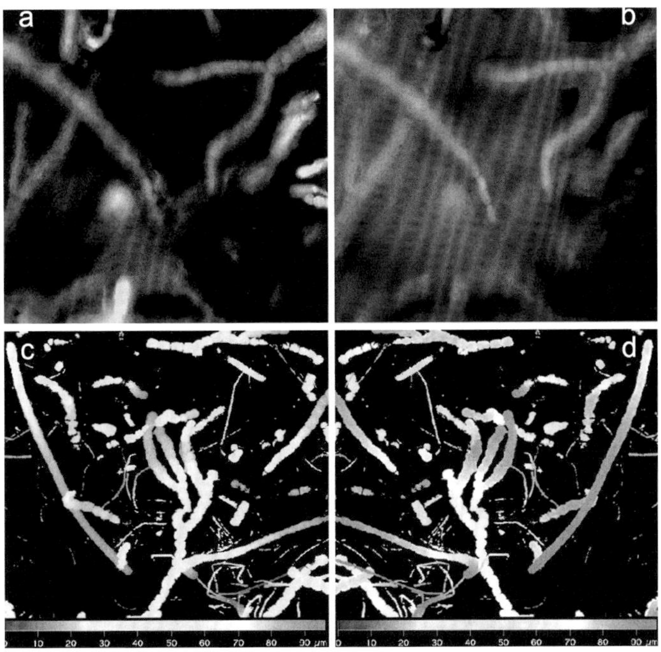

[2] CLSM of intact biofilms after labeling with FITC-Concanavalin A. The filamentous cyanobacterium *L. osteophila* was imaged using 488 (blue) and 543 nm (green) excitation to observe cell autofluorescence (a) and EPS localization (b). The 2-D reconstruction of the biofilm allowed an understanding of the depth distribution of filaments within the biofilm thickness in front (c) and rear (d) view.

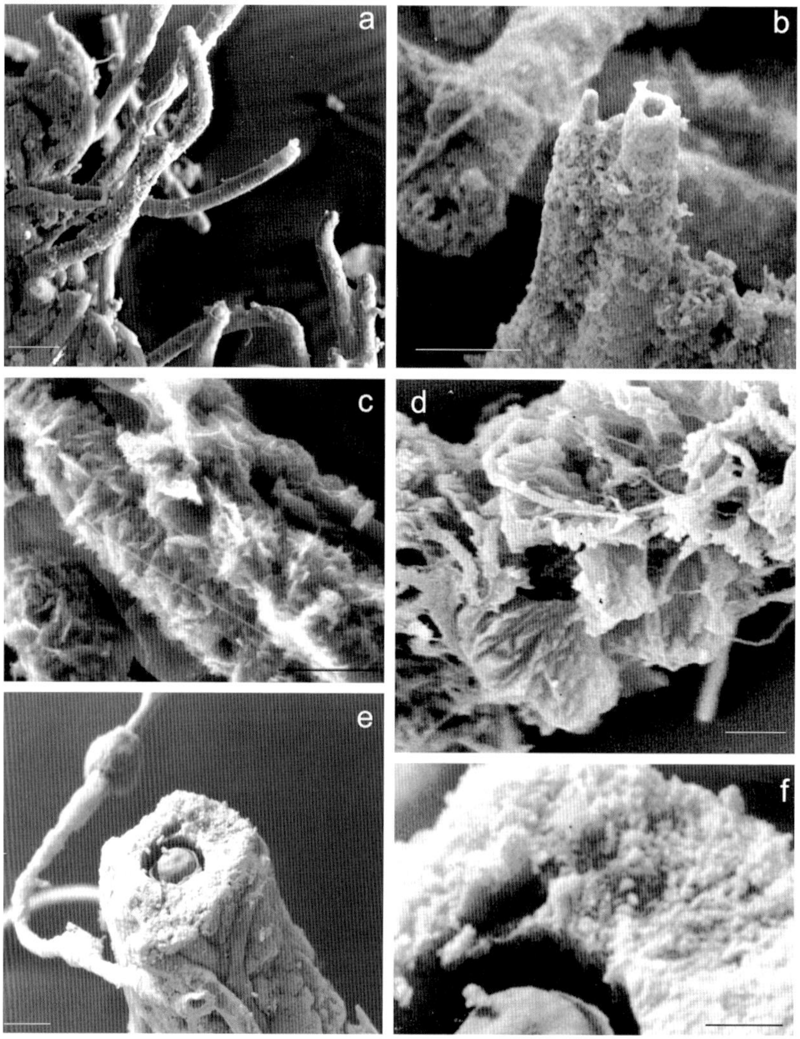

[3] *S. julianum* (a–d) and *L. osteophila* (e, f). Carbonate precipitates were observed either in hydrated (c, d) or conventionally fixed (a, b, e, f) specimens using ESEM and SEM, respectively. Note the different shape of crystals deposited on the surface or within the polysaccharide layers. Bars: (a)(b) = 10 μm, (c) = 5 μm, (e) = 2 μm, (d)(f) = 1 μm.

identical. In fact, although the greatly diverse morphology of many groups of cyanobacteria allows for their visual identification, it cannot be assumed that morphologically similar cyanobacteria, e.g., the widespread oscillatorialean *Leptolyngbya* species, do have the same genotype. It will be, therefore, essential to match genotypes with phenotypes of terrestrial cyanobacteria inhabiting hypogea, and molecular techniques are now available and are being used for phototrophic microbial communities (Ward and Castenholz, 2000).

THE ROLE OF BIOPOLYMERS Apart from biodiversity, there is another crucial aspect to be considered in the assessment of the biodeteriorative action of microbes on monuments. It involves understanding the function of biofilms and of the metabolic machinery that leads to the transformation of the lithic substrata and to a consequent change of the monument ecosystem. New approaches are, therefore, focusing on these aspects in an attempt to provide the information needed for the establishment of biodeterioration control and prevention strategies.

Microbial strategies for adhesion to stone surfaces seem generally to be based on the production and secretion at the cell surface of mucilaginous compounds. Exopolymeric substances secreted by the microorganisms (glycocalyx, sheath or envelope) act as adhesives that allow the cells to stick to the substratum, and their adhesive properties contribute to the formation and cohesion of biofilms. A polysaccharide matrix containing cell debris and significant amounts of inorganic material adsorbed from the substrate makes up most of the cyanobacterial films that will capture airborne particles, bacteria, and spores, thus increasing the masking effect, while transforming and corroding the surfaces. In addition, the anionic nature of exopolymers [FIG. 1C, D] keeps the fibrous matrix highly hydrated, strongly adsorbs cations and dissolved organic molecules from the minerals, and stabilizes dust particles. Calcium ions can be easily extracted and precipitated on polysaccharide sheaths of cyanobacteria as calcium carbonate [FIGS. 1, 3B–F, 4D–F], and, to a greater extent, macronutrients as nitrogen and phosphorus are mobilized from the substrata and metabolized or stored within the cells. [FIG. 4B] Moreover, the photosynthetic activity that sustains the production of new biomass into the ecosystem supports the development of numerous populations of heterotrophic bacteria and fungi that graze on cyanobacterial and algal exopolysaccharides, metabolites, and cell debris [FIG. 4 A, C], thereby increasing the deterioration of stone by synergistic action. Accordingly, studies on the phototrophic microbial communities present in the Roman catacombs have been undertaken in order to characterize the nature of exopolysaccharides produced by cyanobacteria (Albertano et al., 2000b; Alber-

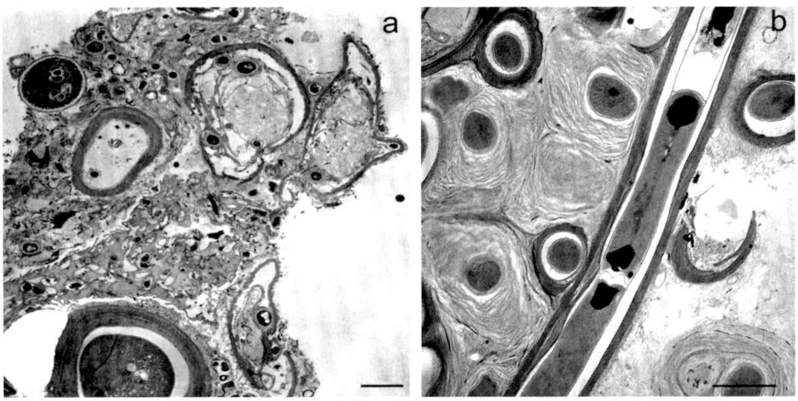

[4] TEM micrographs of biofilms built up by different cyanobacteria on lithic faces of Roman catacombs. Thin section showing interactions between phototrophic cyanobacteria and microalgae and heterotrophic bacteria interspersed among mineral particles and cell debris (a). A biofilm detail with colonies of Chroococcalean cyanobacteria, and a longitudinal section of one *Leptolyngbya* filament with large N-reserve granules of cyanophycin at the cross walls (b). Lower part of a biofilm characterised by empty cyanobacterial sheaths and cell walls of moss protonemata (c). Longitudinal section of one *Fischerella* filament with a thin calcified sheath (d). Cross sections of the calcified layer deposited on the outermost polysaccharide fibrils of the sheath in *S. julianum* (e, f). Note the positive reaction of a glycoprotein layer in the sheath (in black) after staining with phosphotungstic acid in HCL (e). All bars = 2 μm, except (f) = 1 μm.

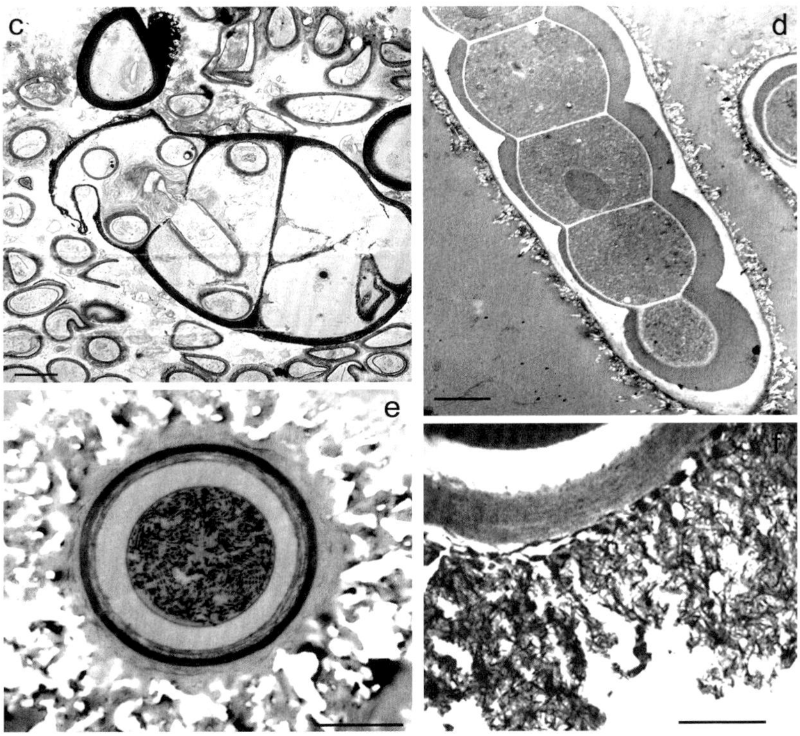
c
d
e
f

tano and Bellezza, 2001; Bellezza et al., 2003), as well as to assess their ability to make use of the available lighting in those sites and their adaptation to different light conditions (Albertano et al., 1991a; Bruno and Albertano, 1999; Bruno et al, 2001).

NONDESTRUCTIVE ASSESSMENT OF CYANOBACTERIAL ACTIVITY The increasing cultural, artistic, and religious importance gained recently by Roman hypogea has boosted interest in new in-situ noninvasive approaches of monitoring and assessing damage and biodecay processes. From this point of view, microsensors could be used to study changes in composition of chemical species mobilized upon stone surfaces during microbial metabolism, in order to evaluate the biodeteriorating activity of microorganisms on the colonized surfaces. The advantages of using microsensors include facility of assemblage, good signal stability, sensitivity, and reproducibility (Albertano and Compagnone, 1999).

The growth of cyanobacterial films, as a result of their photosynthetic and respiratory activity, can induce more or less pronounced variation of the chemical parameters that characterize the microhabitats, and that might possibly cause deleterious effects on the colonized substrata. Amperometric oxygen microsensors were applied to measure photosynthesis and respiration in phototrophic biofilms that develop in Roman catacombs in order to provide new insights on the microbial community function and biogeochemical fluxes (Compagnone et al., 1999).

Scytonema julianum and *Loriella osteophila* [FIGS.1–4], the two filamentous heterocystous cyanobacteria widespread in Roman necropolises and catacombs, can be very dangerous species because of their ability to precipitate carbonate crystals on their polysaccharide sheath and, therefore, to mobilize calcium from calcareous surfaces. The simultaneous recording of curves of photosynthesis at increasing irradiance (P/I) and the application of potentiometric microelectrodes for the measurements of pH have shown that the photosynthetic activity of these species can shift pH values to a sufficient extent to induce precipitation of calcium carbonate (Albertano et al., 2000c, d). In addition, the more recent development of K^+ microelectrodes showed that a decrease of potassium concentration occurred during light exposure in strains of *S. julianum,* possibly due to an active uptake sustained by the photosynthetic activity of these cyanobacteria, while no appreciable variation of soluble calcium due to the metabolic activity was observed in natural and artificial biofilms of the same species.

WHAT CONSERVATION STRATEGY IS SAFE FOR ROMAN HYPOGEA? Current research under the aegis of the EU project CATS is trying to develop and integrate physical and biotechnological methods of limiting microbial growth on valuable archaeological surfaces, and to apply analytical methods to monitor the presence and extent of the damage caused by phototrophic and heterotrophic microorganisms. Pilot studies will investigate the possibility of using new lighting systems providing wavelengths poorly used by cyanobacterial photosynthesis. This could drastically decrease the growth of cyanobacteria and, therefore, the quantity of organic matter available to the associated heterotrophic populations.

In addition to the physical approach, newly identified biomolecules related to iron metabolism and cell-to-cell signaling pathways could be checked for their ability to interfere with bacterial and, especially, cyanobacterial metabolism by removing factors indispensable to microbial development. Finally, the application of nondestructive microsensor methodologies to the analysis of biodeteriorative processes, and the understanding of the interactions occurring in situ among microorganisms and the surrounding physicochemical environment are currently recommended as the best approach for the assessment of conservation strategies at these archaeological sites.

ACKNOWLEDGMENTS The author wishes to thank Dr. Palma Mattioli of the University of Rome "Tor Vergata" for her skillful assistance in image improvement. This work was supported by the EU program Energy, Environment and Sustainable Development, under the CATS Project, contract EVK4-CT-2000-00028, the Research National Council of Italy, C.N.R.–Progetto Finalizzato Beni Culturali grant number 01.00472.PF36, and the Italian Ministry of University and Research, M.I.U.R.–Cofin 2001.

REFERENCES

ALBERTANO, P., 1993. Epilithic algal communities in hypogean monuments environment. *Giornale Botanico Italiano*, 127, 385–392.

ALBERTANO, P., 1997. Elemental mapping as tool in the understanding of microorganisms—substrate interactions. *Journal of Computer-assisted Microscopy*, 9, 81–84.

ALBERTANO, P., and BELLEZZA, S., 2001. Cytochemistry of cyanobacterial exopolymers in biofilms from Roman hypogea. *Nova Hedwigia*, 123, 501–518.

ALBERTANO, P., and COMPAGNONE, D., 1999. Ultrastructural and analytical approaches to the study of stone microbial communities. In: Monte, M., (Ed.), *Eurocare–Euromarble Proceedings EU 496/8*, CNR Editions, Rome, pp. 89–93.

ALBERTANO, P., and URZI, C., 1999. Structural interactions among epilithic cyanobacteria and heterotrophic microorganism in Roman hypogea. *Microbial Ecology*, 38, 244–252.

ALBERTANO, P., KOVÁCIK, L., and GRILLI CAIOLA, M., 1994. Preliminary investigations on epilithic cyanophytes from a Roman Necropolis. *Archiv für Hydrobiologie, Algological Studies*, 75, 71–74.

ALBERTANO, P., KOVÁCIK, L., MARVAN, P., and GRILLI CAIOLA, M., 1995. A terrestrial epilithic diatom from Roman Catacombs. In: Marino, D., and Montresor, M., (Eds.), *Proceedings of the Thirteenth International Diatom Symposium*, Biopress Limited, Bristol, pp. 11–21.

ALBERTANO, P., LUONGO, L., and GRILLI CAIOLA, M., 1991a. Influence of different lights on mixed cultures of microalgae from ancient frescoes. *International Biodeterioration*, 27, 27–38.

ALBERTANO, P., LUONGO, L., and GRILLI CAIOLA, M., 1991b. Observations on cell structure of micro-organisms of an epilithic phototrophic community competing for light. *Nova Hedwigia*, 53, 369–381.

ALBERTANO, P., LUONGO, L., and GRILLI CAIOLA, M., 1991c. Ultrastructural investigations on algae deteriorating Roman frescoes. In: Baer, N., Sabbioni, C., and Sors, A.I, (Eds.), *Science, Technology and European Cultural Heritage*, Butterworth-Heinemann Ltd, Oxford, pp. 501–504.

ALBERTANO, P., BARSANTI, L., PASSARELLI, V., and GUALTIERI, P., 2000a. A complex photoreceptive structure in the cyanobacterium *Leptolyngbya sp. Micron*, 31, 27–34.

ALBERTANO, P., BRUNO, L., BELLEZZA, S., and PARADOSSI, G., 2000b. Polysaccharides as a key step in bio-erosion. In: Fassino, V., (Ed.), *Proceedings of 9th International Congress on Deterioration and Conservation of Stone*, Vol. 1, Elsevier, The Netherlands, pp. 425–432.

ALBERTANO, P., BRUNO, L., D'OTTAVI, D., MOSCONE, D., and PALLESCHI, G., 2000c. The effect of photosynthesis on pH variation in cyanobacterial biofilms from Roman catacombs. *Journal of Applied Phycology*, 12, 379–384.

ALBERTANO, P., BRUNO, L., MOSCONE, D., D'OTTAVI, D., and PALLESCHI, G., 2000d. Evaluation of cyanobacterial impact on stone surfaces in Roman hypogea by using microsensors. In: Guarino, A., (Ed.), *Science and Technology for the Safeguard of Cultural Heritage in the Mediterranean Basin*, Vol. 2, Elsevier, Paris, pp. 701–703.

ARIÑO X., HERNANDÉZ-MARINÉ, M., and SAIZ-JIMENEZ, C., 1997. Colonization of Roman tombs by calcifying cyanobacteria. *Phycologia*, 36, 366–373.

ASCASO, C., WIERZCHOS, J., SOUZA-EGIPSY, V., DE LOS RIOS, A., and DELADO RODRIGUES, J., 2002. In situ evaluation of the biodeteriorating action of micro-organisms and the effects of biocides on carbonate rocks of the Jeronimos Monastery (Lisbon). *International Biodeterioration and Biodegradation*, 49, 1–12.

BELLEZZA, S., PARADOSSI, G., DE PHILIPPIS, R., and ALBERTANO, P., 2003. *Leptolyngbya sp.* strains from Roman hypogea: cytochemical and physico—chemical characterisation of exopolysaccharides. *Journal of Applied Phycology*, in press.

BRUNO, L., and ALBERTANO, P., 1999. Photoacclimation of sciaphilous epilithic cyanobacteria isolated from Roman hypogea. *Archiv für Hydrobiologie, Algological Studies*, 94, 89–103.

BRUNO, L., PIERMARINI, S., and ALBERTANO, P., 2001. Characterisation of spectral emission by cyanobacterial biofilms in the Roman Catacombs of Priscilla in Rome (Italy). *Nova Hedwigia*, 123, 229–236.

COMPAGNONE, D., DI CARLO, V., BRUNO, L., ALBERTANO, P., and PALLESCHI, G., 1999. Development of oxygen microsensor for monitoring cyanobacterial photosynthesis in Roman hypogea. *Analytical Letters*, 32, 213–222.

DORMIEDEN, T., GORBASHINA, A.A., and KRUMBEIN, W.E., 2000. Patina (physical and chemical interactions of sub—aerial biofilms with objects of art). In: Ciferri, O., Mastromei, G., and Tiano, P., (Eds.), *Microbes and Art, The Role of Microbial Communities in the Degradation and Protection of Cultural Heritage*. Plenum Press, New York, pp. 105–120.

ERLICH, H.L., 1996. How microbes influence mineral growth and dissolution. *Chemical Geology*, 132, 5–9.

HERNANDÉZ-MARINÉ, M., ASENCIO, A.D., CANALS, A., ARIÑO, X., ABOAL, M., and HOFFMANN, L., 1999. Discovery of population of the lime-incrusting genus *Loriella* (Stigonematales) in Spanish caves. *Archiv für Hydrobiologie, Algological Studies*, 94, 121–138.

HERNANDÉZ-MARINÉ, M., and CANALS, T., 1994. *Herpyzonema pulverulentum* (Mastigocladaceae) a new cavernicolous atmophytic and lime-encrusted cyanophyte. *Archiv für Hydrobiologie, Algological Studies*, 75, 123–136.

HOFFMANN, L., 2002. Caves and other low-light environments: aerophitic photoautotrophic microorganisms. In Bitton, G., (Ed.), *Encyclopedia of Environmental Microbiology*. John Wiley & Sons, New York, pp. 835–843.

TOMASELLI, L., LAMENTI, G., BOSCO, M., and TIANO, P., 2000. Biodiversity of photosynthetic micro-organisms dwelling on stone monuments. *International Biodeterioration and Biodegradation*, 46, 251–258.

URZÌ, C., and ALBERTANO, P., 2001. Studying phototrophic and heterotrophic microbial communities on stone monuments. In: *Methods in Enzymology* 336, Academic Press, San Diego, CA, pp. 340–355.

VIDELA, H.A., GUIAMET, P.S., and GOMEZ DE SARAVIA, S., 2000. Biodeterioration of Mayan archaeological sites in the Yucatan Peninsula, Mexico. *International Biodeterioration and Biodegradation*, 46, 335–341.

WARD, D.M., and CASTENHOLZ, R.W., 2000. Cyanobacteria in geothermal habitats. In: Whitton, B.A., and Potts, M., (Eds.), *The Ecology of Cyanobacteria*. Kluwer Academic Publishers, Dordrecht, pp. 37–60.

WARSCHEID, TH., and BRAAMS, J., 2000. Biodeterioration of stone: A review. *International Biodeterioration and Biodegradation*, 46, 343–368.

C. URZÌ

F. DE LEO

P. DONATO

V. LA CONO

STUDY OF MICROBIAL COMMUNITIES COLONIZING HYPOGEAN MONUMENT SURFACES USING NONDESTRUCTIVE AND DESTRUCTIVE SAMPLING METHODS

University of Messina, Italy

INTRODUCTION An important step in art conservation is the acquisition of information about the state of deterioration of the artifact being studied. It is often necessary to sample significant amounts from a valuable object in order to clarify whether (micro-)organisms are involved, which type are involved, and the relationship that these have with the material itself. Within the framework of the European project "CATS"—[C]yanobacteria [At]tack Rock[s]: Control and Preventive Strategies to Avoid Damage Caused by Cyanobacteria and Associated Microorganisms in Roman Hypogean Monuments—we were involved in specific tasks aimed at answering the following questions:

- Which decay processes are caused by cyanobacteria-based biofilm?
- What can we learn about the diversity and spatial distribution of microorganisms in the biofilm?
- What kind of modification can occur to the chemoorganotrophic microflora when we apply methods to control (reduce or eliminate) the cyanobacterial community?

Due to the artistic and historic value of most of the surfaces affected by biofilm, we had to choose different sampling procedures—nondestructive and/or destructive—depending upon the sampled surface, in order to get valuable information on the microbial communities colonizing those surfaces. In recent years, several nondestructive methods have been proposed to obtain direct or indirect evidence of microbial colonization (e.g., agar fingerprinting, adhesive tape strips, sterile needle sampling) and advantages and disadvantages of these have been listed by Urzì and De Leo (2001). However, destructive sampling methodologies are often unavoidable in order to get information on the pattern, depth, and extent of biodeterioration.

In this paper, we report on the results we have obtained within the framework of the CATS project, carrying out different analyses after using minimally destructive as well as nondestructive sampling methods.

EXPERIMENTAL Samples were taken in the Christian Catacombs of St. Callistus and Domitilla, in Rome, Italy, as well as in the "Zuheros" (Bats) caves in Spain, in heavily microbially colonized areas. The principal type of alterations observed were described as: (1) green patinas; (2) white veils or patinas; and (3) black to gray patinas (observed only in the dark areas). Samples were taken with a sterile scalpel from the biofilm/substrate for culture analyses and for fingerprinting of DNA extraction from microbial communities; while adhesive tape (Fungi –Tape™, Scientific Device Lab. Inc., Glenview, IL) was used mainly for direct microscopic observations and to perform fluorescent in situ hybridization (FISH) directly from the samples.

Biomass Analysis Both cultural and molecular analyses were carried out from the biomass samples. Enumeration and identification of chemoorganotrophic bacteria, fungi, and cyanobacteria were done after inoculation of the suspensions and respective decimal dilutions in suitable growth media for bacteria (BRII), fungi (DRBC), and cyanobacteria (BG11) as described by Urzì and Realini (1998).

Single-strand-conformational-polymorphism technique, or SSCP, was applied for the fingerprint of microbial population, as described by Schwieger and Tebbe (1998). Total genomic DNA from biofilm samples was extracted by using a modified protocol of Zhou et al. (1996) and amplified by universal primers (Com1, forward 5'CAGCAGCCGCGGTAATAC3', 519–536 and primer Com2-Ph [reverse] 5'CCGTCAATTCCTTTGAGTTT3', 907–926) (Life Technologies, Italy). Primer Com2-Ph contained a 5'-terminal phosphate group. PCR products were purified and the digestion of the single strand was carried out using lambda exonuclease (Pharmacia Amersham Biotech, Italy). The samples were subjected to electrophoresis in an 0.6 MDE gel (FMC Bioproducts, Rockland, Maine), and gels were stained with a silver stain system by a protocol recommended by the manufacturer (Promega).

PROBE	TARGET GROUP	PROBE SEQUENCE (5' TO 3')	TARGET SITE[a] (rRNA POSITION)	%FA[b]	REFERENCE
EUB338	*Bacteria*	GCTGCCTCC-CGTAGGAGT	16S (338)	35	Amann et. al., 1995
GPBHGC	Gram positive bacteria with high G+C content	TATAGTTAC-CACCGCCGT	23S (1901)	20	Roller et. al., 1992
Streptomyces probe	*Streptomyces* sp.	GCGTCGAAT-TAAGCCACA	16S (924)[c]	50	Stackebrandt et al., 1991

[a] *E. coli* numbering (Brosius et al., 1981). [b] Percentage formamide (FA) in hybridization buffer.
[c] *Streptomyces ambofaciens position.*

[TABLE 1] Oligonucleotide probes used in this study.

Light and Epifluorescent Microscopy Adhesive tape samples were cut into small pieces. A drop of sterile water or Acridine Orange solution (0.01%) was put between a glass slide and the tape (face up), and a cover glass was placed on top, in order to keep the tape as flat as possible (Urzì and Albertano, 2001). Direct observations of samples were carried out using a light microscope equipped with phase contrast (DML, Leica) and/or a light epifluorescent microscope (Leica Q550 IW) equipped with a mercury lamp (OSRAM HBO 50W) and 450–490 nm excitation filter.

Fluorescent In-Situ Hybridization (FISH) Applied Directly to Adhesive Tape Strips FISH technique is currently used to analyze complex communities, as well as for the in-situ identification of microorganisms and the phylogeny of unculturable microorganisms. It is usually carried out on membranes or on glass. We successfully carried it out directly on adhesive tape strips.

Adhesive tape pieces (0.5 cm × 0.5 cm) were preliminarily fixed at +4°c for 3 hr by adding three volumes of fixation buffer (4% paraformaldehyde in PBS, pH 7.2) in order to avoid enzymatic processes and allow a slight cell permeabilization, and then rinsed in saline phosphate buffer, PBS. The procedure for FISH was the one described by Stahl and Amann (1991). The oligonucleotide probes were labeled with different fluorochromes: FITC (fluorescein isothiocyanate), Cy3 (indocarbocianine), and TRITC (tetramethylrhodamin isothiocyanate) were supplied by Biosense S.r.l. (Italy). The oligonucleotide probes used are listed in TABLE 1.

After hybridization, the adhesive tape pieces were placed on microscope glass slides and mounted in glycerol-based antifading agent (SlowFade® Light Antifade kit, Molecolar Probe), a thin coverslip was placed over the material, and the excess was gently squeezed out.

The slides were observed with a Leica DMR/HCS epifluorescence microscope equipped with a 50W mercury lamp HBO and a different combination of filters at a magnification of 100x with different excitation filters depending upon the fluorochrome used.

Images were made with a photocamera using Kodak Gold Ultra film 400ASA or with a Leica Quantimed COHU high performance CCD camera Q550IW.

RESULTS AND DISCUSSION The approach used to study the microbial community colonizing hypogean surfaces yielded a better understanding of the relationships among the different populations and between microorganisms and substrate. Also, we could compare our results to those obtained by other partners in the aforementioned project (www.idronaut.it/research-project/CATS/cats.html).

The adhesive-tape sampling technique gave valuable information on the microorganisms involved in the biofilm. In fact, assessment of damage carried out through microscopic analyses of adhesive tape samples and culture analyses of biofilm demonstrated that the Roman Catacombs environment differs notably from that of the Zuheros (Bats) caves. In the first case, we could associate the different type of alterations with specific microbial communities, as summarized in the following:

- The green biofilm was based on a microbial community in which phototrophic microorganisms were prevalent. Heterotrophic bacteria, closely associated with algal filaments, were sometimes observed, while free-living bacteria were common in the biofilm. Most were Gram + rod bacteria and cocci. Fungi were rare. The presence of artificial light was clearly connected with the pattern of growth. In association with the biofilm the substrate was softer than in the noncolonized surfaces. A diffuse crystals precipitation, biologically mediated (Albertano and Urzì, 1999) was also observed in the far end of the biofilm margins.
- The white patina was mainly due to the colonization of Gram + rod bacteria, cocci, and especially filamentous bacteria (*Streptomycetes*). Phototrophic microorganisms were barely represented. Fungi were only occasionally present.

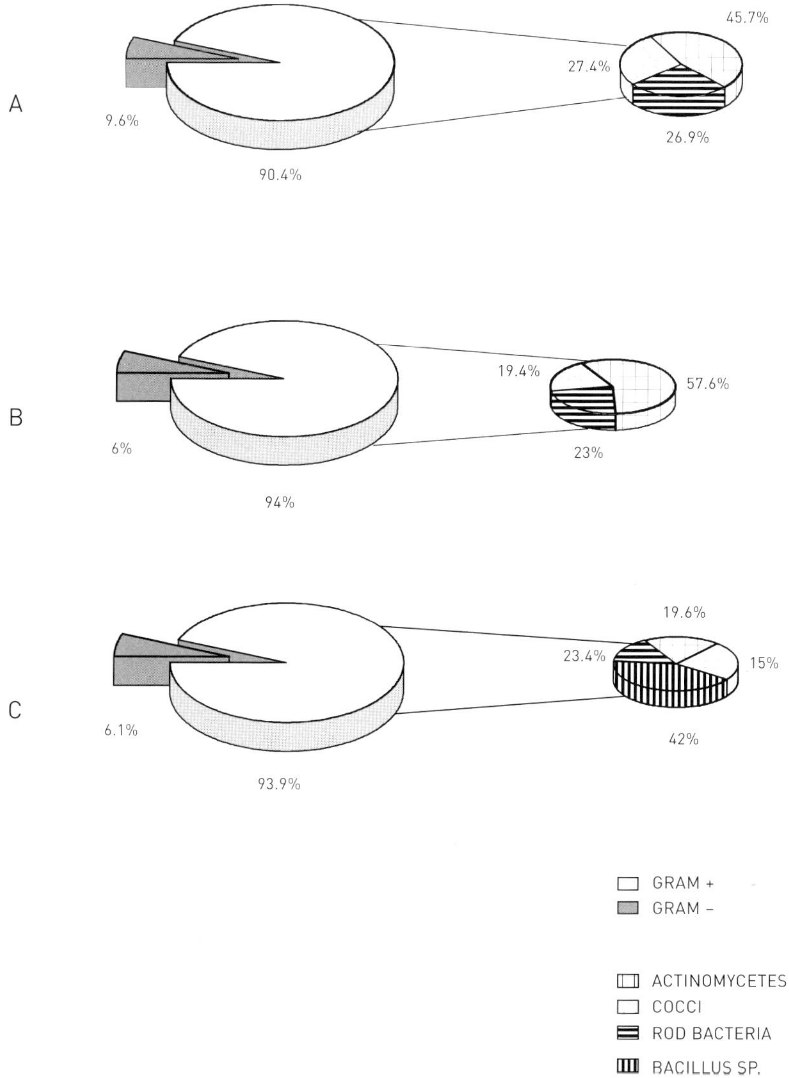

[1] Percentages of chemoorganotrophic bacteria isolated from Roman Catacombs and Zuheros Caves. (a) Domitilla Catacomb; (b) St. Callisus Catacomb; (c) Zuheros Caves.

• Gram + rod bacteria and cocci were isolated from the gray-black alterations (observed only in the dark locations), but in low amounts. Black fungi were occasionally found. No phototrophic microorganisms were isolated.
• In the Zuheros caves, the main type of alteration observed in all sites was a green patina of different color intensity. The analysis of a microbial community living in association with the alterations demonstrated that the green biofilm was almost entirely due to the colonization of phototrophic microorganisms, while the presence of heterotrophic microorganisms (bacteria and fungi) was scarce. Bacteria, mostly Gram + rod bacteria forming spores, cocci, and filamentous bacteria, were isolated infrequently. They were low in number and in diversity.

The bacterial strains isolated (549 strains) revealed a different distribution of bacterial groups, as shown in FIG. 1. In the Roman Catacombs, *Actinomyctes* were prevalent, while in the Zuheros caves spore-forming *Bacillus* were the most common isolates.

Studies of the microbial community by SSCP were very useful in demonstrating differences and similarities among the populations taken from the different biofilms. [FIG. 2] In fact, each band obtained with the SSCP technique corresponded to different genera/species. Identification of microorganisms involved, through DNA extraction from the gel and its sequencing, will be possible and is currently in progress. Comparison of sequences of the strains isolated through cultural methods and those obtained from the SSCP fingerprinting of microbial population will allow us to better understand the dynamics of bacterial populations living in the biofilm.

The FISH technique applied directly on adhesive tape was successful in allowing a better characterization of bacterial communities in biofilm. [FIGS. 3A AND B] The isolation of strains from the growth media was useful in gaining information on the most representative strains and their taxonomy; with that information we could design and choose the appropriate molecular probe to be used. Through FISH and microscopic observation we could not only determine the relationships between phototrophic and heterotrophic microorganisms [FIG. 3A], but we could also get information on the type of bacteria present. [FIG. 3B]

However, the application of the technique was limited by:

• The simultaneous presence of phototrophic and heterotrophic organisms, due to chlorophyll autofluorescence interfering with fluorochrome conjugation to the molecular probe [FIG. 3A]; and
• The choice of the right probe to use with a specific sample; in our case, of the biodeteriogen responsible for the damage.

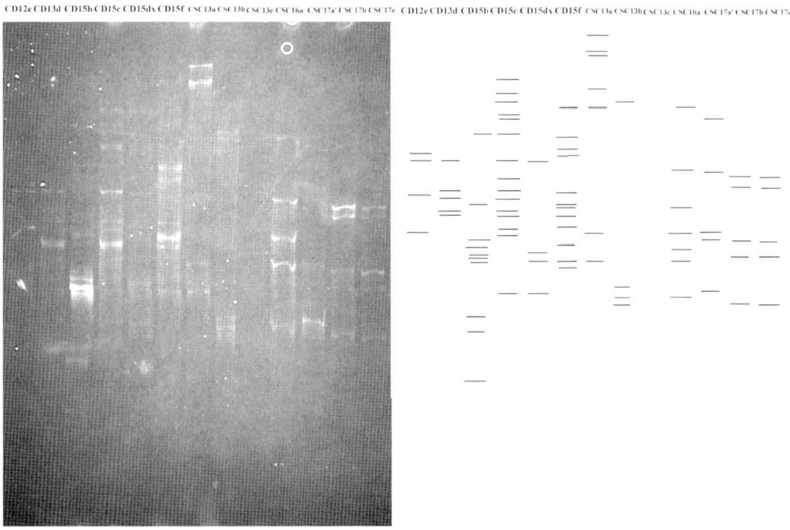

[2] SSCP electrophoretic patterns obtained from biofilm samples taken from the Roman Catacombs. Bands at the same height indicate the same organisms. Comparison of electrophoretic patterns of the microbial communities from the Roman Catacombs showed a greater biodiversity in the Domitilla samples (CD) than in the St. Callistus (CSC) ones.

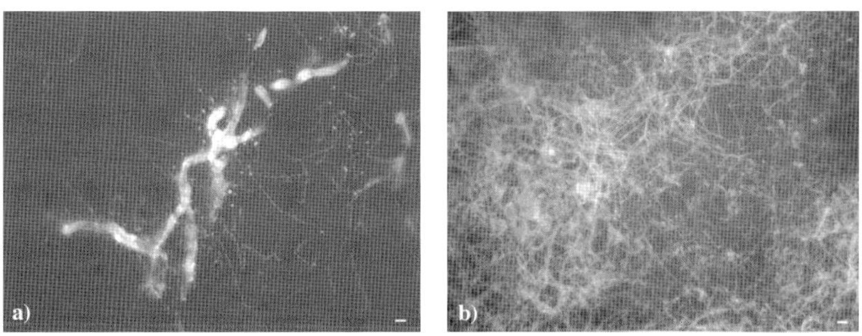

[3] Evidence of microbial community from two different samples taken with adhesive tape strips from the Roman Catacombs of Domitilla (left) and St. Callistus (right). (a) Autofluorescent filamentous algae and bacterial filaments shown by EUB 338 labeled with FITC. (b) *Streptomyces* hyphae labeled with a universal probe EUB 338 labeled with FITC. Bar is 10 μm.

CONCLUSIONS From our results, it is evident that a combination of different sampling methods is needed in order to detect microflora colonizing artistic surfaces, to minimize damage inflicted during the sampling procedure, and to get useful information concerning the microorganisms responsible for biodeterioration. Using this approach, we could answer the first two questions posed by the project—those concerning damage and the microorganisms involved.

The third question will be faced at the end of project (end of the year 2003) and again when sampling of the surfaces is required; a well-planned approach can reduce the number of samples to a minimum.

In fact, the transformations occurring in the biofilm after the application of different wavelengths of light could be easily detected both by FISH analysis (applied directly on the adhesive tape) and by the comparison of total microbial population evidenced by the SSCP technique before and after the treatments.

ACKNOWLEDGMENTS This work was carried out with the financial support of the European Community through the CATS research project (project number EVK4-CT-2000–00028), the Network COALITION (project number EVK4–1999– 20001), the Italian C.N.R. (project 01.00654.PF36), and M.I.U.R. 2000 and 2001.

REFERENCES

ALBERTANO, P., and URZÌ, C., 1999. Structural interactions among epilithic cyanobacteria and heterotrophic microorganisms in Roman hypogea. *Microbial Ecology*, 38, 244–252.

AMANN, R.I., LUDWIG, W., and SCHLEIFER, K.H., 1995. Phylogenetic identification and in situ detection of individual microbial cells without cultivation. *Microbiological Reviews*, 59, 143–169.

BROSIUS, J., DULL, T. J., SLEETER, D. D., and NOLLER, H.F., 1981. Gene organization and primary structure of a ribosomal RNA operon from *Escherichia coli*. *Journal of Molecular Biology*, 148, 107–127.

ROLLER, A., LUDWIG, W., and SCHLEIFER, K.H., 1992. Gram-positive bacteria with a high DNA G+C content are chacterized by a common insertion within their 23S rRNA genes. *Journal of General Microbiology*, 138, 1167–1175.

SCHWIEGER, F., and TEBBE, C.C., 1998. A new approach to utilize PCR-single-strand-conformation polymorphism for 16S rRNA gene-based microbial community analysis. *Applied Environmental Microbiology*, 64, 4870–6.

STACKEBRANDT, E., WITT, D., KEMMERLING, C., KROPPENSTEDT, R.M., and LIESACK, W., 1991. Designation of Streptomycete 16S and 23S rRNA-based target regions for oligonucleotide probes. *Applied and Environmental Microbiology*, 57, 1468–1477.

STAHL, D. A., and AMANN, R., 1991. Development and application of nuclcic acid probes. In: Stackebrandt, E., and Goodfellow, M., (Eds.), *Nucleic Acid Techniques in Bacterial Systematics*, John Wiley and Sons, New York, pp. 205–248.

URZÌ, C., and REALINI, M., 1998. Colour changes of Noto's calcareous sandstone as related to its colonisation by microorganisms. *International Biodeterioration and Biodegradation*, 42, 45–54.

URZÌ, C., and ALBERTANO, P., 2001. Studying phototrophic and heterotrophic microbial communities on stone monuments. In: Doyle, R.J., (Ed.), *Methods in Enzymology*, vol. 336, Academic Press, San Diego, CA, pp. 340–355.

URZÌ, C., and DE LEO, F., 2001. Sampling with adhesive tape strip: an easy and rapid method to monitor microbial colonization on monument surfaces. *Journal of Microbiological Methods*, 44, 1–11.

ZHOU, J., BRUNS, M. A., and TIEDJE, J.M., 1996. DNA recovery from soils of diverse composition. *Applied Environmental Microbiology*, 62, 316–322.

WOOD AND ARCHAEOLOGICAL MATERIALS

R.A. BLANCHETTE

DETERIORATION IN HISTORIC AND ARCHAEOLOGICAL WOODS FROM TERRESTRIAL SITES

University of Minnesota, St. Paul, MN

ABSTRACT Wood buried in terrestrial sites or exposed to the environment is rapidly degraded by microorganisms. In addition to decay by fungi and bacteria, nonbiological processes also deteriorate wood. Archaeological wood that survives for long periods is found in environments that protect it from decomposition. Although decay processes are restricted, these woods are rarely free from attack. This paper reviews the types of deterioration associated with historic and archaeological woods from a number of terrestrial sites, including Anasazi great houses approximately 1000 years old in New Mexico, USA; ancient Egyptian tombs; an 8th century, B.C., tomb found in Tumulus MM (Midas Mound), at Gordion, Turkey; and historic wood from huts built on Ross Island, Antarctica, and Ellesmere Island in the Canadian High Arctic between 1881 to 1911 by the early explorers of the North and South Polar regions. Soft-rot fungi were the major cause of wood decay in all of the woods examined. These fungi tolerated adverse environmental and substrate conditions and were able to cause extensive decay in many of the woods examined. Decay by brown-rot fungi was also found and was often associated with termite or other insect damage. In addition to microbial decay, a chemical defibration of wood was found associated with high salt concentrations. This type of attack occurred in woods from the Antarctic, Anasazi great houses, and ancient Egyptian tombs, and is more prevalent and damaging than was previously realized.

INTRODUCTION Wood and other organic matter will deteriorate over time from microbial decay and nonbiological degradation processes when exposed to the environment. In most temperate and tropical areas, wood decomposes rapidly, leaving little to no evidence of the original substrate. In some environments, however, wood is protected from biological and chemical attack and may remain relatively unaltered. But even in the most extreme environments where microbial growth is restricted and nonbiological deterioration processes are limited, wood can deteriorate. Over long periods, microbes that are able to grow in adverse conditions degrade the wood, and various compounds, such as salts, cause a slow chemical deterioration. Most, if not all, historic and archaeological woods have been affected to some degree by degradation.

Archaeological woods that survive the many forms of biological and chemical deterioration are extremely valuable resources since they can provide important information about past cultures. These woods can also be used to obtain accurate dating of materials from a site by dendrochronology observations (Schweingruber, 1993), determine where objects have originated by tree species identification (Tennessen et al., 2002), and provide information on past environments by examining chemical and biological signatures that have been left in the wood (Blanchette, 2000; Blanchette et al., 1994; Filley et al., 2001).

Most archaeological woods are found buried in saturated soils or peat bogs in a waterlogged condition. The major factor responsible for protecting the wood from decay in these environments is reduced oxygen concentrations (Blanchette, 2000). Oxygen is needed for aerobic microbes to grow, and limited oxygen concentrations impede microbial colonization and decay. It is very rare, however, for oxygen to be completely excluded in these environments and for wood to be completely free from attack. Many investigations have demonstrated that different types of degradation and varied microbial successions occur in wood when it is buried and waterlogged (Bjordal et al., 1999; Blanchette et al., 1990; Singh et al., 1990). Archaeological wood may also be found in other terrestrial environments that are not waterlogged. In these situations, extremely dry environments, cold temperatures, or other types of adverse conditions have limited the rate and extent of wood deterioration taking place. Currently there is little information available on the microorganisms or degradation processes occurring in these situations. This paper provides an assessment of the types of deterioration that have been found in historic and archaeological woods from several different terrestrial (nonwaterlogged) environments and gives insights to better understand the decay processes that affect these woods. Information on wood deterioration is important in helping to develop appropriate conservation methods so that these valuable cultural resources can be preserved. It is also becoming increasingly important for developing successful methods of in situ preservation when wooden objects are reburied or preserved at the excavation site.

MICROBIAL WOOD DECOMPOSITION Moisture is essential for microbial decay to take place, and there are few natural environments where there is no moisture. Even in buried tombs from arid regions some moisture is usually present, allowing selected microbes to grow and progressively degrade the wood. In addition to moisture, many other factors influence microorganism growth, such as temperature, pH, nitrogen, and other nutrients (Eriksson et al., 1990). Favorable conditions allow rapid colonization by fungi and fast decomposition, but unfavorable conditions usually cause other organisms that tolerate the extreme environmental conditions to dominate. The type of wood and presence of extractives within the wood cells also influences decay. Woods from cedar, juniper, cypress, redwood, and oak are more resistant to microbial decay than are pine, birch, beech, aspen, and other woods that have fewer extractives (Zabel and Morrell, 1992).

When microorganisms decay wood, they cause distinct morphological and chemical changes that are signatures of the causal organisms. Examination of the patterns of microbial attack can reveal what type of organism was responsible and provide a great deal of information on the current condition of the wood. Under aerobic conditions, fungi are the primary decomposers of wood. Fungi that cause wood decay have been classified into broad categories of white, brown, and soft rot based on the color and texture of the residual wood after decay (Blanchette, 1998; Eriksson et al., 1990). White-rot fungi are common wood degraders in forest ecosystems and have the capacity to produce extracellular enzymes that degrade all cell wall components. The type of enzymes produced and the sequence of their production govern the amount and extent of cellulose, lignin, and hemicellulose degradation from wood. Some species are very efficient degraders of cellulose while others degrade extensive amounts of lignin without significant cellulose degradation (Blanchette, 1991). Different forms of white rot result when cell wall components are degraded from the wood in varying amounts. Although white-rot fungi are some of the most prevalent wood-decaying organisms in nature, they are not commonly found attacking archaeological wood. Another group, brown-rot fungi, are common wood degraders in forest ecosystems and are frequently found attacking wood in service (Zabel and Morrell, 1992). These fungi preferentially degrade the polysaccharide components of wood and do not degrade appreciable amounts of lignin. During incipient stages of decay, extensive depolymerization of cellulose occurs, resulting in significant losses in wood strength properties. As decay progresses, cellulose and hemicellulose are removed, leaving a highly lignified and recalcitrant woody residue. Brown- and white-rot fungi are taxonomically placed in the Basidiomycota.

Soft-rot fungi are classified as Ascomycota and Deuteromycota and can be distinguished from other decay fungi by decay patterns they produce in wood (Blanchette, 1998; Nilsson et al., 1989). Typically, soft-rot fungi produce cavities that spiral within the secondary wall of wood cells following the microfibrillar orientation of cellulose (Type I attack) (Daniel and Nilsson, 1998). In transverse sections of wood, holes of varying sizes can be observed in the secondary walls, whereas in radial or tangential sections they are elongate cavities, often with pointed ends. In some woods, a different form of attack occurs (Type II attack) and the entire secondary wall is gradually eroded, leaving a relatively intact middle lamella. Soft-rot fungi are often associated with waterlogged woods, and the term soft rot is used because the affected wood surface appears soft in wet environments. Soft rot can also occur in nonwaterlogged conditions and the advanced stages of decay appear brown and crumbly and may look similar to brown-rotted wood. Soft-rot fungi are often found in environments that are not conducive to the growth of brown- and white-rot fungi. Since soft-rot fungi can completely degrade the entire secondary cell wall, they have the capacity to degrade some lignin as cellulose and hemicellulose are metabolized. The highly lignified middle lamella, however, is not degraded and persists even in the most advanced stages of decay.

Bacteria are commonly found in wood and may be associated with decay fungi or act as scavengers to utilize previously decomposed substrates. They may also be primary degraders of wood in wet environments. Some bacteria are restricted in their capacity to degrade only the pit membranes of wood cells, while other types can directly attack the cell wall. Three major groups of wood-destroying bacteria have been identified—erosion, tunneling, and cavitation bacteria. Since these types of bacterial degradation are most prevalent in waterlogged woods and have not been identified from relatively dry terrestrial sites, they will not be discussed in detail. For reviews on bacterial degradation of wood see Blanchette (2000), Blanchette et al. (1990), Daniel and Nilsson (1998), and Singh and Butcher (1991).

DECAY IN ARCHAEOLOGICAL AND HISTORIC WOODS

Wood from Anasazi Great Houses The Anasazi of northwestern New Mexico built great houses approximately 1000 years ago using enormous amounts of wood from a variety of tree species. In Chaco Canyon, it is estimated that tens of thousands of trees were used for roof beams, secondary support materials in roofs, and door and ventilator lintels to build Pueblo Bonito, Pueblo del Arroyo, Chetro Ketl, and other great houses that are now part of the Chaco Culture National Historic Park (Windes and McKenna, 2001). In Aztec Ruins National Monument, located approximately 80 km north of Chaco Canyon, the West Ruin currently has about 6000 pieces of exposed wood (Tennessen et al., 2002). Most excavations of the great houses were done in the late 19th and early 20th century. After these excavations, mud brick walls and a large amount of wood that remained in the structures were exposed to the environment. Although the San Juan Basin is an arid region of the southwestern U.S. receiving only about 20 cm of annual precipitation, exposed wood has deteriorated, and over the past 8 to 10 decades all woods have had some degree of degradation present. Soft-rot fungi caused the major type of decay found in wood from Aztec Ruins and Chaco Canyon great houses. The ends of beams in the great houses showing cracks and what is usually considered to have characteristics of general weathering were found to have typical Type I soft-rot attack. [FIGS. 1A–C] Wood that visually appeared brown with cracks and checks was also found with decay typical of soft-rot fungi. This occurred in samples of Douglas fir, pine, juniper, cottonwood, and aspen. In cottonwood or aspen a Type II attacked was observed. Some brown rot was evident in a few wood samples from these sites and appeared to be associated with dry land termite damage. Since the mud brick and wood have continued to deteriorate in many of the excavated great houses, the U.S. National Park Service is currently reburying large sections of the structures as a method of preservation. The effect of reburial on reactivation of extant fungi in the wood and on the potential for new decay fungi to colonize and attack the wood is not known and needs investigation.

Ancient Egyptian Wood Archaeological woods from different ancient Egyptian tombs in the Nile Valley were examined to determine the major types of deterioration present. A previous publication has provided a detailed assessment of decay in many different archaeological woods from Egypt (Blanchette et al., 1994) and selected examples are used here to compare the various types of deterioration observed with those found in other terrestrial sites. All of the sites are from the arid Nile Valley region, although some moisture was undoubtedly present during the

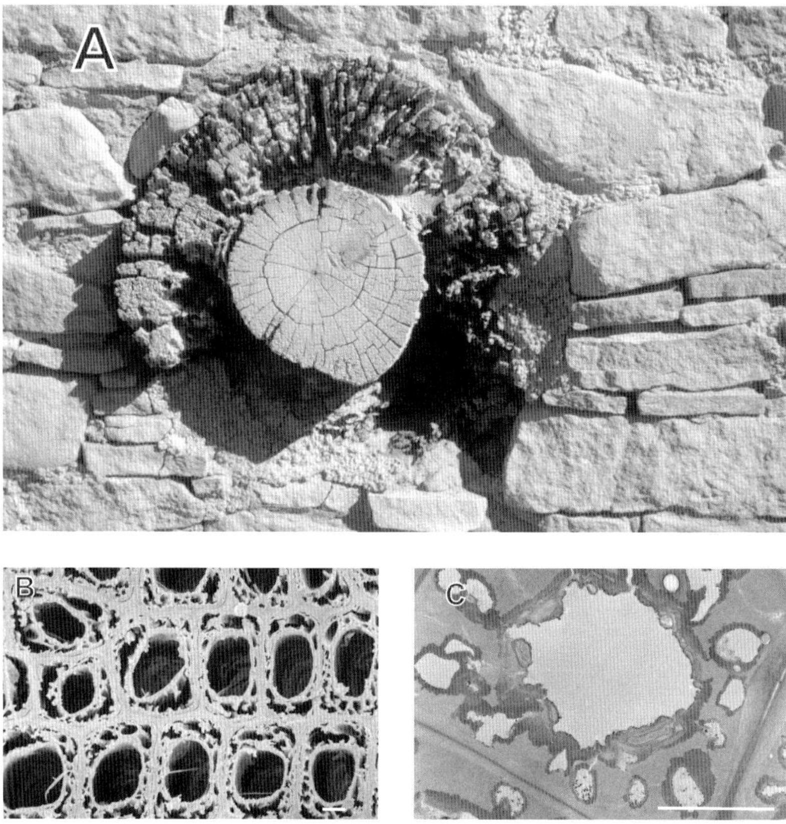

[1] Soft rot in wood from Anasazi great houses. (A) Exposed beam from Chetro Ketl, a great house that is part of the Chaco Culture National Historic Park, New Mexico, showing wood decay. (B) Scanning electron micrograph of a transverse section showing soft-rot cavities within secondary walls of wood cells. (C) Transmission electron micrograph of a transverse section with soft-rot cavities inside the secondary wall. Bar = 5 μm.

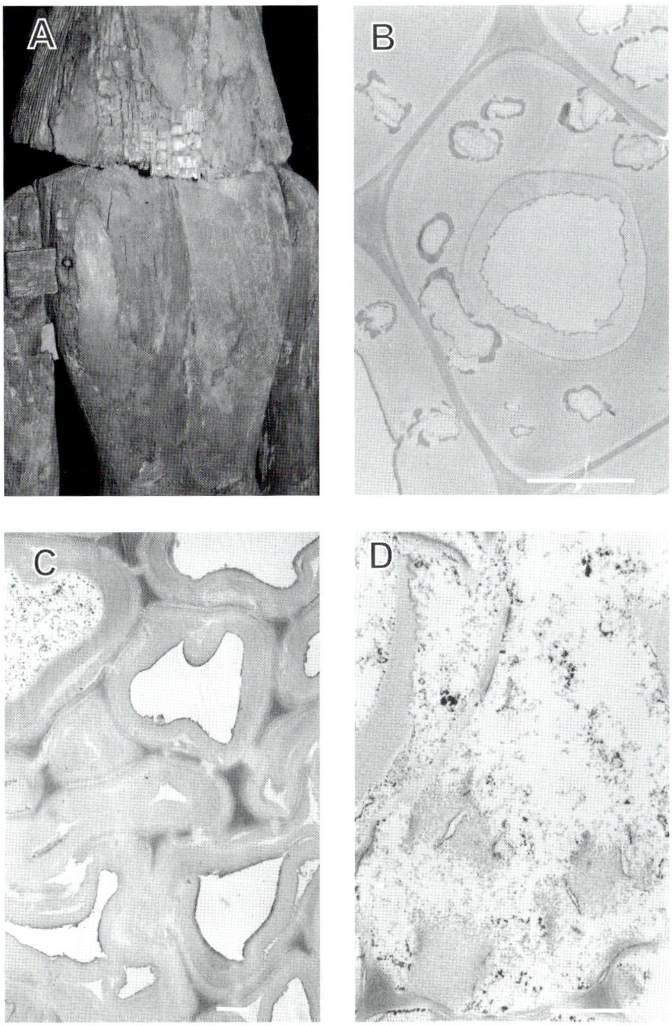

[2] Decay associated with wooden objects from ancient Egyptian tombs. (A) Wooden statue from the tomb of the scribe of Mitry, Saqqara, 2340 B.C. (currently in the Metropolitan Museum of Art, acc. no. 26.2). (B to D) Transmission electron micrographs of transverse sections showing characteristics of soft rot (B) and brown rot (C and D). (B) Soft-rot Type I cavities within the secondary wall. (C) Cells degraded by brown rot fungi are swollen and appear porous. (D) A residual cell structure disintegrated into minute particles after brown-rot attack. The brown-rot fungus degraded the cellulose, leaving severely altered cell walls that had very little strength. Bar = 5 µm.

time the wood was in the various tombs. Evidence of insect damage on some of the objects provides further evidence that the tombs were not completely sealed off from outside influences. In wooden statues from Saqqara (2340 B.C.), two types of decay were found. [FIGS. 2A–D] A brown rot was present in the lower parts of some objects in areas that appeared to be in contact with the floor and lower walls of the tomb chamber. The decay caused extensive damage and resulted in severe weakening of wood strength. Transverse sections of the affected wood showed swollen and porous cell walls that were extremely weak. The entire cell wall often collapsed and disintegrated into minute particles. This decay represented a very advanced stage of brown rot with little cellulose remaining. The residual cell walls were made up of only a loose matrix of lignin that maintained little of the original wood's strength properties. The decay, however, was localized to certain parts of the object, with relatively sound wood, or wood with soft rot, present in other areas. In these wooden statues and other objects examined, brown-rotted wood was associated with sites of termite or other insect damage.

Soft rot (Type I) was also evident in many objects examined. [FIG. 2A] Visually, the decay resembled brown rot and appeared brown with checking occurring in the dry wood; however, microscopic examination indicated that a soft-rot fungus caused the decay. Various stages of soft rot were observed, with some cells exhibiting extensive cavity formation while others had only a few cavities within the secondary walls. [FIG. 2B]

The environment of some tombs was conducive to brown rot attack by basidiomycetes, indicating that sufficient moisture was present and conditions were appropriate for this type of decay fungus to colonize and degrade the wood. The associated insect damage indicates that the tomb and burial site were not free from outside influences, and termites and other insects entering the tomb could have vectored the decay fungi. The ability of insects to enter the tomb suggests that avenues for additional moisture may also have been present. The presence of soft rot on upper regions of statues [FIG. 2A] or other wooden objects not in direct contact with the tomb floor suggests that micro-environmental conditions in these areas may not have been conducive to brown-rot degradation, and fungi tolerant of reduced levels of moisture (soft-rot fungi) were most prevalent.

Wood from Tumulus MM Tumulus MM (Midas Mound), built in the 8th century B.C., is thought to be the tomb of the Phrygian King Midas. An inner wooden funeral chamber was made from pine and cedar timbers and an outer tomb structure consisting of large-diameter juniper logs surrounded it (Young, 1981). After

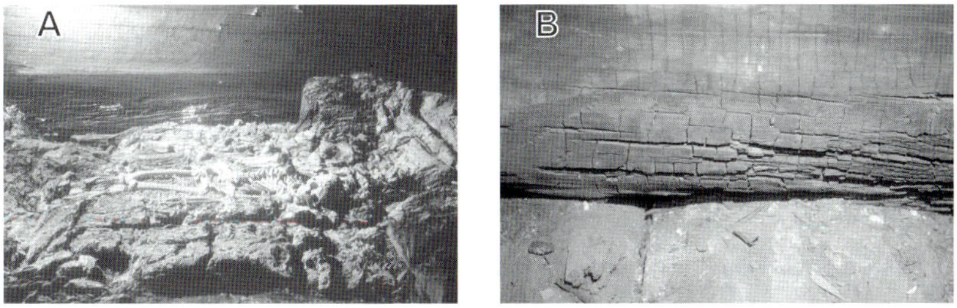

[3] Wood decay in the Tumulus MM Tomb, Gordion, Turkey. (A) The king's remains can be seen on the severely degraded cedar coffin within the wooden tomb chamber. (B) The lower wall and floor show decay by soft-rot fungi. The decayed wood appeared brown with numerous cracks and checks.

the king was placed in the tomb it was covered by limestone rubble, a dome of clay, and 53 m of limestone-rich earth. When excavated, the wooden tomb was found to contain magnificent furniture, a coffin with human remains, and many other cultural properties (Simpson and Payton, 1986).

Investigations of decay in the wooden tomb structure, furniture, and coffin indicated that only a soft-rot type of attack was present throughout the tomb. [FIGS. 3A AND B] The decay was extremely advanced, and micromorphological examination showed Type I soft rot within cell walls of all woods examined. [FIGS. 4A–D] In transverse sections, the secondary wall contained an extraordinary number of cavities. [FIG. 4A] Often cavities were observed to coalesce, forming larger voids in the cell walls. The numerous holes present in the secondary walls resulted in very weak cells that were found to easily fragment. [FIGS. 4B AND D] In longitudinal sections, the soft-rot cavities were often clearly evident. [FIG. 4B] The furniture made primarily from boxwood and walnut also had cell walls with Type I soft-rot attack. The degradation, however, had progressed to such an advanced stage that most of the secondary wall was often completely removed, and decay resembled that of Type II soft rot. The environmental conditions within the tomb over the past 2700 years were dry, but leachate from alkaline water seeped into the tomb chamber through the limestone overburden. This moisture could have come from clay placed over the tomb when the tumulus was built or from other natural sources. Apparently, the conditions within the tomb were not conducive to white- and brown-rot fungi, but soft-rot fungi, able to tolerate the low moisture and alkaline conditions, caused considerable degradation (Blanchette and Simpson, 1992). Recently it was shown by stable nitrogen isotope analyses that the soft-rot fungus probably started to degrade the tomb at the coffin area and mobilized the king's highly ^{15}N-enriched

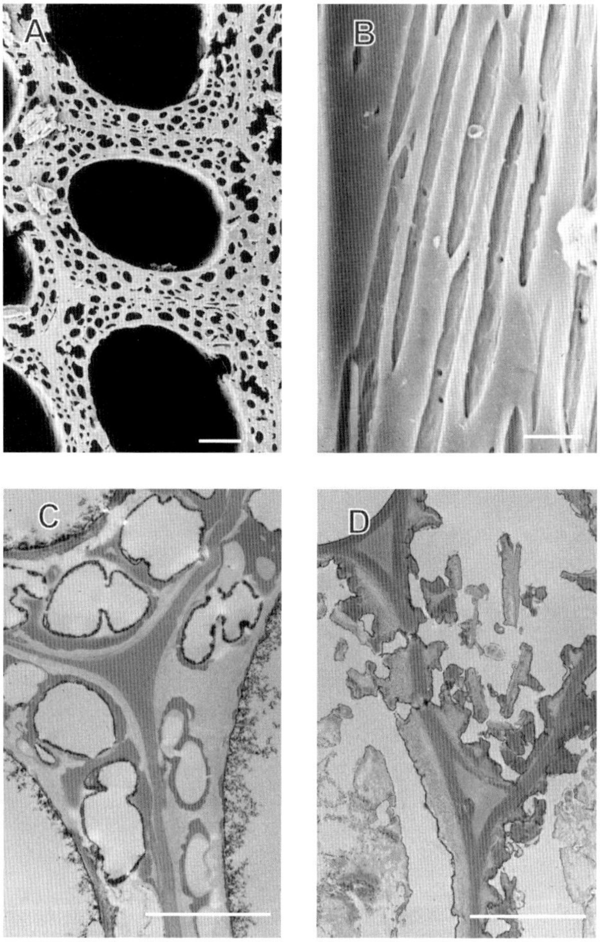

[4] Decay by soft rot in wood from the coffin and tomb structure found within Tumulus MM. Scanning (A and B) and transmission electron micrographs of transverse (A, C, and D) and radial sections (B). (A) Advanced stages of soft rot in wood from the coffin showed extraordinarily numerous cavities within the secondary wall. (B) Cell wall degradation by Type I soft rot showing the elongated cavities with pointed ends. (C and D) Wood from the tomb floor with cavities in the secondary wall. Little of the cell wall remained unaffected between the cavities, and the extremely weak decayed cells easily fragmented. Bar = 5 μm.

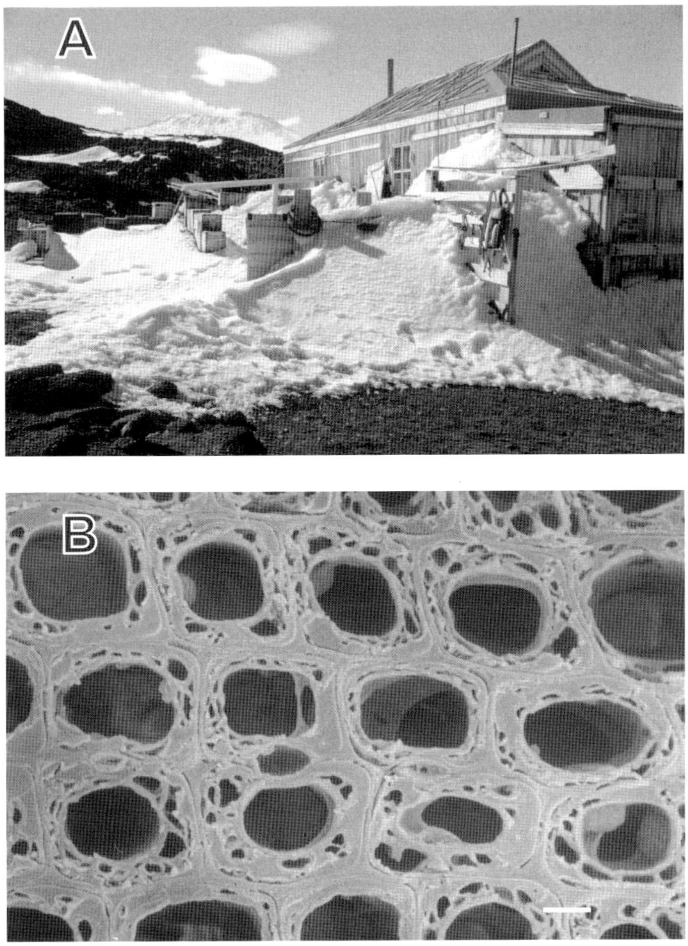

[5] Decay in Cape Royds hut, Ross Island, Antarctica. (A) Historic hut at Cape Royds built by Ernest Shackleton in 1908 during the Nimrod expedition. (B) Soft rot in wood from the Cape Royds hut with cavities within the cell walls. Bar = 5 μm.

nutrients to continually colonize and decay wood throughout the tomb (Filley et al., 2001). The redistribution of the king's nitrogen, high in ^{15}N due to his diet that had been rich in meat, could be identified in samples taken from the entire tomb. In areas of the tomb sampled that were not decayed, ^{15}N values were extremely low. The soft-rot fungus effectively recycled the nutrients of the king and transported them from the wooden coffin [FIG. 3A] into the floor and wall timbers of the tomb [FIG. 3B] over long periods of time. The extensive amount of decay found in the massive cedar coffin, cedar timbers of the tomb floor, and juniper logs (all woods with natural resistance to decay) suggests that the fungus was active for hundreds of years within the tomb, and one major decay episode, possibly caused by one soft-rot fungus, was responsible for the deterioration (Filley et al., 2001).

Arctic and Antarctic Historic Woods Three huts, Discovery Hut (built by Robert F. Scott and his crew in 1901), Cape Royds Hut (erected by Ernest Shackleton's Nimrod expedition in 1908; [FIG. 5A]), and the Cape Evans Hut (built by Scott's Terra Nova expedition in 1911), were used for sheltering men and equipment for several years during scientific investigations and attempts to reach the South Pole. The Cape Evans hut was also used by the Ross Sea Party in 1914–17, which was part of Shackleton's Imperial Trans-Antarctic expedition. These sites were abandoned once the expeditions were over, leaving the buildings and thousands of artifacts behind. Today these historic sites from Antarctica's Heroic Era provide a remarkable view into the past. Furniture, food stores, scientific apparatus, and other items are left in the huts as a reminder of what everyday life was like for the early explorers.

However, during the past few decades, serious wood deterioration has become evident in the huts and artifacts, causing concern about the long-term preservation of these historic sites (Harrowfield, 1995). It is a great misconception that the polar cold and dry climate can protect organic material from decomposition (Blanchette et al., 2002; Hughes, 2000). In actuality, significant deterioration takes place over time in the Ross Sea Region. Examination of wood in contact with the ground at these historic sites indicated that wood decay was occurring. Microscopic examination showed that a soft-rot attack was present. [FIG. 5B] Isolations made on culture media from the wood revealed that *Phialophora*-like species were associated with the degradation. The only type of biological decay found was a soft rot causing Type I attack. At the locations of the huts, the environment is above 0°c for only a few weeks in the austral summer, and high concentrations of salts are present in soils and in the historic woods. The extreme polar environment influences the microorganisms that are able to persist in this ecosystem. The *Phialophora*-like organisms are able to tolerate the adverse conditions and cause a slow degradation of the historic wood over time.

In the Canadian High Arctic, the early explorers of the North Polar Region also built huts during their explorations. In 1881, an American expedition led by Adolphus Greely built a large building called Fort Conger about 500 miles from the North Pole on Ellesmere Island (Dick, 2001). This was part of the International Polar Year efforts to study the Arctic regions. Fort Conger was the American headquarters for exploration of the interior of Ellesmere Island and the coast of northern Greenland, and for the collection of scientific information throughout the region. The Greely expedition did not get relief shipments as expected, and 19 of the 25 men in the expedition died of starvation and exposure. Robert Peary also used this site as a base camp for his campaigns to reach the North Pole. In 1900, he refitted Fort Conger from a large building into a smaller group of wooden huts for use during his attempts to reach the pole (Dick, 2001). Today, the huts remain with a tremendous amount of historic wood and other artifacts still present at the site. Wood decay is also prevalent and samples obtained from the wood of Fort Conger and the huts built by Peary indicated that soft-rot fungi were the only type of wood-degrading fungi present. All woods examined had a Type I soft-rot attack. Environmental conditions in the Arctic are also extreme, with a very short summer season and high salt concentrations occurring in soils and wood along the coastal regions. These conditions limit the types of organisms that can attack the wood with only specific soft-rot fungi, able to tolerate the adverse conditions, as the dominant decay-causing organism present.

Nonbiological Deterioration In addition to the biological degradation found at these different archaeological and historic sites, nonbiological degradation was also found. The occurrence of this chemical deterioration of wood is much more common than previously realized. The decay is characterized by a fuzzy, defibrated appearance on wood surfaces suggestive of mechanical abrasion (Blanchette et al., 2002). [FIGS. 6A AND B] Surface wood cells are detached and may be removed by wind or snowmelt, exposing the underlying cells that are then subjected to the same defibration process. Over long periods of time, the wood is gradually eroded away. In Antarctica, defibration is associated with wood containing high salt concentrations. [FIG. 6A] The wood absorbs moisture and salts from snowmelt and the salts precipitate at the evaporative surface. As evaporation occurs, exceedingly high concentrations of salts accumulate in the wood. In polar regions where rainfall does

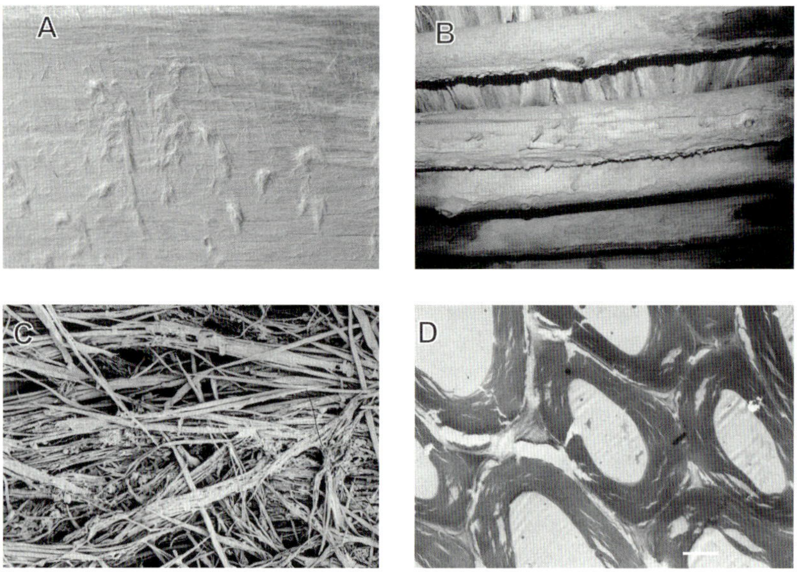

[6] Nonbiological degradation of wood. (A) Exterior wallboard from Robert F. Scott's Cape Evans hut in Antarctica showing severe defibration of wood from high salt concentrations. Wood cells are degraded and detached from the wood. (B) Difibration in roof beams from Pueblo del Arroyo, an Anasazi great house in Chaco Culture National Historic Park, New Mexico. (C) Scanning electron micrograph of chemically deteriorated wood from Cape Evans hut in Antarctica showing wood cells separating at the middle lamella and detaching from adjacent cells. (D) Transmission electron micrograph of a transverse section from the mask of a coffin lid from an ancient Egyptian tomb at Abydos, 1570–1293 B.C. (currently in the Museum of Fine Arts, Boston, acc. no. 01.7431) showing salt deterioration that has affected the wood cells. Cells are delaminated by degradation of the middle lamella; secondary walls also show deterioration with cracks and fissures present within the remaining secondary wall layers. Detached wood fibers in C are approximately 50 to 80 µm wide. Bar in D = 5 µm.

not occur, salts are not leached from the wood. The salts corrode the wood, causing a chemical deterioration of the middle lamella region of wood cells. The selective attack on the middle lamella causes a separation of wood cells, and individual cells detach from the wood surface. [FIG. 6C] When wet, the affected area appears as if the wood surface has been pulped.

Although the mechanism of chemical degradation taking place is not well understood, the decay appears similar to the attack on wood found in temperate regions on preservative-treated marine pilings, in wood from warehouses that store deicing salts and fertilizer, and on the internal surfaces of boats and cooling towers (Johnson et al., 1992; Wilkins and Simpson, 1988). In these situations the damage is thought to be caused by the growth of salt crystals within wood cells. The results presented here indicate that cell separation is due to a chemical attack on the middle lamella causing cells to break away from adjacent cells. This type of deterioration has also been found in woods from Anasazi great houses at Chaco Culture National Historic Monument and Aztec Ruins National Monument [FIG. 6B], as well as in some woods from ancient Egyptian tombs. In wood of the great houses, roof beams protected from direct rain absorbed runoff water containing dissolved salts. Evaporation resulted in salt accumulation on the beam surfaces. Over time these surfaces were deteriorated by chemical action of the salts, causing the entire surface of the affected beams to become defibrated. [FIG. 6B] Often, the roof beams were in a protected location and a large layer of detached wood fibers remained on the wood surfaces.

Similar surface defibration was found on Egyptian woods. These woods, located in tombs where moisture apparently carried alkaline substances into the wood, had localized areas of defibrated cells. Transmission electron micrographs of the deteriorated woods showed that the middle lamella was destroyed and cells detached from one another. [FIG. 6D] In addition, the secondary wall also appeared to separate. Parallel series of separations occurred within the secondary wall layers, indicating that all parts of the woody cell wall, including the cellulose-rich secondary wall layers, were affected by the chemical attack. [FIG. 6D] Differences in appearance of the surface deterioration occurring in the Egyptian woods were most likely due to the much longer time of burial, to the more direct contact with alkaline substances, or to the different types of salts as compared to those found in the Antarctic historic huts or the Anasazi great house woods.

CONCLUSIONS Environmental conditions that exist in areas where archaeological wood is protected from rapid decay exclude most wood-destroying fungi. However, microorganisms that can tolerate these extreme conditions grow and may degrade the wood. In the wooden objects examined in these studies, soft-rot fungi were the major wood-destroying organisms found. They appear to dominate in woods subjected to a variety of conditions, including the limited moisture of buried tombs, high salt concentrations, and the extreme conditions of the cold polar environment. In temperate and tropical ecosystems, white- and brown-rot fungi are the major decay fungi that cause rapid degradation of wood, and soft-rot fungi are not frequently encountered. However, soft-rot fungi do occur and can be found in naturally decay-resistant woods, preservative-treated woods, and waterlogged woods. In addition to these selected situations where soft rot is found, these fungi also appear common in archaeological woods from dry, alkaline, or cold terrestrial sites. The presence of some brown rot in the ancient Egyptian woods and the Anasazi great houses suggests that, when conditions are favorable, brown-rot fungi may cause substantial decay in buried objects or wood exposed to the environment. The association of brown rot with termite and other insect damage indicates that specific conditions may influence the introduction of brown-rot fungi into the tombs. Sufficient moisture and nutrients also have to be available at these locations, and salts or other substances that inhibit growth of basidiomycetes must be limited.

Although soft-rot fungi are the major cause of decay in archaeological wood from terrestrial sites, little is known about the biology and ecology of these fungi. Since many sites, such as Tumulus MM, support the growth of only soft-rot fungi, and exceedingly advanced stages of decay can be found, these sites provide an excellent opportunity to study this unusual form of wood decay.

The defibration of wood due to salt deterioration processes was found frequently in the woods examined at these different terrestrial sites. Now that the characteristics of this form of chemical damage are known, and methods to distinguish it from other types of deterioration have been demonstrated, there will undoubtedly be more reports on its occurrence. Additional investigations of the mechanism of chemical attack on wood are needed and further characterization of the varying types of damage should be carried out.

Archaeologists, conservators, and other scientists working with excavated archaeological woods have the opportunity to gain significant information about past cultures by studying wood. The use of wood to understand site chronology, past environments, deterioration processes, and microbial biodiversity in extreme environments strongly suggests that archaeological wood should not be neglected. Information on the chemical and morphological condition of the woods is also of prime importance so that appropriate objects conservation methods can be selected, and if in-situ preservation is required, this type of information will help in the selection of procedures that will ensure successful preservation of the materials.

ACKNOWLEDGMENTS The author thanks Benjamin Held, Joel Jurgens, Kory Cease, and Andre Abad for assistance with many aspects of this research. A large number of cooperators helped to provide samples and their expertise. For research on wood from Tumulus MM at Gordion, Turkey, this includes Elizabeth Simpson, The Baird Graduate Center for Studies in the Decorative Arts, New York; G. Kenneth Sams, Department of Classics, University of North Carolina, Chapel Hill; and Timothy Filley, Department of Earth and Atmospheric Sciences, Purdue University; for work with the ancient Egyptian woods, Robert Koestler, Metropolitan Museum of Art, New York; and Pamela Hatchfield, Museum of Fine Arts, Boston. Wood from the historic huts of the Ross Sea region of Antarctica was investigated in collaboration with the Antarctic Heritage Trust, Christchurch, New Zealand; the New Zealand Antarctic Institute; and Roberta Farrell, University of Waikato, New Zealand; Wood from the Canadian High Arctic was investigated in collaboration with Quttinirpaaq National Park, Parks Canada; and the Nunavut Research Institute. Wood studies from Chaco Culture were done in cooperation with the Chaco Culture National Historic Park and Aztec Ruins National Monument, New Mexico; and The Getty Conservation Institute, Los Angeles. This material is based in part upon work supported by National Science Foundation grant 9909271.

REFERENCES

BJORDAL, C.G., NILSSON, T., and DANIEL, G., 1999. Microbial decay of waterlogged wood found in Sweden applicable to archaeology and conservation. *International Biodeterioration and Biodegradation*, 29, 381–398.

BLANCHETTE, R.A., 1991. Delignification by wood-decay fungi. *Annual Review of Phytopathology*, 29, 381–398.

BLANCHETTE, R.A., 1998. A guide to wood deterioration caused by fungi and insects. In: Dardes, K., and Rothe A., (Eds.), *The Conservation of Panel Paintings*. Getty Conservation Institute, Los Angeles, pp. 55–68.

BLANCHETTE, R.A., 2000. A review of microbial deterioration found in archaeological wood from different environments. *International Biodeterioration and Biodegradation*, 46, 189–204.

BLANCHETTE, R.A., and SIMPSON, E., 1992. Soft rot and wood pseudomorphs in an ancient coffin (700 B C) from Tumulus MM at Gordion Turkey. *International Association of Wood Anatomists Bulletin*, 13, 201–213.

BLANCHETTE, R.A., NILSSON, T., DANIEL, G., and ABAD, A., 1990. Biological degradation of wood. In: Rowell, R.M., and Barbour, R.J., (Eds.), *Archaeological Wood: Properties, Chemistry, and Preservation, Advances in Chemistry Series*, Vol. 225. American Chemical Society, Washington, DC, pp. 141–174.

BLANCHETTE, R.A., HAIGHT, J.E., KOESTLER, R.J., HATCHFIELD, P.B., and ARNOLD, D., 1994. Assessment of deterioration in the archaeological wood from ancient Egypt. *Journal of the American Institute of Conservation*, 33, 55–70.

BLANCHETTE, R.A., HELD, B.W., and FARRELL, R.L., 2002. Defibration of wood in the expedition huts of Antarctica: an unusual deterioration process occurring i n the polar environment. *Polar Record*, 38, 313–322.

DANIEL, G.F., and NILSSON, T., 1998. Developments in the study of soft rot and bacterial decay. In: Bruce, A., and Palfreyman, J.W., (Eds.), *Forest Products Biotechnology*. Taylor and Francis, London. 326 pp.

DICK, L., 2001. *Muskox Land: Ellesmere Island in the age of contact*. University of Calgary Press, Calgary. 615 pp.

ERIKSSON, K.E., BLANCHETTE, R.A., and ANDER, P., 1990. *Microbial Degradation of Wood and Wood Components*. Springer-Verlag, Berlin, 407 pp.

FILLEY, T.R., BLANCHETTE, R.A., SIMPSON, E., and FOGEL, M.L., 2001. Nitrogen cycling by wood decomposing soft-rot fungi in the "King Midas tomb" Gordion, Turkey. *Proceedings of the National Academy of Sciences*, 98, 13346–13350.

HARROWFIELD, D.L., 1995. *Icy Heritage: Historic Sites of the Ross Sea Region*. Antarctic Heritage Trust, Christchurch. 88 pp.

HUGHES, J., 2000. Ten myths about preservation of historic sites in Antarctica and some implications for Mawson's huts at Cape Denison. *Polar Record*, 36, 117–130.

JOHNSON, B.R., IBACH, R.E., and BAKER, A.J., 1992. Effect of salt water evaporation on tracheid separation from wood surfaces. *Forest Products Journal*, 42, 57–59.

NILSSON, T., DANIEL, G., KIRK, T.K., and OBST, J.R., 1989. Chemistry and microscopy of wood decay by some higher Ascomycetes. *Holzforschung*, 43, 11–18.

SCHWEINGRUBER, F.H., 1993. *Trees and Wood in Dendrochronology: Morphological, Anatomical, and Tree-ring Analytical Characteristics of Trees Frequently Used in Dendrochronology*. Springer-Verlag, Berlin, 402 pp.

SIMPSON, E., and PAYTON, R., 1986. Royal wooden furniture from Gordion. *Archaeology*, 39, 40–47.

SINGH, A.P., and BUTCHER, J.A., 1991. Bacterial degradation of wood cells: a review of degradation patterns. *Journal of the Institute of Wood Science*, 12, 143–249.

SINGH, A.P., NILSSON, T., and DANIEL, G.F., 1990. Bacterial attack of *Pinus sylvestris* wood under near-anaerobic conditions. *Journal of the Institute of Wood Science*, 11, 237–249.

TENNESSEN, D., BLANCHETTE, R.A., and WINDES, T.C., 2002. Differentiating aspen from cottonwood in prehistoric wood from Chacoan great houses. *Journal of Archaeological Science*, 29, 521–527.

WILKINS, A.P., and SIMPSON, A., 1988. Defibring of roof timbers. *Journal of the Institute of Wood Science*, 11, 121–125.

WINDES, T.C., and MCKENNA, P.J., 2001. Going Against the Grain: Wood Production in Chacoan Society. *American Antiquity*, 66, 119–140.

YOUNG, R.S., 1981. *Three Great Early Tumuli. The Gordian Excavations Final Reports*. Vol. 1. The University Museum. Philadelphia, PA. 326 pp.

ZABEL, R.A., and MORRELL, J.J., 1992. *Wood Microbiology, Decay and its Prevention*. Academic Press, Orlando, 476 pp.

P. HOFFMANN

DEGRADATION PATTERNS IN WATERLOGGED WOOD AND THE TWO-STEP PEG TREATMENT FOR LARGE ARCHAEOLOGICAL FINDS: THE CASE OF THE BREMEN COG

Deutsches Schiffahrtsmuseum, Bremerhaven, Germany

INTRODUCTION Archaeological wood is a term used by conservators for wood and wooden objects excavated by archaeologists. It is normally wet, waterlogged, or wet and frozen. Only in rare situations can wood survive in a dry state—and then it has to be very dry: The Pharaoh's ships buried at the foot of one of the Giza pyramids are such an exception.

Waterlogged soils, sediments, or deposits can provide low-oxygenated, or anoxic environments, where the biological degradation of the organic material is restricted to very-slow-working specialist microbes, or is impossible at all. Archaeological wood will nearly always have been attacked by soft-rot fungi, and/or bacteria. A pure chemical degradation has not yet been conclusively demonstrated, although hydrolysis of carbohydrates seems probable over long periods of time. In a waterlogged environment the wood substance consumed by microorganisms is substituted by water. The degraded piece of wood keeps its original dimensions. However, the degraded wood has lost mechanical strength, and its properties are seriously altered. On drying, the wood shrinks heavily, or even collapses; splits develop, as well as cross-grain cracks; and it warps and distorts.

Conservators try to prevent these catastrophes. [FIG. 1] We try to transform the water-filled object to an air-dry object that can sit in a showcase or on the shelf of a museum's store and still look like it did when it was excavated. We have a number of methods at our disposal (e.g., Grattan and Clarke, 1987), even in the case of boats and ships that do not fit into a lab handily, and the question is which method is best for the given object.

MACROSCOPIC DEGRADATION PATTERNS A piece of archaeological wood normally has a soft and obviously degraded surface. Pricking with a pointed knife or needle will indicate if there is a core of harder tissue in the timber, and how thick the soft degraded layer is. The heavily degraded layer may not be of uniform thickness; degradation often proceeds in pockets and along cracks in the timber. To get an idea of the degradation pattern, the depths of easy penetration with the "needle test" are taken at many points and marked in a sketch of the timber or object.

Saw-cuts across the timbers give a clearer image: Areas of heavy degradation are cut without leaving a fibrous surface, as less degraded areas do. [FIG. 2] In addition, in most wood species there are distinct color differences. In most cases there is a clear front between heavily and only slightly degraded tissue. Sometimes areas of heavy degradation are irregularly distributed over the cross section. And sometimes the timber is uniformly and heavily degraded throughout.

MICROSCOPIC DEGRADATION PATTERNS Histological staining techniques, the use of polarized light, and fluorescence techniques permit a differentiation between degraded and nondegraded wooden cells (Hoffmann and Jones, 1990). With the light microscope, the visual evidence is confirmed. Quite often, there is a distinct front separating heavily degraded tissue from nonaffected, sound tissue. The areas of totally degraded cells are in direct contact with the areas of sound cells. In other cases there is a transition zone in which sound and totally degraded cells, or cell clusters, are mixed. On the microscopic level, archaeological wood seems to consist only of quite heavily degraded cells and nondegraded cells, arranged in varying patterns. [FIG. 3] Wood that seems to be of an intermediate state of degradation when its density or chemical composition is analyzed, consists in reality of a mixture of nondegraded and heavily degraded cells. Only sometimes does one observe cells in the process of being degraded: The cell wall seems to be transformed from sound to completely degraded in one step, the process progressing from the cell lumen into the cell wall.

ULTRASTRUCTURAL DEGRADATION PATTERNS The microbial attack on archaeological wood deposited in a waterlogged environment leaves distinct degradation pat-

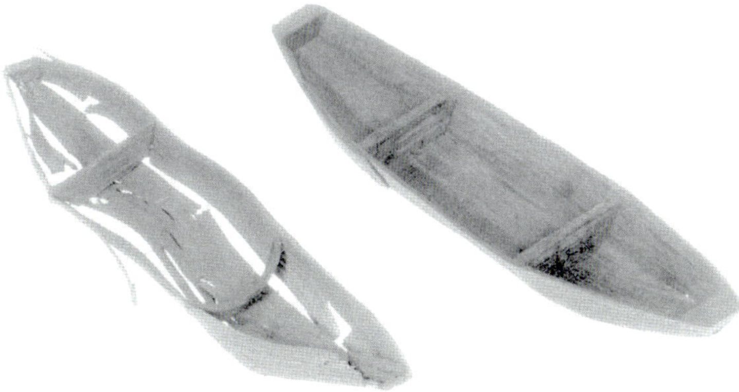

[1] Two model boats built from archaeological wood and air-dried. Left: without PEG stabilization; right: with PEG stabilization. Photo: E. Laska, Deutsches Schiffahrtsmuseum.

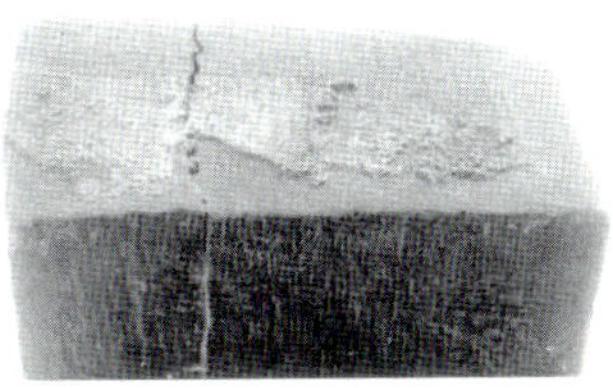

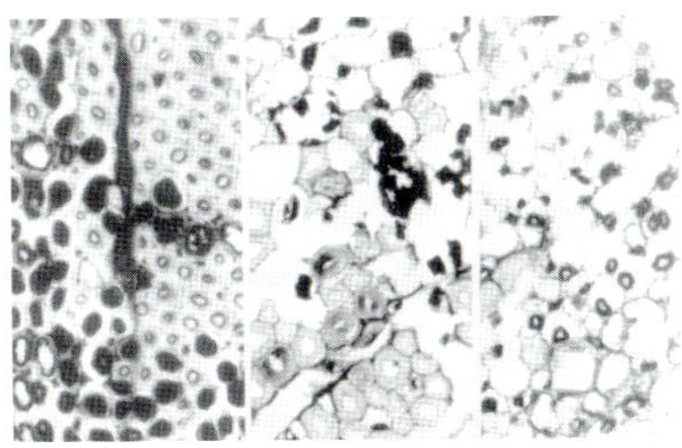

[2] Cross section of a keel plank from a Roman river barge from the Rhine; the wood is heavily degraded except for a small central area. Photo: author. [3] Degradation patterns on the cellular level. Left: degradation front between degraded and nondegraded areas. Middle: mosaic of degraded and nondegraded cell clusters. Light micrographs: author.

terns in the wood cell walls specific to the groups of active microbes (e.g., Blanchette et al., 1990). The different degradation patterns have one characteristic in common: Where the microbes have attacked, the chemically complex cell wall structure has completely disintegrated. Outside of the area of total degradation the cell wall appears unaltered and in its original state. [FIG. 4]

The ultrastructural analysis confirms the light microscopic evidence. Archaeological wood can be described as consisting of two wood qualities—nondegraded tissue with sound cell walls, and heavily degraded tissue with next to no cell walls at all, only the system of middle lamellae having survived. During the degradation of a cell the cell walls again consist of nondegraded portions, and totally degraded portions.

DEGRADATION PATTERN AND THE TWO-STEP PEG TREATMENT FOR STABILIZATION

The concept of archaeological wood being composed of two wood, or cell wall qualities only, arranged in different ways, allows for a simple classification. A waterlogged timber can be nondegraded with only a negligible surface layer of degraded cells; the timber can consist of two qualities, in most cases a layer of heavily degraded tissue enveloping a core of nondegraded wood; or the timber may be heavily degraded throughout.

Traditionally, heavily degraded wood is satisfactorily stabilized against shrinkage by bulking it with a high molecular weight polyethylene glycol (PEG) (e.g., Grattan and Clarke, 1987). The PEG is introduced into the residual wood structure by diffusion from a heated aqueous solution, and on cooling and drying the wood the PEG solidifies and braces the frail tissue against shrinkage.

For the late 4th century large extended logboat from Yverdon in Switzerland, for example, its wood being predominantly heavily degraded, a tank impregnation with PEG 3000 led to a very good stabilization ([FIG. 5]; Kaenel, 1994).

Most large wooden archaeological finds, however, such as boats and ships, consist of timbers combining two wood qualities, heavily and only slightly degraded tissue. High molecular weight PEG cannot penetrate into nondegraded tissue and cell walls, and thus will not stabilize such wood (Hoffmann, 1984: Young and Sims, 1989). The Bremen Cog, a huge medieval merchant ship built A.D. 1380, the ship I have worked with the past 20 years, consisted of two-quality timbers; mostly a rather thin degraded layer enveloped the little degraded bulk of the wood. This wood would shrink about 12 % across the grain upon drying if it was not stabilized.

When I joined the Deutsches Schiffahrtsmuseum to head the conservation of the Cog, I developed a two-step PEG treatment designed to stabilize two-quality

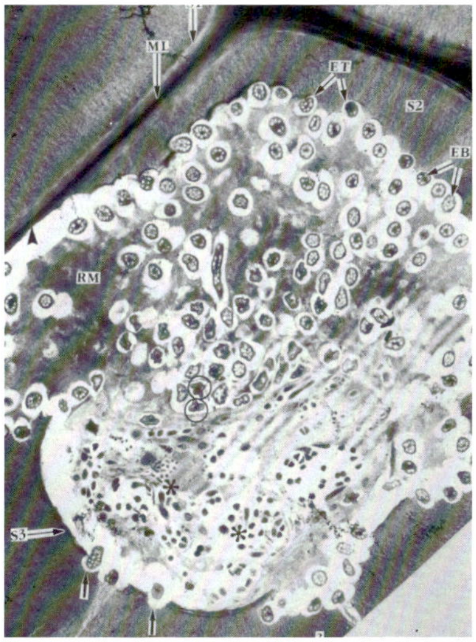

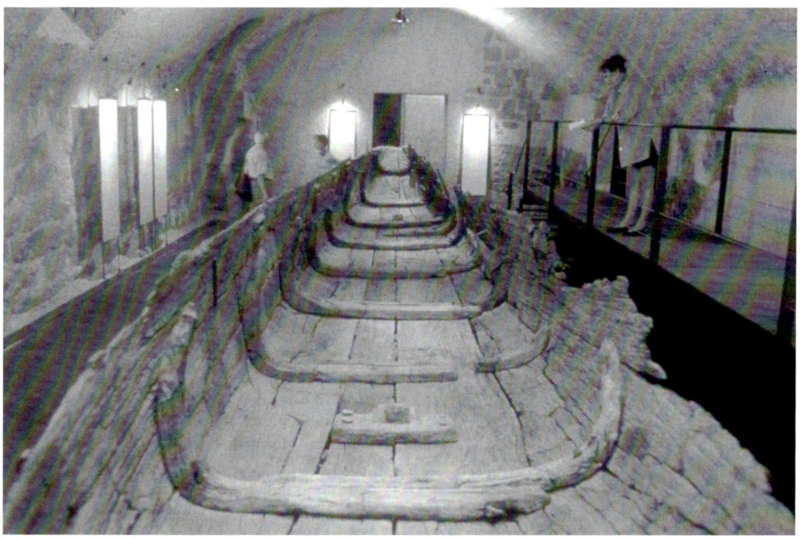

[4] Erosion bacteria degrading the walls of a wood cell. EB = erosion bacteria; ET = erosion troughs; ML, S_1, S_2, S_3 = cell wall layers; RM = residual material; * = secondary degrading bacteria. Electron micrograph: A. Singh, T. Nilsson, G. Daniel, Swedish Agricultural University. [5] The extended logboat from Yverdon successfully stabilized. Photo: G. Kaenel, Musée Cantonal d'Archéologie et d'Histoire, Lausanne, Switzerland.

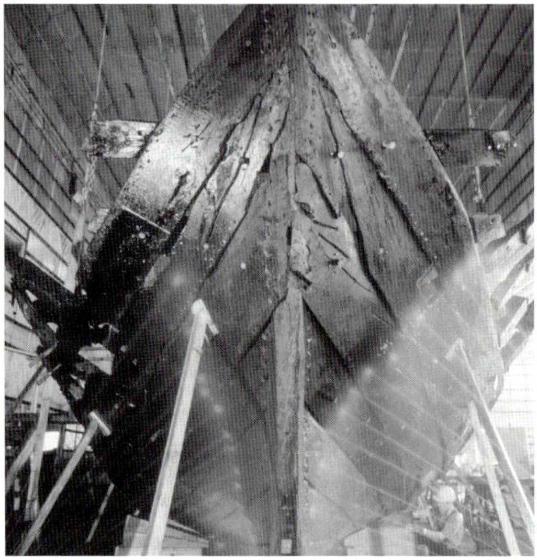

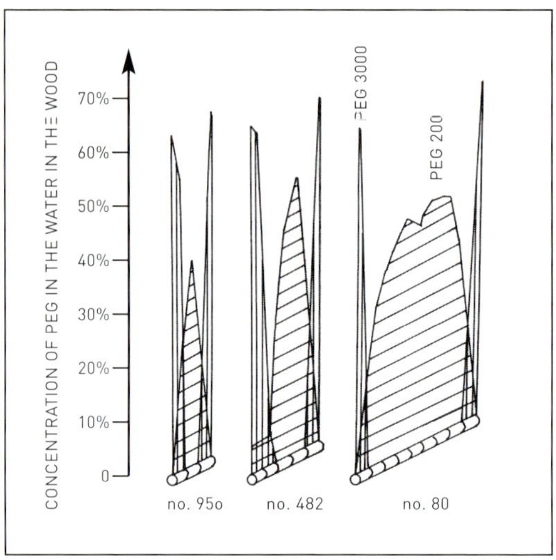

[6] The timbers of the Bremen Cog are kept wet during the reconstruction of the ship. Photo: G. Meyerdierks, Deutsches Schiffahrtsmuseum. [7] PEG distribution in core samples taken at the end of the stabilization treatment.

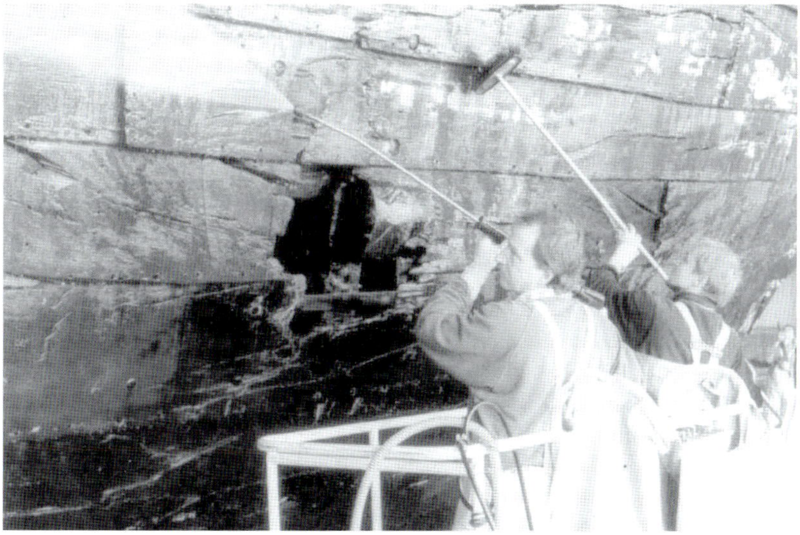

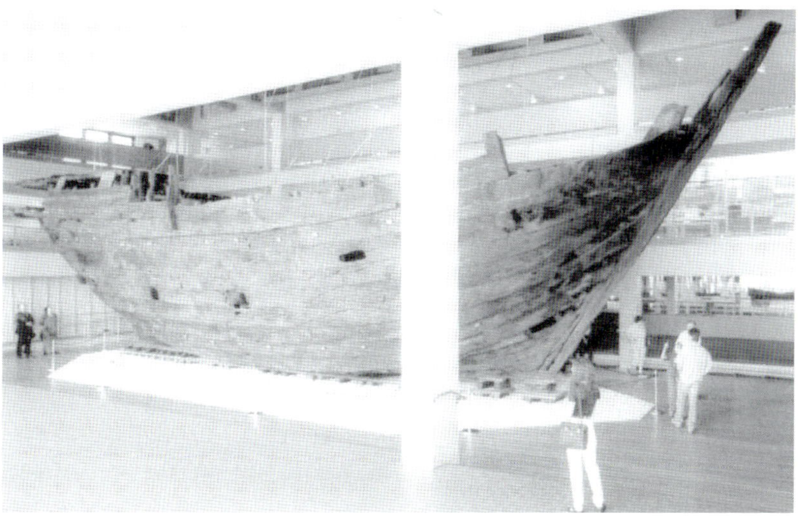

[8] Cleaning off surplus PEG 3000 with a steam jet. Photo: author. [9] The Bremen Cog in the Deutsches Schiffahrtsmuseum. Photo: E. Laska, DSM.

timbers: In a first bath with low-molecular-weight PEG 200, these small molecules diffuse through the heavily degraded wood and into the little degraded tissue. They enter the intact cell walls, and substitute for parts of the water there. On drying the wood, the nonvolatile PEG remains in the cell walls and keeps them in their waterlogged, original shape. The timber thus holds its dimensions. In a second bath, high molecular weight PEG 3000 fills and bulks the heavily degraded parts of the timbers (Hoffmann, 1986).

The Bremen Cog had been salvaged in pieces and was then reconstructed from its wet timbers. [FIG. 6] A tank was constructed around the hull and filled with 800 m^3 of water. PEG 200 was added at a slow rate of 5% per year—owing to limited funds—to an end concentration of 40%. The progress of the impregnation was monitored once a year by analyzing 5-mm Ø core samples taken from representative timbers. When the concentration of PEG in the wood had reached equilibrium with the treatment solution, we exchanged the bath for a solution of high-molecular PEG 3000. The concentration in this bath started at 60%, and was raised to 76% after one year. Two years later the timbers showed the desired PEG distribution: a high concentration of only PEG 3000 in the outer decayed wood, and a sufficient amount of PEG 200 in the inner portions of the timbers. [FIG. 7] We ended the treatment, and a local shipyard carefully dismantled the tank (Hoffmann, 2001).

With a steam jet, brushes, sponges, and spatulas we cleaned the ship of surplus PEG. [FIG. 8] After drying for two years now, the average shrinkage of the timbers measures about 2.5%. The two-step treatment has suppressed 84% of the shrinkage expected for nonstabilized wood of this quality.

Today the Bremen Cog stands as the central exhibit in our museum. She looks a pleasant dark brown color, the wood appears dry, and no traces of the PEG inside are visible. [FIG. 9] The result of this very long conservation experiment has so far met our hopes and expectations.

REFERENCES

BLANCHETTE, R.A., NILSSON, T., DANIEL, G., and ABAD, A. 1990. Biological degradation of wood. In: Rowell, R.M., and Barbour R.J., (Eds.), *Archaeological Wood, Advances in Chemisty*, Series 225, American Chemical Society, Washington, DC, pp. 141–176.

GRATTAN, D.W., and CLARKE, R. 1987. Conservation of waterlogged wood. In: Pearson, C., (Ed.), *Conservation of Marine Archaeological Objects*, Butterworths, London etc., pp. 164–206.

HOFFMANN, P. 1984. On the stabilisation of waterlogged oak wood with PEG/Molecular size versus degree of degradation. In: *Proceedings of the 2nd ICOM Waterlogged Wood Working Group Conference*, Grenoble, CETBGE/Grenoble, pp. 195–116.

HOFFMANN, P. 1986. On the stabilisation of waterlogged oak wood with PEG.II. Designing a two-step treatment for multi-quality timbers. *Studies in Conservation*, 31, 103–113.

HOFFMANN, P. 2001. To be and to continue being a cog: the conservation of the Bremen Cog of 1380. *International Journal of Nautical Archaeology*, 30 (1), 129–140.

HOFFMANN, P., and JONES, A.M. 1990. Structure and degradation process for waterlogged archaeological wood. In: Rowell, R. M., Barbour, R.J. (Eds.), *Archaeological Wood, Advances in Chemistry* Series 225, American Chemical Society, Washington, DC, pp. 35–66.

KAENEL, G. 1994. PEG conservation of a Gallo-Roman barge from Yverdon-les-Bains (Canton of Vaud, Switzerland). In: *Proceedings of the 5th ICOM Group on Wet Organic Archaeological Materials Conference*, Portland, Ditzen, Bremerhaven, pp. 143–166.

YOUNG, G.S., and SIMS, R. 1989. Microscopical determination of peg in treated wood—the effect of distribution on dimensional stabilization. In: Mac Leod, I. D., (Ed.), *Proceedings of the ICOM Working Groups on Wet Organic Archaeological Materials and Metals Conference*, Fremantle, W.A. Museum, Perth, pp. 109–140.

E. SIMPSON

THE CONSERVATION OF THE WOODEN OBJECTS FROM GORDION, TURKEY: METHODS FOR THE TREATMENT OF DRY ARCHAEOLOGICAL WOOD

University of Pennsylvania Museum of Archaeology and Anthropology, Philadelphia, PA,
and The Bard Graduate Center for Studies in the Decorative Arts, New York, NY

EXCAVATIONS AT GORDION AND EARLY TREATMENT OF THE WOODEN ARTIFACTS A large collection of rare and unusual wooden objects was excavated at the site of Gordion, capital of the kingdom of Phrygia, between the years 1950 and 1973. Excavations were carried out by the University of Pennsylvania Museum of Archaeology and Anthropology under the direction of Rodney S. Young. At least 37 pieces of ornate wooden furniture and 56 wooden objects were recovered from three royal tombs—Tumulus P, Tumulus MM, and Tumulus W, excavated in 1956, 1957, and 1959, respectively. In addition, many carbonized wood fragments were found in the "destruction level" of the City Mound. The Gordion finds are now recognized as the most important well-preserved collection of wooden artifacts from the ancient Near East.

Date of the Gordion Wooden Objects New C-14 dates for organic samples from the City Mound establish a date for the destruction level in the late 9th century B.C.; Tumulus W and Tumulus P are approximately contemporary. Tumulus MM, the largest mound at the site, is now dated to the second half of the 8th century B.C., the period of the Phrygian King Midas or his immediate predecessor (Simpson, 2001).

Condition of the Wood Many of the wooden objects were well preserved inside the tomb chambers, which had been covered over with huge mounds of earth (tumuli). The high clay content of the earth had kept the moisture levels inside the tombs fairly constant and low enough to inhibit major degradation. Woods used for the Gordion artifacts include boxwood, juniper, walnut, yew, maple, oak, pine, and cedar.

The condition of the various objects was dependent on the types of wood used (some, such as boxwood, proved highly resistant to decay), on the degree of protection afforded by the position of the objects in the tomb chambers, the proximity of the wood to metals and organic substances in the burials, and the exposure of the wood to moisture that had entered the tombs, either at the time of burial or during the course of excavation (Spirydowicz et al., 2001; Blanchette and Simpson, 1992).

While much of the wood survived in relatively good condition, some of the objects had degraded substantially. Studies have shown that soft rot was an important factor in the damage that the wood sustained (Blanchette et al., 1991; Blanchette and Simpson, 1992; Nelson et al., 1995).

Early Attempts at Conservation The wooden objects were studied briefly, were drawn, and underwent conservation treatment after their excavation. Two different treatments were used: (1) Most of the wood was coated with a solution of Alvar, a polyvinyl acetal resin, dissolved in acetone. (2) A second method of treatment was used for three inlaid pieces of furniture from Tumulus MM—the inlaid table and two serving stands. These pieces were first soaked in a solution of alcohol and water in an attempt to dry the wood. The objects were then immersed in a bath of wax and gasoline, in the hope that the solution would enter the pores of the wood. Unfortunately, the wax remained largely on the surface, affording little added strength and attracting dust and dirt (Simpson and Spirydowicz, 1999; Spirydowicz et al., 2001).

THE GORDION FURNITURE PROJECT After study and treatment, the wood was placed in storage at Gordion and in the Museum of Anatolian Civilizations, Ankara. Information about the wooden objects from Tumulus W, P, and MM was published posthumously by Young in the first volume of the Gordion excavations final reports (Young, 1981). During the preparation of Young's volume for publication, numerous errors were found in the field drawings of the wooden furniture, and a new study of the pieces was begun in 1978 by Elizabeth Simpson.

Subsequent examination of the artifacts in Turkey showed that the wood was extremely fragile and in need of further conservation. A project to retreat the wood

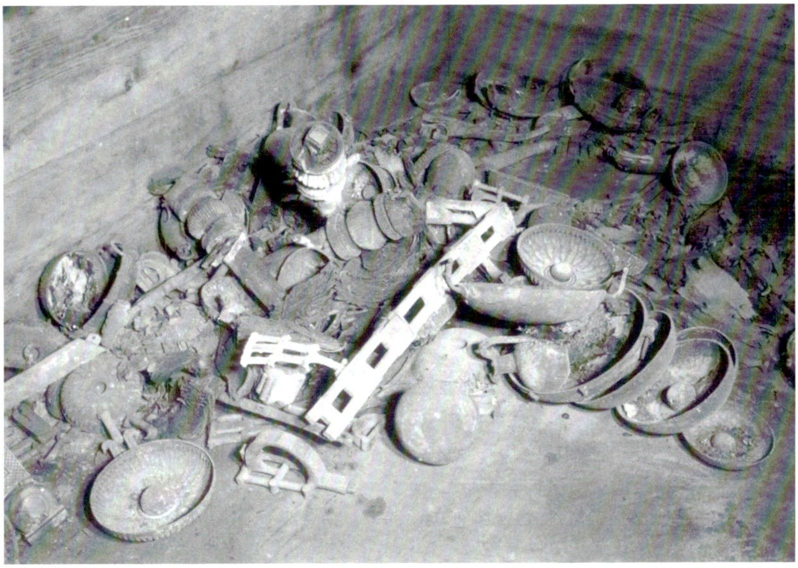

[1] The inlaid table, Tumulus MM, Gordion, as found in the tomb in 1957.

was begun in 1981 under the auspices of the University of Pennsylvania Museum, directed by Simpson, and with the participation of Robert Payton, then conservator for the British Institute of Archaeology in Ankara. In the summer of 2001 the conservation project was completed after 21 years of field work in Turkey, the last 12 years under the supervision of the project's senior conservator, Krysia Spirydowicz, professor of objects conservation at Queen's University, Kingston, Ontario.

The Inlaid Table from Tumulus MM The first piece to be retreated was the marvelously elaborate inlaid table from Tumulus MM (Simpson, 1983; Payton, 1984; Spirydowicz, 1996; Simpson and Spirydowicz, 1999). The table had been made from 40 separate components, most of which had fallen to the floor in their original positions [FIG. 1], providing important clues for the later reconstruction. [FIG. 2] The table's three curved legs terminated in lion-paw feet, and carved struts rose from the feet to support the corners of the square frame. The frame was made from four pieces of wood, carved as a series of interconnected inlaid panels. From the panels of the frame rose 18 inlaid struts that supported the tabletop; the corner struts took the form of handles, indicating that the table was portable. The whole construction was secured by dowels in an internal bracing system. The tabletop was made of walnut; the legs, frame, dowels, and struts were boxwood; and the inlay was juniper.

Most of the boxwood elements had survived intact, along with their juniper inlay, although a few pieces were warped or degraded. By contrast, the walnut tabletop was preserved only in fragments. Some of the damage that had occurred was the result of water seepage into the tomb prior to its excavation in 1957: A drilling campaign, utilizing a water-cooled drill bit, had been carried out in 1955 and 1956 in order to locate the chamber within the tumulus. The condition of the wood had been further compromised by its initial treatment in the wax-and-gasoline mixture. When the pieces of the table were examined in 1981, they were black and gritty, due to the dirt that had become embedded in the wax on the wood's surface.

For the new treatment, the wood was first cleaned of the dark, waxy deposit with cotton swabs soaked in the solvent toluene. It was then consolidated under vacuum in a solution of polyvinyl butyral (Butvar B-98), dissolved in a mixture of toluene and ethanol. After consolidation, the wood was removed from the solution, wrapped in polyethylene, and allowed to dry slowly over several months. The wood was then unwrapped, and a stronger solution of Butvar B-98 was applied to reinforce cracked or damaged areas. Finally, the surface was cleaned of excess consolidant.

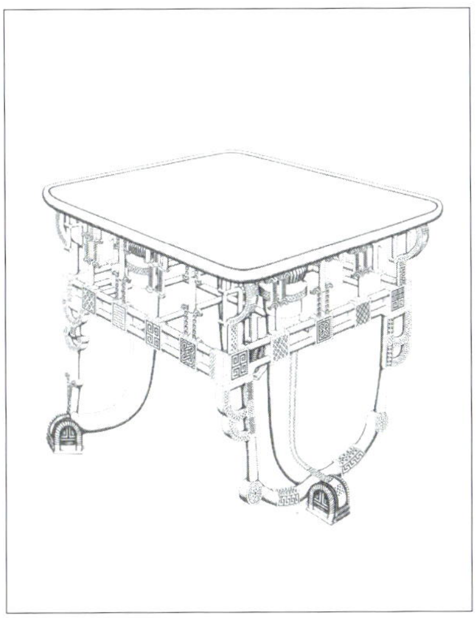

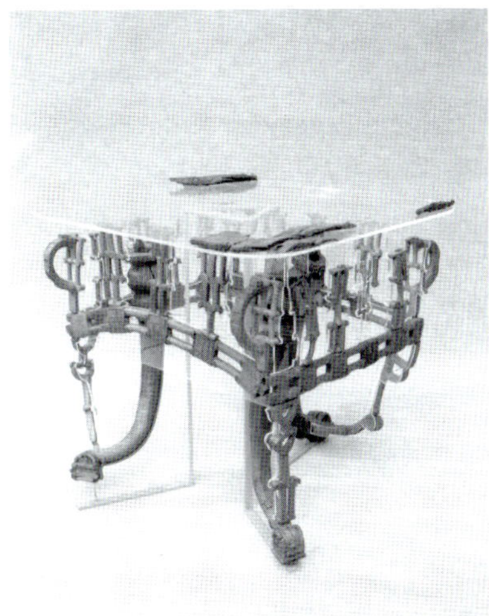

[2] Reconstruction drawing of the inlaid table from Tumulus MM, front view. Elizabeth Simpson, 1985.
[3] Inlaid table, Tumulus MM, view of the back and left sides. Reconstructed for display in the Museum of Anatolian Civilizations, Ankara, 1989.

The treatment was a success, lightening the color of the wood and strengthening it considerably. The pieces could finally be handled safely and the drawing and photography of the table completed. The inlaid table was reconstructed on a Plexiglas mount [FIG. 3] for display in the Museum of Anatolian Civilizations (Simpson and Payton, 1986; Simpson and Spirydowicz, 1999).

Aspects of the Treatment Program The wooden objects that had been treated with Alvar at the time of excavation required a somewhat different approach. During retreatment, the Alvar was removed from the surface of the wood with acetone, aided by the application of gel poultices consisting of a solution of Klucel HF in ethanol or an ethanol-acetone mixture (Simpson and Spirydowicz, 1999; Spirydowicz et al., 2001). The wood was then consolidated with Butvar B-98 according to the method used for the inlaid table.

This method, refined over the next several years by Payton, Spirydowicz, and other members of the conservation team, has been used for most of the wooden artifacts from Gordion (Spirydowicz, 1996; Simpson and Spirydowicz, 1999). Among the pieces that were effectively saved by this treatment method are the two inlaid serving stands and eight plain tables from Tumulus MM, and the inlaid serving stand, bed, carved tables, and numerous small wooden objects from Tumulus P.

During the course of these treatments, the procedures and concentration of the consolidant (Butvar B-98) were optimized according to the results of investigations carried out in collaboration with Robert Blanchette, Arno Schniewind, and other scientists (Spirydowicz et al., 2001). The process has proved highly effective for the consolidation of dry archaeological wood.

Bronze and Butvar: The "Screen" from Tumulus W Certain pieces required unusual measures, among them the so-called screen from Tumulus W, which was recovered as a mass of broken wood fragments. Carved in openwork geometric patterns and covered with bronze studs [FIG. 4], this was apparently once a magnificent serving stand. The wood had degraded, and many of the surviving studs were held in place only by means of the Alvar coating applied at the time of excavation. The stand was removed from the Gordion depot, where it had been stored since 1959, and taken to Ankara for retreatment, a process that would take eight years of painstaking work to finish.

The wood was first cleaned of the Alvar coating [FIG. 5], a difficult and time-consuming procedure because of the extremely fragile condition of the wood. The fragments were cleaned with the aid of solvent gel poultices. Removal of the Alvar

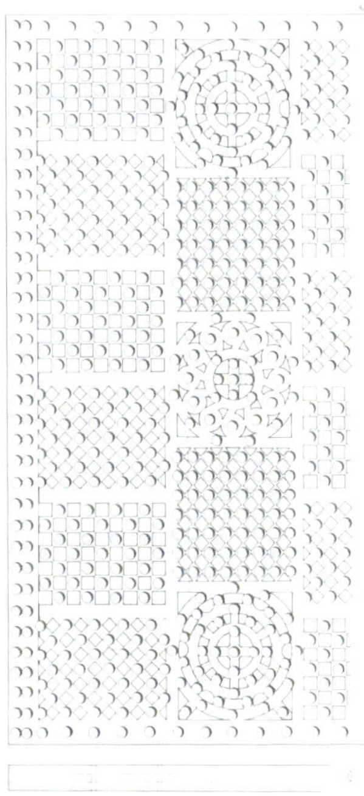

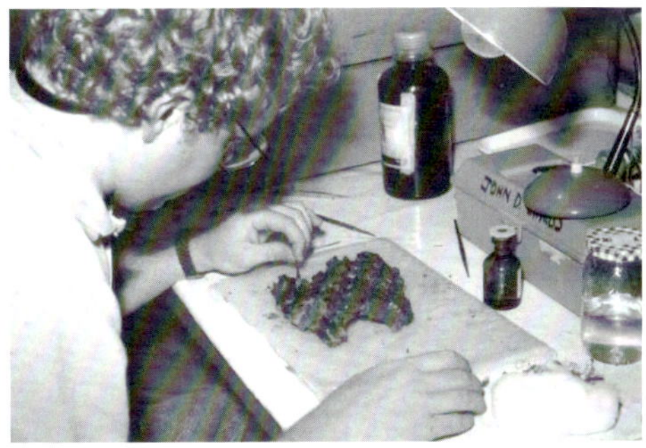

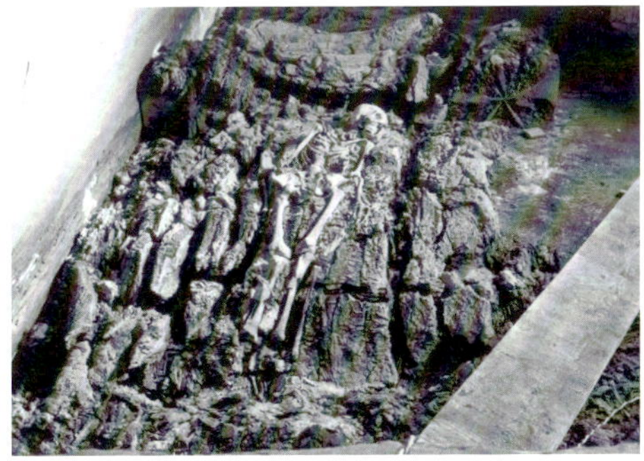

[4] Reconstruction drawing of the "screen" from Tumulus W, Gordion. Elizabeth Simpson, 1999.
[5] John Childs of the conservation team cleaning a fragment of the Tumulus W "screen," 1990.
[6] Remains of the king on his collapsed log coffin, Tumulus MM, Gordion, as found in 1957.

from the bronze studs necessitated scraping with a scalpel under magnification. The studs were found to have bronze disease, which they had developed during their years in storage at Gordion. The corrosion was removed, and loose studs were treated with Benzotriazole and coated with a solution of Acryloid B-72 to inhibit further corrosion.

Many of the bronze studs were still attached to the wooden fragments and were therefore subject to the same consolidation treatment to be used for the wood. Tests conducted to determine the effect of Butvar B-98 on the attached studs indicated that the acidic nature of the consolidant actively promoted corrosion of the bronze. Other consolidants were tested, but because of the lack of data concerning their long-term effects, and because of our past success with Butvar for degraded wood, Butvar was retained, although with a different means of application. A method was developed that allowed the wood to be soaked in the consolidant without immersing the studs in the solution. Following the wood's consolidation, the attached studs were recleaned, treated with Benzotriazole, and coated with the Acryloid B-72 solution. The Tumulus W "screen" was mounted for storage on Plexiglas in 1998.

Unconsolidated: The King's Coffin from Tumulus MM A few of the Gordion wooden objects have remained unconsolidated. The king's coffin [FIG. 6] was not conserved because of the extremely degraded condition of the wood; no method has yet been found to allow the piece to be treated safely and effectively. The coffin, which was made from a huge cedar log, cut in half and hollowed out, survives now only in a few large fragments. Studies carried out on this piece have yielded crucial information about the anatomy and identification of the wood and have revealed new and important data relating to soft-rot deterioration (Blanchette and Simpson, 1992; Filley et al., 2001).

The dead king had lain in state in his open log coffin during a funeral ceremony held outside the tomb before the burial. After the ceremony, his body was lifted out of the coffin, the coffin parts were disassembled and placed in the tomb, and the king was finally lowered into the coffin, as found by the excavators 2700 years later (Simpson, 1990; Simpson, 2001). Research has led to a reconstruction of the king's funeral and even the items on the menu at the funerary banquet (McGovern et al., 1999; Simpson, 2001).

Special Studies, Storage, and Display Special studies have revealed the species of woods used (Aytuğ, 1986; Aytuğ, 1988; Aytuğ and Görcelioğlu, 1988; Aytuğ and Pehlivan, 1989), information about the wood's deterioration (Blanchette et al., 1991;

Blanchette and Simpson, 1992; Filley et al., 2001; Nelson et al., 1995), techniques of Phrygian furniture construction (Simpson, 1999), and previously unrecognized aspects of the design and function of these important objects (Simpson, 1988).

Four major pieces of furniture have been reconstructed on Plexiglas mounts for display in the Museum of Anatolian Civilizations, installed in custom display cases made in Turkey (Simpson and Spirydowicz, 1999). Steel storage cabinets were fabricated in the U.S., made in modules, shipped to Turkey, and assembled in the Gordion storage room of the Ankara Museum. The wood has been placed in the drawers of the storage cabinets, set into ethafoam bedding cut to conform to the shapes of the pieces (Simpson and Spirydowicz, 1999). The Gordion wooden objects are thus accessible to scholars and scientists for study in safe, secure storage.

The effective collaboration of the conservators, scientists, archaeologists, and other specialists participating in the Gordion Furniture Project has produced much new information about the craft and technology of ancient woodworking, contributed to the study of wood science, advanced the field of wood conservation, and ensured the preservation of nearly 100 precious Phrygian wooden artifacts for the future.

REFERENCES

AYTUĞ, B., 1986. Kral Midas'la gömülen mobilyalar (The furniture buried with King Midas). *Arkeometri Ünitesi Bilimsel Toplantı Bildirileri* VI, Tübitak, Ankara, 3–11.

AYTUĞ, B., 1988. Le mobilier funéraire du roi Midas I. In: Hackens, T., et al. (Eds.), *Wood and Archaeology*, Acts of the European Symposium, Louvain-la-Neuve, October 1987, PACT 22, Strasbourg: Council of Europe, 357–368.

AYTUĞ, B., and GÖRCELIOĞLU, E., 1988. Gordion'daki "P" Tümülüsü'nde bulunmuş mobilyalar (Furniture found in Tumulus P at Gordion). IV. *Arkeometri Sonuçları Toplantısı*, Kültür Bakanlığı, Ankara, 49–59.

AYTUĞ, B., and PEHLIVAN, S., 1989. Gordion'daki "P" Tümülüsü'nde bulunmuş mobilyalar -II- ve bazı odun eşyalar (Furniture found in Tumulus P at Gordion -II- and some wooden objects), Aksay Ünitesi, Bilimsel Toplantı Bildirileri I, Tübitak, Ankara, 123–148.

BLANCHETTE, R., CEASE, K., ABAD, A., KOESTLER, R., SIMPSON, E., and SAMS, G., 1991. An evaluation of different forms of deterioration found in archaeological wood. *International Biodeterioration* 28, 3–22.

BLANCHETTE, R., and SIMPSON, E., 1992. Soft rot and wood pseudomorphs in an ancient coffin (700 B.C.) from Tumulus MM at Gordion, Turkey. *Bulletin of the International Association of Wood Anatomists* 13 (2), 201–213.

FILLEY, T., BLANCHETTE, R., SIMPSON, E., and FOGEL, M., 2001. Nitrogen cycling by wood decomposing soft-rot fungi in the "King Midas tomb," Gordion, Turkey. *Proceedings of the National Academy of Sciences* 98 (23), 13346–13350.

MCGOVERN, P., GLUSKER, D., MOREAU, R., NUNEZ, A., BECK, C., SIMPSON, E., BUTRYM, E., EXNER, L., and STOUT, E., 1999. A funerary feast fit for King Midas. *Nature* 402, 863–64.

NELSON, B., GONI, M., HEDGES, J., and BLANCHETTE, R., 1995. Soft-rot fungal degradation of lignin in 2700 year old archaeological woods. *Holzforschung* 49, 1–10.

PAYTON, R., 1984. The conservation of an eighth century B.C. table from Gordion. In: Brommelle, N., et al. (Eds.), *Adhesives and Consolidants*, International Institute for Conservation of Historic and Artistic Works, London, 133–137.

SIMPSON, E., 1983. Reconstructing an ancient table: the "Pagoda" Table from Tumulus MM at Gordion. *Expedition* 25 (4), 11–26.

SIMPSON, E., 1988. The Phrygian artistic intellect. *Source* 7 (3/4), 24–42.

SIMPSON, E., 1990. Midas' bed and a royal Phrygian funeral. *Journal of Field Archaeology* 17, 69–87.

SIMPSON, E., 1999. Early evidence for the use of the lathe in antiquity. In: Betancourt, P., et al. (Eds.), *Meletemata: Studies in Aegean Archaeology Presented to Malcolm H. Wiener*, Université de Liège and University of Texas at Austin, 781–785.

SIMPSON, E., 2001. Celebrating Midas: contents of a great Phrygian king's tomb reveal a lavish funerary banquet. *Archaeology* 54 (4), 26–33.

SIMPSON, E., and PAYTON, R., 1986. Royal wooden furniture from Gordion. *Archaeology* 39 (6) 40–47.

SIMPSON, E., and SPIRYDOWICZ, K., 1999. *Gordion Wooden Furniture: The Study, Conservation, and Reconstruction of the Furniture and Wooden Objects from Gordion, 1981–1998*, Museum of Anatolian Civilizations, Ankara.

SPIRYDOWICZ, K., 1996. The conservation of ancient Phrygian furniture from Gordion, Turkey. In: Roy, A., Smith, P. (Eds.), *Archaeological Conservation and Its Consequences, International Institute for Conservation of Historic and Artistic Works*, London, 166–171.

SPIRYDOWICZ, K., SIMPSON, E., BLANCHETTE, R., SCHNIEWIND, A., TOUTLOFF, M., and MURRAY, A., 2001. Alvar and Butvar: the use of polyvinyl acetal resins for the treatment of the wooden artifacts from Gordion, Turkey. *Journal of the American Institute for Conservation* 40, 43–57.

YOUNG, R., 1981. *Three Great Early Tumuli: The Gordion Excavations Final Reports*, volume I, The University of Pennsylvania Museum of Archaeology and Anthropology, Philadelphia.

B.W. HELD[1]
R.A. BLANCHETTE[1]
J.A. JURGENS[1]
S. DUNCAN[2]
R.L. FARRELL[2]

DETERIORATION AND CONSERVATION ISSUES ASSOCIATED WITH ANTARCTICA'S HISTORIC HUTS

1 University of Minnesota, St. Paul, MN 2 University of Waikato, Hamilton, New Zealand

ABSTRACT During Antarctica's heroic era, explorers used Ross Island as a starting point for expeditions to the South Pole and other regions of Antarctica. Two huts built by Robert F. Scott in 1902 and 1911, and one built by Sir Ernest Shackleton in 1908, were used to escape the extreme Antarctic environment during their explorations. Large quantities of supplies brought to sustain the expeditions for several years were stored in and around the huts. Once the expeditions were completed, the huts and unused supplies were abandoned. Antarctica's dry, cold environment has provided some protection to the huts and artifacts, but significant deterioration has occurred over the past 9 to 10 decades and serious issues concerning preservation must be addressed if the huts are to be successfully preserved.

Biological degradation of wood is common in the huts. A soft-rot type of decay was found to be attacking wood in contact with the ground at several locations. The fungal species responsible for this decay are being characterized and studied. Actively growing fungi have been observed and isolated from clothing, leather, wood, foodstuffs, and other artifacts within the huts. Environmental data loggers placed at various locations throughout the three huts have revealed that during the austral summer, temperatures rise above freezing and relative humidity within the structures is often well over 80%, providing conditions conducive to mold growth.

The impact of visitors on relative humidity inside the huts was also studied. Nonbiological deterioration is evident in the huts, resulting in a defibration of wood surfaces. This unusual corrosive type of chemical degradation is caused by salts that affect the lignified regions of the woody cell wall, resulting in cell separation. Snowmelt carrying dissolved salts is absorbed by the wood and evaporated at the surface. Elemental analyses have shown that very high concentrations of salts are present in affected woods. Wind abrasion has also caused significant erosion of the huts' exterior wood.

Wood treatments are being studied in field tests at three locations on Ross Island, Antarctica, to determine their effectiveness as possible protectants for the exterior of the huts. Results from this work are providing important insights into how wood deteriorates in the polar environment and are creating a foundation of information to initiate conservation plans to preserve these historic sites for future generations.

INTRODUCTION In January of 1902 Robert F. Scott's Discovery expedition built the first of three huts in the Ross Sea Region of Antarctica. [FIG. 1] Modeled after an Australian outback structure, the 9.3 × 9.1-m hut was prefabricated and built at Hut Point on Ross Island to be used as living quarters and to store supplies. Explorations of the region were made, and scientific information was collected during the expedition. Sir Ernest Shackleton, who served with Scott during the Discovery expedition, mounted his own "British Antarctic Expedition" and returned to Antarctica with hopes of being the first to the South Pole. Shortly after arriving on Ross Island in early 1908 he built a hut 13 km north of Hut Point at Cape Royds. [FIG. 2] Situated adjacent to a large Adele penguin rookery, it is the smallest of the three huts studied and measures 7 × 5.8 m. The hut was used for several years during the expedition and was vital for survival while the expedition members readied themselves for the long journey to the pole. Having fallen 97 miles short of reaching the pole, Shackleton returned to England, leaving the hut and all materials associated with the expedition at the site. Scott returned to Antarctica in 1910 on the ill-fated Terra Nova expedition. He built a large hut, 15.5 × 7.5 m, at Cape Evans. [FIG. 3] This hut was the most spacious of the three huts on Ross Island and had a stable attached for the ponies that were brought from England for hauling supplies.

Since each of the expeditions left quickly when the relief ships arrived, the huts and thousands of artifacts were left behind. The dry cold conditions of the Antarctic environment have provided some protection from decay for the huts, but after

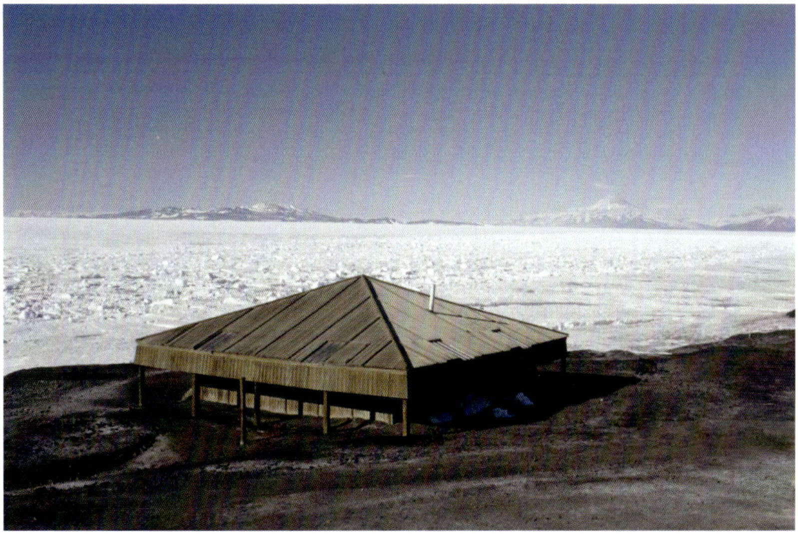

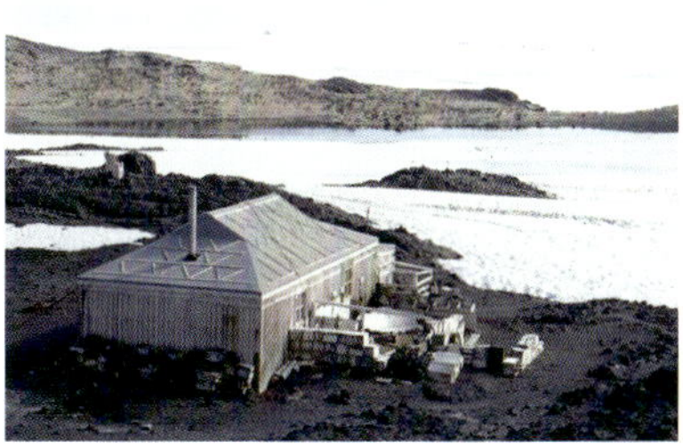

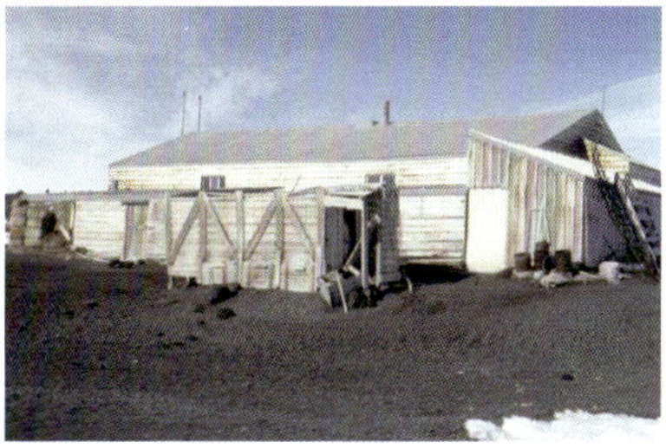

[1] Robert F. Scott's 1902 Discovery expedition hut in the Ross Sea region of Antarctica. [2] Sir Ernest Shackleton's 1908 hut 13 km north of Hut Point at Cape Royds, Ross Island. [3] Scott's 1910 Cape Evans hut from his Terra Nova expedition.

[TABLE 1] Wood identifications made from the Ross Island historic huts and associated artifacts

WOOD ID	LOCATIONS	
	INTERIOR	EXTERIOR
CAPE EVANS HUT		
Abies	Scott's table, supply boxes	NW siding, stables roof slats
Betula	Supply boxes (plywood)	Supply boxes (plywood)
Picea	Main table, darkroom, supply boxes	NE siding, stables siding, latrine, cross, sledge NW corner
Pinus (*sylvestris*)	Ceiling, floor, walls	NE, SE siding, fuel boxes, latrine
Pseudotsuga menziesii	Not found	Replaced siding on hut above stables
CAPE ROYDS HUT		
Abies	Supply box in kitchen	Doghouse
Betula	Supply boxes (plywood)	Supply boxes (plywood)
Picea	Ceiling, floor, walls, supply boxes	Siding, fence in stables, supply boxes
Pinus (*sylvestris*)	Ceiling beams, facia	Facia, meteorological station
Ulmus	Not found	Supply box in garage
DISCOVERY HUT		
Picea	Supply boxes, wall repair	Not found
Pinus (*sylvestris*)	Floor, walls, ceiling	Siding, roof, veranda
Pseudotsuga menziesii	Beams, floor repair	Beams, veranda corner posts, corner siding, replaced siding N side veranda

9–10 decades serious deterioration has occurred, and these problems must be addressed if the huts are to be preserved for the future.

Nonbiological types of deterioration are the most common forms of deterioration associated with the huts and artifacts. One very prevalent type of attack is due to high salt concentrations that corrode the wood surfaces. This chemical attack has only recently been described and the conditions for its occurrence elucidated (Blanchette et al., 2002). Wind erosion is another very serious form of nonbiological deterioration that occurs commonly on exterior woods, especially those with a southerly exposure. Biological deterioration can also be commonly found in association with the wood of the huts and artifacts that is in contact with the ground. A soft-rot type of decay has been found attacking woods in the stables area at the Cape Royds hut and in wooden artifacts at various locations at the historic site. The interiors of the huts have been affected by surface molds. Little is known about the fungi that can colonize and degrade wood in the Antarctic environment, and this investigation was done to further investigate their role in wood deterioration in the historic woods. This study also determined the wood species used for the construction of the huts and various wooden artifacts found at the historic sites. This information is essential for conservation efforts if any of the deteriorated woods need to be replaced.

MATERIALS AND METHODS Minute wood samples were taken from inconspicuous locations on the huts and other artifacts for fungal isolations, salt deterioration evaluations, elemental analysis, and wood identification, and they were stored in sterile containers until they were later studied in the laboratory. This work was done in cooperation with the Antarctic Heritage Trust and under Antarctic Conservation Act permit numbers 2001–015 and 2002–001. Elemental analysis of defibrated wood was carried out by using multi-elemental inductively coupled plasma atomic emission spectroscopy (Blanchette et al., 1994; Munter and Grande, 1981; U.S. Environmental Protection Agency, 1979, 1983).

Interior and exterior temperature and relative humidity data were taken using HOBO® H8 Pro data loggers. Data loggers were calibrated to take readings of temperature and RH every hour year round. The data loggers were downloaded during field visits to the huts in 2001 and 2002. The number of data loggers used in the Cape Evans, Cape Royds, and Discovery huts are 6, 5, and 4, respectively. Data loggers were placed throughout the huts at varying heights.

Fungal isolations were made by aseptically removing small segments from wood samples and placing them in selective media: 1.5% Difco malt extract agar, basidiomycete select agar (Worrall, 1991), and a 1.5% Difco acidified malt agar. Wood samples were also prepared for scanning electron microscopy using previously described methods (Blanchette et al., 2002).

Wood identifications were made by taking small sections from wood and examining them with light microscopy for anatomical characteristics (Hoadley, 1990; Panshin and Zeeuw, 1970).

RESULTS

Wood Identification A large variety of different woods were used to construct the huts. [TABLE 1] The major woods used for Cape Evans and Discovery huts were *Pinus sp.* with characteristics of a hard pine. The species is a hard pine, likely *P. sylvestris,* as the wood originated from Europe. This wood was used for the floors, walls, and ceilings of the huts. Cape Royds hut is constructed primarily out of *Picea.* All plywood supply crates were identified as *Betula sp.* at all locations. Douglas fir is another important species and was used for the beams and corner posts on the veranda of Discovery hut. Boards that were replaced during conservation work many years ago on a section of the veranda siding are also Douglas fir.

Salt Deterioration Salt deterioration can be commonly found affecting the huts and associated artifacts. Affected wood has a fuzzy appearance when dry due to the separation of surface fibers. [FIGS. 4 AND 5] Under wet conditions affected surfaces of wood have the appearance of having been pulped. A severe defibration is taking place on the exterior siding of the stables on Cape Evans hut. [FIG. 6] Snow with elevated amounts of salts accumulates on the roof of the stables and melts as temperatures rise in the austral summer, causing melt water with significant amounts of sodium and chloride ions to drip onto the side wallboards. [TABLE 3] This salt solution cascades down the siding and is absorbed by the wood. The salts accumulate in the wood and deterioration occurs. The surface wood fibers become defibrated and detach from one another. Boards replaced by conservators on the hut as recently as 1991, from above the stables area, are already showing significant defibration when observed in 2002.

Cape Royds hut and Discovery hut are also affected by salt deterioration. At Cape Royds, the damage is especially severe along the north side in the stables area. Some of the most severe damage found on the interiors of the huts is in the porch of Cape Royds hut, in which the ceiling and wall boards are extensively bleached and defibrated. Table 2 shows concentrations of sodium ions of this wood exceed 50,000 ppm. Many wooden supply boxes located on the south and east sides of the

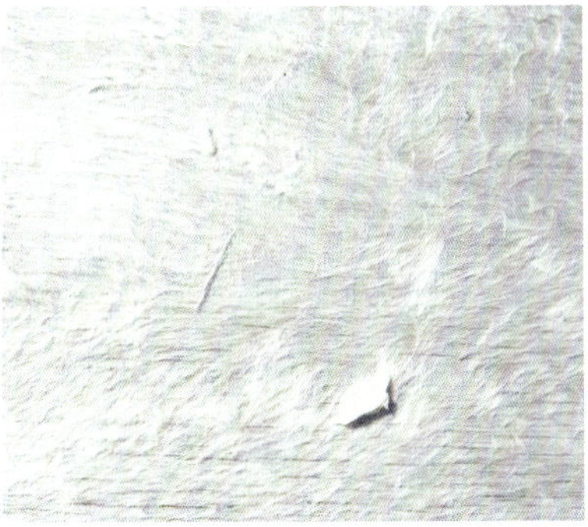

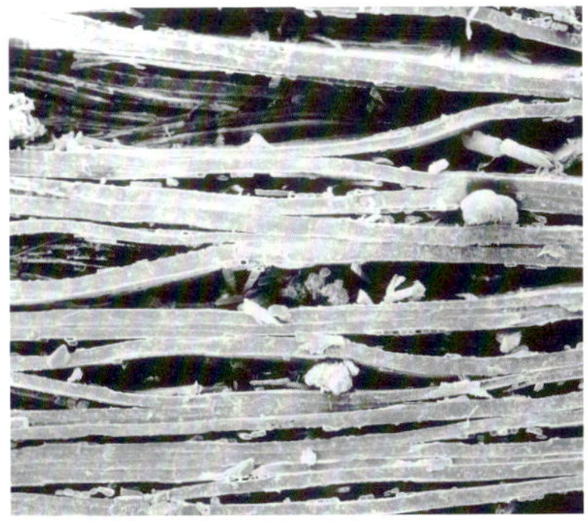

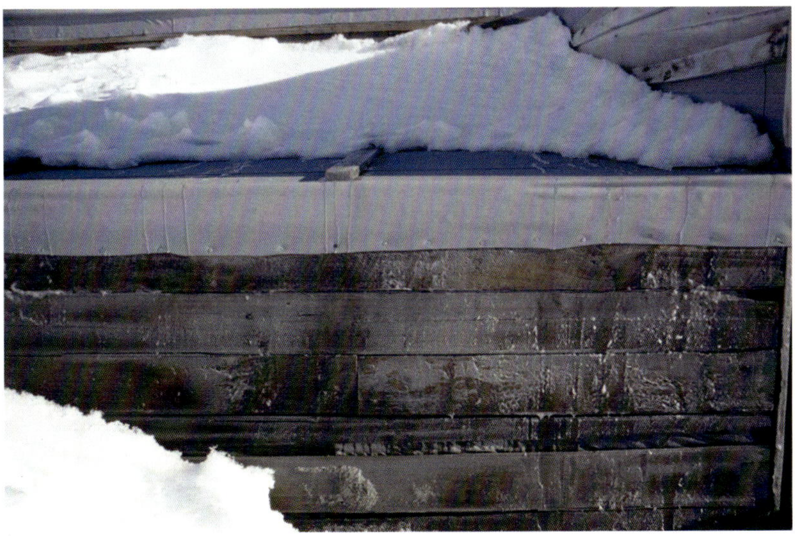

[4, 5] Salt-deteriorated wood showing a fuzzy appearance when dry due to the separation of surface fibers.
[6] Exterior siding of the stables of the Cape Evans hut showing severe defibration.

[TABLE 2] Elemental analysis (ppm) of wood from the Ross Island historic huts. Adapted from Blanchette et al., 2002.

LOCATION	MATERIAL	P	K	CA	MG	MN	AL	FE	NA
Cape Evans	Exterior defibrated wood	132	4355	6379	8331	49	558	431	98,650
Cape Royds	Porch defibrated ceiling	132	2828	7451	6091	71	647	649	52,970
Discovery Hut	Exterior defibrated wood	547	7070	4250	10,728	78	3285	2602	246,321
Sound modern wood	Construction board	102	527	403	133	51	5	10	7

[TABLE 3] Chloride and sodium concentrations in melt water from the Cape Evans stables roof. Elements in ppm. Adapted from Blanchette et al., 2002.

SAMPLE	CL	NA
Roof melt water	12,283	5,736
Old snow on roof	626	395
New snow on roof	34	23

hut are also damaged. Supply boxes made from plywood are not only defibrated but have also become delaminated. Wood sampled from beneath the veranda at Discovery hut had the highest concentration of salts from all woods sampled, with sodium concentrations of 246,321 ppm. [TABLE 2]

The huts themselves are not the only woods being affected by salt deterioration. Artifacts made from wood and exposed to salts and moisture are also affected. Wooden crates, doghouses, sledges, and even nonwooden objects such as rattan pony snowshoes, are being deteriorated. Salts also cause significant corrosion of metals at the historic huts.

Wind Erosion Wind erosion can be identified in many locations of the huts and associated artifacts. High-velocity winds originating from the South Pole carry airborne particles that cause a sand-blasting effect on the exposed wood. Therefore most of the significantly eroded areas are those that face south. Wood is not eroded uniformly by wind. [FIG. 7] Windborne particles erode the highly lignified, thick-walled latewood cells at a slower rate than they do the thin-walled earlywood cells, leaving affected wood with an uneven, furrowed appearance.

Three test panels, one at each hut location, have been set up to test possible treatments for use on the huts. Blocks of pine and spruce were treated with four silicon-based treatments and an oil-based paint similar to the type that was originally used on Discovery hut. Wood blocks have been assessed and more time is needed to complete the study. Significant erosion is taking place on the control blocks after only two years of exposure.

Biological Deterioration Minute samples from many areas suspected of being decayed were collected and examined. A unique form of decay has been found degrading wood in contact with the ground. It is described as a soft-rot type of decay causing cavities to form in the cell walls. Isolations have indicated that the decay is caused by *Phialophora*-like fungi. Figure 8 shows extensive cavities caused by the soft-rot fungi located in the secondary walls of the wood cells. [FIG. 8] Affected wood was from the stables area at Cape Royds hut. This area has significant amounts of penguin guano due to penguins gathering in this location for short periods of time. Pure cultures of the decay fungi have been obtained and are being characterized and studied further. Cultures similar to those causing soft rot have been isolated from many other woods at all three hut locations.

In addition to decay fungi, extensive mold growth has been found in Cape Evans hut. [FIG. 9] The fungal growth is most common on wood near or on the floor and

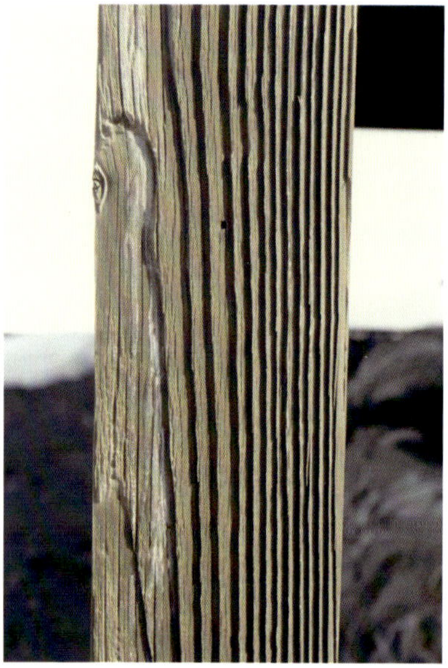

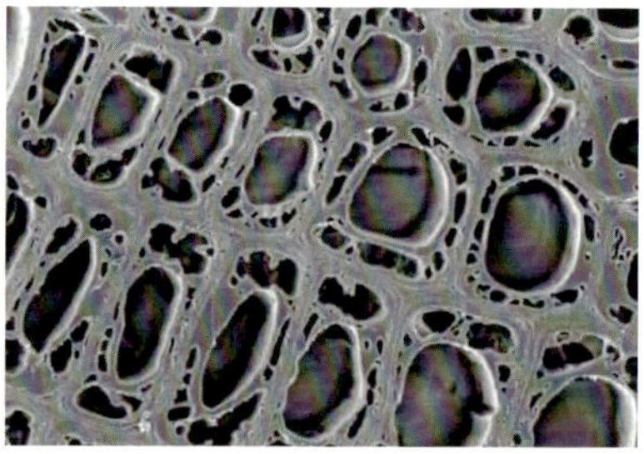

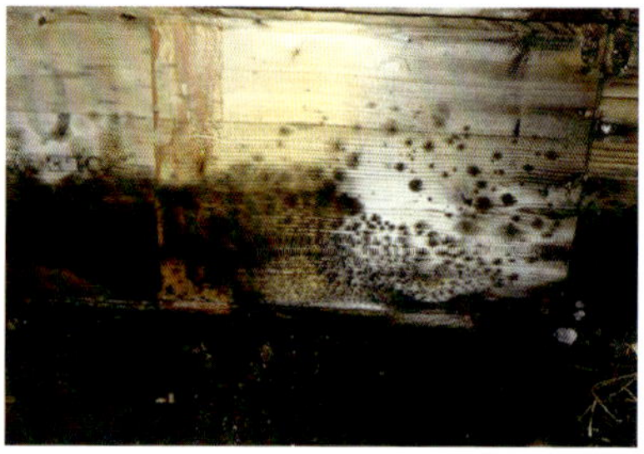

[7] Wood eroded by wind shows an uneven erosion. [8] Extensive cavities caused by the soft-rot fungi located in the secondary walls of the wood cells. [9] Extensive mold growth has been found in Cape Evans hut, most commonly on wood near or on the floor and often in areas with limited air circulation.

often in areas with limited air circulation. Frost accumulating at the junction of the wall and floor several feet away is probably contributing increased moisture to the area as temperatures rise and the ice melts, enabling the mold growth to occur. Cultures of the mold have been obtained and identified as *Cladosporium spaerospermum* and *C. cladosporioides.*

Environmental Monitoring Temperature and relative humidity have been collected for over a two-year period. A graph showing temperature and RH for two years in Discovery hut is shown in figure 10. [FIG. 10] The RH over time is quite elevated, although the temperature rises above freezing only for a short period during the austral summer months. Relative humidity in the huts is elevated and is often above 80%. Figure 11 shows data for a one-month period from December to January 2002 inside Cape Evans hut, where the average RH is 84% and the maximum relative humidity observed was 93%. [FIG. 11] The average temperature for the same period is 1.4°C and reaches a maximum of 7°C. These conditions are conducive to mold growth within the hut.

Hundreds of visitors from cruise ships or scientists carrying out research in the area enter the huts every season. The impact on the interior environment of the hut from visitors has not been previously determined. Using visitor data, several days were selected when large groups entered Cape Royds; the goal was to determine their possible effect on relative humidity. On Jan. 8, 2001, 66 visitors entered the hut. [FIG. 12] Four visitors entered the hut three days prior to Jan. 8, and three visitors entered the hut three days after the large group. Forty-five visitors entered the hut on Feb. 14, 2001, and there were no other visitors in the days prior to or after Feb. 14. [FIG. 13] Both events show no apparent impact on RH from the visitors entering the hut. Instead, the RH appears more reflective of the temperature, with increases in RH following increases in temperature.

DISCUSSION

Wood Identification Woods used in the construction of the huts and wooden artifacts have been identified. [TABLE 1] For successful conservation of the historic huts the identification of the wood will help any restoration efforts to provide the appropriate wood species that was originally used. During this investigation it was found that some woods recently used in restoration work were not the same type used in the original structure. Harrowfield (1981) reports the general assumption that Discovery hut was made of Douglas fir (*Pseudotsuga menziesii*), while our investigations found Douglas fir was used for some main timbers and corner posts of

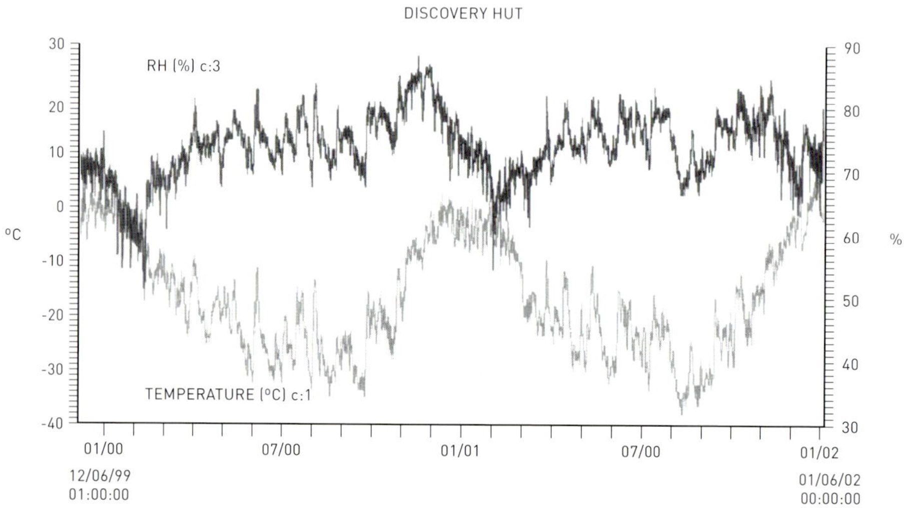

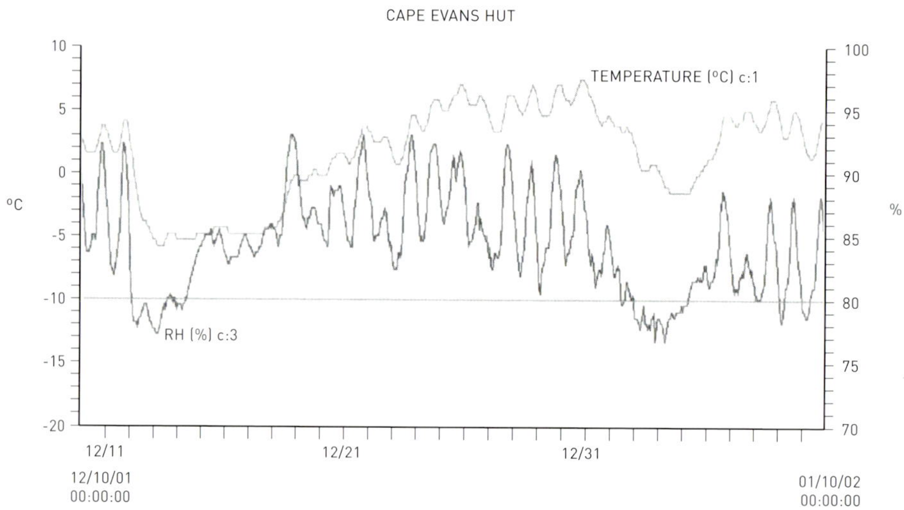

[10] Graph of relative humidity and temperature in Discovery hut over a 2-year period. [11] Graph showing relative humidity and temperature in Cape Evans hut over a one-month period indicating high relative humidity and temperatures above 0°c. The horizontal line indicates 80% relative humidity.

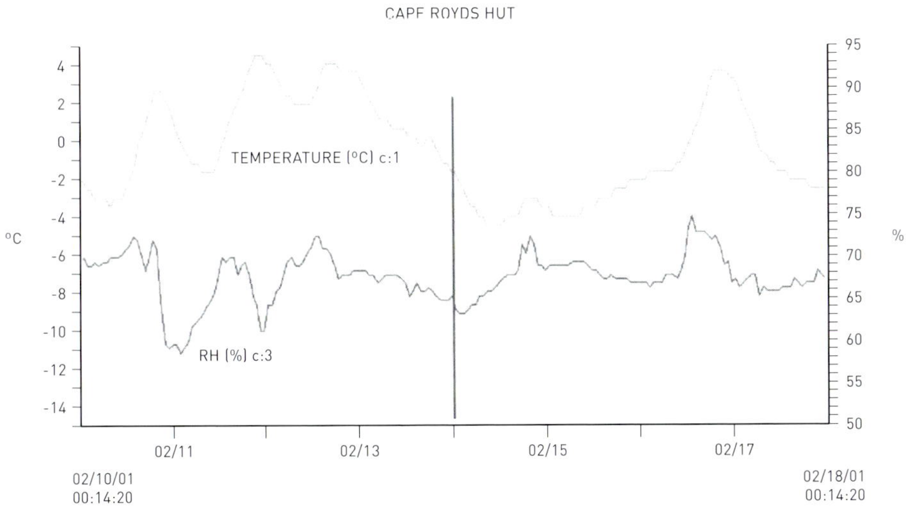

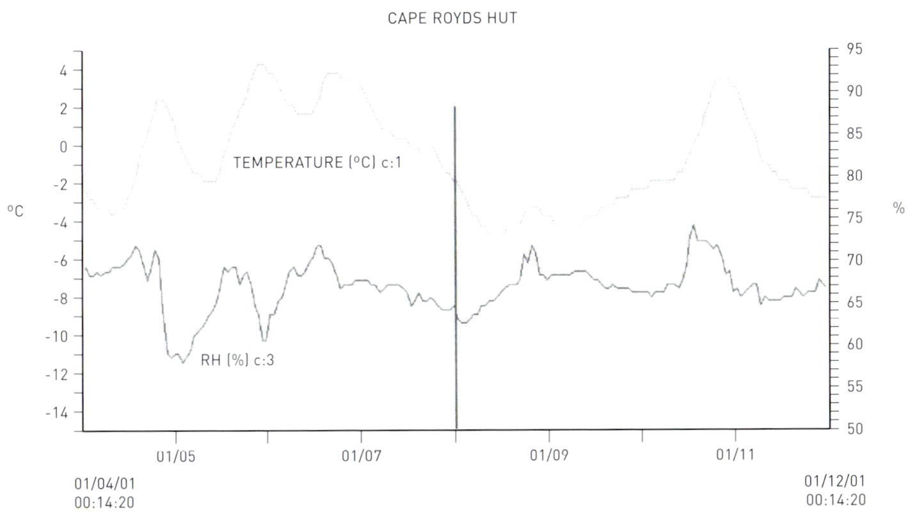

[12] Graph of relative humidity and temperature in Cape Royds hut. The vertical line indicates a day in which 66 visitors entered the hut. [13] Graph of relative humidity and temperature in Cape Royds hut. The vertical line indicates one day in which 45 visitors entered the hut.

the veranda, and pine (*Pinus sp.* similar to *P. sylvestris*) was used for exterior siding, interior walls, floor, and ceiling. The misconception that the wood was Douglas fir was likely the reason for conservators using Douglas fir in 1964 to replace lost wall boards from the north side of the veranda. The original boards of the veranda still present on the hut are *Pinus sp.* Cape Evans hut has also undergone restoration and in 1991 some of the deteriorated siding on the northwest- or sea-side of the hut was replaced. Douglas fir was also used for this wood but the original woods remaining on the hut indicate the exterior boards were made of pine. Nearly all of the exterior wallboards sampled from Cape Evans are pine. At Cape Royds the hut has been thought to be made from fir (Harrowfield, 1981), but our investigations have found the wood is spruce (*Picea sp.* and most likely Norway spruce). Pony snowshoes previously thought to be bamboo (Harrowfield, 1981) are actually rattan. The identification of most of the major woods used in the huts is now confirmed [TABLE 1] and conservators can use this information for selecting the appropriate woods for any future restoration work.

Salt Deterioration Wood exposed to high salt concentrations can be severely deteriorated. In more temperate regions, a defibration of surface fibers has been identified on preservative-treated marine pilings, salt and fertilizer warehouses, and chemical plants (Johnson et al., 1992; Wilcox et al., 1991). In a study by Parameswaran (1981), the changes occurring in wood used from a potash fertilizer storehouse were similar to the alkali pulping process that takes place during papermaking. While the process appears to be similar, salt defibration associated with the historic huts occurs at a much slower rate. On Ross Island, salts commonly migrate to the surface of the soil, forming thin crusts (Campbell and Claridge, 1987). These salts are then blown and accumulate in snowdrifts. Salts from the Ross Sea also contribute aerosols of salt solutions that occur in snow and ice around the historic huts.

Although the exact process of salt deterioration in wood is not fully understood, it is apparent that the high salt concentrations cause a chemical reaction to take place in which the hemicellulose and lignin in the middle lamella are degraded (Blanchette et al., 2002). This is exhibited in affected woods by a defibration of surface fibers, giving the wood surface a fuzzy appearance. There are many locations at all three huts that are affected by salt deterioration. All of these locations involve moisture absorption from melt water in pools on the ground, or by the melting of snow from the roofs directly on the huts or artifacts. To best protect the huts against this problem, excess moisture accumulation must be kept from being in contact with historic woods. Drainage can be improved in some areas adjacent to the huts where water tends to accumulate when snow melts. Also, melt water with high concentrations of dissolved salts drips onto the sides of some areas of the huts. Extending the drip line of the roof several inches out over the wall could avoid melt water coming in contact with the wood. Additional research is needed to fully understand this complex process of wood defibration from different types of salts in the polar environment as well as to find better methods to protect wood from the damage that salt causes.

Wind Erosion Wind erosion is a serious problem for the historic huts and artifacts, which are exposed to strong southerly winds. Research on wind movements around the historic hut at Cape Adare suggests that wind barriers may be effective at protecting the historic woods (Harrowfield, 1996). Wood protectants are also being tested as possible treatments to slow the process of wind erosion (author's unpublished data). The treatments will be evaluated at the end of the five-year study period. Further research is needed to identify coatings that may prove effective against nonbiological effects and are acceptable for use on the historic woods.

Biological Deterioration Soft rot is the only type of wood decay found at the historic sites investigated. The same species of soft rot that was isolated from decayed wood at Cape Royds has also been isolated from many other samples taken from wood in ground contact at Cape Evans and Discovery hut. It is interesting that these fungi seem to be common in the Antarctic environment but rarely in other parts of the world. There is limited knowledge about the biology and physiology of these unusual fungi, and their role in the Antarctic ecosystem is unknown (Vincent, 1998; Vishniac, 1992). The location in which soft rot occurs appears to be associated with an area where penguin guano accumulates. The increased nitrogen and elevated pH may influence the growth and decay rates of these fungi. This also may be a reason why soft rot is prevalent in wood from Cape Royds but the same fungi are not causing wood decay at other sites. Isolations of similar species have been made from other locations in Antarctica, indicating these fungi are likely to be endemic to the continent (Vincent, 1988; Sun et al., 1978).

The soft-rot fungi have only several weeks each year during which temperatures are high enough (above 0°c) to attack the wood. It is most likely that the decay would be much more extensive if temperatures were not limiting fungal activity. *Phialophora* species occurring in temperate areas have also been known to tolerate toxic substances such as copper, chrome, and arsenic in preservative-treated timbers (Daniel and Nilsson, 1988; Wang et al., 1990). Further research could provide information into factors responsible for their tolerance and give insight into potential methods of control of these decay fungi.

Environmental Monitoring Relative humidity within the huts is elevated for much of the year. Of greatest concern is the time during the austral summer in which both RH is high and temperatures are above freezing. A reduction of RH during these periods may limit the growth of surface molds. Also, air circulation in the huts is currently poor. Increasing circulation in problem areas where mold growth occurs may be beneficial. It does appear that visitors to the huts have no significant effect on RH inside the huts. [FIGS. 12 AND 13] Temperature and RH data will continue to be collected for several years to confirm these results.

CONCLUSIONS The research presented here has provided valuable information on the deteriorating effects influencing the Ross Island historic huts. The species of wood that the huts and associated wooden artifacts are made of have been identified, and the types of nonbiological and biological deterioration commonly associated with the huts have been identified.

The major forms of nonbiological deterioration are salt defibration and wind erosion. Salt solutions are repeatedly absorbed by the historic woods, which precipitate water at evaporative surfaces, causing high concentrations of salts to build up. Extensive defibration results, caused by a degradation of the middle lamella region of wood cells, which becomes weakened and deteriorated, causing cells to separate. Wind erosion over the past 9–10 decades has caused significant damage. Most affected is wood that is unprotected from the strong southerly winds, which carry windborne particles that slowly erode the wood surface. Earlywood is preferentially eroded at a faster rate than latewood, apparently due to lower amounts of lignin and thinner cell walls in the earlywood zone.

Biological organisms are also affecting the huts. An unusual type of soft rot has been found attacking wood in contact with the soil at Cape Royds hut. *Phialophora*-like fungi isolated from the decayed woods are being characterized and studied for their decay capacity. These fungi have also been commonly isolated from all of the historic structures and many of the wooden artifacts located outside the huts. Active mold growth caused by *Cladosporium sp.* has also been found growing on wood in the Cape Evans hut. Mold growth appears to be exacerbated by conditions of high relative humidity and temperatures above 0°c during the austral summer months. Environmental monitoring of temperature and relative humidity has been initiated and will continue over the next several years. At present, the relative humidity is elevated above 80% for much of the summer months, and visits by large numbers of people (from cruise ships) do not appear to have a significant effect on the internal environment of the huts. The results provided from these investigations should serve as a resource for conservators involved in the preservation of these historic sites, allowing the huts to survive long into the future.

ACKNOWLEDGMENTS We thank Antarctic Heritage Trust and their conservators for cooperation with these studies, and Joanne Thwaites for her assistance at the historic huts. The authors also thank the New Zealand Antarctic Institute and personnel of Scott base for their assistance in carrying out this research. This material is based upon work supported by the National Science Foundation grant 9909271.

REFERENCES

BLANCHETTE, R.A., HELD, B.W., and FARRELL, R.L., 2002. Defibration of wood in the expedition huts of Antarctica: an unusual deterioration process occurring in the polar environment. *Polar Record*, 38, 313–322.

BLANCHETTE, R.A., HAIGHT, J.E., KOESTLER, R.J., HATCHFIELD, P.B., and ARNOLD, D., 1994. Assessment of deterioration in archaeological wood from ancient Egypt. *Journal of the American Institute of Conservation*, 33, 55–70.

CAMPBELL, I.B., and CLARIDGE, G.G.C., 1987. *Antarctica: soils weathering processes and environment.* New York, Elsevier. 368 pp.

DANIEL, G.F., NILSSON, T., 1988. Studies on preservative tolerant Phialophora species. *International Biodeterioration*, 24, 327–335.

HARROWFIELD, D.L., 1981. *Sledging into history*. Macmillan, Auckland, 118 pp.

HARROWFIELD, D.L., 1996. The role of wind in the destruction of an historic hut at Cape Adare in Antarctica. *Polar Record*, 32, 3–18.

HOADLEY, R.B., 1990. *Identifying Wood.* Taunton Press, New Town, CT, 233 pp.

JOHNSON, B.R., IBACH, R.E., and BAKER, A.J., 1992. Effect of salt water evaporation on tracheid separation from wood surfaces. *Forest Products Journal*, 42, 57–59.

MUNTER, R.C., and GRANDE, R.A., 1981. Plant tissue and soil extract analyses by ICP-AES. In: Barnes, R.M., (Ed.), *Developments in Atomic Plasma Spectrochemical Analyses*, Heydon and Sons, Philadelphia, pp. 653–673.

PANSHIN, A.J., and ZEEUW, C., 1970. *Textbook of Wood Technology*, 3rd edition, McGraw Hill, New York.

PARAMESWARAN, N., 1981. Micromorphology of spruce timber after long-term service in a potash store house. *Holz als Roh und Werkstoff*, 39, 149–156.

SUN, S.H., HUPPERT, M., and CAMERON, R.E., 1978. Identification of some fungi from soil and air of Antarctica. *Terrestrial Biology* 3. Antarctic Research Series v.30, 26 pp.

U.S. ENVIRONMENTAL PROTECTION AGENCY, 1979. *Methods for chemical analysis of water and wastes.* Cincinnati, OH: Environmental Monitoring and Support Laboratory, Office of Research and Development (Method 325.2; EPA-600 4–79–020).

U.S. ENVIRONMENTAL PROTECTION AGENCY, 1983. *Methods for chemical analysis of water wastes.* Cincinnati, OH: Environmental Monitoring and Support Laboratory, Office of Research and Development (Method 200.7; EPA-600 4–79–020).

VINCENT, W.F., 1988. *Microbial ecosystems of Antarctica.* Cambridge University Press, Cambridge. 304 pp.

VISHNIAC, H.S., 1992. Biodiversity of yeasts and filamentous microfungi in terrestrial Antarctic ecosystems. *Biodiversity Conservation*, 5, 1365–1378.

WANG, C.J., WANG, K., and ZABEL, R.A., 1990. *Identification manual for fungi from utility poles in the eastern United States.* American Type Culture Collection, Rockville, Maryland, 356 pp.

WILCOX, W., BOTSAI, E., and KUBLER, H., 1991. *Wood as a building material: A guide for designers and builders.* John Wiley & Sons, New York, 215 pp.

WORRALL, J.J., 1991. Media for selective isolation of hymenomycetes. *Mycologia*, 83, 296–302.

J.A. JURGENS[1]
R.A. BLANCHETTE[1]
D.N. CARLSON[2]

EVALUATING THE WOODEN REMNANTS OF THE TEKTAŞ BURNU SHIPWRECK

1 University of Minnesota, St. Paul, MN 2 Institute of Nautical Archaeology, College Station, TX

ABSTRACT A small vessel thought to be a Greek merchant ship sank to the floor of the Aegean Sea near Tektaş Burnu, Turkey, between 440 and 425 B.C., or shortly thereafter. Scattered among the many remaining artifacts were a few small wooden remnants of the ship's hull and equipment. These remains were investigated to determine the type of wood used in the construction of the ship and to obtain information on the degradation processes that have occurred in the wood over the past millennia. Several small segments of wood were all that remained of the ship, and 12 of these woods have been identified—as *Pinus* (5 samples), *Quercus* (3 samples), and *Ulmus* (3 samples), with one very small sample remaining unidentified. Micromorphological studies indicated the woods were severely degraded and lacked structural integrity. In all samples examined, the secondary wall layers of the wood cells were extensively degraded with minute holes and tunnels present among disrupted cell wall material. The middle lamella between cells remained, and was not eroded. The deterioration in the secondary wall layers has characteristics indicating that bacterial degradation was the major type of microbial decay. The advanced stage of decay and changes that have occurred from environmental influences and secondary scavenging bacteria were found to often mask the decay of the primary bacteria that attacked the wood. Erosion and tunneling bacteria followed by secondary scavenging bacteria most likely were involved with the deterioration observed. Decay by erosion bacteria is characterized by depletion of cellulose and hemicelluloses in the secondary wall, leaving only a lignin-rich framework of the middle lamella. Tunneling bacteria attack the secondary wall, producing small tunnels, and may penetrate and degrade the middle lamella to a limited extent. In sediment-covered, waterlogged woods where oxygen is restricted, both erosion and tunneling forms of bacterial degradation may be found. The ship's wood that remained was infiltrated with copper and other metal corrosion products that apparently helped to inhibit microbial degradation and aid in the wood's preservation. Identification of the wood species found in the shipwreck has provided information on the types of woods used for building this ship and the characterization of decay has documented the current condition of the wood and elucidated the type of microbial degradation occurring in this waterlogged environment.

INTRODUCTION Wood provides a wealth of information when it is found at an archeological site. Wooden artifacts can give great insight into past civilizations, cultures, technology, and environmental conditions (Blanchette, 1995). However, wood is often lost to microbial decay processes in terrestrial and marine environments, unless conditions that suppress these organisms prevail. All wood-degrading microorganisms have general requirements for survival that include: moisture, oxygen, nutrients, favorable temperature, suitable pH, and a nontoxic substrate (Blanchette and Hoffman, 1994; Eaton and Hale, 1993). If any of these requirements are not provided, their development is greatly reduced or nonexistent. In terrestrial conditions where these requirements are met, brown- and white-rot fungi are typically responsible for rapid degradation of wood (Blanchette and Hoffman, 1994). However, in marine environments where oxygen is often limited, soft-rot fungi and bacteria are the dominant organisms that attack wood, causing a relatively slow decay (Daniel and Nilsson, 1997; Singh and Butcher, 1991). It is also very important to understand the degradation and chemical properties of archaeological woods if conservation and preservation procedures are to be successful (Blanchette, 2000). This paper reviews the current understanding of microbial decay of archeological wood from marine environments and investigates the wooden remnants of the Tektaş Burnu shipwreck.

Fungi that cause soft-rot decay in wood belong to the *Ascomycota* and *Deuteromycota* phyla. They cannot compete with the brown- and white-rot fungi and are typically found in woods that have limited access to oxygen, or are found in an environment of moisture extremes that will not support the basidiomycetous decay fungi. Soft-rot fungi are capable of two forms of decay: type 1 or cavity formation, and type 2, a cell wall erosion type of attack. Bacteria need high moisture contents and are most common in waterlogged environments even when oxygen is limited. Wood-degrading bacteria are often divided into three major groups: erosion bacteria, tunneling bacteria, and cavitation bacteria, based on the type of attack within the various cell wall layers. Pit-degrading and scavenging bacteria are two other groups that degrade wood or wood residues, but these have not been studied extensively. The pit-degrading bacteria concentrate their attack on bordered pits and related structures, and scavenging bacteria are considered secondary degraders that gain their required resources from residual material generated by primary degraders. These organisms may work individually or together to slowly degrade historically important woods in marine environments.

WOOD-DEGRADING ORGANISMS FROM MARINE ENVIRONMENTS

Erosion Bacteria In marine environments one of the most predominant types of wood degradation appears to be caused by erosion bacteria (Singh and Kim, 1997). This type of cell wall attack starts as conical troughs that initiate from the cell lumen and move toward the middle lamellae. The bacteria appear to preferentially degrade cell wall layers that have a considerable amount of cellulose and hemicellulose, and to avoid regions of the cell with high lignin contents. The secondary cell wall, consisting of the S_1, S_2, and S_3 layers, is the largest region of the cell wall and the layer most degraded. [FIG. 1] Erosion bacteria appear to enter the cell through the lumen by penetrating the S_3 layer to access the remaining layers of the secondary wall. Although the bacteria penetrate localized areas of the S_3, they do not fully degrade this layer. The S_3 layer is often absent in areas of the wood cell wall where extensive degradation of the S_2 layer has occurred and erosion troughs have coalesced into larger voids. It is not apparent, however, whether this is caused by direct bacterial degradation or a loss of structural support from the underlying S_2 layer after it has been degraded (Daniel and Nilsson, 1997). Within the secondary wall, the bacteria tend to follow the cellulose microfibril orientation. After advanced decay by erosion bacteria, the secondary wall is riddled with holes, but the middle lamella remains unaltered. Wood decay by bacteria requires some oxygen for degradation to progress, but erosion bacteria apparently can tolerate near anaerobic conditions since they can be found in sediment-covered, waterlogged woods (Daniel and Nilsson, 1997).

Tunneling Bacteria Unlike erosion bacteria, tunneling bacteria do have the ability to degrade lignin and the middle lamella region of the cell wall. They enter wood cells through the lumen, and have also been observed entering the secondary wall via the chambers of bordered pits when the lumen is full of extractives (Singh and Butcher, 1991). Once the cell wall has been penetrated, the bacteria readily degrade all layers of the secondary wall and often do not follow the orientation of cellulose microfibrils in the S_2 layer. In the secondary wall, the bacteria produce small tunnels that are similar in diameter to the bacterium. As they degrade the cell wall, concentric bands of residual materials from degradation and extracellular slime are left behind. [FIG. 1] If conditions remain favorable, decay progresses and the small tunnels eventually coalesce, forming larger areas of degradation. Secondary bacterial scavengers often grow and utilize the residual cell wall components left by the tunneling bacteria. There are a few areas of the wood cell that appear to be immune to degradation by tunneling bacteria—the corner regions of the middle lamellae (Singh and Butcher, 1991). Although evidence of tunneling and erosion bacteria can often be found within the same cell wall, tunneling bacteria seem to require greater oxygen concentrations than erosion bacteria do (Björdal, 2000).

Cavitation Bacteria As the name implies, this type of bacterial attack produces cavities primarily in the S_2 layer. Similar to the erosion and tunneling bacteria, cavitation bacteria appear to enter the cell wall by boring a small hole in the S_3 layer, but they do not degrade large areas of this layer. After removing the cellulose and hemicellulose in the S_2 layer, the thin S_3 appears to collapse because of reduced support. Studies have also shown that the cavities created by these bacteria are not limited to the S_2 and can extend into the S_1 layer (Singh and Butcher, 1991). The cavities are often associated with bordered pits, implying that the bacteria gain entry to the cell through the pit chamber. Cavities have also been observed in areas adjacent to the bordered pits, suggesting that direct penetration of the cell wall is also possible (Singh and Butcher, 1991). The cavities are generally angular and often diamond-shaped, with the long axis of these cavities running either parallel to or perpendicular to the long axis of the cell. [FIG. 2] This type of wood-degrading bacteria does not appear to have the ability to degrade the middle lamellae. The oxygen requirement for cavitation bacteria is not yet known, but they do not appear to be common in extremely oxygen-depleted environments.

Pit-Degrading Bacteria Although cavitation bacteria appear to gain entrance into the wood cell though bordered pit chambers, and tunneling bacteria are able to degrade pit membranes, there is a type of bacteria that seem to selectively degrade only the pit membrane and not the cell wall (Burnes et al., 2000; Singh, 1997). They

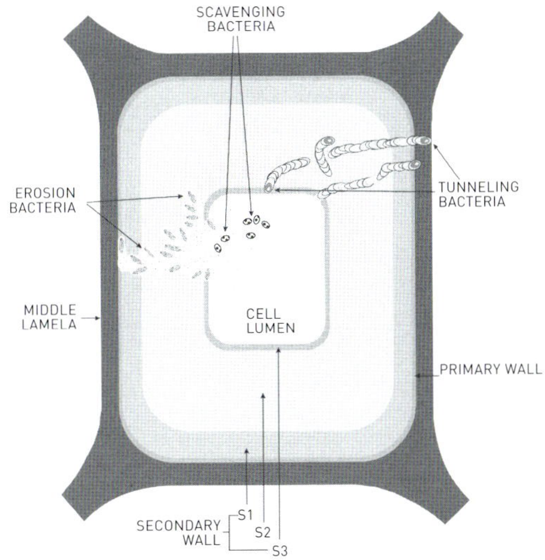

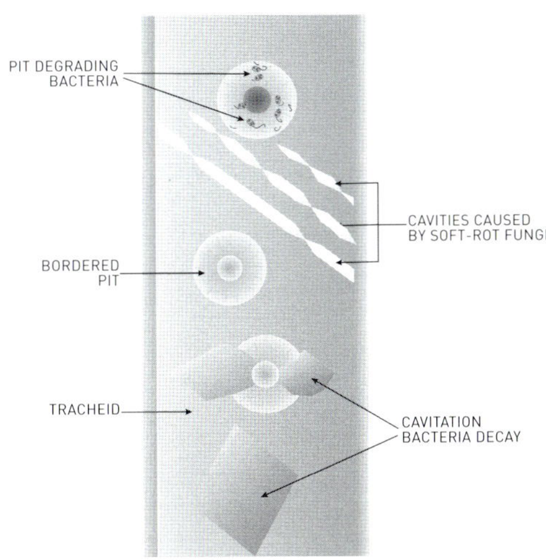

[1] Diagrammatic representation of a wood cell and the sites where different bacteria act. [2] Cavities created by bacteria are generally angular and often diamond-shaped, with the long axis of these cavities running either parallel to or perpendicular to the long axis of the cell.

appear to preferentially degrade the nonlignified, pectin-rich region of the pit membrane including the margo and torus, which are primarily composed of cellulose. These pit-degrading bacteria accumulate in the pit chamber and attach to the microfibrils of the margo. [FIG. 2] They have the ability to completely destroy the margo, leaving the torus partially intact, or may cause a complete dissolution of both structures.

Scavenging Bacteria and Environmental Influences Scavenging bacteria and environmental influences often mask degradation patterns created by wood-degrading bacteria (Blanchette, 1995). Scavenging or secondary bacteria degrade the residual materials left after degradation by tunneling and erosion bacteria. [FIG. 1] They are not found in areas where primary bacterial degradation is occurring, and appear to rely on degraded wall components and not intact cell wall material (Singh and Butcher, 1991). These organisms and the influence of environmental factors on woods that have been exposed to marine environments for long periods often remove or alter residual material, creating large voids in the wood cell and making positive identification of primary degraders difficult. Scavenging bacteria appear to tolerate near anaerobic conditions, similar to those of erosion bacteria.

Soft-Rot Fungi Soft- rot decay by fungi can be divided into two different categories: type 1 or cavity formation, and type 2 or erosion attack, both of which can be observed in hardwoods and softwoods. The type 1 form of decay is characterized by angular cavities in the S_2 layer that follow the cellulose microfibril orientation when viewed longitudinally. [FIG. 2] These chains of cavities with conical ends are often seen as round holes or voids when viewed in transverse sections. The fungus enters the wood cell wall through the lumen by creating a small-bore hole in the S_3 layer. Once the hyphae are aligned with microfibrils, growth stops and cavity formation is initiated. The cavities expand for a period and hyphal growth begins again with the production of a proboscis hypha, which may extend from one or both ends of the cavity. As hyphal growth stops, a new cavity begins, creating a chain of cavities that are connected by angular ends. This type of decay primarily affects the S_2 and S_1 layers, with minimal or no damage to the S_3 (Eaton and Hale, 1993; Khalili et al., 2001). Soft-rot fungi do not degrade the middle lamella.

A complete erosion of the secondary wall and a slight modification of the middle lamella in advanced stages of degradation characterize the type 2 form of soft-rot decay (Daniel and Nilsson, 1997). There are differences between the erosion of secondary wall layers of hardwoods and softwoods (Eaton and Hale, 1993). The erosion process in hardwoods often consists of troughs of different sizes, giving

SAMPLE #	DESCRIPTION	IDENTIFICATION	DESCRIPTION
Tektaş Lot 389 Hull remains (?)	Softwood	*Pinus sp.* (Pine)	Bacterial degradation
Tektaş Lot 394 Hull remains (?)	Hardwood	*Quercus sp.* (Oak) (White oak group)	Bacterial degradation
Tektaş Lot 831.01 Peg	Softwood	*Pinus sp.* (Pine)	Bacterial degradation
Tektaş Lot 867.01 One of 11 fragments from atop lead anchor stock core	Hardwood	*Ulmus sp.* (elm)	Bacterial degradation
Tektaş Lot 868.01/TK 252 Wood from anchor stock core 868	Hardwood	*Ulmus sp.* (elm)	Bacterial degradation
Tektaş Lot 902/TK 251 Wood from anchor casting bolt 867	Hardwood	*Ulmus sp.* (elm)	Bacterial degradation
Tektaş Lot 912/TK 270 Found with lead casting bolt, thought to be part of either anchor shaft or anchor tine	Softwood	*Pinus sp.* (Pine)	Bacterial degradation
Tektaş Lot 989.01	Softwood	*Pinus sp.* (Pine)	Bacterial degradation
Tektaş Lot 1025 Damaged fragment has nail hole and copper corrosion products	Softwood	*Pinus sp.* (Pine)	Bacterial degradation and copper oxidation products
Tektaş Lot 1032 Thought to be dunnage or wickerwork found intermingled with small fish bones	Hardwood	Unknown, very small diameter, odd hardwood stem, very decayed	Appears to be dunnage, not a structural component
Tektaş Lot 1166 One of four small fragments from beneath lead anchor stock core 1152	Hardwood	*Quercus sp.* (Oak)	Bacterial degradation
Tektaş Lot 1168/TK 303 Wood beneath anchor stock core 1151	Hardwood	*Quercus sp.* (Oak) (White oak group)	Bacterial degradation

[TABLE 1] Wood sample identification from the Tektaş Burnu shipwreck with comments on sample location and type of degradation found.

the degraded wood a striped appearance. The decay involves a degradation of all secondary cell wall layers, starting at the lumen and moving toward the middle lamellae. In softwoods the fungus enters the wood cell through the S_3 layer, but does not appear to degrade it. The decay progresses into the cellulose-rich S_2 layer, but has only a minimal effect on the S_1 and no effect on the middle lamellae. Soft-rot fungi appear to have a limited tolerance to low oxygen. However, when diffusible oxygen is readily available in marine environments, soft rot is often the dominant wood-degrading microorganism (Daniel and Nilsson, 1997). Some soft-rot fungi and most of the above-mentioned wood-degrading bacteria do have the ability to degrade wood treated with preservatives (CCA, CCB, etc.) in wet environments (Daniel and Nilsson, 1997; Singh and Butcher, 1991). The degree to which these organisms tolerate inhibitory chemicals varies, with tunneling bacteria having the greatest tolerance and soft-rot fungi being the least tolerant of microorganisms tested (Daniel and Nilsson, 1997; Eaton et al., 1989). Rarely does a single type of wood-degrading microorganism colonize and decay wood in marine environments; rather, a combination of attacks affects the wood. This can often make identification of degrading organisms difficult and mask previous decay patterns.

Marine Crustaceans and Borers The deterioration of wood in marine environments is often greatly influenced by marine crustaceans and borers (Cragg et al., 1999). In aquatic systems that support their existence, they may cause severe damage to wood in a short period. Water temperature, salinity, and oxygen availability are some of the most important factors controlling the distribution of these invertebrates, and if levels of any of these requirements drop too low their activities are inhibited (Cragg et al., 1999). The high salinity and relative warmth of the Aegean Sea supports large populations of wood-borers, which often preferentially attack woods that have been degraded by marine fungi and bacteria (Eaton and Hale, 1993; Cragg et al., 1999).

TEKTAŞ BURNU SHIPWRECK In September 1996 the Institute of Nautical Archaeology research vessel, Virazon, discovered a shipwreck about 40 meters below the surface of the Aegean Sea, while doing an annual survey for wrecks off the Turkish coast. The site is located on the southwest shore of Turkey, at Tektaş Burnu, or the "One Rock Cape." [FIG. 3] Shortly after its discovery three amphoras were recovered, identified, and subsequently dated to between 440 and 425 B.C. (Carlson, 1999). Excavation of the Classical Greek ship was not initiated until the summer of 1999 and was complete in the summer of 2001. [FIG. 4] The vast majority of arti-

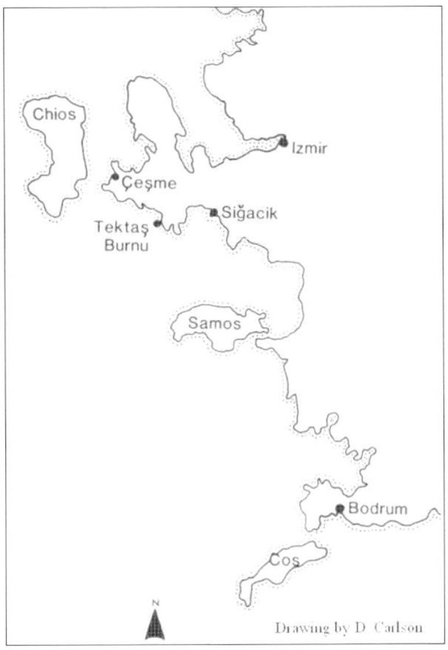

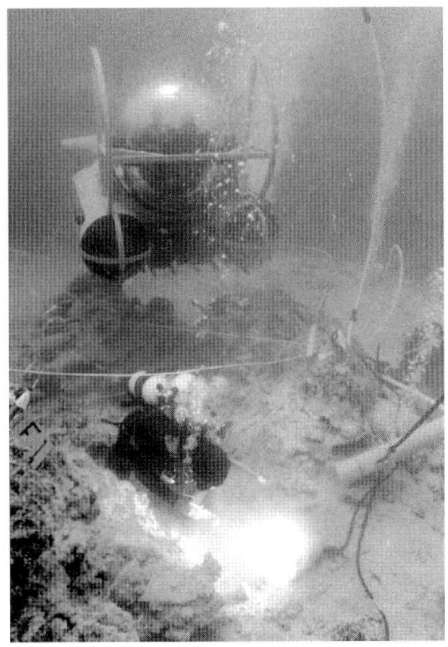

[3] The shipwreck site is located on the southwest shore of Turkey, at Tektaş Burnu, or the "One Rock Cape." [4] Underwater excavation of the Classical Greek ship.

facts were amphoras, with more than 250 recovered from the site. However, many cups, bowls, oil lamps, cooking pots, and other ceramic items were found mixed among the amphoras (Carlson, 2001). Two of the most interesting and historically important finds at the wreck were associated with rectangular lead bars and cooper nails. The lead bars were cores from wooden anchor stocks and represent the earliest example of what is referred to as a Type IIa anchor (Trethewey, 2001). The cooper nail size and clenched shape provided evidence that the ship's wooden hull was nailed to its internal frame, which again represented the earliest documented use of this type of construction (Bass and Platt, 2002).

Materials and Methods Wood samples were obtained from the Institute of Nautical Archeology, College Station, Texas. All samples went through a gradual desalination process shortly after their removal from the shipwreck site and were transported in fresh water to the University of Minnesota. Small segments of the original samples were prepared for scanning electron microscopy by the following procedure. Segments were cleaned in deionized water, infiltrated with O.C.T. (Tissue Tek®) embedding medium under vacuum, frozen to –20°c, and sectioned with a Cryo-cut freezing microtome. All sections were then washed in water and dehydrated through an ethanol series and critical-point-dried. Small segments of the historic woods were also taken for elemental analysis. These consisted of representative samples from the three different woods identified and a piece of modern pine and oak analyzed for comparison. The analysis was carried out by multi-elemental inductively coupled plasma atomic emission spectroscopy as described by Blanchette et al. (2002).

Wood Identification of the Tektaş Burnu Shipwreck Although the ship's hull, internal frame, anchors, and undoubtedly portions of its cargo were made from wood, only small remnants of wood remain. The ship is thought to have come to rest between two large rocks, keeping it above the sea floor and away from sediments. If the ship had been covered in sediments, oxygen would have been limited and more of the hull would most likely have remained. Samples of the wood were identified to their genus and are listed in Table 1, along with additional comments provided by the divers who recovered them. [TABLE 1] Five of the samples are pine (*Pinus sp.*) [FIG. 5A], three are oak (*Quercus sp.*) [FIG. 5B], three are elm (*Ulmus sp.*) [FIG. 5C], and one very small piece remains unidentified. The oak and pine appear to have been used in the hull construction and the elm may have been used as the anchor stock. Most of the woods were associated with either the lead cores from the anchor or

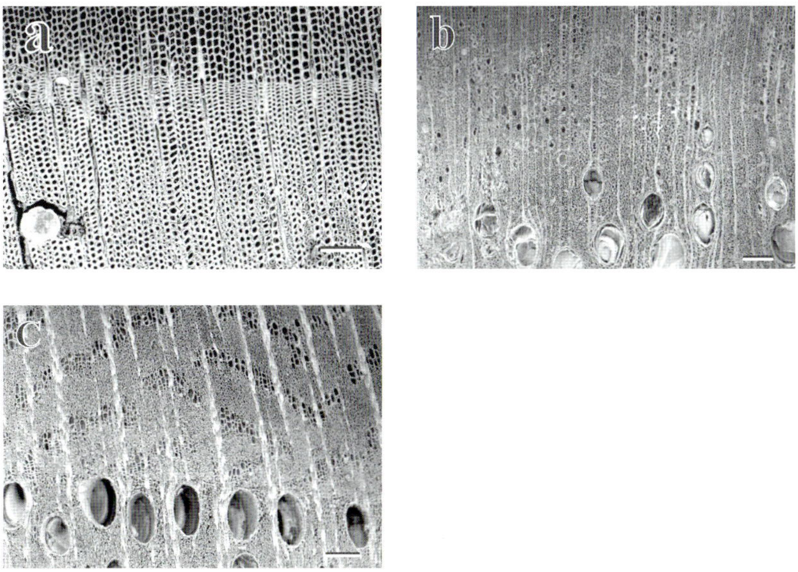

[5] Examples of the wood found from the wreck: (a) pine (*Pinus sp.*); (b) oak (*Quercus sp.*); and (c) elm (*Ulmus sp.*).

the copper hull nails, and were dark in color or had obvious metal corrosion products on their surface. In the preparation of samples for both microscopy and elemental analysis it became apparent that the woods were not just affected by metals on the surface, but that corrosion products had infiltrated the wood. The samples found near the lead cores had black streaks or were black throughout the entire piece, and wood found near copper nails or with nail holes in it was green, similar to the color of oxidized copper.

Wood Decay in the Tektaş Burnu Shipwreck Evidence of marine borers and extensive decay was present in all samples, with the oak exhibiting the greatest alteration to its anatomy, and the pine showing the least. Figure 6b is a transverse section of oak displaying a honeycombed appearance common to highly degraded woods found in marine environments. [FIG. 6B] The secondary wall of the fiber cells has large voids that appear to have been caused by a coalescing of cavities caused by erosion bacteria. Holes in the middle lamellae also indicate tunneling bacteria caused the decay. Little evidence of residual material generated by these primary degraders is evident in the wood cell walls, suggesting that scavenging bacteria and environmental influences have affected the degraded cells. The S_3 layer has also been altered and appears to have collapsed into the cell lumen, probably because the underlying S_2, which provides structural support to the very thin S_3 layer, was severely degraded. The only layer that has remained relatively intact is the highly lignified middle lamellae.

Degradation in fiber cells from the elm wood sample is very similar in appearance to that of the oak. [FIG. 6C] The degradation patterns indicate that erosion bacteria caused the holes, and that they often coalesced in the secondary wall, creating large voids. Small-diameter holes in the middle lamellae and vessel walls verify that tunneling bacteria were also present in these cells. The concentric bands and extracellular slime produced by tunneling bacteria and the granular matrix created by erosion bacteria during their decay processes are also absent in the elm sample, and scavenging bacteria and environmental influences were most likely responsible for degrading this material.

The pinewood exhibited the least amount of decay, when compared to the other woods. [FIG. 6A] The wood does not have the large voids caused by extensive erosion or tunneling bacteria decay. The small-diameter holes in the secondary wall and middle lamellae suggest tunneling bacteria were present, and the existence of larger-diameter holes in the secondary wall, and characteristics of the degradation, indicate that erosion bacteria were responsible for some of the degradation in this wood as well.

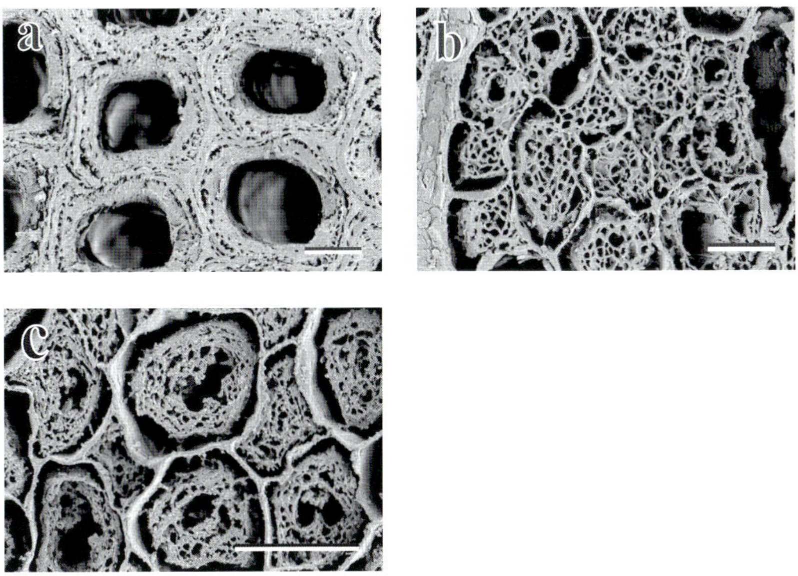

[6] Examples of the degree of decay for the three types of wood found in the wreck: (a) Pinewood exhibited the least amount of decay, when compared to the other woods; (b) oak and (c) elm exhibited the greatest alteration to their anatomy.

Element Analysis of Tektaş Burnu Shipwreck Wood Although the wood of the Tektaş Burnu shipwreck suffered extensive degradation, a few remnants survived not only microbial decay, but attack by wood-boring marine invertebrates. It is possible that sediments gradually covered the woods, eliminating oxygen needed for the wood-degrading organisms to persist. However, the large concentrations of metal ions [TABLE 2] present in the woods most likely had an inhibitory effect on microbial growth. Since decay was found in the wood cell walls of the samples, degradation had occurred before the infiltration of metals into the wood. The decay probably made the wood more permeable, allowing the metal corrosion products to easily penetrate it. Microbial degradation probably occurred for many years after the ship had sunk, and once metal ion concentrations increased the decay stopped and the woods were protected from further attack. Wood without metals apparently continued to decay and was completely degraded.

The importance of analyzing element concentrations in archaeological woods taken from marine environments has recently been highlighted with the discovery of high acidity in woods recovered from the Swedish warship Vasa (Sandström et al., 2002). The ship was removed from Stockholm harbor, where it had rested for 333 years, and was placed in a museum after extensive treatment of its waterlogged timbers. In recent years the pH of some of the timbers had values below 2, and acid hydrolysis of the wood has become a major concern. It was determined that sulfur products, produced by sulfate-reducing bacteria, reacting with corroded iron bolts used to hold the ship together, had generated sulfuric acid. The authors stress the importance of removing metal ions such as iron and copper to inactivate the reaction and make conservation of the wooden artifacts successful.

CONCLUSIONS Wood-degrading organisms in marine environments are often responsible for the loss of archaeological wood. However, when requirements for their existence are not met, degradation stops. The woods recovered from the Tektaş Burnu shipwreck exhibited extensive degradation caused by erosion, tunneling, and scavenging bacteria as well as marine borer activity, but they did persist when essentially all other woods were lost. The presence of heavy metals and toxic elements found in the woods appears to be responsible for the preservation of these few remaining wood remnants. An understanding of the decay and elements present within archaeological woods is required if conservation methods are to be successful on these important wooden cultural properties.

[TABLE 2] Elemental analysis of woods from the Tektaş Burnu shipwreck and two modern woods used for comparison. Elements in ppm.

Element	*Ulmus sp.* (Tektaş Lot 867.01)[a]	*Pinus sp.* (Tektaş Lot 1025)[a]	*Quercus sp.* (Tektaş Lot 1166)[a]	Modern *Pinus sp.*	Modern *Quercus sp.*
Aluminum (Al)	471	95	4048	5	6
Arsenic (As)	108	98	72	[b]	[b]
Boron (B)	77	69	525	2	34
Cobalt (Co)	10	414	31	0	[b]
Copper (Cu)	800	90647	46	1	8
Iron (Fe)	2397	9406	4039	10	45
Magnesium (Mg)	4636	2360	6934	133	34
Manganese (Mn)	11	18	6	51	28
Sodium (Na)	2514	8432	10275	7	32
Nickel (Ni)	109	1597	110	0	0
Lead (Pb)	7980	1896	22324	1	1
Vanadium (V)	925	534	130	0	[b]
Zinc (Zn)	156	237	74	2	6

[a]Numbers assigned to the samples when collected.
[b]Element not measured.

REFERENCES

BASS, G.F., and PLATT, C., 2002. *Golden Age Treasures.* National Geographic, pp. 102–117.

BJÖRDAL, C.G., 2000. Waterlogged archaeological wood; biodegradation and its implications for conservation. Doctoral thesis. ISSN 1401-6230, ISBN 91-576-5876-5.

BLANCHETTE, R.A., 1995. Biodeterioration of archeological wood. *CAB Biodeterioration Abstracts,* 9, 113–127.

BLANCHETTE, R.A, 2000. A review of microbial deterioration found in archeological wood from different environments. *International Biodeterioration,* 46, 189–204.

BLANCHETTE, R. A., HELD, B.W., and FARRELL, R.L., 2002. Defibration of wood in the expedition huts of Antarctica: An unusual deterioration process occurring in the polar environment. *Polar Record,* 38, 313–322.

BLANCHETTE, R.A, and HOFFMANN, P., 1994. Degradation processes in waterlogged archaeological wood. In: Hoffmann, P., (Ed.), *Proceedings of the 5th ICOM Group on Wet Organic Archaeological Materials Conference,* Bremerhaven, Germany: International Council of Museums, pp. 111–142.

BURNES, T.A., BLANCHETTE, R.A., and FARRELL, R.L., 2000. Bacterial biodeterioration of extractives and patterns of bordered pit membrane attack in pine wood. *Applied Environmental Microbiology,* 66, 5201–5205.

CARLSON, D.N., 1999. The 1999 excavation season at Tektaş Burnu, Turkey. *INA Quarterly,* 26, 3–8.

CARLSON, D.N., 2001. The 2000 excavation season at Tektaş Burnu, Turkey. *INA Quarterly,* 28, 3–8.

CRAGG, S.M., PITMAN, A.J., and HENDERSON, S.M., 1999. Development in the understanding of the biology of marine crustaceans and in methods of controlling them. *International Biodeterioration,* 43, 197–205.

DANIEL. G., and NILSSON, T., 1997. Developments in the study of soft rot and bacterial decay. In: Bruce, A., and Palfreyman, J.W., (Eds.), *Forest Products Biotechnology.* Taylor & Francis, New York.

EATON, R.A., and HALE, M.D.C., 1993. *Wood decay, pests and protection.* Chapman and Hall, New York.

EATON, R.A., AMPONG, F., BARNACLE, J., BEESLEY, J., BULTMAN, D., COOKSON, L., CRAGG, S., DE PALMA, J., GAMBETTA, A., HENNINGSSON, B., LEVI, M., LEVY, C., NILSSON, T., and ORLANDI, E., 1989. An international collaborative marine trial to investigate the effect of timber substrate on the efficacy of CCA and CCB wood preservative. *Material und Organismen,* 24, 51–79.

KHALILI, S., NILSSON, T., and DANIEL, G., 2001. The use of soft rot fungi for determining the microfibrillar orientation in the S2 layer of pine tracheids. *Holz Roh Werkstoff,* 58, 439–447.

SANDSTRÖM, M., JALLLEHVAND, F., PERSSON, I., GELIUS, U., FRANK, P., and HALL-ROTH, I., 2002. Deterioration of the seventeenth-century warship *Vasa* by internal formation of sulphuric acid. *Nature,* 415, 893–897.

SINGH, A.P., 1997. Initial pit borders in *Pinus radiata* are resistant to degradation by soft rot fungi and erosion bacteria but not tunneling bacteria. *Holzforschung,* 51, 15–18.

SINGH, A.P., and BUTCHER, J.A. 1991. Bacterial degradation of wood cells: a review of degradation patterns. *Journal of the Institute of Wood Science,* 12, 143–157.

SINGH, A.P., and KIM, Y.S., 1997 Biodegradation of wood in wet environments. *The International Group on Wood Preservation*, Document No: IRG/WP 97–10217.

TRETHEWEY, K., 2001. Lead anchor-stock cores from Tektaş Burnu, Turkey. *International Journal of Nautical Archaeology*, 30, 109–114.

TREATMENT AND PREVENTION

J. SUZUKI[1]
R.J. KOESTLER[2]

VISUAL ASSESSMENT OF BIOCIDE EFFECTS ON JAPANESE PAINT MATERIALS

1 Museum of Fine Arts, Boston, MA 2 The Metropolitan Museum of Art, New York

ABSTRACT This paper reports on the results of experiments to test for visual changes to paint systems using a statistical binary procedure after biocidal treatments, and a comparison of the visual data with those from a spectrophotometer. Four biocides were tested: two gases—argon (Ar, suffocant for fungi and insects) and nitrogen (N_2, suffocant for insects); and two liquid fungicides—the architectural antimicrobial D/2 (Prosoco, Inc., Lawrence, KS), and an experimental pine emulsion (Tampere University of Technology, Finland). Four paint sets were subjected to the biocides: watercolor, acrylic color, egg tempera, and Japanese paint, making for a total of 52 sample sets. The visual assessment procedure provided information on tone and hue change, gloss change, and formation of precipitate. The results indicated that argon and nitrogen gas had no visual effect on any pigments; D/2 changed parts of each sample set; and the experimental pine resin caused such a major change that it was dropped from testing. The spectrophotometric results had better than 92% agreement with the visual technique.

INTRODUCTION Virtually all art materials are susceptible to microbial attack. When it occurs, an immediate response is needed to minimize the damage to the material. This response may include application of a biocidal treatment. In order to make an informed selection of what biocidal treatment to use, it is important to have some idea of the side effects of any treatment. A technique to assess for visual change in paint samples was presented in Koestler et al. (1993). This technique has been used to assess the changes to four water-based paints: watercolor, acrylic color, egg tempera, and Japanese paint, after treatment with four different biocides —two gases—argon and nitrogen; and two liquids—an architectural biocide, D/2, and an experimental pine resin extract.

MATERIALS AND METHODS The watercolor and acrylic color used were proprietary brands (for watercolor, Winsor and Newton, England; and for acrylic color Liquitex, USA). Egg tempera was made with dry pigments (Holbein, Japan), then ground into the medium of 1:1 egg yolk and water. The dry pigments for Japanese paint were ground into the solution of Japanese animal skin glue, Sanzenbon-nikawa (all pigments were Japanese products). Each paint had 12 colors, making for a total of 52 samples sets. [TABLE 1]

Supports Watercolor and acrylic color were applied on watercolor paper (Fabrian, cellulose 100%, Cartiere Miliani Fabriano, Italy). Egg tempera was applied on a poplar plywood panel of 3-mm thickness. The panel was sized with 7% rabbit skin glue solution and prepared with gesso ground (for the lower six layers, gesso grosso; for the upper four layers, gesso sottile with 7% rabbit skin glue solution). Japanese paint was applied on Japanese paper (Kumohada-mashi, a hemp and Kozo fiber 100%, sized with animal skin glue, Uematsu Co., Japan).

Paint Application and Dryness Each paint was applied in 2.5-cm-wide strips on each support. And each paint was diluted with water in the usual painting method, and applied in about six layers with a soft-flat brush as evenly as possible. Triplicate strips were prepared for controls and each experimental sample. Control and experimental samples were stored under identical light conditions before and after treatment. Both were hung on a wall, in dim light, where they dried for about 10 months.

Biocides and Application Four biocides were tested: two gases—argon (Ar, suffocant for fungi and insects) and nitrogen (N_2, suffocant for insects); and two liquid fungicides—the architectural antimicrobial D/2 (Prosoco, Inc., Lawrence, KS), and an experimental pine emulsion (Tampere University of Technology, Finland). Argon and nitrogen gas have been used as insect suffocants while argon has recently been

[TABLE 1] Paints tested

WATERCOLOR	ACRYLIC COLOR[1]	EGG TEMPERA[2]	JAPANESE PAINT[3]	
			English Name	Japanese Name
Chinese White	Zinc White	Zinc White	Chalk	*Gofun*
Titanium White	Titanium White	Titanium White	White Lead	*Enpaku*
Alizarin Crimson	Alizarin Crimson	Rose Madder	Synthetic White	*Iwashiro*
Cadmium Red	Cadmium Red	Cadmium Red	Synthetic Pink	*Iwamomo*
Terre Verte	Chromium Oxide Green	Terre Verte	Light Red	*Taisha*
Winsor Emerald	Emerald Green	Oriental Green	Vermilion	*Shu*
Prussian Blue	Prussian Blue	Prussian Blue	Red Lead	*Entan*
Cobalt Blue	Cobalt Blue	Cobalt Blue / Deep	Verdigris	*Rokusho*
Cadmium Yellow	Cadmium Yellow	Cadmium Yellow	Synthetic Verdigris	*Shin Iwarokusho*
Raw Sienna	Raw Sienna	Raw Sienna	Azurite	*Iwagunjo*
Burnt Umber	Burnt Umber	Burnt Umber	Gold	*Ao-Kindei*
Ivory Black	Ivory Black	Ivory Black	Silver	*Jun-Gindei*

[1] Support paper was also tested. [2] Support wood board and gesso were also tested. [3] Japanese paper was also assessed.

shown to be effective on some fungi (Tavzes et al., this volume).

The argon and nitrogen gases were commercial grade (minimum 99.995% pure, actual measured value of oxygen about 10 ppm). Samples treated with Ar or N_2 gas were enclosed in Marvel Seal 360 (a heat-sealable aluminized Mylar) and brought down to an oxygen level under 100 ppm. Packets of oxygen scavenger (Freshpax D-2000 cc, Multisorb Technologies, Buffalo, NY) were placed inside each bag to maintain the low oxygen level. The humidity level was about 55%. Triplicate samples were kept in this environment for 1 month.

Triplicate sample strips of each paint sample were sprayed with the D/2, from a distance of about 3 inches, until the complete surface was covered (two to three spurts of the applicator bottle). The pine emulsion was tested on canvas and found to severely stain the canvas brown, so further testing with this product was not conducted.

Evaluation Methods Visual color change was assessed by two trained observers with the unaided eye and with a light microscope using a statistical binary procedure (as per Koestler et al., 1993). A 2.5 × 3 cm grid was superimposed on each sample, and divided into a 5 × 6 matrix, and five sites were selected by a random number technique and examined separately by the two observers. Samples were compared to corresponding control areas. Each observer looked at 5 of 30 boxes for each of the three paint samples per test. A total of 15 out of 90 boxes were looked at by each observer, percentages were calculated separately for each paint sample/observer, and then the two percentages were averaged to yield the final estimate for the area affected (see Koestler et al., 1993, for details). Changes were assessed in the following categories: visual color change (broken into tone and hue), change in gloss, and presence of any precipitate. Along with each set of binary statistical data is an impression of overall percentage change of the surface.

Instrumental evaluation of changes resulting from treatment was carried out with a Minolta CM-2002 spectrophotometer. The color values using CIE 1976 L*a*b* of control and experimental samples were measured before and after treatment. The same random field design as was used in the visual method described above was used for the instrumental method, with minor modification. A total of three areas, measured three times each, for each of the three experimental paints (27 measurements/test paint) were averaged to yield the raw data for the color calculations. The diameter of the measuring beam was 0.3 mm. Tests were performed first on samples of the paint to assess for structural color differences and orientation of the sample. These were done by predefining one direction of the samples as 0°, measuring, as above, then rotating the sample 45° at a time so that a total of eight directions (the four cardinal and four sub-cardinal directions) were measured. Final tests were run at 0° and 90°; however, only the 0° results are presented here. The visual tone results for black or white samples were compared to ΔE^*_{ab} values, and the visual hues of color samples were compared to ΔH^*_{ab}.

RESULTS AND DISCUSSION Argon and nitrogen gas had no visual effect on the pigments. The spectrophotometric readings were also negative. The experimental pine sol resin was tested on one sample and the results were so immediately disfiguring that no further testing was carried out with this material.

The D/2 results were the most interesting; full results from the visual assessment are summarized in TABLES 2–5. Normally, it is not recommended that a water-based biocide be used on the test materials used in this study. But there is really no choice in the U.S. market, as D/2 is the only registered biocide available for use on anything related to art (in this case masonry), and since this study was interested in testing and developing the assessment methodology, its inclusion is warranted.

The visual color change has been divided into tone and hue. This methodology is a refinement of that used in Koestler et al. (1993), which had just referred to color change. Assessing tone yields information on the lightness or darkness of the pigment, while hue gives information on the similarity of color. There are many ways to describe color differences; Berns (2000, pp. 73–74) recommends using ΔL^*, Δa^*, and Δb^* for neutral tones such as blacks, whites, and grays. For chromatic samples he recommends the use of ΔH^*_{ab}, ΔL^*, and ΔC^*_{ab}. A summary of the spectrophotometric data is given in TABLES 6–9. The CIELAB formulae used in this paper are:

$$\Delta E^*_{ab} = [(\Delta L^*)^2 + (\Delta a^*)^2 + (\Delta b^*)^2]^{1/2}, \text{ for black, white, and gray samples;}$$

and

$$\Delta H^*ab = [(\Delta E^*_{ab})^2 - (\Delta L^*)^2 - (\Delta C^*_{ab})^2]^{1/2}, \text{ for chromatic samples.}$$

In general, D/2 caused significant changes in tone and hue in three out of four color sets (acrylic being the exception). Gloss changes occurred in all four paint sets. Sometimes the change in gloss was accompanied by an apparent precipitation on the surface. On some samples white particles (visible with light microscopy) remained on the surface, and on others there seemed to be a dried emulsion.

The acrylic sample set had the least change of the four with D/2. All of these changes were alterations in the gloss (an increase in all four cases). It is possible that the change in gloss on acrylic samples could have been countered by rinsing.

[TABLE 2] Summary of visual observations on acrylic samples treated with D/2

	VISUAL COLOR CHANGE				CHANGE IN GLOSS		PRECIPITATION	
	TONE		HUE					
	RANDOM FIELD	OVERALL IMPRESSION	RANDOM FIELD	OVERALL IMPRESSION	RANDOM FIELD	OVERALL IMPRESSION	RANDOM FIELD	OVERALL IMPRESSION
Support Paper	0%	0%	0%	0%	0%	0%	0%	0%
Zinc White	0%	0%	0%	0%	0%	0%	0%	0%
Titanium White	0%	0%	0%	0%	7%	67% G	0%	0%
Alizarin Crimson	0%	0%	0%	0%	0%	0%	0%	0%
Cadmium Red	0%	0%	0%	0%	0%	0%	0%	0%
Chromium Oxide Green	0%	0%	0%	0%	0%	0%	0%	0%
Emerald Green	0%	0%	0%	0%	30%	50% G	0%	0%
Prussian Blue	0%	0%	0%	0%	3%	17% G	0%	0%
Cobalt Blue	0%	0%	0%	0%	0%	0%	0%	0%
Cadmium Yellow	0%	0%	0%	0%	0%	0%	0%	0%
Raw Sienna	0%	0%	0%	0%	0%	0%	0%	0%
Burnt Umber	0%	0%	0%	0%	0%	0%	0%	0%
Ivory Black	0%	0%	0%	0%	15%	100% G	0%	0%
Total % change	0%		0%		4%		0%	

G: glossy

[TABLE 3] Summary of visual observations on Japanese paint samples treated with D/2

	VISUAL COLOR CHANGE				CHANGE IN GLOSS		PRECIPITATION	
	TONE		HUE					
	RANDOM FIELD	OVERALL IMPRESSION	RANDOM FIELD	OVERALL IMPRESSION	RANDOM FIELD	OVERALL IMPRESSION	RANDOM FIELD	OVERALL IMPRESSION
Japanese Paper	0%	0%	0%	0%	0%	0%	0%	0%
Chalk (*Gofun*)	13%	17% D	3%	83% Y	0%	0%	0%	0%
White Lead (*Enpaku*)	0%	0%	0%	0%	0%	0%	0%	0%
Synthetic White (*Iwa-shiro*)	0%	0%	0%	83% Y	0%	0%	0%	0%
Synthetic Pink (*Iwa-momo*)	20%	67% L	23%	17% Y	0%	0%	0%	0%
Light Red (*Taisha*)	0%	0% L	0%	0%	0%	17% M	0%	0%
Vermilion (*Shu*)	0%	0%	0%	0%	0%	0%	0%	0%
Red Lead (*Entan*)	20%	67% L	7%	50% Y	3%	100% UE	7%	17%
Verdigris (*Rokusho*)	7%	33% L	7%	17% Y	0%	17% M	0%	0%
Synthetic Verdigris (*Shin Iwa-rokusho*)	37%	83% L	27%	67% Y	0%	0%	0%	0%
Azurite (*Iwa-gunjo*)	3%	17% D	0%	0%	0%	0%	0%	0%
Gold (*Ao-Kindei*)	0%	0%	0%	0%	0%	0%	0%	0%
Silver (*Jun-Gindei*)	3%	67% L	3%	67% Y	0%	0%	0%	0%
Total % change	8%		5%		<1%		<1%	

L: light, D: dark, Y: yellowed, R: reddish, M: matte, UE: uneven

[TABLE 4] Summary of visual observations on watercolor samples treated with D/2

	VISUAL COLOR CHANGE				CHANGE IN GLOSS		PRECIPITATION	
	TONE		HUE					
	RANDOM FIELD	OVERALL IMPRESSION	RANDOM FIELD	OVERALL IMPRESSION	RANDOM FIELD	OVERALL IMPRESSION	RANDOM FIELD	OVERALL IMPRESSION
Chinese White	0%	0%	0%	0%	0%	0%	0%	0%
Titanium White	0%	0%	0%	0%	0%	0%	0%	0%
Alizarin Crimson	13%	33% L	0%	0%	0%	0%	0%	0%
Cadmium Red	0%	17% L	0%	0%	0%	0%	0%	0%
Terre Verte	0%	17% L	0%	0%	0%	0%	0%	0%
Winsor Emerald	17%	50% L	3%	0% Y	0%	0%	17%	100%
Prussian Blue	10%	33% D	23%	67% Y	0%	0%	0%	0%
Cobalt Blue	40%	67% D	0%	0%	17%	100% UE	0%	0%
Cadmium Yellow	0%	0%	0%	0%	0%	0%	0%	0%
Raw Sienna	37%	67% L	0%	0%	0%	0%	13%	50%
Burnt Umber	13%	67% L	3%	0%	0%	0%	0%	0%
Ivory Black	33%	33% D	0%	0%	7%	33% UE	0%	0%
Total % change	**14%**		**2%**		**2%**		**3%**	

L: light, D: dark, Y: yellowed, UE: uneven

[TABLE 5] Summary of visual observations on egg tempera samples treated with D/2

	VISUAL COLOR CHANGE				CHANGE IN GLOSS		PRECIPITATION	
	TONE		HUE					
	RANDOM FIELD	OVERALL IMPRESSION	RANDOM FIELD	OVERALL IMPRESSION	RANDOM FIELD	OVERALL IMPRESSION	RANDOM FIELD	OVERALL IMPRESSION
Support Wood Board	0%	0%	0%	0%	0%	0%	0%	0%
Gesso Ground	20%	50% D	17%	33% Y	0%	0%	0%	0%
Zinc White	50%	100% D	53%	100% Y	0%	0%	0%	0%
Titanium White	37%	33% D	30%	83% Y	0%	17% UE	0%	0%
Rose Madder	0%	0%	0%	0%	0%	0%	7%	50%
Cadmium Red	0%	0%	0%	0%	0%	0%	3%	33%
Terre Verte	57%	100% L	3%	17% Y	3%	0% G	0%	0%
Oriental Green	0%	0%	0%	0%	7%	83% UE	0%	0%
Prussian Blue	37%	67% D	0%	0%	13%	100% UE	23%	67%
Cobalt Blue / Deep	63%	100% D	0%	0%	23%	83% UE	27%	67%
Cadmium Yellow	0%	0% D	0%	0%	0%	100% UE	0%	33%
Raw Sienna	90%	100% D	37%	50% R	0%	67% UE	0%	67%
Burnt Umber	60%	100% D	0%	0%	3%	50% UE	30%	83%
Ivory Black	17%	83% L	0%	0%	20%	100% UE	53%	100%
Total % change	**31%**		**10%**		**5%**		**10.0%**	

L: light, D: dark, Y: yellowed, R: reddish, G: glossy, UE: uneven

	DE*Lab FOR B&W	DH*ab FOR COLOR SAMPLES
Support Paper		
Zinc White	0.28	
Titanium White	0.20	
Alizarin Crimson		0.08
Cadmium Red		0.30
Chromium Oxide Green		0.08
Emerald Green		0.31
Prussian Blue		**2.87**
Cobalt Blue		0.02
Cadmium Yellow		0.43
Raw Sienna		0.17
Burnt Umber		0.18
Ivory Black	0.66	

[TABLE 6] Summary of spectrophotometry data for acrylic samples treated with D/2

		DE*Lab FOR B&W	DH*ab FOR COLOR SAMPLES
Japanese Paper			
Chalk	*Gofun*	1.00	
White Lead	*Enpaku*	0.06	
Synthetic White	*Iwa-shiro*	0.70	
Synthetic Pink	*Iwa-momo*		0.20
Light Red	*Taisha*		0.08
Vermilion	*Shu*		0.30
Red Lead	*Entan*		0.20
Verdigris	*Rokusho*		0.10
Synthetic Verdigris	*Shin Iwa-rokusho*		0.11
Azurite	*Iwa-gunjo*		0.36
Gold	*Ao-Kindei*	0.90	
Silver	*Jun-Gindei*	0.69	

[TABLE 7] Summary of spectrophotometry data for Japanese paint samples treated with D/2

	DE*Lab FOR B&W	DH*ab FOR COLOR SAMPLES
Chinese White	0.90	
Titanium White	0.62	
Alizarin Crimson		1.30
Cadmium Red		0.05
Terre Verte		0.72
Winsor Emerald		0.82
Prussian Blue		2.31
Cobalt Blue		1.06
Cadmium Yellow		1.65
Raw Sienna		0.32
Burnt Umber		0.13
Ivory Black	1.24	

[TABLE 8] Summary of spectrophotometry data for watercolor samples treated with D/2

	DE*Lab FOR B&W	DH*ab FOR COLOR SAMPLES
Support Wood Board		
Gesso Ground	1.13	
Zinc White	0.72	
Titanium White	0.70	
Rose Madder		0.21
Cadmium Red		0.02
Terre Verte		0.70
Oriental Green		0.34
Prussian Blue		0.80
Cobalt Blue / Deep		1.32
Cadmium Yellow		0.74
Raw Sienna		3.26
Burnt Umber		0.92
Ivory Black	1.52	

[TABLE 9] Summary of spectrophotometry data for egg tempera samples treated with D/2

This is not something that is usually recommended with paper, and it was not done in this study. It should be noted that D/2 is meant for use on masonry surfaces. It is possible that if it were used on acrylic painted masonry surfaces, if rinsed, it would cause little or no change to the paint.

Egg tempera has a ground, so it is believed that the D/2 soaked into it and browned it slightly. Nine samples had their gloss changed, but not in a consistent direction as was the case with the acrylic.

Japanese paint has a coating on the surface, but it is not enough to inhibit the D/2 from soaking in and changing the appearance of the paint. The effect was mostly a lightening and yellowing of the samples, and only three had their gloss changed.

The visual observations overall showed that D/2 was detrimental to all four paint types. To rank effect from least to most damaging: The acrylic was damaged the least, with a 4% average gloss change; then Japanese paint (8% average tone change and 5% average hue change); then watercolor (14% average tone change and 2% average hue change); and finally egg tempera was the most damaged by the D/2 with a 31% average tone change and a 10% average hue change.

Comparison of the visual observations with the spectrophotometric results yields agreement in as many as 48 out of 52 (92%) sample sets. It all depends upon the value for ΔE or ΔH that is accepted as the cutoff point for measuring a real difference. There is no absolute cutoff point for when a real color difference (ΔE or ΔH) is measured. It depends upon the surface roughness. Tiano et al. (this volume) on rough stone surfaces uses a ΔE of 3, while Koestler et al. (1993) on oil paint surfaces had discussed a ΔE of 1. On smooth painted surfaces it may be that a lower value is correct, but on uneven surfaces (or if uneven distribution of pigment occurs) a higher value may be more meaningful. If we assume a ΔE (or ΔH) of 1 as signifying a real difference, four samples gave an instrumental reading greater than that *and* these did not agree with the visual observations. If we move the cut-off point up to a ΔE or ΔH of 3, then *no instrumental reading said there was a change when the visual said there wasn't.* There are still four visual changes that the instrument did not see that have to be explained.

In the four remaining cases that did not agree, it was the visual observations that measured a change and the spectrophotometer that did not. These differences may be explained as follows: Japanese paint samples synthetic pink and synthetic verdigris are made of small particles of pigment, and it is possible that the spectrophotometer's probe size (0.3 mm) relative to the pigment particle sizes gave a higher variability in readings between control and treatment. In the case of the egg tempera, two pigments, both white (zinc white and titanium white) were clearly changed in tone and hue; these samples also had darkened, yet the spectrophotometer did not measure a change. This may be due to the fact that a small precipitate from the D/2 was on the surface. Visual examination probably compensated for the presence of the white particles and noted the overall change in the surface. The spectrophotometer may have "seen" the white particles and averaged them in with the darkened surface to give an unchanged reading.

In general, it was felt that the trained conservators' eyes, when coupled with the statistical binary assessment method, yielded truer results than did the instrumental technique, but they were close. (This assumes, of course, that both techniques didn't produce an error in the same direction.) The binary statistical assessment procedure permits a relatively rapid low-tech procedure to assess the effects of a treatment on a painted surface. The refinement of Koestler et al. (1993) of dividing color change into tone and hue permitted the tone of black and white samples to be compared to color calculation values that reflected just a lightening and darkening effect, and the hue of color samples to be compared to the color calculations that reflected mostly color changes. This seems to give a greater degree of agreement between visual and instrumental data. (Had DH^*_{ab} rather than the ΔE^*_{ab} been used for the black and white samples, three data sets would have fallen out of agreement with the visual observations.)

One other point worth noting concerns the meaning of the ΔE, or color difference value and visual observations. Billmeyer and Saltzman (1981, p. 109), in their influential book on principles of color chemistry (2nd edition) state:

> "It is our belief that color measurement is the same whether the eye alone is used, or the eye and other instruments are used in combination. The instruments are aids to the eye. Their function...is to express the color of materials or color differences in terms of what the eye perceives.... It does not supercede the trained eye...in many cases it may become a substitute for years of experience in the training period."

Berns (2000, p. 130), in the 3rd edition to the Billmeyer and Saltzman classic, states:

> "We recognize that all existing color-difference calculations do not correlate perfectly with visual judgments, even under the most carefully controlled experimental conditions. Thus, built into any instrumental color control system is visual validation. In cases where the instrumental measurement is close to a color tolerance limit, visual observations are required. Ideally, they should be performed by a number of observers."

Despite the power and utility of the visual binary statistical procedure, it does not eliminate the need for an instrumental approach to color assessment. Not many art objects have an appropriate color control samples that can be used to assess color change in the object over time. Indeed, the control is usually the previous state of the same object under consideration. This necessitates having some instrumental technique that can record the current color state of an object's surface that can then be compared to the object at some later time to assess for change. The essential aspect of the color measurement is that the conditions of measurement be clearly defined to make future comparisons meaningful. In future testing we will try to examine the error bar of ΔE's or ΔH's associated with unevenly distributed paint surfaces and determine how to relate this to real art painting surfaces when a control sample doesn't exist.

ACKNOWLEDGMENTS We would like to thank Masako Koyano of the Art Conservation Lab, in Tokyo, and Betsy Court of Balboa Art Conservation Center, San Diego, for the pigments used for the sample paints. We also thank Kim Kotary and Daniel L. Koestler for assistance with data manipulation, Nancy Britton for many helpful discussions, and Nobuko Kajitani for the use of the Metropolitan Museum of Art Textile Conservation Department's Minolta spectrophotometer. We are especially indebted to Hisae Suzuki for her hours of support on this project. United States support for the senior author was kindly provided by the Foundation for Cultural Heritage, Tokyo, Japan.

REFERENCES

BERNS, R.S., 2000. *Billmeyer and Saltzman's Principles of Color Technology*. 3rd Edition. John Wiley and Sons, New York.

BILLMEYER, F.W., and SALTZMAN, M., 1981. *Principles of Color Technology*. 2nd Edition. John Wiley and Sons, New York.

KOESTLER, R.J., PARREIRA, E., SANTORO, E.D., and NOBLE, P., 1993. Visual effects of selected biocides on easel painting. *Studies in Conservation*, 38, 265–273.

TAVZES, C., POHLEVEN, J., POHLEVEN, F., and KOESTLER, R.J., 2003. Anoxic Eradication of Fungi in Wooden Objects. In: Koestler, R.J., Koestler, V.R., Charola, A.E., and Nieto-Fernandez, F.E., (Eds.), *Art, Biology, and Conservation: Biodeterioration of Works of Art*. The Metropolitan Museum of Art, New York, pp. 426–439.

TIANO, P., BRACCI, S., and RESCIC, S. 2003. Biomediated Calcite Precipitation for the Reinforcement of Monumental Stones. In: Koestler, R.J., Koestler, V.R., Charola, A.E., and Nieto-Fernandez, F.E., (Eds.), *Art, Biology, and Conservation: Biodeterioration of Works of Art*. The Metropolitan Museum of Art, New York, pp. 486–497.

C. TAVZES[1*]
J. POHLEVEN[1]
F. POHLEVEN[1]
R.J. KOESTLER[2]

ANOXIC ERADICATION OF FUNGI IN WOODEN OBJECTS

1 University of Ljubljana, Slovenia 2 The Metropolitan Museum of Art, New York
* Corresponding author

ABSTRACT This work examined the effects of low-oxygen conditions, achieved with either argon or nitrogen gas, on the viability of the wood-decay fungi *Antrodia vaillantii, Coniophora puteana, Gloeophyllum trabeum, Serpula lacrymans,* and *Trametes versicolor* cultivated on infested wood samples. The fungal cultures were exposed to low oxygen concentrations, below 20 ppm, for one to five weeks in hermetically sealed vessels or gas-impermeable double-enveloped plastic enclosures.

After treatment, CO_2 production of cultures was measured to determine if mycelia regenerated. In addition, the ability of hyphae to resume growth was also assessed by transferring the treated blocks to fresh nutrient agar. The effect of the anoxic conditions on the mycelia of treated blocks was expressed as either a complete absence of regrowth of the fungus or as a delay in the mycelia's recovery.

Results clearly showed that argon was more effective than nitrogen gas at reducing or eliminating fungal activity. Anoxic treatment by either argon or nitrogen gas did not affect *T. versicolor* cultures in the time span of the experiment. *G. trabeum* was only slightly affected, and *A. vaillantii* mycelial cultures proved to be less sensitive to anoxic treatment than those of *S. lacrymans* and *C. puteana,* which lost their viability completely after one month of treatment.

INTRODUCTION Objects of cultural and historical heritage are irreplaceable and should therefore be protected from damage caused by insects and wood-decay fungi. Wooden objects may be attacked by various wood-boring insects, e.g., the furniture beetle or the powderpost beetle, which can, over long periods, cause severe damage. The furniture beetle, in particular, likes to attack wood that has been infested by fungi. Wood-attacking fungi can decay wooden objects very rapidly and in a short period completely destroy them.

The anoxic treatment method has been developed to eliminate insect infestations with no side effects to the object (Gilberg, 1991; Koestler, 1992, 2001). This approach has proven very successful, especially when coupled with measurement of insect respiration (Koestler et al., 2000; Stusek et al., 2000).

Using an FTIR system for CO_2 detection, even a single organism in a 10-liter sealed enclosure can be detected (Koestler et al., 2000). Since an infestation in an art object generally involves many more insects, their presence can be readily detected in volumes greater than 10 liters. Measurement of respiration is also important in demonstrating the effectiveness of any treatment, anoxic or chemical.

Another important capability of the FTIR respiration system is that by measuring the loss of respiration activity after anoxic treatment, accurate length of treatment (LOT) can be established for different insects, their life stages, object types and sizes, and temperature and humidity conditions. In practice it has been found that LOT values for insects in art objects sometimes differ significantly from those given in published studies of insects in artificial situations, e.g., in jars.

Anoxic treatment for insects is typically performed using argon or nitrogen gas. Valentin et al. (1992) have shown that argon is 25–50% faster at killing insects than is nitrogen. They speculated that the argon dried out the insects faster than nitrogen did, and that loss of moisture was the reason for the faster deaths with argon. A more plausible explanation is that since the argon-treated insects die faster they have more time to desiccate than do nitrogen-treated ones. Argon, being heavier than oxygen, sinks to the bottom of the enclosure and pushes any remaining oxygen out of the object and to the top of the enclosure. Nitrogen, on the other hand, is lighter than oxygen and rises to the top, while the oxygen sinks to the bottom of the enclosure, and remains within the object longer than under argon.

Eradicating a fungal infestation is more difficult than eliminating insects. Dry wood is nearly completely protected against germination of fungal spores, but if the object is already infested with fungal mycelia, drying will not always eradicate the infestation. For example, the dry-rot fungus *S. lacrymans* cannot be eliminated by drying, and a more active procedure should be employed (Pohleven, 2000).

Virtually all of the procedures to eradicate fungi involve chemical treatment that can potentially damage the art, and that can be dangerous for conservators and the environment. Since the anoxic treatment has proven to be safe and successful for insect control, it was decided to test its effectiveness for fungal control too.

While there have been a few physiological studies on the effect of different oxygen concentrations on growth and survival of fungi, and on the extent of decay that they cause, none have specifically addressed using low-oxygen environments to kill them. Jensen (1967) reported that fungal biomass production was significantly lower at an oxygen concentration of 15% (as compared to the normal oxygen concentration of 20.9%) and that it ceased in the absence of oxygen. Otherwise, if the oxygen concentration was kept at a fixed level and the carbon dioxide concentration was varied, the amount of dry weight produced decreased as carbon dioxide concentrations increased. The ability of several fungal species to degrade wooden blocks was severely retarded when oxygen content was as low as 1% or CO_2 levels were raised to 10% (Highley et al., 1983). On the other hand, delignification was not affected by an atmospheric CO_2 content of 14%, and was only slightly retarded at 7% of O_2 in the bulk gas phase, concentration inside the wood particles could be lower (Reid, 1985). The ability of fungi to survive in an environment with a low oxygen concentration is best described by Scheffer (1986). There it was shown that the fungi that survived the longest were more often heart-rot rather than the sap- and products-decay fungi. Some of the fungi were even able to survive in sealed vessels for more than two years. However, most of the fungal species studied died within three months, and cultures of two of the species died within a week.

None of the above-cited authors used pure argon or nitrogen gas to replace the oxygen and carbon dioxide atmosphere in experimental vessels. Tavzes et al. (2001) have shown that some species of brown-rot fungi lost significant viability when grown in pure culture medium (PDA) under extremely low oxygen concentrations generated with argon or nitrogen gas. Optimization of the asphyxiation method might shorten the time needed for eradication of these fungi. In insects, argon gas has been shown to be more effective than nitrogen in killing insects—25 to 50% faster (Valentin et al., 1992). The aims of the current research were to ascertain whether fungi in wood blocks responded the same way as those same fungi in nutrient medium (PDA), and to see whether argon was more effective than nitrogen as the replacement gas in anoxic eradication of fungi. If the process reported by Tavzes and co-workers (2001) could be improved upon, then it could be useful in treating fungal-infested art objects.

MATERIALS AND METHODS

Organisms, Media, Material, and Inert Gases Strains of wood-decay basidiomycetes *Antrodia vaillantii* (DC. ex Fr.) Ryv. (*A.v.4*), *Coniophora puteana* (Schum. ex Fr.) Karst. (*C.p.*), *Gloeophyllum trabeum* (Pers.) Murr. (*G.t.2*), *Serpula lacrymans* (Wulf. ex. Fr.) Bond. (*S.l.5*), and *Trametes versicolor* (L. ex Fr.) Pilat (*T.v.*) were obtained from the fungal collection of the Department of Wood Science and Technology, Biotechnical Faculty, University of Ljubljana, Slovenia. The mycelia were grown on potato dextrose agar (PDA) (Difco Laboratories, Detroit, MI). Wood blocks used in the experiment were Standard SIST EN 113, dimensions $1.5 \times 2.5 \times 5$ cm (longitudinal direction) Norway spruce (*Picea abies*) sapwood and beech (*Fagus sylvatica*) (ECS, 1989).

Ventilation lids were made using Metalwork valves (Metalwork, Italy), Manesmann-Rexroth (Germany) adaptors, and Rectus (Germany) connectors. All of the tubes in the system were made of polyurethane plastic. To improve sealing, silicon vacuum paste (silicon high vacuum grease medium, Merck, Germany) was used on all connections and contact surfaces.

Argon (99.98% pure) and nitrogen (99.999% pure) gas were used to generate the anoxic environments within the experimental jars and plastic enclosures, and synthetic air (20.5% O_2 and 79.5% N_2) was used to perform respiration measurements.

Equipment for O_2 and CO_2 Concentration Measurements Two devices were used for measuring the concentration of gases within the experimental jars. One of them was supplied by Art Care International, Inc. (Orangeburg, NY) for measurements of O_2 concentrations in the range from 0 to 2000 ppm (with a relative accuracy of 2%). It consisted of a membrane pump, an oxygen perfusion sensor, and a digital display. The other one was produced by ECHO d.o.o. (Slovenia), and consisted of a membrane pump (flow rate 0.5 l/min), an electrochemical oxygen sensor that measures in the range from 0 to 25% (absolute accuracy 0.01%), an IR carbon dioxide sensor (0–3000 ppm, accuracy 5 ppm) and a 16-bit A/D converter for computerized data acquisition. The ECHO system was a closed-circuit system and permitted measurements of changes in O_2 and CO_2 concentrations over time.

Design of the Experiment Fungi were inoculated in jars onto previously autoclaved PDA. Mycelia were allowed to grow in a normal atmosphere; i.e., jars were covered with punctured lids for one week (apertures were filled with sterile cotton to prevent contamination). In that time mycelia overgrew approximately two-thirds of the nutrient medium diameter and formed vital and fast-growing mycelia. Steam-sterilized beech wood blocks were put onto *T. versicolor* cultures and spruce samples onto *A. vaillantii, C. puteana, G. trabeum,* and *S. lacrymans* cultures (on a plastic net over the cultures) for eight weeks. During that period the blocks were completely colonized. Infested wood blocks were carefully removed from underlying mycelia and put either in empty experimental jars or double-enveloped plastic enclosures, and were subjected to asphyxiation. The jars, sealed with ventilation lids, were flushed either with argon or nitrogen gas, until the concentration of oxygen in the exhaust air was below 10 ppm. The double-enveloped plastic enclosures used a heat-sealable, triple-layer EVA-nylon-EVA material. The inner bag had two valves inserted into it to allow gas exchange to the required level (less than 20 ppm). To reduce potential oxygen leakage into the inner bag, an outer bag was constructed around it and then flushed for ten minutes with argon. Oxygen scavenger was placed inside the outer bag, which was then heat-sealed.

All of the inoculations, re-inoculations, exchanges of the lids, and sealing of the inner bags were performed in a laminar flow hood, in sterile conditions. The temperature in growth chambers was maintained at 25°C throughout the experiment.

Observations on the jars were performed over a five-week period. Four experimental groups were formed. The first group was tested after one week of exposure to anoxic conditions; the second, third, and fourth after two, three, and five weeks, respectively. Each group contained six experimental jars. Three of them contained four wood blocks infested with *A. vaillantii, C. puteana,* or *T. versicolor* and then treated with argon gas. The other three were treated with nitrogen gas and contained three wood blocks infested with their respective fungal strain.

After the period of anoxic treatment, wood blocks were removed from experimental jars and placed onto sterilized PDA in jars covered with punctured lids (apertures were filled with sterile cotton to prevent contamination). Mycelial cultures infesting the wood blocks were incubated in normal O_2 conditions in a growth chamber. Visual monitoring for possible growth of hyphae was done every day, for at least four weeks.

Five double-enveloped bags were constructed, one for each tested species (*A. vaillantii, C. puteana, G. trabeum, T. versicolor,* and *S. lacrymans*). Five infested wood block samples of each species were placed in their respective bag, flushed with argon as described above, and incubated for four weeks in a growth chamber. After the four-week treatment, recovery of respiration was measured. To measure viability of the fungus, the bags were opened and the blocks were placed in experimental jars containing fresh, sterilized PDA. Ventilation lids were placed on each jar. Jars were flushed with synthetic air and the concentrations of O_2 and CO_2 were measured for five minutes using the ECHO measuring apparatus. Then the valves on the jar lids were closed and the cultures incubated for one hour. After the incubation period, another five-minute measurement was performed. A CO_2-respiration value was calculated as the difference between the average values obtained during the first five-minute reading and the final one. After respiration measurements, samples were tested for regeneration as described above for samples treated in jars.

RESULTS

Mycelial Cultures, Anoxically Treated in Experimental Jars Figure 1 shows regeneration results of *A. vaillantii* cultures from wood blocks asphyxiated with either argon or nitrogen in experimental jars. [FIG. 1] It should be noted that the higher the bar in the graph the longer it took for the culture to recover. Bars that reach the top of the graph represent cultures that did not regenerate in the timeframe of the experiment. As can be seen in the figure, argon-treated samples started to lose viability earlier and to a greater extent than the ones treated with nitrogen. Argon treatment had an effect on the viability of the cultures as early as the second week, and by the fifth week 50% of the cultures failed to regenerate. Nitrogen treatment did not exhibit any effect after three weeks and by the fifth week only 33% of the cultures did not regenerate.

Figure 2 shows that after just one week of treatment with argon, *C. puteana* cultures were not able to regenerate, whereas treatment with nitrogen caused only a short-term loss of viability even after five weeks of treatment. [FIG. 2]

Treatment with either argon or nitrogen had no apparent effect upon the viability of *T. versicolor* cultures infesting wood blocks within the timeframe of the experiment. All of the cultures regenerated within two days after removal from treatment jars, the same response as exhibited by the controls.

Mycelial Cultures, Anoxically Treated in Double-Enveloped Plastic Enclosures Three of the brown-rot fungi, *A. vaillantii, C. puteana,* and *S. lacrymans,* were unable to regenerate after four weeks of asphyxiation in argon. The fourth brown-rot fungus, *G. trabeum,* commenced growth three days after being placed in ideal growth conditions. Cultures of the white-rot fungus, *T. versicolor,* were also unaffected by anoxic treatment and resumed growth as soon as control cultures did ([FIG. 3], lower section).

The upper part of figure 3 summarizes the data for one hour of CO_2 production, measured immediately after removal from anoxic conditions. The cultures of the three brown-rot species that did not regenerate did not have any detectable CO_2

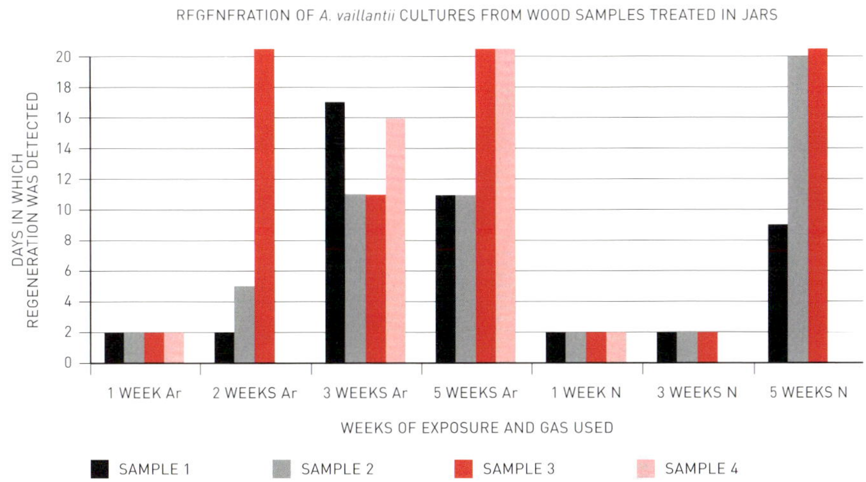

[1] Regeneration of *A. vaillantii* cultures, grown on Norway spruce sapwood blocks and treated with either argon or nitrogen gas for selected period. If the bar showing days in which regeneration was observed exceeds 20 days, it means that this culture did not regenerate.

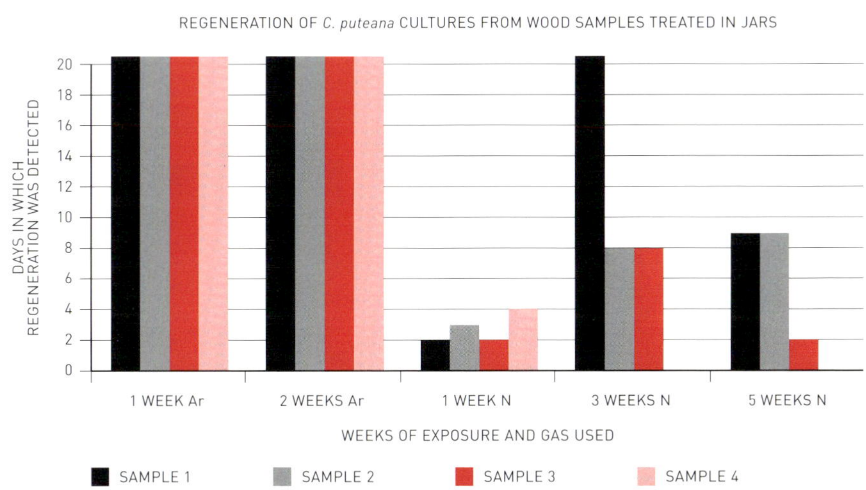

[2] Regeneration of *C. puteana* cultures, grown on Norway spruce sapwood blocks and treated with either argon or nitrogen gas for selected period. If the bar showing days in which regeneration was observed exceeds 20 days, it means that this culture did not regenerate.

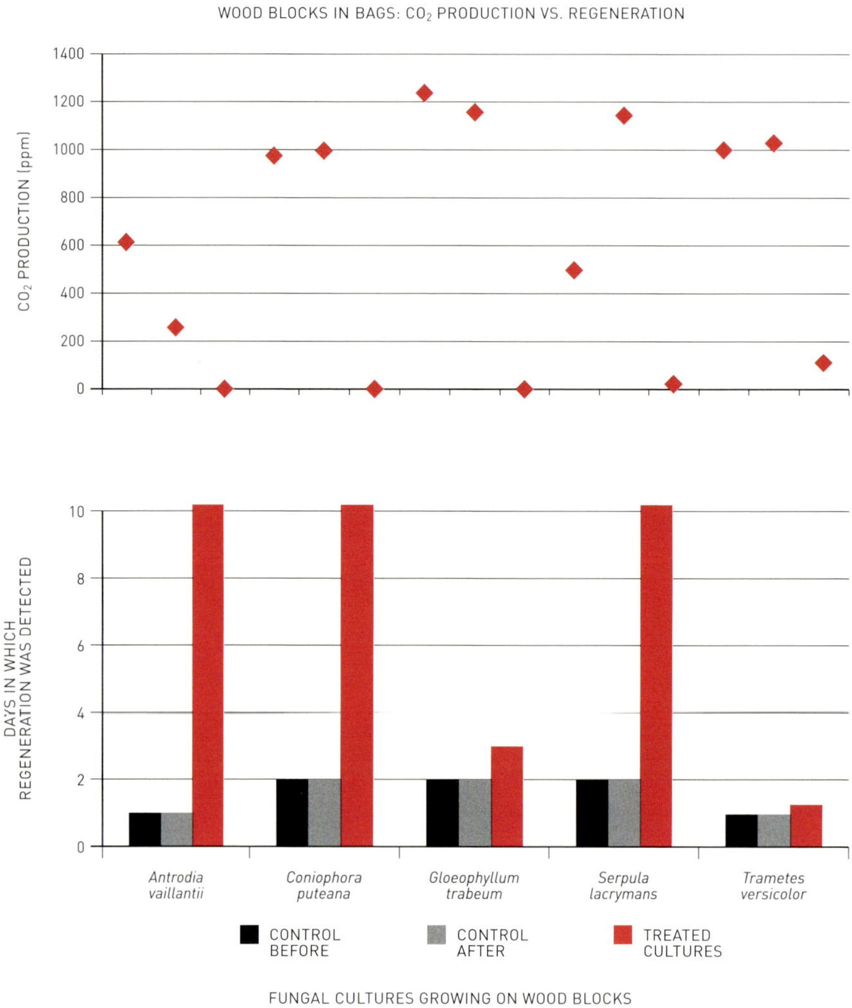

[3] Regeneration and CO_2 production of *A. vaillantii, C. puteana, G. trabeum, S. lacrymans,* and *T. versicolor* cultures, cultivated on wood blocks and asphyxiated for four weeks in double-enveloped enclosures. If the bar showing days in which regeneration was observed exceeds ten days, it means that this culture did not regenerate.

production, especially when compared to their respective controls. The white-rot fungus that recovered quickly gave CO_2 readings four to ten times greater than those of any of the brown-rot fungi. The fourth brown-rot fungus, *G. trabeum*, which took three days to show detectable growth, did not have any CO_2 production immediately after removal from the anoxic environment.

DISCUSSION The experimental results clearly demonstrate that at least some fungal infestations in wood can be eradicated by anoxic treatment within four weeks. Furthermore, there was a clear difference in treatment effectiveness between argon and nitrogen gas, with argon being able to eradicate *C. puteana* infestations in wood blocks completely while nitrogen did not.

Prior work by Tavzes et al. (2001) on suffocation of fungal cultures on PDA with either argon or nitrogen did not show a significant difference in effectiveness between the two gases. That work showed that even in 16 weeks of treatment at least some samples of the species tested were still viable. The reason was believed to be that the fungi could absorb readily accessible nutrients from the PDA medium. However, this is not the case in the wood blocks because there the fungi had to secrete oxygen-requiring enzymes in order to break down this heterogeneous and complex polymer and thus derive the needed nutrients. In the absence of oxygen, fungi may not produce enough energy to create and/or secrete these enzymes.

A. vaillantii and especially *C. puteana* cultures infesting wood blocks started to lose viability much earlier and the effect was much more pronounced if argon, rather than nitrogen gas, was used to create an anoxic environment in experimental jars. That difference was also observed for insect control in wooden objects, where argon gas proved to have higher efficiency (Koestler, 1993, 1996) because its higher specific weight enabled it to penetrate into wood cell lumens and push out the oxygen trapped within. Results on suffocation of fungal cultures on PDA showed that there was no significant difference in the loss of viability of treated mycelia regardless of which gas—argon or nitrogen—was used in a treatment (Tavzes et al., 2001). The amount of oxygen dissolved in water within agar media is obviously not high enough to help fungi to defy the adversity of the oxygen-depleted atmosphere. Therefore, one has to be careful when making direct analogies between experiments on nutrient growth media and treatment of fungi-infested wood.

Asphyxiation of fungi-infested wood blocks in plastic enclosures was the procedure most similar to practical conservation measures taken for insect control. Since argon has proven to have higher efficiency than nitrogen in lowering the viability of fungal cultures infesting wood blocks (Tavzes et al., 2002), it was decided to use only this gas for the test with plastic enclosures. For this refined experiment, five fungal species, well-known destroyers of wooden art and historic objects, known to differ in their sensitivity to anoxia (Scheffer, 1986), were chosen.

Three of the brown-rot fungi, *A. vaillantii, C. puteana,* and *S. lacrymans,* died within a month of exposure to anoxia achieved with argon. [FIG. 3] In the experiments described by Scheffer (1986), these same or similar species responded in an analogous manner. In general, the period of time needed to kill the fungi was shorter in our tests than in Scheffer's study. This is probably because in our experiments O_2 was eliminated rapidly, putting the fungi under immediate stress rather than waiting as Scheffer did for them to consume the residual O_2 in the vessels. In the case in which the fungi did not die there was a measurable effect, either in a lengthening of time needed to recover (e.g., *A. vaillantii* in jars) or in lowered CO_2 respiration reading (*G. trabeum* in bags). Scheffer indicated that *T. versicolor* survived 6–12 months and *G. trabeum* 3–6 months. This is borne out in our study, as *T. versicolor* recovered in one day and *G. trabeum* in three. Another indication of *T. versicolor* being more resilient than *G. trabeum* is seen in CO_2 production measured immediately after exposure to normal air (119 vs. 1 ppm). The results clearly show a species-specific ability of fungi to survive low oxygen concentrations. Therefore to ensure effective use of the argon-anoxic procedure it is necessary to identify the species infesting the object. Rapid identification of species would be important to speed up the conservation effort on an art object. Molecular biology techniques of DNA identification (spores or hyphae) offer the promise of doing this (Adair et al., 2002; Di Bonaventura et al., 2002).

Dry wood is nearly completely protected against infestation (germination) by fungal spores, but if the object is already infested with fungal mycelia, drying will not always eradicate the infestation. If all cells in the fungal mycelium that are causing damage to a wooden object die, and the object is maintained properly afterwards and stored in dry conditions that prevent recurrence of an infestation, the treatment is completed and there is no more threat to its integrity and appearance (Pohleven, 2000). Anoxically treated cultures of *S. lacrymans* completely lost their viability after a month of exposure. Since *S. lacrymans* infestations of wood are the most obstinate and cannot be eradicated with drying, this bodes well for further development of anoxic treatment for control of fungal infestations of wooden art objects. Future studies will focus on testing more species and assessing the effectiveness of the argon-anoxic procedure for treating fungus-infested art and cultural heritage objects.

ACKNOWLEDGMENTS The authors thank Dr. Josip Korosec from The Institute for the Protection of Cultural Heritage of Slovenia, and Mrs. Elizabeta Benko-Mächtig from that Institute's Restoration Center, for their technical support and for sharing their experience in practical application of anoxic treatment procedures for preserving fine art objects.

We are also grateful to Messer Slovenia for supplying us with gases and technical equipment necessary for our experimental work.

REFERENCES

ADAIR, S., KIM, S. H., and BREUIL, C., 2002. A molecular approach for early monitoring of decay basidiomycetes in wood chips. *FEMS Microbiology Letters*, 211, 117–122.

DI BONAVENTURA, M.P., DESALLE, R., EVELEIGH, D.E., BALDWIN, A., and KOESTLER, R.J., 2003. Studies of fungal infestations of Tiffany's drawings: Limits and advantages of classical and molecular techniques. In: Koestler, R.J., Koestler, V.R., Charola, A.E., and Nieto-Fernandez, F.E., (Eds.), *Art, Biology, and Conservation: Biodeterioration of Works of Art*. The Metropolitan Museum of Art, New York, pp. 94–109.

EUROPEAN COMMITTEE FOR STANDARDIZATION, 1989. Wood preservatives; Determination of the toxic values against wood destroying basidiomycetes cultured on agar medium, *EN* 113. Brussels, p. 14.

GILBERG, M., 1991. The effects of low oxygen atmospheres on museum pests. *Studies in Conservation*, 36, 93–98.

HIGHLEY, T.L., BAR-LEV, S. S., KIRK, T.K., and LARSEN, M.J., 1983. Influence of O_2 and CO_2 on wood decay by heartrot and saprot fungi. *Phytopathology*, 73, 630–633.

JENSEN, J.F., 1967. Oxygen and carbon dioxide affect the growth of wood decaying fungi. *Forest Science*, 13, 304–389.

KOESTLER, R.J., 1992. Practical aspects of nitrogen and argon gas fumigation procedures for insect control in museum objects. *Preprints of the Second International Conference on Biodeterioration of Cultural Property*. Yokohama, Japan. Oct. 5–8, 1992, pp. 96–98.

KOESTLER, R.J., 1993. Insect eradication using controlled atmospheres, and FTIR measurement for insect activity. *ICOM 10th Triennial Meeting*, Washington, DC Vol. II, pp. 882–886.

KOESTLER, R.J., 1996. Anoxic treatment for insect control in panel paintings and frames with argon gas. *American Institute of Conservation Paintings Specialty Group Postprints*. AIC, 1717 K Street, NW, Suite 301, Washington, DC 20006, pp. 61–72.

KOESTLER, R.J., 2001. Detecting and Controlling Insect Infestation in Fine Art. In: Stevenson, C.M., Lee, G., and Morin, F.J., (Eds.), *Pacific 2000*, Proc. 5th International Conference on Easter Island and the Pacific. Easter Island Foundation, Los Osos, CA, pp. 541–545.

KOESTLER, R.J., SARDJONO, S., and KOESTLER, D.L., 2000. Detection of insect infestation in museum objects by carbon dioxide measurement using FTIR. *International Biodeterioration and Biodegradation*, 46, 285–292.

POHLEVEN, F., 2000. The threat from fungi to wooden items of our cultural heritage. In: Benko Mächtig, E., (Ed.), *RES—Wood in the Process of Restoration*, Publications of the Restoration Centre of the Republic of Slovenia, The Restoration Centre of the Republic of Slovenia, Ljubljana, Slovenia, pp. 25–30.

REID, I.D., 1985. Biological delignification of aspen wood by solid-state fermentation with the white-rot fungus *Merulius tremellosus*. *Applied and Environmental Microbiology*, 50, 133–139.

SCHEFFER, T.C., 1986. O_2 requirements for growth and survival of wood-decaying and sapwood–staining fungi. *Canadian Journal of Botany*, 64, 1957–1963.

STUSEK, P., POHLEVEN, F., and CAPL, D., 2000. Detection of wood boring insects by measurement of oxygen consumption. *International Biodeterioration and Biodegradation*, 46, 293–298.

TAVZES, C., POHLEVEN, F., and KOESTLER, R.J., 2001. Effect of anoxic conditions on wood decay fungi treated with argon or nitrogen. *International Biodeterioration and Biodegradation*, 47, 225–231.

TAVZES C., POHLEVEN, F., JENISEK, M., and KOESTLER, R.J., 2002. Effect of asphyxiation on wood decay fungi treated with argon and nitrogen gas. 33rd Annual Meeting of International Research Group on Wood Preservation, IRG Documents 2002, IRG Secretariat, Stockholm, IRG/WP 02–10452.

VALENTIN, N., ALGUERO, M., and MARTIN DE HUAS, C., 1992. Evaluation of disinfection techniques for the conservation of polychrome sculpture in Iberian museums, In: *International Institute for Conservation. Conservation of the Iberian and Latin American Cultural Heritage*, pp. 165–167.

V. BULLOCK

MICROBIOLOGICAL IMPURITY OF CONSERVATION TREATMENT WATERS

University of Western Sydney, Australia

ABSTRACT In a survey in Sydney, Australia, conservators rated the microbiological purity of treatment water as a high priority. However, analysis showed bacteria to be present at high levels in all waters tested. Materials conservation appears to have adopted the water purification technologies of industry but not the monitoring methodologies. An examination of international water quality standards and industrial case studies leads to the suggestion that conservation-specific water standards be developed.

INTRODUCTION Materials conservation, like other sectors and industries, is sensitive to the quality of its water supply. In 2000, under a cooperative project, leading materials conservation laboratories in Sydney, Australia, filled out a questionnaire about water-related priorities and practices and also permitted analyses of their laboratory waters. This paper analyzes the findings of that project, the Sydney Conservation Laboratories Water Assessment 2000 (henceforth "SCLWA 2000"; Bullock, 2001), particularly with regard to microbiology. Sydney is home to some of Australia's oldest and most important cultural institutions, and conservation practices and facilities are well developed.

SCLWA 2000 SCLWA 2000 had three aims:
1. To compare the quality of tap and purified waters at eight leading materials conservation laboratories in Sydney;
2. To rate the water delivery systems that are being used with cultural materials in terms of quality, cost, maintenance, and suitability; and
3. To make findings and collected data available to all participating organizations as an aid to their decision making in the areas of water delivery systems and the wet treatment of cultural materials.

The methods were a range of water analyses (including pH, conductivity, ion chromatography, inductively coupled plasma mass spectrometry, suspended solids, and heterotrophic plate count) and a questionnaire. The questionnaire gathered information on usage, maintenance, suitability, costs, and decision-making concerning tap and pure waters in each laboratory. Choice of water type used at different treatment stages with different material types was also recorded. Questionnaires were forwarded to all participating laboratories and a 100% return rate was achieved. The conservator(s) most familiar with laboratory waters and purification equipment filled out the tick-box questionnaire. Pure water comparisons were made with international standards (ASTM, 1997; ISO, 1987; and British Pharmacopoeia Commission, 2000); tap water comparisons were made with official "Sydney Water" test results and ARMCANZ/NHMRC drinking water standards. Standards consulted for sampling, collection, and analyses were: Standards Australia, AS2031.2–1987; AS/NZS 5667; AS 4276–1995; and Franson, 1998.

Sydney drinking water derives from partially protected catchments and is processed using direct filtration (no sedimentation step and Fe_3Cl^- is the coagulant) as the catchment waters are typically colored waters of low turbidity. Typical measurements from the government utility "Sydney Water" for conductivity are 181–209 (S/cm µmicroSiemens/cm) and, for pH, 7.7–8.1 (Sydney Water, 1996), according well with international potable water standards. Three different local systems supplied water to the conservation laboratories from Warragamba dam.

WATER STANDARDS AND EFFECTS OF PURIFICATION Standards are achieved by consultation between industry, professional organizations, and university and government departments. This is equally the case for potable and high-purity water standards, as well as many other non-water-related fields. "Standards Australia" has created some of its own water standards (including collection procedures), but for high-purity waters, overseas standards are used. International standards (ASTM, 1997; ISO, 1987) for reagents or analytical laboratory use recognize a three-four-fold classification system. Pure water may range between 0.056 (ultrapure) and 5.0 μS/cm and pH 5 and 8. (Chlorides, sodium, total silica, and total organic carbon measure are also specified.) Where bacterial levels need to be controlled, ASTM overlays a further threefold classification, A–C. Acceptable heterotrophic plate count and endotoxin levels for Type A are 10/1000 mL and <0.03, respectively; for Type B, 10/100 mL and 0.25; and for Type C, 100/10 mL. In addition, pharmacopoeias specify microbiological purity levels required for water for injection, dialysis, etc.

Salts, metals, and some organic compounds (such as chlorine and fluoride) are permitted in or added to drinking water by authorities for the purposes of public health because they are found to reduce the prevalence of bacterial pathogens. Most pathogens derive from human intestines and secondarily from other warm-blooded animals. Very few pathogens are capable of living free in the environment for extended periods (Cliver and Newman, 1987). However, it has been found that municipal water treatment and storage processes may actually assist the growth of "nuisance organisms" on reservoir walls and in distribution mains, and that this effect can be extended by further attempts at purification. (Nuisance organisms may include procaryotic bacteria, such as planktonic and benthic cyanobacteria; iron, manganese, and sulfur bacteria; actinomycetes and fungi; and eucaryotic organisms such as algae, crustacea, and protozoa (ARMCANZ/NHMRC, 1996.) Most often, local water purification aims to remove toxic organic/inorganic contaminants. Purified water is thus left with no residual toxicity to microbiological contamination. Further, the employment of large surface area filters and/or resin beds, also a source of nutrition, encourages the propagation of nuisance organisms and resultant local bacterial and fungal production. In turn these may maintain populations of crustacea and protozoa. Autolytic breakdown of any of these organisms results in the release of cellular compounds that affect the color, turbidity, taste, and odor of water (McFeters et al., 1993, and ARMCANZ /NHMRC, 1996).

SCLWA 2000—SELECTED QUESTIONNAIRE RESULTS Questionnaire responses in SCLWA 2000 showed that water quality testing in conservation laboratories was limited. Testing was generally omitted for each water type (i.e., tap, distilled, deionized, and reverse osmosis), due to lack of time for testing and lack of familiarity with the tests. Those respondents conducting tests tested for pH and/or conductivity. Where the company supplying the purifier was used for maintenance, this was often arranged in response to occasional problems. Two laboratories scheduled annual servicing.

Microbiological content was ranked together with low ion content and consistency of quality by all six respondents to this part of the survey as a high priority in choosing water purification systems. However, no laboratory had established a way of determining the microbiological status of a water. Apart from lack of time and understandable lack of familiarity with very precise microbiological testing methods such as the heterotrophic plate count, conservation laboratories are not clean enough for traditional microbiological investigations. For microbiological investigations, it is inevitable that a microbiologist attends to take samples and then prepares specimens and conducts plate counts in a relatively sterile microbiological laboratory. The complexity and cost of proper microbiological testing deters conservators.

With regard to choice of water types used in conservation treatment, it was found that both tap and pure waters were used. Tap water was used in the early stages of treatment with metal (3 respondents), and ceramics/stone (3), but also with glass (1), wood (1), paper (1), paintings (1), and large engineering (1). Tap water was used in the later stages of treatment with large engineering objects only. None used tap water for rinsing objects. From the results of the questionnaire a clear shift to pure water usage in the later stages of object treatment and rinsing was noted, although ethnographic/natural history, textiles, and photographic and electronic media were treated with pure water only. Deionization was the most commonly used water purification technology used. Distilled water was used at a number of laboratories, but one laboratory used only distilled water for each treatment stage (early/late/rinse) of the ten material categories. Where there was access to only one type of purified water, this tended to be used at all three treatment stages, except when tap water was used in early treatment stages (as described above).

Where more than one type of purified water was available, quite different individual approaches were noted. In the case of textiles, one laboratory chose to use deionized water at all three stages of treatment, while another started with deionized and moved to distilled water for later stage treatment and rinsing. A third laboratory started treatment of metal with tap water, finalized treatment with distilled, deionized, or reverse osmosis water, but rinsed only with distilled water. A fourth laboratory chose to use only distilled water for the treatment of electronic media, despite having reverse osmosis and deionized waters available. In one case deionized and distilled water were regarded as entirely interchangeable.

SCLWA 2000—SELECTED ANALYTICAL RESULTS Scientific analyses showed that water purification in the conservation laboratories surveyed reduced conductivity, anions, metals, pH, and suspended solids; it also redistributed and slightly reduced cations. Conductivity and pH are the simplest shorthand indicators of the purity of a water. However, it should be noted that pH is regarded as an unreliable and enigmatic indicator with truly pure waters, due to the extremely low levels of cations/anions present. Measurement of tap waters during the survey produced conductivities in the range of 186–209 µS/cm and with pH predominantly 6–7. Conservation laboratory purification of Sydney tap water was found to reduce conductivities by between 40% and 99.9% to 0.9–122 µS/cm and pH from 6–7 to 5. Only two waters classified as technically pure according to received organic/inorganic definitions (i.e., conductivity in the range 0.056 µS/cm-5.0 µS/cm). No particular form of purification was shown to be consistently purer than another.

Two microbiology tests were conducted in relation to SCLWA 2000. Test A was a traditional heterotrophic plate count on waters from eight conservation laboratories. [FIG. 1] Results showed that bacteria were present at similar high numbers in both tap and purified waters and that none classified as technically pure according to received microbiological definitions (i.e., <10 colony forming units per mL). Numerically the results showed that purified water registered more colony forming units than tap water; however, according to standard microbiological method the very high numbers of colony forming units that were grown nullified these results. Test B was conducted on commercially guaranteed pure deionized and reverse osmosis waters using Millipore® HPC and Yeast & Mould Samplers. Results showed that deionized water contained more bacteria than reverse osmosis waters and that the Millipore® Samplers provided an accurate and easy way to test the microorganic health of a water, perhaps suitable to the limited resources, time availability, and laboratory conditions of the conservator. Although the colonies grown from the commercial pure waters were very few (i.e., <1 colony forming unit per mL and within specification for pharmaceutically pure water), the results accord well with the theory that deionized water supports microorganisms due to the use of resins with large surface areas.

MICROBIOLOGY IN CONSERVATION Practicing conservators know the theory and reality of the effect of dissolved and undissolved organics and inorganics on cultural materials, but relatively little about the role of microorganisms. Where conservators have published on microbiology, it has most often been with regard to fungi. Fungi (i.e., yeasts, molds) and bacteria may be found in water and air—both are endemic. Bacterial by-products (i.e., enzymes on which viruses, protozoa, and fungi feed) can be particularly harmful and insidious, but certain enzymes may also be harnessed in conservation treatment. In cases of the deterioration of internationally significant items of cultural heritage, conservators are increasingly seeking the expert input of microbiologists for accurate diagnoses. In such investigations microbiologists typically present conservators with full lists of microorganisms present. Depending on the level of the conservator's microbiological education and residual funding, this information can influence conservators' decisions as to which biocides and traditional techniques to use in conservation treatment. However, some microbiologists are calling for microbiological investigation that goes beyond the provision of data to the elucidation of the mechanisms of deterioration and greater analysis of symbiotic relationships between materials and microbes, as well as between microbial species present, at cultural sites or on items of cultural heritage (Koestler et al., 1994). Further, it is being suggested that microbiologists work toward interpreting this information to suggest specific microbiological treatment solutions (enzymes or competitive strains) that are more delicate, environmentally friendly, and effective than the standard conservation treatment techniques (May et al., 2003).

WATER QUALITY IN INDUSTRY SCLWA 2000 included case studies of pure water management in three sensitive industries in Australia: pharmaceuticals, photographics, and electronics. The U.S. space program was also investigated. Both the U.S. space program and electronics (semi-conductor) appear to set particularly high water-quality standards. In 1998 a revision of high-quality water systems at the U.S. space program was prompted by the Genesis mission, which required ultraclean surfaces to collect "atoms spewed out by the Sun and return them to Earth for detailed study" (Hango et al., 2000). They were also looking to provide a clean environment for the examination and storage of rocks from the moon, meteorites from Antarctica, and dust from comets and asteroids.

Certain industries are required by law to conform to published water quality standards. For example, pharmaceuticals must demonstrate (microbiological) hygiene levels indicated by designated pharmacopoeias. International water quality standards are generally revised at five-year intervals. Each standard revision becomes more stringent and it can be a large task for companies to upgrade systems to deliver the higher sensitivity levels required by revised standards, e.g., parts per billion rather than parts per million measurement of contaminants. Other industries, for example, electronics, are not required to conform to standards legally, but use them as baseline guides. Marketing edge and technological advances cause some industries using pure waters to develop their own in-house water quality

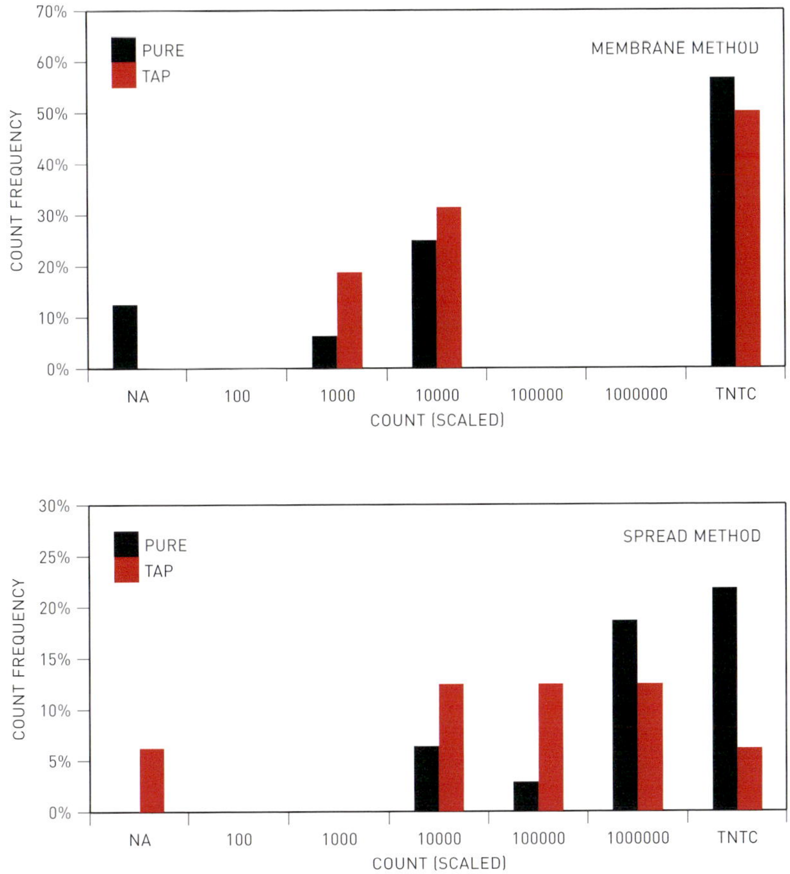

[1] Heterotrophic plate count of tap and purified waters from eight conservation laboratories. Count, in colony forming units (cfu), is plotted according to tap or purified water origin, averaged across duplicates and scaled up to represent 100 mL. Two preparation techniques were employed, and graphed results are presented in terms of these: spread/pour plate (0.1 mL) and membrane filtration (1.0 and 10 mL). Incubated for 6 days at 25°c on R2A (low nutrient) media. N=21 for pure water; N=24 for tap water.

guidelines, which evolve much more quickly and appropriately to their circumstances. Similar to official standards and guidelines, in-house high-quality water guidelines include diagnostic parameters and testing procedures, measurement frequency, and recommended actions to take when water quality goes out of specification—in extreme cases, plant shutdown.

While some industries (fossil fuel power plants, petrochemical, paper and pulp industries) may be primarily interested in preventing boiler feed water from scaling or corroding plant equipment, microbiological issues are important in all plants using water purification equipment as continuously used monitoring equipment (i.e., continuous conductivity, daily monitoring of oxygen, chlorides, and sulfates to ppb levels) must be sanitized to enable accurate measurements. In-house tailoring of water quality is also advised at different operating localities within single organizations, which may have different regional feed water, wastewater, and condensate requirements. It is also found that superficially identical water purification systems regularly display significant differences. Eradication of microorganisms is now an industry focus equivalent to the filtration of suspended and dissolved organic and inorganic contaminants. In-house guidelines give industries a better guarantee that treated water will not contribute to defective products or damage plant equipment, while at the same time ensuring that the treatment system is operating efficiently and economically.

IMPLICATIONS FOR CONSERVATION If bacteria are deleterious to organic and inorganic end-users and products in sensitive industries/sectors, what are the implications for the conservation of cultural materials? What is the mechanism of interaction between aged cultural materials and different water types? What makes conservators choose between different water types? Does treatment with bacteria-laden water predispose cultural materials to long-term stability risks? For example, would such a washed object be more inclined to deterioration in a humid storage environment? The case of the symbiotic relationship between iron and sulfur reducing bacteria is by now well known to marine conservators, after identification by engineering research. Are there other symbiotic relationships between materials and bacteria that are as yet unproven? It might be reasonable to assume that organic materials would be at greater risk from bacterial attack than inorganic, but the iron/sulfur case is a clear exception to this.

In conservation we are not bound by legal standards outside of those used to safeguard the environment and our health, and we do not publicly define many technical details of our work. Commonly we have limited funds with which to assess existing work practices, and as our prime responsibility is to an employer with its own broader objectives, our time is often spent on matters other than conservation operational matters. As a result, issues such as water quality may only be informally discussed. In addition, we have been bound comparatively loosely by professional codes of ethics and practice; however, discussions concerning accreditation have "returned focus to the standards and core functions of conservation" (Henderson and Dollery, 2001).

CONCLUSION It is clear that there is a disjunction in conservation between the desire to minimize the contamination of objects under treatment by using purified water, and the reality of achieving truly pure water. It is apparent that conservators must embrace monitoring methodologies from industry along with the water-purifying hardware already in place, if indeed the goal is to produce pure water for conservation treatment. However, while it may seem logical to follow the lead of industries that invest heavily in water research and development, the encouragement to use pure, even ultrapure water, may actually be inappropriate to conservation. Perhaps less ionically pure water is better suited to the treatment of some cultural heritage materials or particular degrees of deterioration. If microbiological contamination is found not to predispose cultural materials to extra risk, it would be far less costly to use nonsterile water. Ionic and microbiological water purity come at high monetary cost and constant vigilance, both of which are difficult for conservation to achieve. It may be that conservation's trusting approach to water quality has been fortuitous. It would be better, however, to research the mechanism of interaction between water and deteriorated cultural materials, and, in particular, the effect of bacteria on cultural materials. Like others, we should work to establish a water standard specific to our unique requirements through our professional bodies and through standards organizations, as water is conservation's most commonly used solvent.

ACKNOWLEDGMENTS Thanks are due to the participating conservation laboratories in Sydney and to Fort Dodge Australia for the use of their facilities. Thanks also to Millipore Australia for the donation of one box of samplers. I acknowledge and appreciate the help of Dr. Christopher Jenkins in the preparation of this paper, and Dr. Richard Thomas and the Australian National Maritime Museum in the preparation of the poster on this subject presented at the ABC2002 Conference.

REFERENCES

AMERICAN SOCIETY FOR TESTING AND MATERIALS, 1997. Standard Specification for Reagent Water D1193–91 In: *Annual Book of ASTM Standards*, Section 11, Water and Environmental Technology, 11.01. West Conshohocken, Pennsylvania, pp. 112–114.

AGRICULTURE AND RESOURCES MANAGEMENT COUNCIL OF AUSTRALIA AND NEW ZEALAND/NATIONAL HEALTH AND MEDICAL RESEARCH COUNCIL, 1996. *Australian Drinking Water Guidelines*, ARMCANZ/NHMRC, Canberra, pp. 2–16.

BRITISH PHARMACOPOEIA COMMISSION, 2000. *British Pharmacopoeia 2000*. The Stationery Office, London.

BULLOCK, V.M., 2001. Sydney Conservation Laboratories Water Assessment 2000. M. Applied Science Research Reports [Unpubl.], University of Western Sydney, Kingswood, Australia.

CLIVER, D.O., and NEWMAN, R.A., (Eds.), 1987. Drinking Water Microbiology—Special Issue of the Journal of Environmental Pathology, *Toxicology and Oncology*, 7 (5/6) May–Aug. 1987. International Society for Environmental Toxicology and Cancer and Committee on the Challenges of Modern Society (NATO/CCMS), p. 38.

FRANSON, M-A-H., (Man. Ed.), 1998. *Standard Methods for the Examination of Water and Wastewater (20th Edition)*. American Public Health Association/American Water Works Association/Water Environment Federation, Washington, DC, pp. 9–1 to 9–40.

HANGO, R.A., WARREN, J., ALLTON, J., and STANSBERY, E., 2000. Special Applications—High Purity Water Used for Moon and Mars Specimen Curation. In: *Ultrapure Water*, October. Tall Oaks Publishing, Littleton, pp. 75–78.

HENDERSON, J., and DOLLERY, D., 2000. Growing Pains—The Development of a Conservation Profession in the UK. In: Roy, A., and Smith, P., (Eds.), *Tradition and Innovation: Advances in Conservation*. Melbourne Congress 10–14 October 2000. IIC, London, p. 88.

INTERNATIONAL ORGANIZATION FOR STANDARDIZATION, 1987. Water for analytical laboratory use—Specification and test methods. *ISO* 3696: 1987 (E). ISO, Switzerland, pp. 1–5.

KOESTLER, R.J., BRIMBLECOMBE, P., CAMUFFO, D., GINELL, W., GRAEDEL, T., LEAVENGOOD, P., PETUSHKOVA, J., STEIGER, M., URZI, C., VERGES-BELMIN, V., WARSCHEID, T. (1994). How do environmental factors accelerate change? In: Krumbein, W.E., Brimblecombe, P., Cosgrove, D.E., and Staniforth, S., (Eds.), *The Science, Responsibility, and Cost of Sustaining Cultural Heritage*, John Wiley & Sons, New York, pp. 149–163.

MCFETERS, G.A., BROADAWAY, S.C., PYLE, B.H., and EGOZY, Y., 1993. Distribution of Bacteria within Operating Laboratory Water Purification Systems. In: *Applied and Environmental Microbiology*, 59, 1410–1415.

MAY, E., PAPIDA, S., and ABDULLA, H., 2003. Consequences of Microbe-Biofilm-Salt Interactions for Stone Integrity in Monuments. In: Koestler, R.J., Koestler, V.R., Charola, A.E., and Nieto-Fernandez, F.E., (Eds.), *Art, Biology, and Conservation: Biodeterioration of Works of Art*. The Metropolitan Museum of Art, New York, pp. 452–471.

STANDARDS AUSTRALIA, 1987. AS2031.2-1987. *Selection of Containers and Preservation of Water Samples for Chemical and Microbiological Analysis Part 2—Microbiology.* Standards Australia, Sydney.

STANDARDS AUSTRALIA, 1995. AS4276-1995. *Water Microbiology.* Standards Australia, Sydney.

STANDARDS AUSTRALIA, 1998. AS/NZ 5667. *Water Sampling.* Standards Australia, Sydney.

SYDNEY WATER, 2000. Sydney Water Analyses for February 2000. [Unpubl. data]

SYDNEY WATER, 1996. *Analysis of Water from Customers Taps.* Sydney Water, Sydney, pp. 1–4.

E. MAY[1]
S. PAPIDA[1]
H. ABDULLA[2]

CONSEQUENCES OF MICROBE-BIOFILM-SALT INTERACTIONS FOR STONE INTEGRITY IN MONUMENTS

1 University of Portsmouth, Portsmouth, UK 2 Suez Canal University, Ismailia, Egypt

ABSTRACT Scanning electron microscopy confirmed the presence of bacteria and biofilm on all substrates, in situ and in the laboratory. Bacterial populations were not favored on substrates of a particular mineralogy in situ or in the laboratory; however, sheltered sites favored the development of a rich, homogeneous biofilm. The presence of salts had a profound effect on the appearance of the biofilm, in situ and in the laboratory experiments, causing contraction, coagulation, and desiccation. Weathering experiments also indicated that salts could also interact destructively with filamentous bacteria, in particular with the genus *Streptomyces,* and cause severe damage to stone. This biodeterioration could be attributed to biofilm production, which may enhance the effects of salt crystallization pressure in the pore system of the stone. The extent of FDA hydrolysis, which is an index of growth, was strongly correlated to changes in stone properties. The additional mechanical effects of actinomycete hyphal penetration on the stone were clearly evident from the SEM analysis. Similar reactions to salts are thus observed by SEM in situ and in the laboratory, and these can be linked to patterns of growth and serious damage to stone.

INTRODUCTION: MICROBES AND STONE DECAY The ability of the stone-colonizing microflora to cover and even penetrate material surface layers by the excretion of organic extracellular polymeric substances (EPS) leads to the formation of complex biofilms in which the microbial cells are embedded (Boivin and Costerton, 1991; Denyer et al., 1993). Microorganisms may degrade stone mechanically, chemically, and aesthetically through complementary metabolic activities and biomineralization processes in these biofilms (extracellular polymeric substances, EPS) and microbial mats (Sand, 1997). This paper is concerned with the environmental and microbiological factors that control the development of biofilms on stones, their interaction with salts, and the possible consequences for damage to the stone. Artificial weathering was used to assess how stone is affected by salt, under wet-dry weathering conditions, in combination with biofilm formation by natural bacterial populations or pure cultures of filamentous actinomycetes. Thus comparative studies were done on natural and laboratory biofilms using scanning electron microscopy (SEM) and these were related to measurements of biofilm components and activity, in order to understand the impact of biofilm on the stone deterioration process.

METHODS

Weathering Experiments Weathering experiments were carried out as described by Papida et al. (2000b), with modifications, in order to study the interactions between mixed populations of stone bacteria or actinomycete pure cultures and salt weathering in the deterioration process of stone.

Stones tested in these experiments were Mansfield dolomite and Portland limestone from the U.K. and Stavros and Psathas limestones from Crete, Greece. The stone was cut into discs of 1.0 cm height × 3.8 cm diameter in order to fit into petri dishes and to allow sound velocity determinations on the stone (Rummel and Van Heerden, 1981). The stone discs were dried at 105°c for 24 hours and, after cooling in a desiccator, the initial weight of each stone was then determined. Porosity was determined according to Franklin et al. (1981).

Bacterial suspensions of mixed microbial populations, or MMP, recovered from Portchester Castle, Hampshire, U.K., or Tell Basta tomb, Zagazig City, Egypt, or cultures of *Streptomyces* and *Nocardia brasiliensis*, previously isolated from damaged limestone (10 ml; 6×10^6 cfu/ml) were used to inoculate each stone type.

Inoculated stones were each divided into four groups: control (organisms alone); sterile distilled water; 0.5 M NaCl; 0.5 M Na_2SO_4. Two series of experiments were conducted, one to study MMP, and the second to assess actinomycetes. For the actinomycete series, stones were incubated in an environmental cabinet (Fisons) where temperature and RH settings were adjusted to daily cycles: 11 hours at 35°c and 70% RH; 2 hours at room temperature (for treatments and measurements);

11 hours at 8°C and 40% RH. Every 25 cycles triplicate samples were removed and analyzed for physical (weight loss, porosity, surface hardness, and sound velocity) and biological (viable counts, SEM, carbohydrate and protein) changes. Carbohydrate content was assayed using the colorimetric method of Dubois et al. (1956), as modified by Pacepavicius et al. (1997).

Assessment of Biofilm by Scanning Electron Microscopy Stone samples (0.5 cm^3) were fixed in 4% glutaraldehyde and held at 4°c overnight. The samples were allowed to attain room temperature, then washed in two changes of distilled water for 15 minutes each. Post-fixation was achieved by submerging the samples in 2% aqueous osmium tetraoxide (OsO_4) for 12 hours at 4°c. The samples were allowed to attain room temperature, then washed in two changes of distilled water for 15 minutes each and dehydrated in an acetone series. Stone samples were dried with a Polaron E 3000 critical point drier, with CO_2 as drying agent. After drying, the samples were mounted on aluminum stubs, secured with carbon glue, and coated with gold in a Polaron E5000 diode sputter coater. The coated samples were examined in a JEOL GSM-T20 scanning electron microscope.

STONE DECAY AND BIOFILMS: IN-SITU OBSERVATIONS The mineralogy and structure of the stone in relation to its capacity to collect water, organics, and particles control its predisposition to biodeterioration, or bioreceptivity (Guillitte, 1995). Bacteria produce EPS mucilages, sheaths and fibrous accumulations, consisting of polysaccharides, proteins, and leaked intracellular polymers to facilitate irreversible adhesion to substrates and protect them against dehydration, temperature changes, and toxic agents (Urzi et al., 1991). This provides an anionic medium with high cation-exchange properties to trap aerosols, dust and nutrients, minerals, and organic compound complexes (Christensen and Characklis, 1990). Extracellular polymeric substance biofilms are nonhomogeneous, thick and dense, largely composed of water, forming a very hydrophilic, hygroscopic gel (Decho, 1994). This takes up water from air and releases it under low RH conditions. Stone moisture is thereby increased while porosity, water-uptake capacity, and evaporation are reduced (Warscheid et al., 1993).

Naturally occurring biofilms were studied using SEM at two locations: Portchester Castle, near Portsmouth in the U.K., on the edge of Portsmouth Harbour; and Knania, the fortified walls of the harbor at Knania in Crete, Greece. Both monuments have stone surfaces exposed directly to marine salt aerosols, as well as walls that are protected from the sea.

SEM Observations Notably rich and homogeneous biofilms, composed mostly of bacterial rods, were observed on substrates coming from sheltered areas of Portchester Castle. [FIG. 1B] Heterogeneous biofilm structures and coagulated cells entangling stone particles were observed [FIG. 1A] on substrates directly exposed to sea salt deposition by means of marine aerosols, often considered a primary factor for mechanical coastal stone decay. Algal cells are also seen in fissures of sheltered substrates (not shown) and could be responsible for mechanical breakdown of stone through expansion.

Biofilms were also observed on all surfaces of the limestone blocks of the Venetian Fortification Walls of Khania. [FIGS. 1C AND D] As found at Portchester Castle, biofilm structures were typically desiccated and filamentous-like on the exposed surfaces [FIG. 1C] but more slimy and homogeneous on the sheltered substrates. [FIG. 1D] Active cells under division and biofilms entangling stone particles were observed. Weathered crystals among desiccated biofilms and weathering patterns in pit forms, directly attributable to bacterial activity, were noted. Salt-exposed stone supported heterogeneous biofilms and cell clumps appeared less slimy and coagulated. Filamentous growth was also present. The stony substrate was highly weathered. Debris of crystals and particles of unknown origin were abundant among microbial consortia.

Biofilms in Situ The observation that salts affected the nature of the biofilm seen at Portchester Castle and Khania is in agreement with Gerdes et al. (1994), whose findings determined that high salinity values might cause flocculant and incoherent mats. Sodium chloride does reduce the viscosity of charged polysaccharides (Christensen and Characklis, 1990) while inorganic salts and multivalent inorganic cations coagulate bacterial cells by neutralizing the charge of individual cells (Daniels, 1980). No direct correlation has been found between the size of bacterial populations and substrate location, although diversity may change (Tayler and May, 1991). However, the SEM data presented in figure 1 taken with evidence of the carbohydrate content of stones from sheltered or exposed areas (May et al., 2000), suggests that biofilm slime production may be influenced, rather than bacterial counts, when salts accumulate. [FIG. 1] In addition, filamentous bacteria were seen. Thus the quantity of biofilm produced may change in response to salt accumulation, and this might adversely affect stone integrity.

BIOFILMS, BACTERIA, AND SALTS: LABORATORY OBSERVATIONS

Biofilms and Salts Salts acting on their own are very important decay agents and

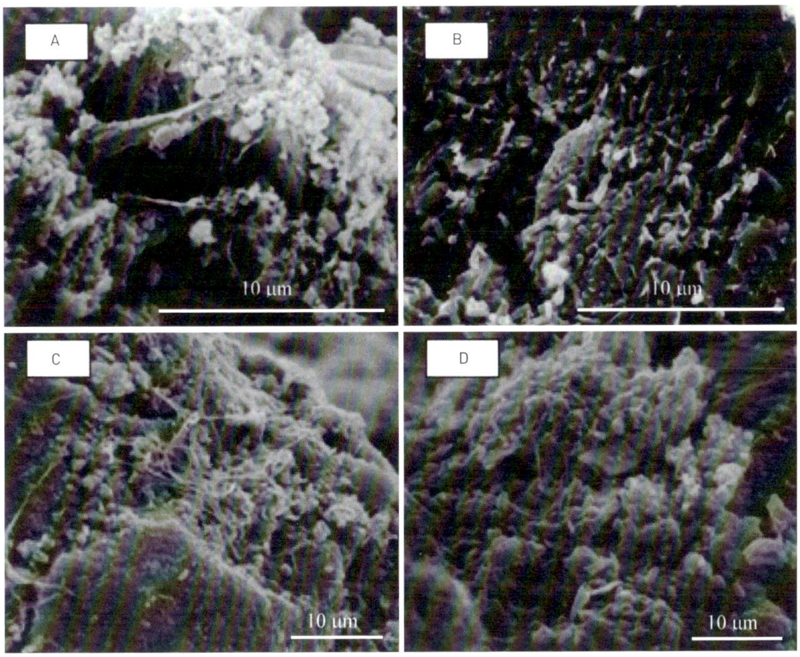

[1] In-situ observations of microbial communities on stone for Portchester Castle showing (a) a salt-exposed site (× 5000); (b) a protected site (× 5000); and, for Knania fortifications, showing (c) a salt-exposed site (× 2000); and (d) a protected site (× 2000). At the protected sites, biofilms are luxuriant and homogenous while those at salt-exposed sites are contracted and flocculant. Filamentous growth is notably present at the salt-exposed sites for both locations, the dimensions of which suggest colonization by actinomycetes.

TREATMENT	% WEIGHT LOSS (after 100 cycles)	
	Without MMP	With MMP
Control	0.02	0.02
Water	0.21	-7.7
NaCl	2.6	3.0
Na_2SO_4	-4.9	-31.1

	Viable count	**0.534**	**0.503**	-0.220	**-0.536**
	Carbohydrate	-0.416	-0.460	0.320	0.472
Streptomyces	Protein	0.143	-0.220	0.195	0.037
	FDA hydrolysis	**0.925**	**0.947**	**-0.936**	**-0.790**
	Viable count	-0.321	-0.359	0.110	-0.277

[TABLE 1] Effect of mixed microbial populations (MMP) on percentage weight loss with Psathas limestone at the end of different treatments (high correlations in bold)

INOCULUM	BIOFILM INDICATOR	CORRELATION VALUES FOR DIFFERENT STONE PROPERTIES			
		Weight loss	Porosity	Sound velocity	Surface hardness
	Carbohydrate	**0.643**	**0.585**	-0.080	-0.344
MMP	Protein	-0.145	-0.184	-0.215	-0.415
	FDA hydrolysis	**0.600**	**0.623**	**-0.790**	-0.453

[TABLE 2] Correlation matrix for biofilm indicators and stone properties during treatment with Na_2SO_4 (high correlations in bold)

can attack stones, mainly mechanically via crystallization, hydration, and thermal expansion in pore spaces during changes in relative humidity and temperature (Rodriguez-Navarro and Doehne, 1999). Suspended particles and salts of (a)biotic origin, serve as nucleation centers for mineral encrustation (Gerdes et al., 1994) and significantly increase biofilm density and affect hydraulic properties (Christensen and Characklis, 1990). Their cohesiveness constantly depends on ionic concentration, temperature, interactions between exopolymer molecules, metal adsorption, and enzymatic hydrolysis (Decho, 1994). In addition, salts formed as a result of microbial metabolism (e.g., nitrifying bacteria) may be stored in nearby biofilms (Wilimzig and Bock, 1995). Salts may also affect EPS properties and reduce the viscosity of charged polysaccharides so that chains are shielded against mutual electrostatic repulsion by contraction from an extended stiff form to a smaller and more flexible one (Smidsrød and Haug, 1971; Christensen and Characklis, 1990).

Consequences of Biofilm-Salt Interactions Weathering simulation tests serve to determine the weathering behavior of natural stones. This approach has been applied in many physical weathering studies (Fitzner and Kalde, 1991; Halsey et al., 1998; Wake and Smith, 1998). Similar methods have been adapted to the field of stone biodeterioration, in order to study the biodeterioration ability of certain organisms under conditions simulating the natural environment (Sand and Bock, 1991; Mansch and Bock, 1998). Furthermore, some studies used this approach to enhance the biodeterioration process by physical or chemical factors (Sterflinger and Krumbein, 1997; Papida et al., 2000a, b).

Papida et al. (2000b) described a series of weathering experiments in which two limestones and a dolomite were subjected to combined microbial and physical weathering simulation cycles, in an attempt to assess the contribution of each agent of decay. Stone discs were exposed to water or solutions of sodium chloride or sodium sulfate under different temperature/wet/dry cycling regimes. Before the weathering cycles, half of the discs were inoculated with mixed microbial populations (MMP), originally recovered from decayed building stone. The presence of MMP greatly accelerated the rates of deterioration of stone of all treatments, measured by weight change and alteration of hydraulic properties of stone. A summary of the weight loss effects on Psathas limestone, induced by treatments in the presence and absence of the mixed microbial inoculum (MMP), is presented in TABLE 1. The MMP were capable of causing a substantial weight loss alone (7.7%), which was greater than that found with Na_2SO_4 (4.9%), but a combination of the two agents more than doubled the additive effect of the salt and the bacteria. Stones showed extensive exfoliation and fissure formation. [FIG. 2] In contrast, for Psathas limestone, NaCl caused a small weight gain, with or without MMP.

In general, when MMP were combined with salts, higher salt and biomass accumulation often caused an initial consolidation effect but eventually resulted in increased mechanical tension. This was actually reflected in a number of properties, i.e., porosity, dry density, and weight, that were subjected to more acute changes than with salts alone. The clay content of stones may promote weathering by MMP, but higher porosity may be protective, allowing space for the deposition of biofilm and crystallized salts. It is possible that, due to incomplete evaporation, salt crystals of both hydrated and dehydrated forms were present at the salt/EPS interface, and water movement between microorganisms and salts under different RH conditions contributed to the mechanical pressures. Thus MMP enhanced stone water content, providing salts with water of hydration over an extended period.

Appearance of Laboratory-Induced Biofilms by SEM All treatments involving MMP revealed the presence of bacteria in biofilms under SEM examination of Psathas limestone. [FIG. 3] Where salts caused mechanical rupture of stones, SEM observations showed dehydrated, contracted, and filamentous-like bacterial biofilms on stones, suggesting desiccation and reduced-viscosity biofilms. The particles observed within biofilms, described by other researchers, have been identified as inorganic salts, carbonates of microbial origin or clay minerals, often serving as nucleation centers for mineral crystallization within crusts (Gerdes et al., 1994; Thierry and Sand, 1995; Rodríguez-Navarro et al., 1997). They significantly increase the density and hydraulic properties of EPS biofilms (Daniels, 1980; Geesey and Jang, 1989). The performance of stones greatly depended on their characteristics and the decay agent/s.

In these laboratory experiments, homogeneous bacterial populations within rich biofilms were recovered on stones supporting MMP alone or with water. [FIG. 3B] When salts were applied to stone [FIG. 3C AND D], biofilm formation resembled more the heterogeneous, desiccated, and filamentous-like types observed in situ (cf. FIG. 1A AND C). For Psathas limestone, weathered crystals could be observed among the biofilms with Na_2SO_4 [FIG. 3D] as well as for NaCl, which induced a highly desiccated, heterogeneous biofilm. [FIG. 3C]

Scanning electron microscopic analysis performed on samples of Mansfield dolomite [FIG. 4] also revealed luxuriant biofilms encasing bacterial cells and entangling stone particles with MMP alone or with water. Mixed microbial populations with NaCl resulted in a rather desiccated biofilm structure on the dolomite

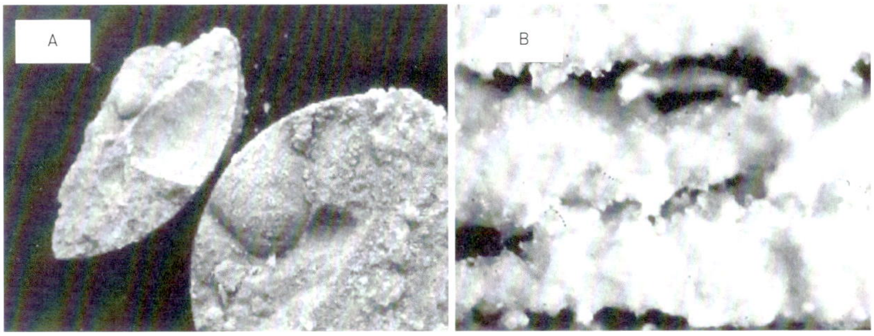

[2] Interaction of bacterial growth and biofilm formation with Na_2SO_4 showing (a) exfoliation and (b) fissure formation at the end of the cycling regime.

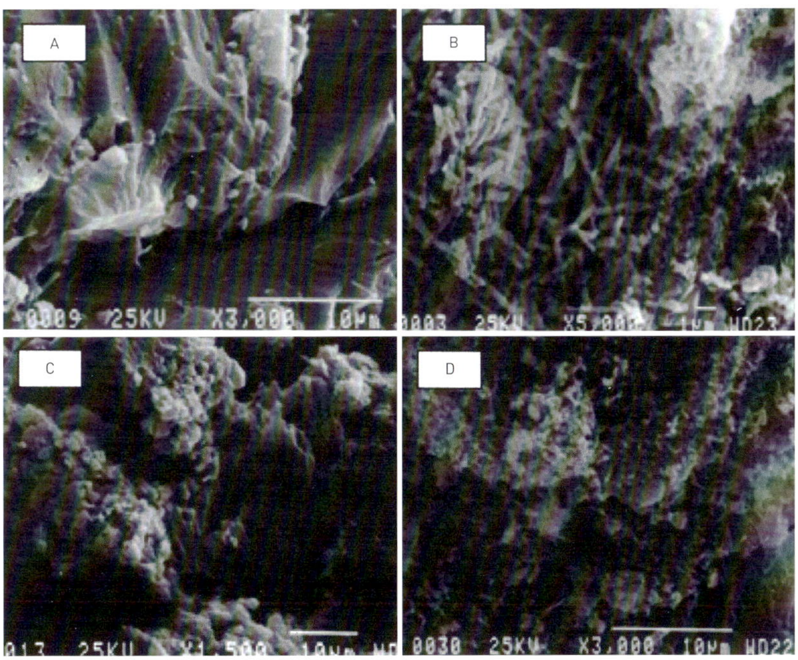

[3] Psathas limestone surfaces under SEM after inoculation with mixed microbial populations and (a) no further treatment (control), (b) water alone, (c) 0.5 M NaCl, and (d) 0.5 M Na_2SO_4, over a 100-cycle regime. Although the control (a) has little visible growth, when water is present (b) a luxuriant biofilm, with clearly defined bacterial rods, is seen. Biofilms formed in the presence of salts (c and d) are discontinuous and flocculant, particularly with NaCl, similar to those observed in situ.

[FIG. 4C] with weathering patterns similar to those attributed to bacterial attack on the limestone. With the most destructive MMP/Na_2SO_4 treatment, the dolomite substrate was seen to be particularly highly weathered [FIG. 4D] and covered by an unknown substance, possibly hydrated or dehydrated Na_2SO_4 crystals or extremely desiccated biofilm.

ACTION OF FILAMENTOUS BACTERIA ON STONE Filamentous microorganisms have been observed on stone surfaces during in-situ studies, and a large range of actinomycetes have been isolated from stone in Egypt at different stages of decay (May et al., 2000). Their presence confirms the findings of Tayler (1991) that they are more common on decayed stone. Although their mode of penetrative growth may represent a threat to stone in buildings, a further series of experiments was done to study the interaction of these different actinomycete species with physical weathering by salts. The actinomycetes used (*Streptomyces* and *Nocardia brasilinsis*) were selected according to a preliminary weight loss screen, as well as because of their tolerance to high salinity. A mixed microbial population (MMP) isolated from deteriorated stones of the Tell Basta tomb in Egypt were also tested in this experiment to compare the effects of pure cultures of actinomycete isolates with MMP on stone under controlled conditions. This biodeterioration experiment aimed to demonstrate the role of actinomycetes in the stone damage under conditions similar to those of the natural environment.

Effects on Stone No detectable weight loss was recorded throughout the experiment in the stones treated with water. In the case of Na_2SO_4, inoculated stones suffered more weight loss than the uninoculated controls. [FIG. 5] After the first 25 cycles, no changes in weight were detected. Thereafter, a gradual increase in weight loss was observed to reach a considerable loss (70–80%) after 100 cycles for both actinomycetes tested. *Streptomyces* was the most aggressive group, causing a weight loss up to 20% more than uninoculated stones that had been treated only with Na_2SO_4. For treatments receiving MMP, only a 7% loss was recorded over the first 50 cycles, but later in the experiment the weight loss reached almost the same level as that with *Streptomyces*. *Nocardia* was more rapid than the other microbes tested but eventually it produced an intermediate weight loss of 10% more than the uninoculated stones.

In contrast to weight loss, little difference in porosity over the first 50 cycles was seen for all treatments with or without microorganisms. [FIG. 6] However, after 50 cycles a remarkable increase in porosity was found in the stones inoculated with

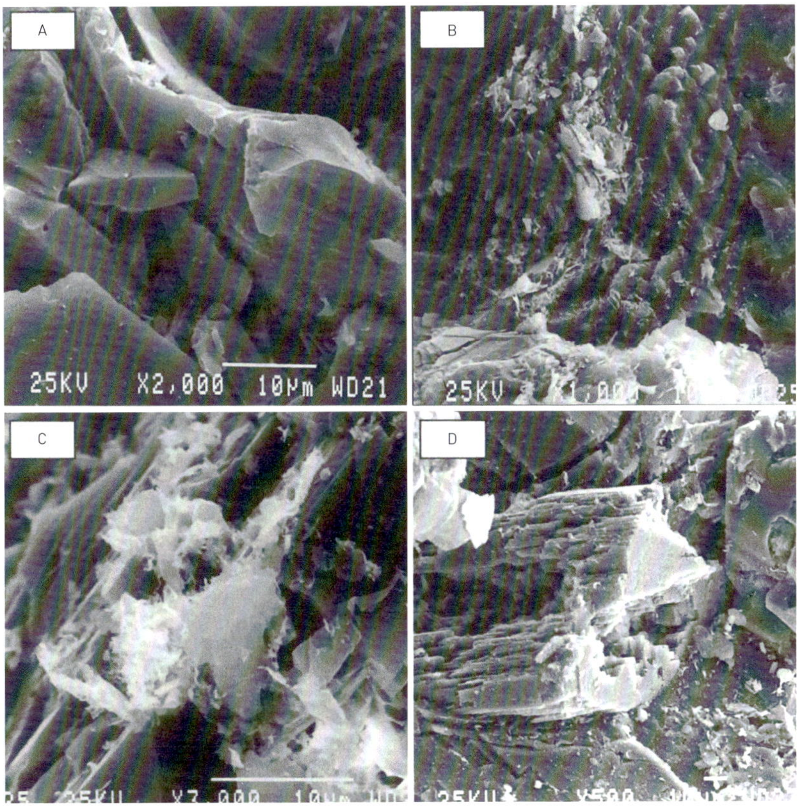

[4] Mansfield dolomite surfaces under SEM after inoculation with mixed microbial populations and (a) no further treatment (control), (b) water alone, (c) 0.5 M NaCl, and (d) 0.5 M Na_2SO_4 over a 100-cycle regime. Extensive, homogenous biofilms occur when water is present (b). NaCl induces obvious contraction of the slime (c) and with Na_2SO_4, there is considerable etching of the mineral surfaces, which are covered in a discontinuous layer of biofilm.

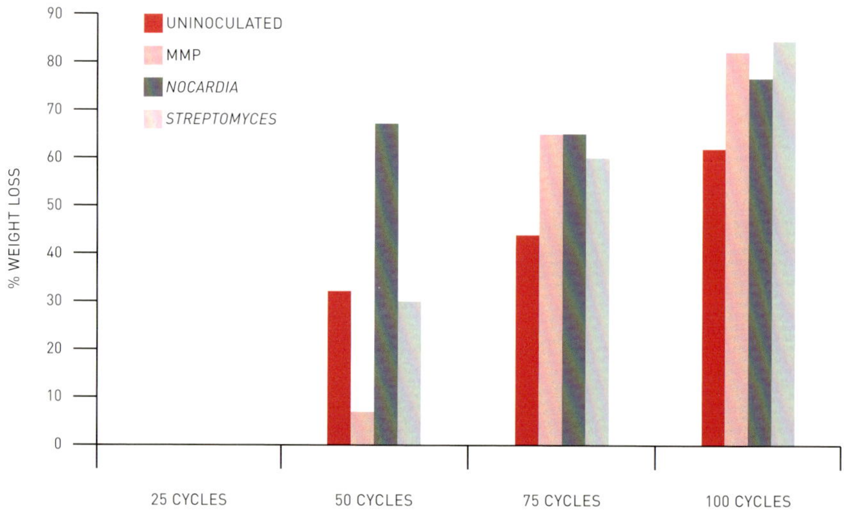

[5] Changes in weight during the cycling regime after treatment with Na_2SO_4 and different microorganisms.

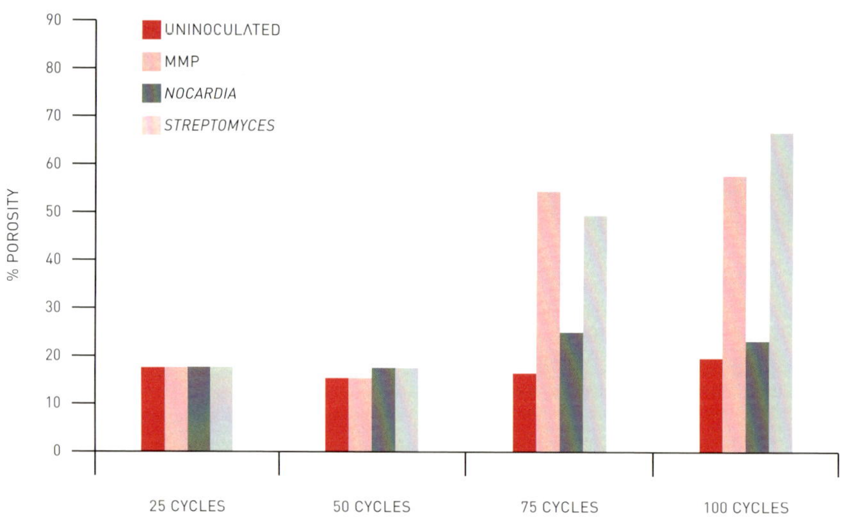

[6] Changes in porosity during the cycling regime after treatment with Na_2SO_4 and different microorganisms.

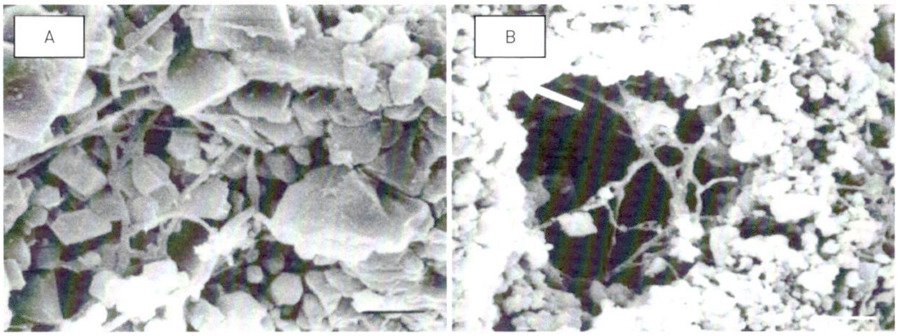

[7] SEM showing actinomycete growth (a) extending hyphae through the limestone structure and (b) bridging pores with hyphal growth and depositing biofilm material around the pore (arrowed). The hyphal filaments are seen to incorporate mineral material as they bridge the pore.

Streptomyces, reaching over 70% after 100 cycles. Mixed microbial populations caused a smaller increase of about 60%. For *Nocardia*, it was much lower, at about 25%, and this was only marginally higher than the uninoculated stones treated with Na_2SO_4 alone, which showed little change in porosity throughout the weathering experiment (16–18%).

Observations on Actinomycete Biofilms Mechanical damage of stones by hyphal penetration of actinomycetes was evident from the SEM analysis. All treatments involving actinomycetes revealed the presence of an extended web of hyphae. [FIGS. 7A AND B] These hyphae were extensively branched and penetrated the stone material, producing patches of biofilm that covered the stone particles. The biofilm accumulated in and around the stone pores. Those stones treated with actinomycetes alone or actinomycetes plus water supported growth for a longer period than those treated with salts, and the presence of hyphae was evident at the end of the experiment. Some hyphae had interacted with salt crystals and covered them with a slime film.

The examination of the biofilm using periodic acid staining (PAS) technique was carried out through different stages of the weathering experiment. The staining intensity increased from light stains in small confined areas in the first stages of the experiment to deep stains in large areas at the end of the 100 cycles. The stones treated with Na_2SO_4 showed a totally stained surface at the end of the experiment.

The mycelial nature of actinomycetes gives them a greater capacity to penetrate the stone if it is friable. This may damage the stone directly as well as indirectly by increasing the surface area of biofilm production, which further enhances the stone damage. The penetration of the substrates by actinomycetes hyphae is related to their ability to excrete a wide range of enzymes (McCarthy and Williams, 1992; Williams, 1985). Tayler (1991) found clear evidence of erosion troughs around margins of actinomycetes colonies attached to stone disc surfaces as observed by SEM. Actinomycetes have also been reported to promote the release of iron from minerals (Kalinowski et al., 2000). Their mycelia were usually mixed with saline crystals in deteriorated paintings (Bassi and Giacobini, 1973).

Correlations of Actinomycete Growth with Stone Properties for Na_2SO_4 The high damage potential of sodium sulfate is a well-documented observation that has been discussed by many authors (Evans, 1970; Cooke, 1979; Robinson and Williams, 1982, Goudie et al., 1997). This potential for damage has led to the widespread use of this salt for testing the durability of building materials (Rodriguez-Navarro and Doehne, 1999). The destructive effect of Na_2SO_4 is mainly based on its high volume expansion when it is crystallized to $Na_2SO_4 \cdot 10H_2O$ during hydration.

During a weathering simulation with Na_2SO_4, actinomycetes and MMP biofilm production was measured in terms of total carbohydrate and protein production, population activity was assessed by the capacity to hydrolyze FDA, and population size was also estimated. The correlation matrixes for these factors and stone properties (namely weight loss, porosity, sound velocity, and surface hardness) are shown in Table 2 for MMP and *Streptomyces sp.* [TABLE 2]

The high weight loss recorded in stones inoculated with either MMP or actinomycetes and treated with Na_2SO_4 was in almost all cases correlated positively with microbial activity measured as FDA hydrolysis. The relationship was especially strong for the *Streptomyces sp.* For MMP, there were strong correlations for carbohydrate content and viable count, which did not occur with the actinomycetes. This may reflect the different types of growth present (and the difficulty of accurately assessing actinomycete growth using viable count procedures). For both, sound velocity showed strong negative correlation with FDA hydrolysis. The analysis of variance of these data showed a high significance difference ($P<0.05$) between the sound velocity measurements of stones inoculated with either actinomycetes or MMP alone and uninoculated stone during the weathering experiment. The viable count for MMP was also strongly correlated with a reduction in surface hardness. However, the difference between presence and absence of microorganisms on stone for both MMP and *Streptomyces sp.* was found to be significant in terms of surface hardness ($P = 0.006$ and 0.01, respectively). The results thus suggest that changes in stone with *Streptomyces* are strongly linked to its growth (measured as FDA activity), and perhaps to the type of growth, while for MMP, changes appear to be linked to both growth *and* biofilm production.

The strong negative correlation observed between FDA activity and sound transmission velocity for both actinomycetes and MMP on stone, indicates the impact of microbial growth on stone integrity. This correlation may be due to the consolidating impact of biofilm on pores and internal spaces during the experiment, as debris or organic matter fills the pores. The transmission velocity of sound waves through limestone was eventually reduced as continued biofilm accumulation, in association with salt expansion, leads to discontinuities and obstacles for the propagation of the waves.

In these experiments pure cultures of actinomycetes were usually more active than MMP in damaging the stone. This could be attributed to the lack of competition that provided a greater opportunity to maximize metabolic activities and consequently cause more damage to stone. However, there were differences between the genera tested. The capacity of the genus *Streptomyces* to cause more damage to stone than *Nocardia* under these experimental conditions may be due to higher biofilm output by *Streptomyces,* as determined by protein and carbohydrate content. Another possible factor is the nonfragmented nature of hyphae in *Streptomyces,* which may give them more strength to penetrate the stone and also increase the surface area for biofilm production. In contrast, the hyphae of *Nocardia* are fragmented.

This investigation shows that the presence of actinomycetes or MMP greatly accelerated the deterioration caused by salts to the limestone, which resulted eventually in the highest material loss. This confirms the findings reported by Papida et al. (2000b) for MMP. The high losses in stone material were in almost every case correlated positively with microbial activity on the stone, but this was especially so in the stones inoculated with *Streptomyces.* This recent work thus suggests an active role for actinomycetes in stone deterioration. Weak negative correlations between actinomycete protein and carbohydrate production and stone porosity may indicate the role of extracellular polymeric substances (EPS) in blocking or reducing pore size, and this may accelerate the rate of stone destruction by salt crystallization. Stones with a high proportion of micropores were reported to be more susceptible to salt decay than those with high proportions of large pores (Rodriguez-Navarro and Doehne, 1999). The mechanical stress induced by biofilm EPS may result from the shrinking and swelling of the colloidal biogenic slimes inside the pore system and even within the mineral grains. The effective alterations in pore size distribution may cause changes in the circulation of moisture and further enhance the chemical dissolution of the stone (Koestler et al., 1985; Viles et al., 1997) and accelerate mineral loss (Barker and Banfield, 1996).

CONCLUSIONS Scanning electron microscopy confirmed the presence of bacteria and biofilm on all substrates, in situ and in the laboratory. Bacterial populations were not favored on substrates of a particular mineralogy in situ or in the laboratory. Although factors other than grain and porosity size control microbial populations in situ, SEM observations indicated a possible preference of microorganisms for extracellular polymeric substance (EPS) production in protected pore niches. Furthermore, interconnected pores may allow groundwater and bacteria to be transported through the stone matrix, although biofilm production within stones may eventually prevent any further penetration. While EPS production is not a selective advantage in the laboratory, total carbohydrate levels in stone discs treated experimentally with MMP were similar to or higher than in-situ levels, a phenomenon possibly encouraged by the environmental conditions provided. Weathering experiments also indicate that actinomycetes, in particular the genus *Streptomyces,* could cause severe damage to stone. This damage affects stone properties in various ways, causing material loss, reduction in surface hardness, or increase in stone porosity, all of which combine to produce a reduction in the velocity of sound transmission through the stone. This biodeterioration could be attributed to biofilm production, which may enhance the effects of salt crystallization pressure in the pore system of the stone. The extent of FDA hydrolysis, which is an index of growth, was strongly correlated to changes in stone properties. The additional mechanical effects of actinomycete hyphal penetration on the stone were clearly evident from the SEM analysis. While no correlation between bacterial populations and total carbohydrate levels has previously been noted in situ, in the laboratory bacterial populations were correlated positively to damage, especially under water stress conditions, with or without salts. The SEM observations show that, in these laboratory studies, salts interact with biofilm and produce patterns of growth and damage similar to those found in situ.

REFERENCES

BARKER, W.W., and BANFIELD, J.F., 1996. Biologically versus inorganically mediated weathering reactions: relationships between minerals and extracellular microbial polymers in lithobiontic communities. *Chemical Geology*, 132, 55–69.

BASSI, M., and GIACOBINI, C., 1973. Scanning electron microscopy: a new technique in the study of works of art. *International Biodeterioration and Biodegradation*, 9, 57–68.

BOIVIN, J., and COSTERTON, J.W., 1991. Biofilms and biodeterioration. In: Rossmoore, H.W., (Ed.), *Biodeterioration and Biodegradation*. Elsevier Applied Science, New York, pp. 53–61.

CHRISTENSEN, B.E., and CHARACKLIS, W.G., 1990. Physical and chemical properties of biofilms. In: Characklis, W.G., and Marshall, K.C., (Eds.), *Biofilms*. Wiley-Interscience, pp. 93–130.

COOKE, R.U., 1979. Laboratory simulation of salt weathering processes in arid environments. *Earth Surface Processes*, 4, 347–359.

DANIELS, S.L., 1980. Mechanisms involved in sorption of microorganisms to solid surfaces. In: Bitton, G., and Marshall, K.C., (Eds.), *Adsorption of Microorganisms to Surfaces*. Wiley-Interscience, pp. 7–58.

DECHO, A.W., 1994. Molecular-scale events influencing the macroscale cohesiveness of exopolymers. In: Krumbein, W.E., Peterson, D.M., and Stahl, L.J. (Eds.), *Biostabilisation of Minerals*. Verlag, pp. 138–148.

DENYER, S.P., HANLON, G.W., and DAVIES, M.C., 1993. Mechanisms of microbial adherence. In: Denyer, S.P., Gorman, J., and Sussman M., (Eds.), *Microbial Films: Formation and Control*. Blackwell Scientific Publications, Oxford, pp. 13–28.

DUBOIS, M., GILLES, K.A., HAMILTON, J.K., REBERS, P.A., and SMITH, F., 1956. Colorimetric method for determination of sugars and related substances. *Analytical Chemistry*, 28, 350–6.

EVANS, I.S., 1970. Salt crystallisation and rock weathering: A review. *Revue de Geomorphologie Dynamique*, 19, 153–177.

FITZNER, B., and KALDE, M., 1991. Simulation of frost-thaw cycle and salt-weathering nature adapted material tests. In: Auger, F., (Ed.), *Proceedings of the International Symposium on the Deterioration of Building Materials*, La Rochelle, France, pp. 103–114.

FRANKLIN, J.A., VOGLER, U.W., SZLAVIN, J., EDMOND, J.M., and BIENIAWSKI, Z.T., 1981. Suggested methods for determining water content, porosity, density, absorption and related properties and swelling and slake durability index properties. In: Brown E.T., (Ed.), *Rock characterisation testing and monitoring. The Commission on Testing Methods*, International Society for Rock Mechanics. Pergamon Press, pp. 71–94.

GEESEY, G. G., and JANG, L., 1989. Interactions between metal ions and capsular polymers. In: *Metal Ions and Bacteria*. pp. 325–357.

GERDES, G., DUNAJTSCIK-PIEWAK, K., RIEGE, H., TAHER, A.G., KRUMBEIN, W.E., and REINECK, H.E., 1994. Structural diversity of biogenic carbonate particles in microbial mats. *Sedimentology*, 41, 1273–1294.

GOUDIE, A.S., VILES, H.A., and PARKER, A.G., 1997. Monitoring of rapid salt weathering in the central Namib desert using limestone blocks. *Journal of Arid Environments*, 37, 581–598.

GUILLLETTE, O., 1995. Bioreceptivity: a new concept for building ecology studies. The Science of the Total Environment, 167, 215–220.

HALSEY, D.P., MITCHELL, D.J., and DEWS, S.J., 1998. Influence of climatically induced cycles in physical weathering. *Quarterly Journal of Engineering Geology*, 31, 359–367.

KALINOWSKI, B.E., LIERMANN, L.J., GIVENS, S., and BRANTHLY, S.L., 2000. Rates of bacteria-promoted solubilisation of Fe from minerals: A review of problems and approaches. *Chemical Geology*, 169, 357–370.

KOESTLER, R.J., CHAROLA, A.E., WYPYSKI, M., and LEE, J.J., 1985. Microbiologically induced deterioraton of dolomitic and calcitic stone as viewed by scanning electron microscopy. In: Furlan, V., (Ed.), *Proceedings of the 5th International Congress on Deterioration and Conservation of Stone*, September 1985, Lausanne. Presses Polytechniques Romandes, Lausanne, pp. 617–626.

MANSCH, R., and BOCK, E., 1998. Biodeterioration of natural stone with respect to nitrifying bacteria. *Biodegradation* 9, 47–64.

MAY, E., PAPIDA, S., ABDULLA, H., TAYLER, S., and DEWEDAR, A., 2000. Comparative studies of bacterial populations on stone in temperate and semi-arid climates. In: Ciferri, O, Tiano, P., and Mastromei, G., (Eds.), *Of Microbes and Art, the role of microbial communities in the degradation and protection of cultural heritage*. Kluwer/Plenum, New York, pp. 49–62.

MCCARTHY, A.J., and WILLIAMS, S.T., 1992. Actinomycetes as agents of biodeterioration in the environment- a review. *Gene*, 115, 189–192.

PACEPAVICIUS, G., LAU, Y.L., LIU, D., OKAMURA, H., and AOYAMA, I., 1997. A rapid biochemical method for estimating biofilm mass. *Environmental Toxicology and Water Quality Journal*, 12, 97–100.

PAPIDA, S., MURPHY, W., and MAY E., 2000a. The use of sound velocity determination for the non-destructive estimation of physical and microbial weathering of limestones and dolomites. In: Fassina, V. (Ed.), *Proceedings of the 9th International Congress on Deterioration and Conservation of Stone*. Elsevier, Amsterdam, pp. 609–617.

PAPIDA, S., MURPHY, W., and MAY E., 2000b. Enhancement of physical weathering of building stones by microbial populations. *International Biodeterioration and Biodegradation*, 46, 305–317.

ROBINSON, D.A., and WILLIAMS, B.G., 1982. Salt weathering of rock specimens of varying shape. *Area*, 14, 293–299.

RODRÍGUEZ-NAVARRO, C., SEBASTIÁN, E., and RODRÍGUEZ-GALLEGO, M., 1997. An urban model for dolomite precipitation: authigenic dolomite on weathered building stones. *Sedimentary Geology*, 109, 1–11.

RODRÍGUEZ-NAVARRO, C., and DOEHNE, E., 1999. Salt weathering: Influence of evaporation rate, supersaturation and crystallisation pattern. *Earth Surface Processes and Landforms*, 24, 191–209.

RUMMEL, F., and VAN HEERDEN, W.L., 1981. Suggested methods for determining sound velocity. In: Brown E.T., (Ed.), *Rock Characterisation Testing and Monitoring*. Published for the Commission on Testing Methods, International Society for Rock Mechanics, Pergamon Press, pp. 105–110.

SAND, W., 1997. Microbial mechanisms of deterioration of inorganic substrates—a general mechanistic overview. *International Biodeterioration and Biodegradation*, 40, 183–190.

SAND W., and BOCK, E., 1991. Biodeterioration of mineral materials by microorganisms —biogenic sulfuric and nitric acid corrosion of concrete and natural stone. *Geomicrobiology Journal*, 9, 129–138.

SMIDSRØD, O., and HAUG, A., 1971. Estimation of the relative stiffness of the molecular chain in polyelectrolytes from measurements of viscosity at different ionic strengths. *Biopolymers*, 10, 1213–1227.

STERFLINGER, K., and KRUMBEIN, W.E., 1997. Dematiaceous fungi as a major agent for biopitting on Mediterranean marbles and limestones. *Geomicrobiology Journal*, 14, 219–230.

TAYLER, S., 1991. Composition and activity of bacterial populations found on decaying stonework. Ph.D. Thesis, University of Portsmouth.

TAYLER, S., and MAY, E., 1991. The seasonality of heterotrophic bacteria on sandstone from ancient monuments. *International Biodeterioration*, 28, 49–64.

THIERRY, D., and SAND, W., 1995. Microbially Induced Corrosion. In: Marcus, P., and Oudar, J., (Eds.), *Corrosion Mechanisms in Theory and Practice*. Series in *Corrosion Technology*, 8, 457–498. Marcel Dekker Inc.

URZI, C., LISI, S., CRISEO, G., and PERNICE, A., 1991. Adhesion to and degradation of marble by a *Micrococcus* strain isolated from it. *Geomicrobiology Journal*, 9, 81–90.

VILES, H.A., CAMUFFO, D., FITZ, S., FITZNER, B., LINGQUIST, O., LIVINGSTON, P., MARAVELAKI, N.V., SABBIONI, C., and WARSCHEID, T., 1997. What is our knowledge of the mechanisms of deterioration and how good are our estimates of rates of deterioration? In: Baer, N.S., and Snethlage, R., (Eds.), *Saving Our Architectural Heritage: The Conservation of Historic Stone Structures*. John Wiley and Sons, New York, pp. 95–112.

WARKE, P.A., and SMITH, B.J., 1998. Effects of direct and indirect heating on the validity of rock weathering simulation studies and durability tests. *Geomorphology*, 22, 347–357.

WARSCHEID, T., BECKER, T.W., BRAAMS, J., GEHRMANN, C., KRUMBEIN, W.E., PETERSEN, K., and BRUGGERHOFF, S., 1993. Studies on the temporal development of microbial infection of different types of sedimentary rocks and its effect on the alteration of the physico-chemical properties in building materials. In: Thiel, M.-J., (Ed.), *Conservation of Stone and Other Materials*. RILEM, vol. 1, pp. 303–310.

WILIMZIG, M., and BOCK, E., 1995. Endangerment of mortars by nitrifiers, heterotrophic bacteria and fungi. In: Bousher, A., Chandra, M., and Edyvean, R., (Eds), *Proceedings of the 9th International Biodeterioration and Biodegradation Symposium*, September 1993. University of Leeds, UK, pp. 195–198.

WILLIAMS, S.T., 1985. Streptomycetes in biodeterioration, their relevance, detection and identification. *International Biodeterioration*, 21, 201–209.

C. BARABESI

E. TAMBURINI

G. MASTROMEI*

B. PERITO

MECHANISMS OF MICROBIAL CALCIUM CARBONATE PRECIPITATION

University of Florence, Florence, Italy

* Corresponding author

ABSTRACT Calcium carbonate crystal deposition is a very diffuse phenomenon that has been described for many microorganisms living in different environments. Nevertheless, the actual molecular mechanisms by which bacteria foster calcium carbonate precipitation are not known. A better understanding at the molecular and genetic levels could improve our knowledge of the process and be useful for biomineralization applications such as conservation of calcareous monumental stones. In this work we report preliminary data obtained with two experimental approaches. The genetic approach, based on the analysis of *Bacillus subtilis* mutants impaired in calcium carbonate crystal production, suggested a link between calcium precipitation and intermediate(s) of fatty acids metabolism. To investigate the role of cell structures we developed an in-vitro test. In this test dead bacterial cells induce calcium carbonate precipitation; moreover, cells from active calcinogenic strains showed a much higher and/or faster production of $CaCO_3$ crystals than did less active strains.

INTRODUCTION Biomineralization is the process through which organisms are involved in mineral formation, as a result of cellular activity that fosters the necessary physical and chemical conditions for such a formation and growth to take place

(Ben Omar et al., 1997). Organisms able to precipitate minerals are present in all major groups, from bacteria to Chordata. However, the way in which these biominerals are formed can be very different among the different taxonomic groups. The majority of the biominerals described contain calcium as their main cation and, in terms of their anions, carbonates are among the most abundant forms.

From an evolutionary point of view it can be stated, without any doubt, that biomineralization processes originated from bacterial activity (Ben Omar et al., 1997). The formation of minerals by microorganisms is a very diffuse phenomenon and their geochemical activity is responsible to a great extent for the deposition of minerals throughout the history of Earth (Lowestam, 1981).

Calcium carbonate precipitation by different bacteria is an example of the important geologic role played by bacteria, a concept that has been discussed since the beginning of the past century. In 1903, Nadson showed that calcium carbonate deposits in Lake Veisowe in Karkou, Russia, could have originated from bacterial activity (Ben Omar et al., 1997). In 1914, Drew called attention to the role of denitrifying bacteria in calcium carbonate deposition in Great Bahama sediments, the most significant calcium carbonate deposits in the world (Drew, 1914). This microbial capability has been described in different environments, such as soils, and freshwater and saline habitats; this capability has been related to the formation of marine calcareous skeletons, carbonate sediments, and soil carbonate deposits (Shinano, 1972; Chafetz, 1986; Vasconcelos et al., 1995). It is worth noting that some carbonate rocks, such as travertines and speleothemes, could be bacterially constructed (Chafetz et al., 1991; Folk, 1993). Other studies demonstrated that calcium carbonate precipitation was a major biogeochemical process not restricted to the activity of any specific group of bacteria (Boquet et al., 1973). Different mechanisms were proposed to explain the involvement of bacteria in carbonate production (Ehrlich, 1995). Studies in this field have pointed out the complexity of this phenomenon, one that can be influenced by the environmental physicochemical conditions and that is correlated with both metabolic activity and cell surface structures (Beveridge, 1989; Fortin et al., 1997).

The metabolic pathways involved in bacterial carbonate formation include both autotrophic and heterotrophic pathways (Castanier et al., 2000). In many cases, carbonate crystals seem to start building up on the bacterial cell wall, suggesting that cells act by themselves as heterogeneous crystallization nuclei. In a few steps a kind of cocoon is formed and, subsequently, biomineral assemblages appears to turn into true crystals, either well shaped or poorly organized, which can enclose bacterial cells within the mineral structure (Castanier et al., 1990). The type of crystals formed (vaterite, aragonite, or calcite) depends on both its growing features and the bacterial strain; different bacteria precipitate different types of calcium carbonate, and the most common crystalline forms are either spherical or polyhedric.

In conclusion, calcium carbonate crystal deposition is a very diffuse phenomenon described for many microorganisms belonging to different taxonomic groups and living in different environments. Nevertheless, the actual molecular mechanisms by which bacteria foster calcium carbonate precipitation are not known. In particular, it is not known if the bacterial capability to induce $CaCO_3$ precipitation and mineralization is due to a general capability of the cell wall and/or other external layers (i.e., a diffuse negative charge of functional groups on cell external layers) or if it could be due to specific molecular or supermolecular structures. A better understanding at the molecular and genetic levels could improve our knowledge of the process and be useful for biomineralization applications such as conservation of calcareous monumental stones.

Our aim is to develop a biological system, one that is bacteria-mediated and able to induce $CaCO_3$ mineral formation in the absence of living cells. We are following two approaches:

1. Identification of bacterial genes and products (molecules) involved in the biomineralization process.
2. Identification of bacterial cell structure(s) inducing $CaCO_3$ precipitation.

MATERIALS AND METHODS

STRAINS Test organisms were: *Bacillus subtilis* 168; *Escherichia coli* XL1-Blue; *Bacillus cereus* kindly given us by G. Orial (LRMH-CPP, Champs sur Marne, France); and *Bacillus subtilis* mutant strain BFA 2427 (Perito et al., 2000).

MEDIA AND GROWTH CONDITIONS The media and growth conditions were as follows: B4 medium (0.4% yeast-extract, 0.5% dextrose, 0.25% calcium acetate); LB medium (1% bacto-triptone, 0.5% yeast-extract, 1% NaCl) supplemented with 0.25% calcium acetate; BC medium (0.6% peptone, 0.6% yeast-extract, 0.22% $CaCl_2$, 0.2% KNO_3, 0.7% NaCl, pH 7,7). Media were solidified with 1.5% agar.

All strains were grown at 37°c. For $CaCO_3$ in vitro testing, bacterial cells were grown in the appropriate culture medium and collected during exponential growth phase. Erythromycin 0.3 g/ml was added to *B. subtilis* mutant BFA2427 cultures.

PREPARATION OF KILLED AND DISRUPTED CELLS

Killing by Heat Cells were autoclaved at 121°c for 30 minutes, and then collected by centrifugation (15 minutes at 5000 rpm). The pellet was resuspended in physiological solution (8.5 g l^{-1} NaCl) to a final concentration of 10^8 cells ml^{-1}. Cell suspension was used as described below, and stored at −20°c.

Disruption by Ultrasound Cells were harvested by centrifugation (15 minutes at 5000 rpm). The pellet was resuspended in physiological solution to a final concentration of 10^8 cells ml^{-1}. Cells were treated with 2 mg ml^{-1} of lysosyme for 30 minutes at 37°c, and protoplast formation was observed with an optical microscope. Cell suspension was broken in an ultrasonic processor at 50 W, with six pulses of 20 seconds. The sonicated suspension was used as described below and stored at −20°c.

$CaCO_3$ PRECIPITATION TEST

Test A This assay was described by González-Muñoz et al. (1996). Liquid salt solution (0.1% $CaCl_2$, 0.4% calcium acetate, 0.3% $NaHCO_3$) or solid (1.5% agar added) were used. The medium was poured into a flask or a petri dish and "seeded" with the sample, incubated at room temperature, and observed with a stereomicroscope for several days.

Test B This assay was described by Addadi and Weiner (1985). A solution of 7.5 mM $CaCl_2$ (Sol. A) was prepared. For each sample serial dilutions were prepared in physiological solution starting from the undiluted one (corresponding to a concentration of 10^8 cells ml^{-1}). One milliliter of each dilution was added to 9 ml of Sol. A in a petri dish (about 50 mm in diameter). Crystallization was induced by slow diffusion of ammonium carbonate vapors in a closed desiccator. Incubation was at room temperature and samples were observed with a stereomicroscope for several days.

Test C A solution of 7.5 mM $CaCl_2$, 35.7 mM $NaHCO_3$ (Sol. A) was filtered through a millipore filter (0.45 μm). Samples were prepared as described for Test B. Incubation was in a closed desiccator (without ammonium carbonate efflux) and at room temperature. Observations of $CaCO_3$ crystal formation were made with a stereomicroscope.

CRYSTALS ANALYSIS Crystals produced on solid growth media were collected as described by Tiano et al. (1999). Precipitates from both liquid growth media and salt solution (test A) were collected by filtration through a millipore filter (0.45 μm). Filters were dried at 60°c overnight.

Precipitates obtained by Tests B and C were dried directly in the petri dishes at 60°c overnight. Precipitates were analyzed by spectrophotometry, FTIR, and X-ray diffraction.

RESULTS

IDENTIFICATION OF BACTERIAL GENES INVOLVED IN THE BIOMINERALIZATION PROCESS

The calcinogenic laboratory strain *Bacillus subtilis* 168, whose genome has been completely sequenced, was used. A previous screening of 1190 mutants of *B. subtilis* 168, obtained from the European *B. subtilis* genome project, allowed us to identify six strains that do not form calcite crystals on solid B4 medium. A preliminary analysis of the mutated gene sequences revealed that, in many cases, the putative function of their products is linked to the fatty acid metabolism (Perito et al. 2000).

This analysis suggested that we should focus our attention on two genes, *ysiA* and *ysiB*, located very close to each other on the *B. subtilis* chromosome and belonging to a putative transcriptional cluster of five genes. We are evaluating the possibility that this cluster is involved in the synthesis of an intermediate of fatty acid metabolism, one that is directly or indirectly involved in the precipitation process. To verify this hypothesis we are planning the following experiments in the near future:

- Production, by insertional mutagenesis, of *B. subtilis* mutants, each carrying an inactivated gene of the *ysiB* cluster;
- Testing of the mutants for their $CaCO_3$ precipitation phenotype; and
- Characterization of the mutants to detect the enzymatic activity coded by the studied genes.

IDENTIFICATION OF BACTERIAL CELL STRUCTURE(S) INDUCING $CaCO_3$ PRECIPITATION

Bacillus subtilis 168 and a very active calcinogenic *B. cereus* strain were used.

$CaCO_3$ Crystal Formation by Bacterial Cells The calcinogenic activity of the bacterial strains were tested on solid medium. *Bacillus cereus* and *B. subtilis* 168 colonies produced crystals visible under the stereomicroscope after 5–7 days of incubation on B4 solid medium. *Escherichia coli* XL1-Blue did not produce visible crystals even after 20 days of incubation. *Bacillus subtilis* 168 produced precipitates visible to the

naked eye after 10 days of incubation in both B4 and LB (supplemented with calcium acetate) liquid media. *B. cereus* produced precipitates visible to the naked eye after 18 days of incubation in BC liquid medium.

Crystals produced on each medium were analyzed by FTIR spectrophotometry and X-ray diffraction. As shown in Table 1, precipitates produced by all the strains on a solid medium were formed by calcite. Precipitates produced in liquid medium were formed by calcite for *B. subtilis* 168 and by calcite and vaterite for *B. cereus*. [TABLE 1]

$CaCO_3$ In-Vitro Precipitation Test To identify bacterial cell structure(s) inducing calcite precipitation, we tested whole dead cells from calcinogenic strains as potential crystallization nuclei. We devoted time to working out a suitable in-vitro precipitation test, one that would be simple, inexpensive, and performable under environmental conditions, keeping in mind the intended goals.

The first assay we used (Test A in Materials and Methods) was described by González-Muñoz et al. (1996) to test bacterial membrane fractions. We tested, on the mineral solid medium of Test A, *B. subtilis* 168 sonicated cells collected both during exponential and stationary growth phases. These samples gave negative results.

Living cells from cultures of *B. subtilis* 168 in B4 liquid medium and *B. cereus* in BC liquid medium were inoculated separately in 25 ml of the salt solution of Test A and tested for precipitate production. After 60 days of incubation at 37°c, precipitates were harvested and analyzed by FTIR spectrophotometry and X-ray diffraction. Precipitates from *B. cereus* cells were formed by calcite and vaterite, while no $CaCO_3$ was detected in precipitates from *B. subtilis* cells. [TABLE 1] Therefore, our work showed that Test A detected calcinogenic activity only with living *B. cereus* cells inoculated into the liquid salt solution.

We also used the assay described by Addadi and Weiner (1985) to test the calcinogenic activity of molecules (Test B in Materials and Methods). Test B was used with the following samples: *B. cereus* exponential phase sonicated cells, *B. cereus* exponential phase autoclaved cells, *B. subtilis* 168 exponential phase sonicated cells, and *B. subtilis* 168 exponential phase autoclaved cells. Results are reported in TABLE 2. All samples were positive, with crystal deposition increasing with time. No visible changes occurred after the seventh day of observation.

The salt solution used as a blank formed precipitates, but with the following features: precipitates were visible after one day of incubation but in a lower quantity with respect to those produced by the samples above; after the second day they

SAMPLE	FTIR[1]	X-RAY[2]
B4 solid medium	No $CaCO_3$	
B. cereus B4 solid medium	Calcite	Calcite
B. subtilis 168 B4 solid medium	Calcite	Calcite
BC liquid medium	No $CaCO_3$	$CaCl_2 \cdot 2H_2O$
B. cereus BC liquid medium	$CaCO_3$	Calcite+Vaterite
LB+CaAc[(3)] liquid medium	No $CaCO_3$	–
B. subtilis 168 LB+CaAc[(3)] liquid medium	$CaCO_3$	Calcite
B4 liquid medium	No $CaCO_3$	–
B. subtilis 168 B4 liquid medium	$CaCO_3$	Calcite
B. subtilis 168 in salt solution (test A)	No $CaCO_3$	No $CaCO_3$
B. cereus in salt solution (test A)	$CaCO_3$	Calcite+Vaterite

[1] FTIR spectrophotometry [2] X-ray diffraction [3] Calcium acetate

[TABLE 1] Precipitates produced either on solid or liquid media

	OBSERVATION AFTER DAYS				
SAMPLE	0	1	2	3	7
Blank	–	+	++	++	++
B. cereus sonicated	–	+++	++++	++++++	++++++++
B. cereus autoclaved	–	+++	++++	++++++	+++++++
B. subtilis 168 sonicated	–	+++	++++	++++++	+++++++
B. subtilis 168 autoclaved	–	+++	++++	++++++	+++++++

Blank = no addition.

[TABLE 2] Precipitates obtained from dead bacterial cells harvested during exponential growth phase, Test B

	OBSERVATION AFTER DAYS				
SAMPLE	0	1	2	3	7
Blank	–	+	++	++	++
B. subtilis 168	–	+++	++++	++++++	+++++++
BFA2427	–	+	+	+++	++++
E. coli XL1-Blue	–	+	+	+++	+++

Blank = no addition.

[TABLE 3] Precipitates obtained from autoclaved bacterial cells harvested during exponential growth phase, Test B

	OBSERVATION AFTER DAYS				
SAMPLE	0	1	2	3	7
Blank	–	+	++	++	++
B. subtilis 168	–	+++	++++	++++++	+++++++
B. subtilis 168 1:2	–	+++	++++	++++++	+++++++
B. subtilis 168 1:4	–	+++	+++	++++++	+++++++
B. subtilis 168 1:8	–	++	+++	++++	+++++
B. subtilis 168 1:16	–	+	++	+++	+++

Blank = no addition.

[TABLE 4] Precipitates obtained from serial dilutions of autoclaved cells harvested during exponential growth phase, Test B

SAMPLE	OBSERVATION AFTER DAYS 0	1	3	7
Blank	–	+	+	+
B. subtilis 168	–	+	++	++++
B. subtilis 168 1:2	–	+	++	++++
B. subtilis 168 1:4	–	+	+++	+++++
B. subtilis 168 1:8	–	+	+++	+++
B. subtilis 168 1:16	–	+	+++	+++
B. cereus	–	+	++	+++++
B. cereus 1:2	–	+	++	+++++
B. cereus 1:4	–	+	++	+++++
B. cereus 1:8	–	+	+	+++
B. cereus 1:16	–	+	+	+++
E. coli XL1-Blue	–	+	+	++
E. coli XL1-Blue 1:2	–	+	+	++
E. coli XL1-Blue 1:4	–	+	+	++
E. coli XL1-Blue 1:8	–	+	+	+
E. coli XL1-Blue 1:16	–	+	+	+
BFA2427	–	+	++	+++
BFA2427 1:2	–	+	+++	++++
BFA2427 1:4	–	+	+++	+++
BFA2427 1:8	–	+	++	++
BFA2427 1:16	–	+	+	+

Blank = no addition.

[TABLE 5] Precipitates obtained from serial dilutions of autoclaved bacterial cells harvested during exponential growth phase, Test C did not increase. Moreover, precipitates were diffused while they were predominantly concentrated in the middle of the petri dish with active samples. Such behavior was seen in all the experiments; for this reason samples showing a precipitate comparable to that of the blank were considered negative, and samples showing a larger precipitate, positive.

To confirm that precipitate formation was not aspecific, we tested autoclaved cells of exponentially growing strains with either little activity (*B. subtilis* mutant strain BFA2427) or no activity (*E. coli* XL1-Blue) in crystal production on B4 medium. Results are shown in TABLE 3. In both cases, cells were less active than *B. subtilis* 168 and the kinetics of precipitation were slower.

To verify the dependence of precipitation on cell concentration and to better compare the calcinogenic activity of different samples, serial dilutions of dead *B. subtilis* 168 cells were tested. The results, reported in Table 4, showed that precipitate production progressively decreased at 1:8 and 1:16 dilutions. [TABLE 4]

To eliminate ammonium vapors, used in Test B, we substituted ammonium carbonate with $NaHCO_3$ (Test C in Materials and Methods).

Samples used in Test B were also used in Test C. Moreover, for each sample, dilutions up to 1:16 were tested. Results are shown in TABLE 5. For data interpretation and comparison between Tests B and C, three aspects should be considered:

1. In both tests, a diffused precipitate formed in the blank salt solution, with no increase after one or two days of incubation.

2. In Tests B and C crystals produced by active samples were predominantly concentrated in the middle of the petri dish and increased with time, even if with slower kinetics than in Test B.

3. In both tests, no visible changes in crystal deposition after the seventh day of incubation were observed.

For undiluted samples, reported in Table 5, results were comparable to those obtained in Test B [TABLE 3] except for *B. subtilis* BFA2427, which at the third day was as active as *B. subtilis* 168, even if the final yield was lower. For diluted samples, in some cases intermediate dilutions (1:2 and/or 1:4) seem to give a greater yield than the undiluted ones (see *B. subtilis* 168 1:4 and *B. subtilis* BFA2427 1:2). At higher dilutions (1:8 and 1:16) precipitate production decreases in all samples. At 1:16 dilution *B. subtilis* 168 and *B. cereus* were still able to give precipitates, while *B. subtilis* BFA2427 and *E. coli* were not. This dilution was then useful in discriminating more active from less active strains. *Escherichia coli* dead cells showed a similar trend both in Test B and C and was the less active strain.

Crystals produced in Test C by *B. subtilis* 168, by *E. coli* Xl1-Blue, by *B. subtilis* BFA2427, and by the blank salt solution were analyzed by FTIR spectrophotometry and X-ray diffraction. All of them were formed by calcite.

DISCUSSION The aim of our work was to study calcium carbonate deposition to determine whether this phenomenon is induced by specific molecular or supermolecular surface structures.

The genetic approach was based on the analysis of *B. subtilis* mutant strains impaired in calcium carbonate crystal production. We have preliminary indications of a link between calcium precipitation and intermediate(s) of fatty acid metabolism. We are further investigating this hypothesis by characterizing a specific gene cluster.

The approach of investigating the role of cell structures, independently of the cell metabolism, was based on developing an in-vitro test for dead bacterial cells. The first assay (Test A) could not detect calcinogenic activity with dead cells. It was described by González-Muñoz et al. (1996) to assay the calcite crystallization activity of bacterial membrane fractions. Since crystal formation was detected by observation with the optical microscope, our negative results could be explained by the lower magnification of the stereomicroscope used. On the other hand, our aim was to obtain a crystal deposition easily detectable either at low magnification or with the naked eye.

Test B, described by Addadi and Weiner (1985) to assay $CaCO_3$ crystallization activity of molecules, and our modified Test C proved to be useful for our purposes. Furthermore, Test C appears to be easier to perform under environmental conditions. The preliminary results described here were obtained by visual observation of crystal production; therefore, these assays are not yet a quantitative test.

It is important to note that in both tests dead bacterial cells induced calcium carbonate precipitation. Ben Omar et al. (1995) reported struvite crystallization induced by killed cells of *Mixococcus xanthus* on a suitable solid medium and proposed that dead cells or cell debris could act by themselves as heterogeneous crystallization nuclei. We have not found any paper reporting crystallization of calcite induced by killed bacterial cells. The same authors reported struvite and calcite crystallization induced by cellular membranes of *M. xanthus* (González-Muñoz et al., 1996), using the first assay we performed (Test A). Regarding calcite crystals, they were formed only on a solid mineral medium. They discussed the lack of calcite crystallization in liquid medium to be in agreement with the results of Addadi and Weiner (1985), who found that "calcite crystals are nucleated on stereochemically analogous faces when proteins are adsorbed onto a rigid substrate." In our research, killed bacterial cells were able to induce calcite crystallization in liquid media (Tests B and C). Even if submitted to aggressive treatments (autoclave, ultrasound), they would probably retain the rigid surface of the cell wall and could act as heterogeneous crystallization nuclei.

Moreover, we found that dead cells from active calcinogenic strains showed a much higher and/or faster production of $CaCO_3$ crystals than did less active strains. This result suggests that calcinogenic strains might have a cellular structure, resistant to the methods used to kill cells, able to promote $CaCO_3$ precipitation. These preliminary results are very encouraging for our work to identify cellular fractions able to induce calcium carbonate precipitation, and they support the possibility of restoring monumental stones without using living bacterial cells.

ACKNOWLEDGMENTS We want to thank Dr. Susanna Bracci (CNR, Florence, Italy) for spectrophotometer, FTIR, and X-ray diffraction crystal analyses, and Dr. Geneviève Orial (LRMH-CPP, Champs sur Marne, France), who kindly gave us *Bacillus cereus.* This work was supported by EC grant EVK4-CT2000–00037, "BIOREINFORCE."

REFERENCES

ADDADI L., and WEINER S., 1985. Interactions between acidic proteins and crystals: Stereochemical requirements in biomineralization. *Proceedings of the Natational Academy of Science*, USA 82, 4110–4114.

BEN OMAR, N., MARTÍNEZ-CAÑAMERO, M., GONZÁLEZ.-MUÑOZ, M.T., ARIAS, J.M., and HUERTAS, F., 1995. *Mixococcus xanthus* killed cells as inducers of struvite crystallization. Its possible role in the biomineralization processes. *Chemosphere*, 30, 2387–2396.

BEN OMAR, N., ARIAS, J.M., GONZALEZ-MUNOZ, M.T., 1997. Extracellular bacterial mineralization within the context of geomicrobiology. *Microbiologia*, 13, 161–72.

BEVERIDGE, T.J., 1989. Role of cellular design in bacterial metal accumulation and mineralization. *Annual Review of Microbiology*, 43, 147–171.

BOQUET, E., BORONAT, A., and RAMOS-CORMENZANA, A., 1973. Production of calcite (calcium carbonate) crystals by soil bacteria is a general phenomenon. *Nature*, 246, 527–529.

CASTANIER, S., MAURIN, A., and PERTHUISOT, J.P., 1990. A trial to get dolomite in freshwater. *Geobios*, 23, 121–128.

CASTANIER, S., LE MÉTAYER-LEVREL, G., ORIAL, G., LOUBIÈRE, J.F., and PERTHUISOT, J.P., 2000. Bacterial carbonatogenesis and applications to preservation and restoration of historic property. In: Ciferri, O., Tiano, P., and Mastromei, G., (Eds.), *Of Microbes and Art: The Role of Microbial Communities in the Degradation and Protection of Cultural Heritage*, Plenum, New York, pp. 203–218.

CHAFETZ, H.S., 1986. Marine peloids a product of bacterially induced precipitation of calcite. *Journal of Sedimentary Petrology*, 56, 812–817.

CHAFETZ, H.S., RUSH, P., and UTECH, N.M., 1991. Microenvironmental controls on mineralogy and habitat of $CaCO_3$ precipitates: an example from an active travertine system. *Sedimentology*, 38, 107–126.

DREW, G.H., 1914. On the precipitation of calcium carbonate in the sea by marine denitrifying bacteria. *Carnegie Institute, Washington Publication*, 182, 9–45.

EHRLICH, H.L., 1995. *Geomicrobiology*, 3rd Edition, Marcel Dekker, New York, 719 pp.

FOLK, R.L., 1993. SEM imaging of bacteria and nanobacteria in carbonate sediments and rocks. *Journal of Sedimentary Petrology*, 63, 990–999.

FORTIN, D., FERRIS, F.G., and BEVERIDGE, T.J., 1997. Surface-mediated mineral development by bacteria. *Rev. Mineral.*, 35, 161–180.

GONZÁLEZ-MUÑOZ, M.T., BEN OMAR, N., MARTÍNEZ-CAÑAMERO, M., RODRÍGUEZ-GALLEGO, M., LÓPEZ GALINDO, A., and ARIAS, J.M., 1996. Struvite and calcite crystallization induced by cellular membranes of *Myxococcus xanthus*. *Journal of Crystal Growth*, 163, 434–439.

LOWESTAM, H.A., 1981. Minerals formed by organisms. *Science*, 211, 1126–1131.

PERITO, B., BIAGIOTTI, L., DALY, S., GALIZZI, A., TIANO, P., and MASTROMEI, G., 2000. Bacterial genes involved in calcite crystal precipitation. In: Ciferri, O., Tiano, P., and Mastromei, G., (Eds.), *Of Microbes and Art: The Role of Microbial Communities in the Degradation and Protection of Cultural Heritage*, Plenum, New York, pp. 219–230.

SHINANO, H., 1972. Studies of marine microorganism taking part in the precipitation of calcium carbonate. *Bulletin of the Japanese Society of Science Fisheries*, 38, 825–832.

TIANO, P., BIAGIOTTI, L., and MASTROMEI, G., 1999. Bacterial bio-mediated calcite precipitation for monumental stones conservation: methods of evaluation. *Journal of Microbiological Methods*, 36, 139–145.

VASCONCELOS, C., MCKENZIE, J.A., BERNASCONI, S., GRUJIC, D., and TIEN, A.J., 1995. Microbial mediation as a possible mechanism for natural dolomite formation at low temperatures. *Nature*, 377, 220–222.

P. TIANO
S. BRACCI
S. RESCIC

BIOMEDIATED CALCITE PRECIPITATION FOR THE REINFORCEMENT OF MONUMENTAL STONES

CNR – Istituto per la Conservazione e la Valorizzazione dei Beni Culturali, Florence, Italy

INTRODUCTION Monumental stone decay is a consequence of the weathering action of several physical, chemical, and biological factors, which induces a progressive dissolution of the mineral matrix. In the case of calcareous stones the material, due to calcite leaching, undergoes an increase in porosity and a decrease in its mechanical characteristics (Pèrez et al., 1995). The products applied to monuments for stone conservation, such as ethyl silicate and silicones, present some drawbacks, and so they are not fully satisfactory as safeguards (Torraca, 1976). In fact, restorers involved in monument conservation are extremely cautious when considering the use of synthetic products as consolidants or to protect stone surfaces. Color changes, crust formation, glossy appearance, and substrate exfoliation, together with environmental pollution, are the most frequent drawbacks. These are essentially related to the photochemical modification of these products, which causes a short-lasting action, and to the use of large amounts of organic solvents. Furthermore, the lack of standard methods for measuring treatment effectiveness can lead to an over- or underestimation of the amount of products to be applied. This situation, together with the past indiscriminate application of all sorts of synthetic materials, has resulted in a real hesitation, on the part of both the restorers and the controlling institutions, to use synthetic products. Thus the historic buildings repair and maintenance field needs innovative and specifically tailored conservation strategies and products.

An old method, based on the application of lime-water, has been recently reused to impart a slight water-repellent and consolidation effect, but it creates a new calcite layer with no bond with the substrate (Peterson, 1981; Moncrieff and Hempel, 1970; Price, 1984). A new approach to improve this kind of conservation treatment, based on the use of inorganic compounds, is to enhance the growth of new calcite crystals, inside stone pores, using biomineralization mechanisms. The first application was based on the use of organic matrix macromolecules (OMMs) extracted from the shell of a mollusk (*Mytilus californianus*). These are usually composed of acidic glycoproteins rich in aspartic and glutamic acid (protein moiety) and carboxylate and sulfate groups (polysaccharide moiety). In the presence of calcium the protein moiety tends to partially adopt the β-sheet conformation, and thus has a high affinity for ions and actively participates in the control of the crystal formation. This process has been investigated in depth (Wheeler and Sikes, 1989) and used to control crystallization of inorganic materials in vitro (Addadi and Weiner, 1990). The feasibility study (Tiano et al., 1992) and the results obtained on small limestone samples (Tiano, 1995), using these OMMs have shown a good efficiency of the bio-induced calcite precipitation process with respect to the decrease of the amount of water absorbed and the increase of the superficial strength of the stone material treated. The positive outcomes of such a type of treatment can be attributed to the facts that:

• New calcite crystals behave as composite material affecting the mechanical properties of the crystals itself (they fracture with a conchoidal cleavage instead of the classical rhombohedral one); and

• New crystals remain strictly bound with the original calcite mineral structure to which the OMMs have been preliminarily absorbed.

Notwithstanding the fact that the extraction and purification of the organic matrix from the mollusk shells have been greatly improved in recent years by using an acid charged ion exchange resin (Albeck et al., 1996), the yield is very low, and production is expensive and time consuming. For these reasons synthetic analog polypetides, such as poly-aspartate (PAsp), and the alternating polypeptide poly Aspartate-Leucine (PAsp-Leu), have been experimented with, as they have been shown to have similar crystal nucleating activity in the appropriate microenvironment (Levi et al., 1998).

At the same time another biomineralization treatment has been developed applying living cultures of selected calcinogenic bacteria (Orial et al., 1992). The bacterial activity can control the crystallogenesis of carbonates through environmental conditions, or crystals can start to build up on the bacterial wall. In a few days a kind of cocoon forms, and subsequently biomineral assemblages appear, progressively turning into true crystals—either well shaped or poorly organized—that trap bacteria within the mineral structure (Castanier et al., 1989). This treatment has demonstrated its efficacy (Castanier, 2000) even if the application of heterotrophic viable bacteria, inside a monumental stone, needs some improvements and caution (Tiano et al., 1999).

The aim of this work is to develop and validate a methodology for monumental stone conservation, based on biomediated calcite precipitation processes, that uses a valid and more practical substitute for the natural macromolecules without the application of viable organisms.

MATERIALS AND METHODS

STONE SAMPLES The tests were conducted on specimens (dimensions 5 × 5 × 3 cm) of two calcareous rocks:

- Pietra di Lecce, a soft bioclastic limestone from southern Italy; and
- Marble Gioia, a calcitic marble from a Carrara quarry.

In order to increase their porosity and to enhance the superficial de-cohesion, all stone specimens were artificially weathered with simple thermal fatigue cycles (1 hour at 200°C followed by 1 hour at 20°C, for a total of 8 cycles). The physical characteristics of the stone used are reported in TABLE 1.

BIOINDUCING TREATMENTS

Products

Silk protein (SP) (from silkworm cocoon)	300 mg/l
Organic matrix macromolecules (OMM)	20 μg/ml
Polyaspartate (PAsp)	20–40 μg/ml
Bacillus subtilis autoclaved suspension (BSF+)	10^6 cells/ml
Escherichia coli autoclaved suspension (ECF-)	10^6 cells/ml

(All products were dissolved in a solution of 7.5 mM of $CaCl_2$.)

	AFTER THERMAL CYCLE (mean values)		
	γ (g/cm^3)	γ_b (g/cm^3)	P%
PIETRA DI LECCE	2.71±0.01	1.73±0.01	36.2±0.1
MARBLE GIOIA	2.72±0.01	2.62±0.01	3.5±0.2

γ = real density; γ_b = bulk density; P = total open porosity (He pycnometer).

[TABLE 1] Physical characteristics of stone specimens

TREATMENT	SP+MeOH	OMM	PAsp 20	PAsp 40	BSF+	ECF-	$Ca(HCO_3)_2$
REF							X
T1		X					X
T2			X				X
T3				X			X
T4	X		X				X
T5					X		X
T6						X	X

[TABLE 2] Treatments applied

TREATMENT	MARBLE GIOIA					PIETRA DI LECCE				
	MEAN BEFORE	SD	MEAN AFTER	SD	DELTA	MEAN BEFORE	SD	MEAN AFTER	SD	DELTA
REF1	19.2	2.9	19.6	2.7	0.4	16.3	4.3	16.8	1.7	0.5
REF2	17.4	2.0	18.0	1.9	0.6	19.6	5.5	20.0	1.1	0.4
REF3	23.9	1.1	24.6	3.0	0.7	17.8	4.8	17.4	2.6	−0.4
T1A	41.3	5.9	41.0	3.2	−0.3	15.0	2.9	15.7	1.7	0.7
T1B	28.3	4.1	28.6	3.1	0.3	17.7	2.8	17.2	1.2	−0.5
T1C	30.4	1.9	31.2	3.1	0.8	18.6	4.2	19.2	2.7	0.6
T2A	25.1	2.5	25.8	1.7	0.7	18.1	5.6	15.3	3.6	−2.8
T2B	29.7	4.1	30.4	2.9	0.7	15.7	0.9	16.6	1.9	0.9
T2C	34.6	4.3	34.5	2.8	−0.1	18.5	5.6	17.5	2.6	−1.0
T3A	28.2	7.0	29.0	2.4	0.8	17.0	4.0	15.6	2.5	−1.4
T3B	31.5	6.5	31.8	5.8	0.3	24.8	3.0	21.8	1.8	−3.0
T3C	34.4	2.9	34.2	2.1	−0.2	17.0	6.8	15.6	2.0	−1.4
T4A	37.1	3.5	35.3	1.9	−1.8	17.9	5.3	17.3	2.3	−0.6
T4B	40.2	8.1	38.8	3.7	−1.4	17.3	3.9	17.6	1.9	0.3
T4C	25.9	2.3	24.8	2.5	−1.1	12.0	2.9	12.6	1.6	0.6
T5A	36.9	7.3	35.7	4.2	−1.2	20.7	8.5	18.6	4.3	−2.1
T5B	41.3	8.7	40.0	2.3	−1.3	19.8	6.7	18.4	3.0	−1.4
T5C	36.5	7.9	34.0	4.4	−2.5	19.3	6.7	17.4	2.5	−1.9
T6A	37.6	6.0	37.2	2.3	−0.4	17.1	3.0	16.4	3.0	−0.7
T6B	38.0	6.3	37.9	3.8	−0.1	13.2	2.9	13.9	2.5	0.7
T6C	28.1	6.1	28.0	1.4	−0.1	20.2	8.3	20.6	1.7	0.4

[TABLE 3] Penetration depth (μm) determined on stone samples before and after the treatments

Timing

a. Overnight pre-adsorption of silk solution by samples. These were successively dried for 3 hours and then soaked in methyl alcohol (80% v/v) for 1 hour and dried again for 3 hours.

b. Overnight pre-absorption of the products by stone samples.

c. Drying of the stone samples for 8 hours at room temperature.

d. Incubation of stone samples with a supersaturated solution of calcium bicarbonate ($Ca(HCO_3)_2$) for 24 hours (obtained by fluxing CO_2 in a saturated $CaCO_3$ solution for 1 hour, filtering and again fluxing with CO_2 for ½ hour).

e. Drying of stone samples, with the treated face upward, for 8 hours at room temperature and then in the oven (60°C) for 3 days.

All the treatments reported in Table 2 have been made by capillary absorption using 200 ml of the treating solution. [TABLE 2] The samples were placed, in a glass Petri dish, with the face to be treated in contact with the solution, soaked up to 3 mm, using thin glass spacers. For each treatment a set of three samples was used.

TREATMENT EVALUATION In order to evaluate the efficacy of the treatment the following parameters were measured (in chronological sequence):

Before treatment

a. drilling resistance test;
b. superficial hardness test (Rockwell superficial test);
c. capillary water absorption;
d. color determination.

After treatment

a. color determination;
b. capillary water absorption;
c. superficial hardness test (Rockwell superficial test);
d. drilling resistance test.

– Color Determination

The chromatic coordinates (Lab system, CIELab 1976), were measured by using a Minolta Chroma Meter (Model CR 200) with light source type C (AA.VV., 1986; 1994). The variation induced by treatments has been determined calculating the ΔE.

– Capillary Water Absorption
The determinations of capillary water absorption (AA.VV., 1986) were carried out only on Gioia marble samples before and after treatments. The amount of water absorbed (by weight) in 60 minutes before and after treatment has been determined.

– Rockwell Superficial Test
The instrument (Galileo DG 201) measures the depth of penetration of a penetrator charged with a known load. The penetrator used for this test is a hardened steel ball (diameter 5 mm) and the load applied is 5 kg for Pietra di Lecce samples and 20 kg for Gioia marble ones.

– Drilling Resistance Test
The drilling resistance test was performed using the DRMS (Tiano et al., 2000) system with the following operative conditions:

- Gioia marble – Rotational speed = 600 rpm; penetration rate = 3 mm/min; hole depth = 7 mm.
- Pietra di Lecce – Rotational speed = 300 rpm; penetration rate = 3 mm/min; hole depth = 7 mm.

The drill bit was a polycrystalline diamond end mill (Diaber) with a 5-mm diameter.

RESULTS AND DISCUSSION

Color The changes of color due to the treatment are negligible, being the calculated value of (E lower than 2 for both the marble and Pietra di Lecce samples (detection limit for human eye (E ≥ 3).

Water Capillary Absorption The results of the water capillary test conducted before and after treatment on the marble samples showed no relevant changes for any treatments. The small changes registered (± 1% of initial value) are in the range of the reproducibility of this test.

Superficial Hardness The results obtained are reported in Table 3 (mean values in μm calculated for five determinations on each sample). [TABLE 3]

If we take into account the standard deviation of mean values, calculated for both mechanical tests for different lithotypes, it is not possible to establish any effective increase in the stone cohesion, either superficially or in the first layer, induced by the treatments applied. Only observation of the recorded cohesion profiles, as determined with the DRMS, in the few millimeters from the external

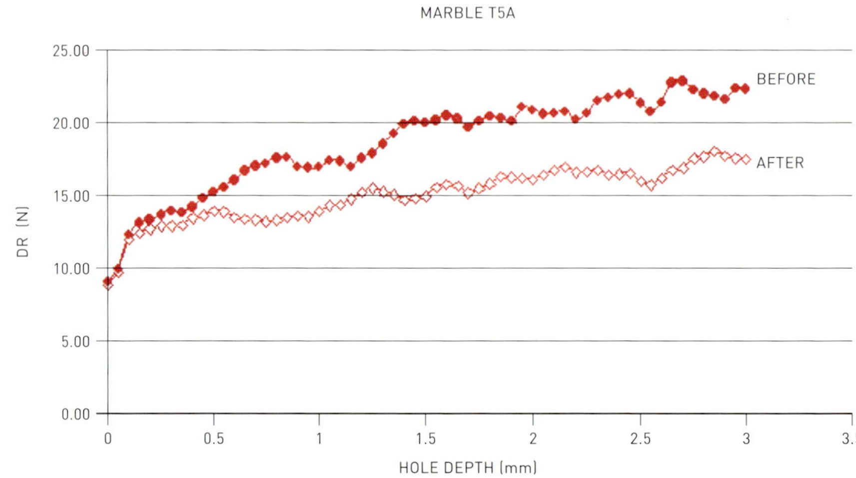

[1] Cohesion profiles determined on marble sample treated with BSF+.

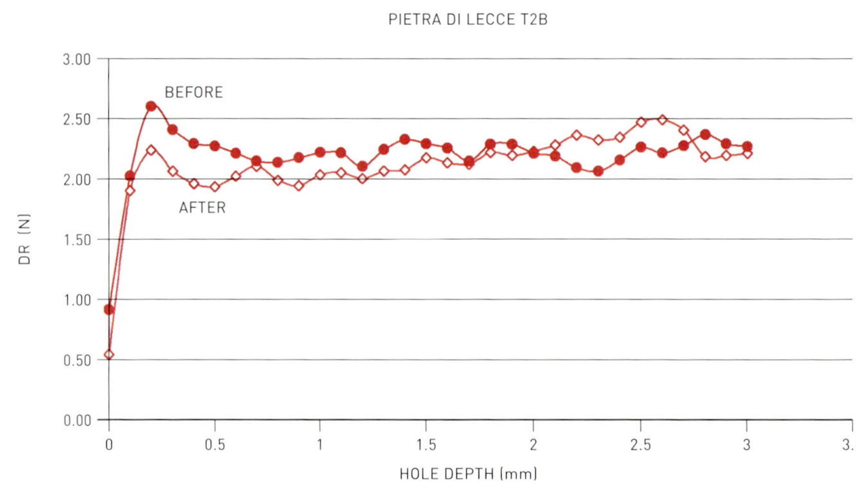

[2] Cohesion profiles determined on Pietra di Lecce sample treated with PolyAsp 20.

TREATMENT	MARBLE GIOIA					PIETRA DI LECCE				
	MEAN BEFORE	SD	MEAN AFTER	SD	DELTA	MEAN BEFORE	SD	MEAN AFTER	SD	DELTA
REF1	21.2	1.1	22.0	0.8	0.7	2.2	0.1	2.1	0.1	0.0
REF2	23.0	0.5	23.3	2.3	0.3	2.3	0.2	2.0	0.2	-0.3
REF3	20.8	0.5	20.2	0.8	-0.6	2.4	0.5	2.2	0.1	-0.2
T1A	15.5	0.7	15.4	0.8	-0.1	2.1	0.2	2.2	0.1	0.1
T1B	23.6	1.2	22.7	1.1	-0.8	2.1	0.2	2.0	0.1	-0.1
T1C	22.6	0.5	20.8	0.2	-1.8	2.2	0.2	1.9	0.2	-0.3
T2A	25.5	0.8	24.7	1.5	-0.8	2.4	0.1	2.3	0.1	-0.1
T2B	20.6	0.3	22.4	0.6	1.8	2.2	0.1	2.2	0.1	0.0
T2C	21.8	0.7	21.9	0.3	0.1	2.2	0.1	2.0	0.1	-0.2
T3A	22.8	0.7	23.8	1.4	0.9	2.2	0.2	2.0	0.1	-0.1
T3B	24.0	1.6	25.2	1.7	1.2	2.0	0.3	1.8	0.2	-0.2
T3C	21.3	0.5	18.5	0.8	-2.8	2.3	0.1	2.2	0.4	-0.2
T4A	20.5	0.5	19.0	1.9	-1.5	1.9	0.1	2.1	0.2	0.2
T4B	21.2	1.1	21.0	0.8	-0.2	2.1	0.1	1.9	0.3	-0.2
T4C	25.6	1.4	22.8	0.8	-2.8	2.1	0.2	2.0	0.4	-0.1
T5A	18.0	3.5	21.9	0.7	4.0	2.2	0.1	2.1	0.0	-0.1
T5B	20.2	0.9	19.7	2.0	-0.5	2.0	0.0	2.0	0.2	0.0
T5C	20.9	0.3	20.7	0.4	-0.2	1.9	0.1	2.2	0.2	0.3
T6A	21.2	0.4	20.1	1.0	-1.2	1.9	0.4	2.2	0.2	0.3
T6B	20.7	0.7	20.7	0.2	0.0	2.1	0.3	2.3	0.1	0.2
T6C	23.5	0.5	23.4	0.2	-0.1	1.9	0.1	2.1	0.2	0.1

[TABLE 4] Drilling resistance (N) determined on stone samples before and after the treatments

surface of some stone specimens gives indications of a little "reinforcing" effect, especially in the case of the treatments T5 (BSF+) and T2 (PAsp20), as is shown in FIGS. 1 AND 2, respectively.

Drilling Resistance The results obtained are reported in Table 4 (mean values in Newtons, calculated for three holes in each sample). [TABLE 4]

CONCLUSIONS It should be taken into account that this is the first time a reinforcing treatment has been evaluated by looking at mechanical properties of the stone in the very first layers of the material. Furthermore, the efficacy of such a treatment is quite difficult to assess because the reinforcing effect should be due to a new calcite formation inside a calcite structure. On the basis of the results obtained, we point out that the proposed evaluation methods seem not to be sensitive enough to measure the reinforcing effect. On the other hand, the new calcite precipitation, even if present, might not be effective enough to influence the mechanical (cohesion) properties of the stone material tested. Further diagnostic methods, such as observation of treated stones in thin sections, and with the SEM, will probably clarify this point. At the same time, the operative conditions of the DRMS and Rockwell test will be adjusted in order to improve their sensitivity.

ACKNOWLEDGMENTS This work was supported by the EU program "Energy, Environment and Sustainable Development," within the framework of the BIOREINFORCE Project (Contract EVK4-CT2000-00037).

REFERENCES

AA.VV., 1986. Assorbimento d'acqua per capillarità. *Documento NORMAL* 11/85, ICR-CNR, (Eds.), Rome.

AA.VV., 1994. Misure colorimetriche di superficie. *Documento NORMAL* 43/93, ICR-CNR, (Eds.), Rome.

ADDADI, L., and WEINER, S., 1990. "Stereochemical and Structural Relations between Macromolecules and Crystals in Biomineralization." In: Mann, S., Webb, J., and Williams, R.J.P., (Eds.), *Biomineralization: Chemical and Biochemical Perspectives.* VCH, FRG, 133–156.

ALBECK, S., WEINER S., and ADDADI, L., 1996. Polysaccharides of intracrystalline glycoproteins modulate calcite crystal growth in vitro. *Chemistry European Journal*, 2, 278–284.

CASTANIER, S., MAURIN, A., and PERTHUISOT, J.P., 1989. Production bactérienne expérimentale de corpuscoles carbonatés, sphéroïdaux a structure fibroradiare. Reflexions sur la définition des ooïdes. *Bull Soc. Géol. France*, 8 (3), 589–595.

CASTANIER, S., LE METAYER-LEVREL, G., ORIAL, G., LOUBIER, J.F., and PERTHUISOT, J.P., 2000. Bacterial carbonatogenesis and applications to preservation and restoration of historic property. In: Ciferri, O., Tiano, P., and Mastromei, G., (Eds.), *Of Microbes and Art—The Role of Microbial Communities in the Degradation and Protection of Cultural Heritage*, Kluwer Academic-Plenum Publisher, pp. 246–262.

LEVI, Y., ALBECK, S., BRACK, A., WEINER, S., and ADDADI, L., 1998. Control over aragonite crystal nucleation and growth: an in vitro study on biomineralization. *Chemistry European Journal*, 4, 389–396.

MONCRIEFF, A., and HEMPEL, K., 1970. Work on the degeneration of sculptural stone. In: *Conservation of Stone and Wooden Objects*, New York Conference, 2nd edition ICC, Ed., London, vol. 1, pp. 103–114.

ORIAL, G., CASTANIER, S., LE METAYER, G., and LOUBIERE, J.F., 1992. The biomineralization: A new process to protect calcareous stone: applied to historic monuments. In: Ttoishi, K, Arai, H., Kenjo, T., and Yamano, K., (Eds.), *Proceedings of the 2nd International Conference on Biodeterioration of Cultural Property*, Yokohama, Japan, pp. 98–116.

PÈREZ, J.L., VILLEGAS, R., VALE, J.F., BELLO, M.A., and ALCADE, M., 1995. Effects of consolidant and water repellent treatments on the porosity and pore size distribution of limestones. In: *Proceedings of the International colloquium Methods of evaluating products for the conservation of porous building materials in monuments.* ICCROM, Rome, pp. 203–211.

PETERSON, S., 1981. Lime water consolidation. In: *Proceedings of the Mortar, Cements and Grouts used in the Conservation of Historic Buildings Symposium*, Rome, ICCROM, pp. 53–61.

PRICE, C.A., 1984. La consolidation du calcaire par compresse de chaux et eau de chaux. In : *Proceedings of Adhesives and Consolidants Meeting*, Paris, ICC, London, pp. 160–162.

TIANO, P., ADDADI, L., and WEINER, S., 1992. Stone reinforcement by induction of calcite crystals using organic matrix macromolecules: feasibility study. In: *The 7th International Congress on Deterioration and Conservation of stone*, Lisbon, Vol 2, pp. 1317–1326.

TIANO, P., 1995. Stone reinforcement by calcite crystals precipitation induced by organic matrix macromolecules. *Studies in Conservation*, 40, 171–176.

TIANO, P., BIAGIOTTI, L., and MASTROMEI, G., 1999. Bacterial Biomediated Calcite Precipitation for monumental Stone Conservation: Methods of Evaluation. *Journal of Microbiological Methods*, 36, 139–145.

TIANO, P., FILARETO, C., PONTICELLI, S., FERRARI, M., and VALENTINI, E., 2000. Drilling force measurement system, a new standardisable methodology to determine the stone cohesion: Prototype design and validation. *International Zeitschrift für Bauinstandsetzen und Baudenkmalpflege* 6, Jahrgang, Aedificatio Publisher, Heft 2, 115–132.

TORRACA, G., 1976. Treatment of stone in monuments—a review of principles and processes. In: Rossi Manaresi, R., (Ed.), *Proceedings of The Conservation of Stone* I, Bologna, pp. 297–315.

WHEELER, A.P., and SIKES, C.S., 1989. Matrix-crystal interaction in $CaCO_3$ biomineralization. In: Mann, S., Webb, J., and Williams, R.J.P., (Eds.), *Biomineralization: Chemical and Biochemical Perspectives.* VCH, FRG, pp. 95–131.

G. ORIAL[1]
TH. VIEWEGER[2]
J.-F. LOUBIERE[3]

BIOLOGICAL MORTARS: A SOLUTION FOR STONE SCULPTURE CONSERVATION

1 Laboratoire de Recherche des Monuments Historiques, Champs sur Marne, France
2 Société GROUX S.A, Tremblay en France, France 3 Société CALCITE BIO CONCEPT, Levallois-Perret, France

ABSTRACT The deterioration of stone monuments generally results in material losses that are traditionally repaired through the use of mortars. The difficulty of this type of intervention on soft stones has resulted in the evolution of restoration techniques toward a new concept: "biological" mortar. This type of mortar, prepared from stone powder and carbonatogenic bacteria, was derived from the biomineralization stone protection technique. After feasibility tests and in-laboratory optimization, it was possible to carry out a first on-site experimentation following the complete restoration of a French monumental portal. The high-technology approach fits the minimum-intervention principle and adheres to present conservation-restoration ethics in France, increasingly aimed toward a minimal-intervention conservation.

INTRODUCTION The deterioration of stones in monuments results in material losses, such as scaling and cracks, that require the use of repair mortars. These are mixtures of binders, aggregates, additives, and water. Traditionally, this type of product is used on the facades of monuments as well as on the statuary. One of the flaws of this type of repair lies in chemical, petro-physical, or mechanical incompatibilities between the stone and the mortar. In particular, the softer the material to be restored, the more delicate the choice of a mortar. The difficulty of intervention on this type of support with classic methods and products (mortars based on a more or less hydraulic binder, epoxy-resin-based glues), made the choices of restoration evolve toward alternative techniques such as "biological" mortar, a new concept developed from stone powder and carbonatogeneous bacteria, derived from the stone protective technique of biomineralization.

Investigations in this field started with in-laboratory feasibility tests. The choice of the bacteria preparation method and the optimizing of the semi-industrial manufacture techniques also represented a non-negligible part in this experimental phase.

Many tests were needed, sometimes with fruitless results, before the first on-site experimentation could be carried out. This was followed by the actual restoration of the western portal of the 12th century Argenton-Chateau church, in western France. This is built with Tuffeau, a soft limestone. The sculptures of this portal had been exposed to the severe climate and were badly damaged. Multiple alterations had affected this precious decor, which urgently required some treatment to be preserved (Bromblet, 1992). This task was part of the vast project of restoring the whole facade. The guideline was the preservation of the statuary through consolidation through the fixing of scales or flaking, as well as the filling up of cracks, rather than its replacement.

STONE PROTECTION TECHNIQUE USING BIOMINERALIZATION In nature, numerous bacteria produce calcium carbonate (e.g., *Bacillus cereus, Bacillus megaterium,* and *Pseudomonas fluorescens*). Such a diversity suggests that the carbonatogenesis process is widespread worldwide (Boquet et al., 1973). It can be found, for example, in France, in the Clamouses underground cave of the Hérault or the L'Ile Adam quarries in the Paris area; in Turkey, in the Pammukale hydrothermal springs; and in Bolivia, in the salted lakes of the Altiplano. Studies have shown that whatever the process used by the bacteria, bacterial carbonatogenesis appears as the displacement of the hydrogenocarbonate balance: $Ca\,(HCO_3)_2 \rightarrow CaCO_3 + CO_2 + H_2O$ (Castanier, 1987).

Carbonatogenesis, in its initial stages, generates an amorphous and greatly hydrated carbonate (magnesium and/or calcium) (Castanier et al., 1999b). It is a precursor of the crystallographic shape recognized in X-ray diffractometry. The use of carbonatogenesis in an industrial setting was developed at the end of the 1980s by a team of researchers at the Pierre and Marie Curie University, in France. The process consists of making calcifying bacteria manufacture a protective layer on the limestone surface. The idea of applying the bacterial process of calcite production to the protection of the stonework, or even historical monuments, was conceived shortly thereafter.

In 1990, a study program aiming to optimize and industrialize the process was started. Although the calcifying power of some bacteria had clearly been established in the laboratory, their viability and the maintenance of their performance in real exterior conditions, as well as their adaptation to an industrial use, were still to be confirmed.

First of all, the calcifying strains, taken from the natural environment, were selected according to their level of carbonate production. This enabled assembling a collection of strains composed of about a hundred bacteria. *Bacillus cereus* was retained as the most efficient bacterium and the most capable of being used industrially. This species identified by the Pasteur Institute of Paris comes from telluric flora and doesn't present any danger to man or to the environment.

Subsequently, studies concentrated on the composition of the feeding solution for these bacteria and on their feeding frequency. Thus, if the liquid must provide the necessary elements for the development of the calcifying bacteria, it must also contain a fungicide, the role of which is to block the parallel and disruptive development of molds present in a latent state on every support, even mineral ones.

Today, the treatment is optimized. It is organized into a phase of sowing onto the stone surface through pulverization of a bacterial medium (mixture of bacteria and feeding solution in an aqueous phase) followed by four daily feedings through the pulverization of a feeding solution (a mixture of organic compounds, mineral salts, and a carbonatogenesis catalyst). The bacteria must be in their exponential phase of growth and the concentration of the medium is valued at about 10^{10} to 10^{12} bacteria per liter.

The feeding phase facilitates carbonatogenesis. Progressively, the bacteria installed in the micro-recesses on the surface of the stone multiply and cling to each other, surrounding themselves little by little with a calcium carbonate cocoon until they form a crystalline structure and an accumulation of superimposed layers. At the beginning of this period, the bacteria remain identifiable through SEM observation. [FIG. 1] After one week, a 4–5-micron-thick calcium carbonate coating that can be called "biocalcin" covers the surface of the material. The fossilized bacterial bodies are not differentiated anymore during SEM examination. At this stage, the newly formed carbonate is still very hydrated. It dries and hardens in the weeks that follow. [FIG. 2]

The first on-site application of this process was carried out about ten years ago, on the Thouars church in central France. This first experimental site has provided, through regular controls, a lot of information on the evolution of outdoor biocalcin (Orial et al., 1992; Le Metayer-Levrel, 1993 and 1996; Le Metayer-Levrel et al., 1997).

The carbonatogeneous bacteria didn't have any difficulty working on an outside facade subjected to daily climatic variations.

SEM observations have confirmed the persistence of the calcite coating and revealed the deterioration that it sustained over the years, similar to that of the stone itself. Thus, biocalcin is sensitive to orientations and shows a differential alteration depending on its localization on the micro-relief of the stone (attacked on the more prominent parts, denser at the bottom of the pores). Biocalcin therefore behaves like a true stone epidermis; it ages at the same time as the stone does, and in the same way. It is estimated that to continue to ensure its protective role, this biocalcin must be renewed every ten years (Le Métayer et al., 1999; Castanier et al., 1999a; Orial, 2000).

PRODUCTION OF BIOLOGICAL MORTARS

PRODUCT CONCEPTION Confronted with the problems of soft stone repair, researchers came up with the idea of a biological mortar. The mortar would be a mixture of aggregate, the stone powder, binder, the bacterial sediment, and the necessary water to form the paste, the feeding solution. This was called "biological" mortar because of the bacterial origin of its binder and because the concept was based on the performance of calcifying bacteria in the formation of an epidermis on stone surfaces. In theory, the bacteria would find the necessary elements for the production of calcium carbonate in the feeding solution and would succeed in binding the grains of stone. Thus, by using a mineral load similar to that of the material to be restored, the biological mortar would evolve in perfect harmony with the treated support, with no difference in color or texture.

The first step of the investigation started with feasibility tests mixing the different components in identical proportions. The result was conclusive; the mortar set. But numerous stages still lay ahead: optimization of the constituents and of their proportions, characterization, definition of implementation specifications (weather fillets, grouts, whitewash, etc.). These studies were connected directly to the restoration of the portal of the Argenton-Château church, where the decision to apply biological mortars had been made.

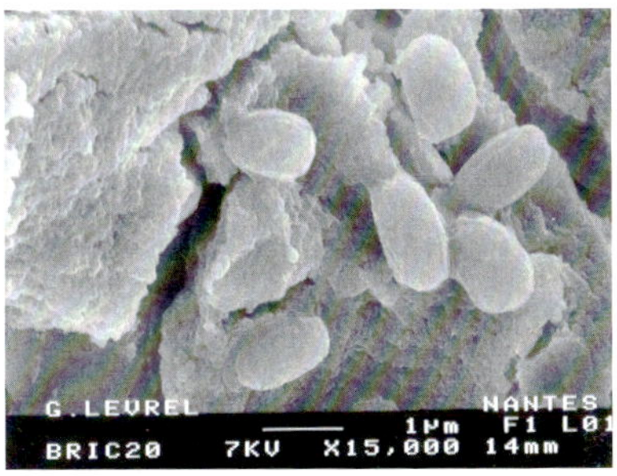

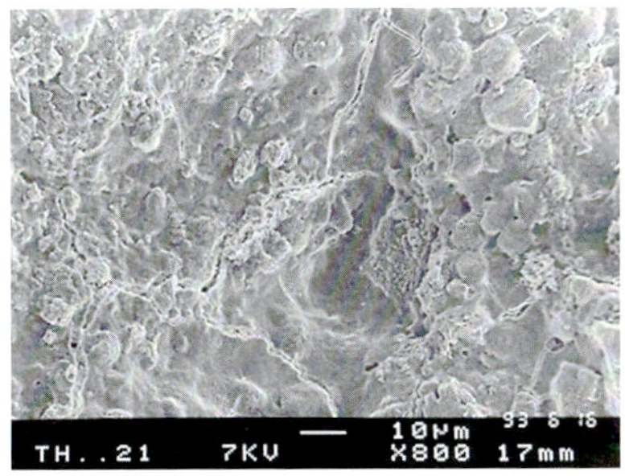

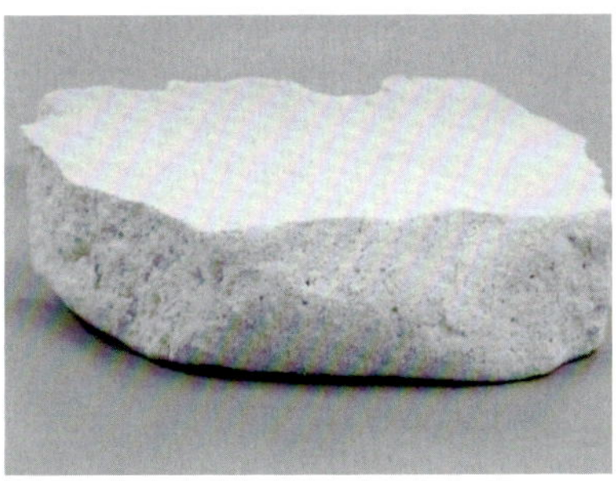

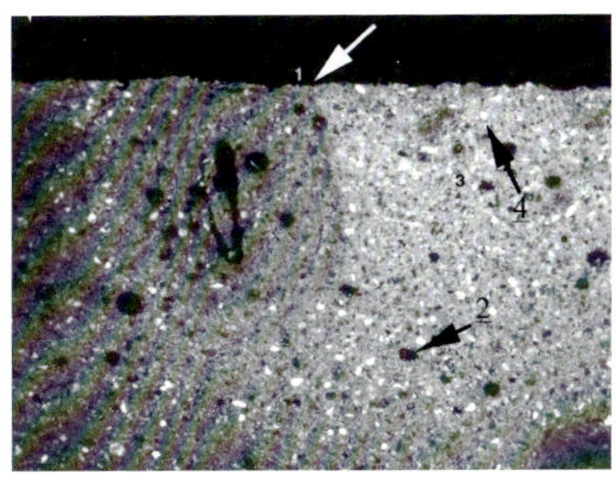

[1] Aspect of the cocoons of $CaCO_3$ surrounding the bacterial bodies. [2] Aspect of the biocalcin coating on the surface of Tuffeau stone. [3] Aspect of a biological mortar sample with a Tuffeau powder base. [4] Optical microscope aspect of cross section of biological mortar: 1 = surface; 2 = pore; 3 = glauconite; 4 = quartz.

IN-LABORATORY ADJUSTMENT Optimization involved the three main components of the mortar:

- The aggregate, i.e., the *stone powder* (most experiments were carried out with Tuffeau powder, constituent stone of the Argenton-Château sculptures);
- Water, in this case *the feeding solution for the bacteria;* and
- The binder, as *bacterial sediment.*

The optimization phase also dealt with the proportions of the different ingredients and the method of preparation (order of incorporation of these ingredients).

THE AGGREGATE: OPTIMIZATION OF THE STONE POWDER GRANULOMETRY A stone powder obtained by stone grinding is formed of dissociated grains of different sizes, with an intergranular space fluctuating according to the size of the grains. This space cannot be too small—otherwise the bacteria cannot access it—nor can it be too large, because the bacteria should be able to fill the empty spaces with their calcium carbonate production. The space between the grains must also allow circulation of the nourishing liquid. Experience with traditional mortar manufacture and the type of granulates used helped in the choice of granulometry. After sifting of the stone powder, granulometries of 40, 160, 250, and 400 μm were tested alone or mixed. For each mortar sample, the criteria for assessment were identical, from the setting of the mortar to the final phase of maturation:

- Presence of cracks
- Ease and homogeneity of setting (to the core of the sample)
- Presence of gas bubbles in connection with bacterial metabolism
- Color and smell

That portion of grains with granulometry between 40 and 160 μm gives the most satisfactory results for the Tuffeau limestone.

FEEDING SOLUTION: OPTIMIZATION OF INGREDIENT CONCENTRATION AND MODE OF FEEDING The feeding solution must provide the necessary elements to stimulate carbonatogenesis: The processes involved in the biomineralization technique are essentially the ammonification of amino acids in aerobiosis and the reduction of nitrates in microaerophilic conditions.

Therefore the liquid has to contain:

- Organic elements, peptone, and yeast extract;
- A source of calcium, a calcium salt;
- A source of nitrate; and

•A specific protein that catalyzes the bacterial functions in the production of calcium carbonate.

During the previous optimization phase of the feeding solution for protective treatment by biomineralization, the in-laboratory experiments had been undertaken in bioreactors with a dosage of dissolved carbonate according to Anderson and Robinson's method (1946). The concentrations had been adjusted in order to get an effective environment for the production of carbonate without excess products, and according to the bacteria used and their concentration.

In the case of mortars, the bacteria used are identical but with a concentration a thousand times higher. It was therefore necessary to adapt the nutritive content to this concentration. The simplest solution would have been to concentrate the first feeding solution. But such a large overload of salts would have been damaging to the production of calcium carbonate. The average balance of the medium was therefore recalculated to increase only certain components, such as organic matter.

After controls and tests, a specific concentrated feeding solution was selected, according to the mass of bacteria. Originally, the first tests of mortars had been based on the operative method of the protective treatment, that is to say, with a repeated feeding for four days following the preparation. It turned out that, in this case, it was difficult to introduce the feeding solution into the core of the mortar after it had begun to set and it was very quickly going to concentrate on the surface of the mortar, thereby provoking colored deposits and developing contaminations. The tests proved that the quantity of feeding solution provided at the time of preparation was sufficient. The feedings that occurred after setting were superfluous, as they were not consumed.

OPTIMIZATION OF BACTERIA PREPARATION

Flocculate Since the beginning of the experiments and applications of the protective treatment, the choice had deliberately been directed toward a bacterial preparation in a lyphilized form, as much for conservation reasons as for transportation. This form of preparation proved to be distinctly less satisfactory than using freshly produced mortars (problems included coloration changes connected to the lyphilization and length of storage) and led to orienting the research toward other means of production and storage of the microorganisms. The objective was to maintain a concentrated biomass, uncontaminated, and effective even after several months of storage.

Work was undertaken then with a specialist in biofixation problems, who suggested a "hyper-concentrated flocculation" technique that would encourage the conservation of the biomass without damaging it. The interesting point of this cell immobilization method is that there is a total preservation of enzymatic activity inside the cell. This "flocculate" was obtained, first through coagulation, then through flocculation, with addition of a powder capable of fixing the microorganisms. This process avoided the centrifugation of the biomass, a delicate operation, and biofixation enabled preservation of activity (Lebesgue, 1979).

Three types of "flocculates" in concentrated pastes were produced:

- With kaolin;
- With pure calcium carbonate; and
- With "Tuffeau powder."

In the required specifications, the produced flocculate had to contain quantities of living bacteria superior to those obtained through reviviscence following lyphilization, i.e., a bacterial concentration at least equal to 10^9 to encourage an optimum production of carbonate.

In reality, the flocculate has a tendency to sediment and the bacteria to float on the surface. Regular bacterial numerations showed a progressive decrease of the concentration in living organisms with time. On the other hand, it turned out that after mixing the flocculates with the mineral load of the mortar itself, the resulting mortars lacked cohesion. This system of cell immobilization was therefore abandoned.

Fresh Bacterial Paste In the very first feasibility tests for mortars, bacteria had been used as bacterial sediment obtained from small laboratory-scale cultures, i.e., using bioreactors that did not exceed 500 ml (which gave a few grams of bacterial residue after centrifugation).

Considering the previous fruitless tests, this method of microorganism preparation seemed the best for application to mortars, but it was imperative to handle large-quantity production problems on a semi-industrial scale.

A developing relationship with a biotechnology company permitted us to prepare biomass in large quantities. This company, equipped with all the necessary equipment, including large bioreactors and a powerful centrifuge, had total expertise in this type of manufacture. After clarifying some points about production parameters (culture medium of the biomass, pH, oxygenation, etc.), specifications were developed and successfully tested:

• A bacterial paste composed of pure *Bacillus cereus* cells in order to avoid uncontrollable and detrimental contaminations;

• A bacterial concentration equal to or over 10^9 per ml; and

• A light beige color additive (according to the Munsell color code).

For the storage of the resulting paste, the biomass freezing technique, which was considered for a moment, was very quickly abandoned because of the risk of a very high loss of organisms upon defrosting. Storage in small pots at a temperature of 3 to 4°C was finally chosen. This allowed their use for conservation purposes for at least 7 weeks, as long as the refrigeration procedure was strictly respected.

From then on, production on a 50-liter bioreactor was started. Biomass concentration through centrifugation enabled recovery of 827 grams of *Bacillus cereus* from 32 liters of fermentation must with a concentration of 3×10^9 viable cells per milliliter. After conditioning and addition of a fungicide (natamycine), bacterial numerations after one day, then after one and two weeks (and finally after several months) showed a good resumption of *Bacillus cereus* growth and the absence of any contamination.

This production made it possible to abandon local laboratory experimentation for full-scale production and on-site use in Argenton.

OPTIMIZATION OF THE MORTAR PREPARATION MODE This optimization involved the balance between weight of bacteria and weight of stone powder, in order to obtain a cohesive mortar.

Tests were carried out by varying the weight of stone powder in relationship to a given quantity of bacteria (one part of bacterial paste for 1, 2, 3, 4, 7, or 9 parts of stone powder). For every sample, a part of feeding solution, containing the useful dose of nutrients, was added, and the final mixture was hydrated as necessary to get an unctuous texture. The samples were placed in molds and were left to dry at ambient temperature.

The evaluation criteria were

- Aspect;
- Color (Munsell code);
- The presence of liquid on the surface;
- The presence of micro-cracks; and
- Resistance to fracture.

The results led to the finalization of the mix: one part of bacterial paste, one part of feeding solution, and two parts of stone powder. [FIG. 3]

FIRST CHARACTERIZATIONS The mortar samples were prepared according to this "recipe" and poured into small molds, with a fragment of Tuffeau stone taken as reference. Other mortars were applied onto a fragment of Tuffeau for examination of texture and mortar/stone contact (Martinet, 1997; Rafai and Martinet, 2002).

From observations using optical microscopy carried out on cross sections perpendicular to the sample surface, it appears that:

• The Tuffeau reference shows a homogeneous finely porous structure. Its matrix is constituted of carbonated clastic calcareous, of micro-calcite, of grains of quartz and glauconite.

• The mortar presents a homogeneous finely micro-porous structure with a few spherical pores, with sizes ranging between 0.1 and 2 mm. The material is essentially made of a micritic matrix coating the micro-grains of calcite, quartz, and glauconite. The contact between grains and matrix seems to form a strong bond. [FIG. 4]

SEM observations brought to light the following findings:

• The Tuffeau reference has a micro-porous microstructure, composed of assemblies of calcium carbonate, quartz, and cristobalite micro-spheres. The assembly is more or less loose and denotes a relatively weak cohesion of the material.

• The mortar presents a micro-porous microstructure with a weak cohesion. The matrix is essentially made of calcium carbonate, of grains of quartz and clusters of biocarbonates. These are well dispersed in the matrix. [FIG. 5]

Porosity measures showed that the mortars had a total porosity lower than that of Tuffeau.

Examination of cross sections of the mortar/stone interface showed that the contact was good and continuous. [FIGS. 6 AND 7]

SYNTHESIS OF THE RESULTS To obtain a biological mortar, one part of *Bacillus cereus* paste with 10^9 cells per milliliter must be mixed with one part of concentrated feeding solution and two parts of stone powder with a granulometry between 40 and 160 µm. The mixing of the bacteria with the stone powder must be carefully carried out in order not to damage the bacterial bodies. This fairly consistent basic mixture can be liquified by adding water, as needed. Thus, one can obtain a range of products that go from mortar to injection grout to very fluid whitewash.

These mortars are well adapted to the repair of Tuffeau, a soft stone of poor durability.

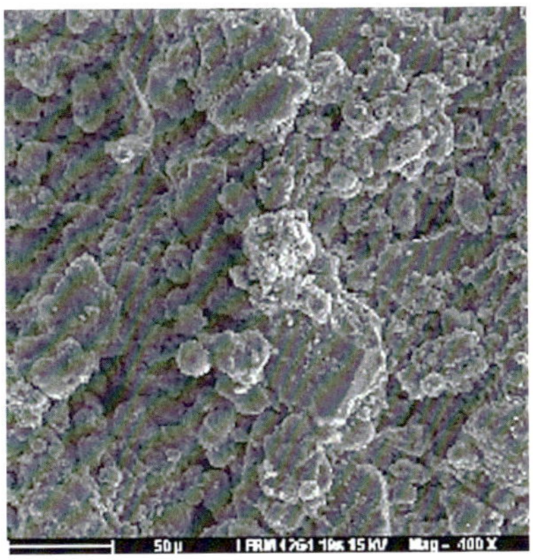

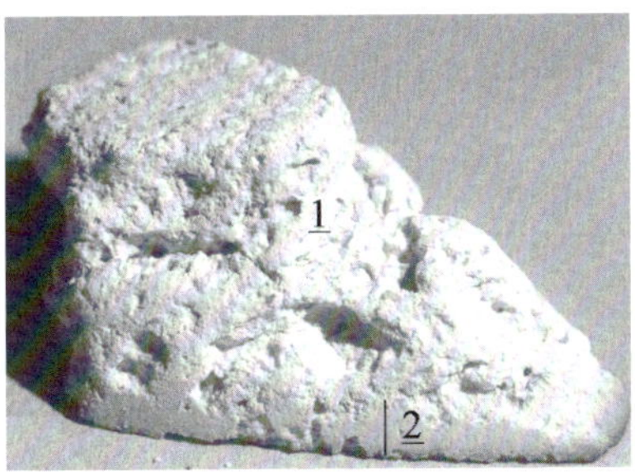

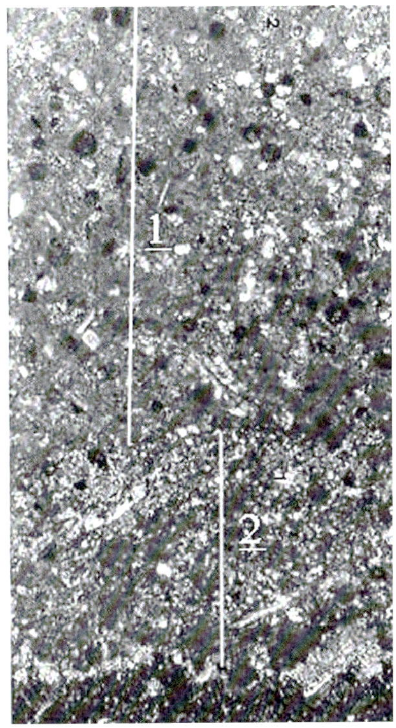

[5] SEM view of biological mortar (cross section). [6] Fragment of Tuffeau stone (1) with a layer of biological mortar (2). [7] Cross section of fragment (fig. 6) showing the similarity of structure and the continuous contact between stone (1) and mortar (2)

RESTORATION OF THE WESTERN PORTAL OF THE ARGENTON-CHÂTEAU CHURCH

THE WORKS The portal is considered to be one of the most remarkable examples of Romanesque sculpture in the Poitou region. The median and lower parts are composed of a big arch with five orders, decorated with figures supported by columns with capitals. On either part of the arches, the surface is decorated with sculpted relief.

Built with granite stone (base and columns) and with Tuffeau (capitals, arches, and sculpted decorations), this portal is exposed to severe weather conditions. The classic morphologies of Tuffeau deterioration can be observed: surface disintegration, failure of adhesion, scalings, surface and in-depth desquamation, and networks of cracks. [FIG. 8] This deterioration leads to material losses, and reduced readability would sooner or later result.

Therefore a restoration campaign was undertaken, and this took place in two phases. The first was the cleaning of the surfaces (elimination of the green biological covering, elimination of the whitewash concealing the original sculpture, and elimination of the concretions and dirt using a laser).

The second phase was devoted to the stone conservation work. It involved consolidation, the laying of weather fillets and micro-repairs with mortar, the application of injection grout, and surface treatment using a whitewash with a light mineral load.

Considering the assessment of the damages and given the knowledge and progress made in laboratory work on biological mortars, it seemed appropriate to combine a traditional method (consolidation with ethyl silicate) and an alternative method (biological mortar) to tackle the problems.

The first applications of biological mortar on site began on test zones, chosen according to the decay facies of the sculpture, and associating two types of intervention:

• Treatment using biological mortar *only* on a very altered sculpture of the right spandril and on an element of the right-side column, and

• Treatment using biological mortar *after* consolidation with ethyl silicate on a sculpted element of the outside arch and on a capital of the right embrasure.

The tests included the laying of weather fillets, in order to secure the scales about to fall, injections in the cracks, and a general surface treatment. After several weeks of aging, and considering the good results obtained, the application of these techniques was extended to the whole portal. Various complementary tests facilitated adjustment of preparation and application conditions, and adaptation to work in the field.

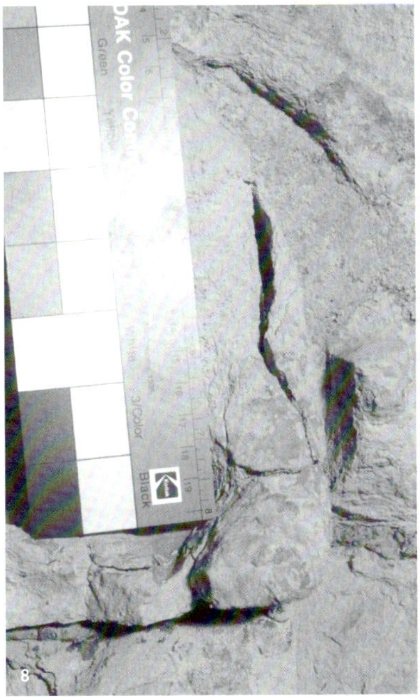

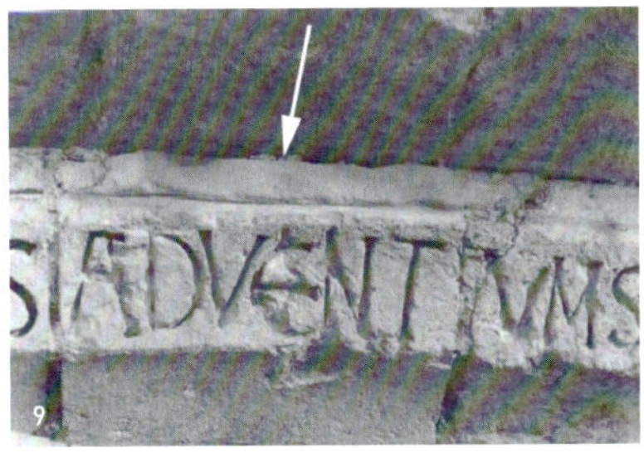

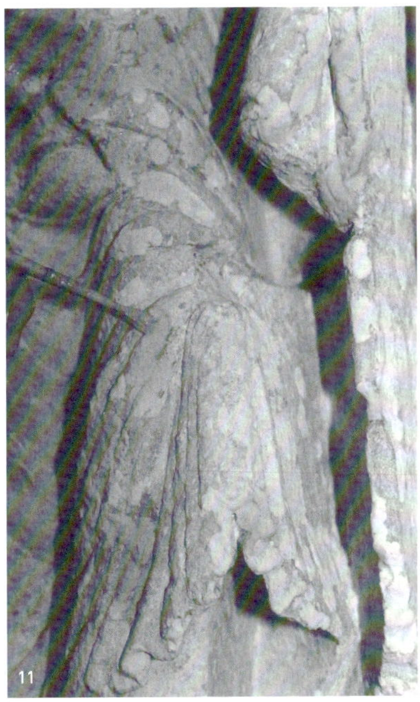

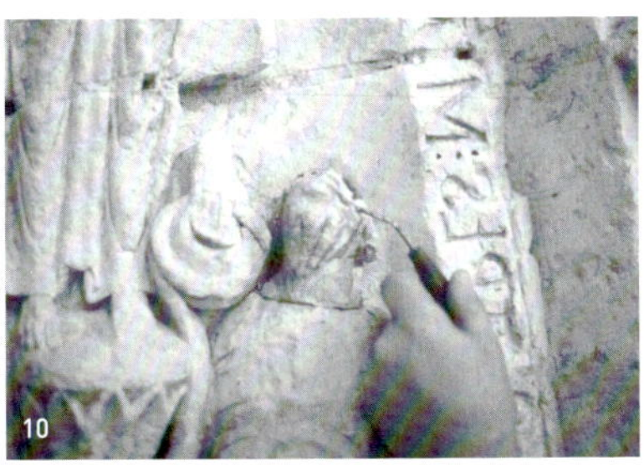

[8] Portal of the Argenton-Château (France) church—example of badly altered sculpture with crack on the shoulder. [9] Example of implementation of thick biological mortar layer. [10] Example of spatula application of weather fillet. [11] Example of whitewash brush application.

Thus, the prepared product must be used within 24 hours in conditions of ambient temperature. Beyond that, adhesion to the support will not be as good, as carbonatogenesis is already extensively in progress.

Product application must be carried out using a spatula, a brush, or a syringe. The choice of the tool is made according to the decay facies, and implementation must be delicate so as not to break the bacterial bodies still alive. [FIGS. 9–12]

It can be estimated that carbonatogenesis is complete four weeks after the mortar is applied. The mortar then is hardened and the definitive formatting of the volumes and of the surface aspects imitating the surrounding original surfaces can be obtained with cutting/leveling tools, small angled rasps and hard-bristle brushes.

Color control depending on the materials to be treated or replaced can be achieved through the addition of mineral pigments, color-stable to light, to the mix at the time of preparation of the mortar, and/or through the application of harmonization patinas on the surface, in water-based solutions, set with a final protective biomineralization treatment.

Forty sculptures, 12 ornaments representing the zodiac figures, and 10 capitals—62 elements altogether—were treated. About 30 of these elements were badly damaged and required a heavy intervention. [FIGS. 13 AND 14]

A portal of this dimension (6 m wide and 4 m high), with this many elements needing treatment, required 5 kg of bacterial paste, which amounted to 20 kg of mortar "base."

The work lasted from June 1999 to March 2000. A visual examination in May 2002, two years after completion, showed that everything was in good condition [FIG. 15]. No disruption was noticeable in spite of strong precipitation at the beginning of the year. A regular scientific follow-up is planned to validate these results.

LESSONS LEARNED FROM THE ON-SITE APPLICATION OF BIOLOGICAL MORTAR Observations made at this experimental site during the intervention enables us to draw some conclusions about this new technique.

The fresh paste kept at 4°C requires a time of reviviscence of about 6 hours before any preparation.

Prior consolidation seems necessary when the support has reached an advanced state of alteration. An in-depth intervention requires the maximum preservation of the stone epidermis.

It appears necessary to work in successive stages, with, first of all, the setting of weather fillets (in layers, if needed, to avoid withdrawals and flaking).

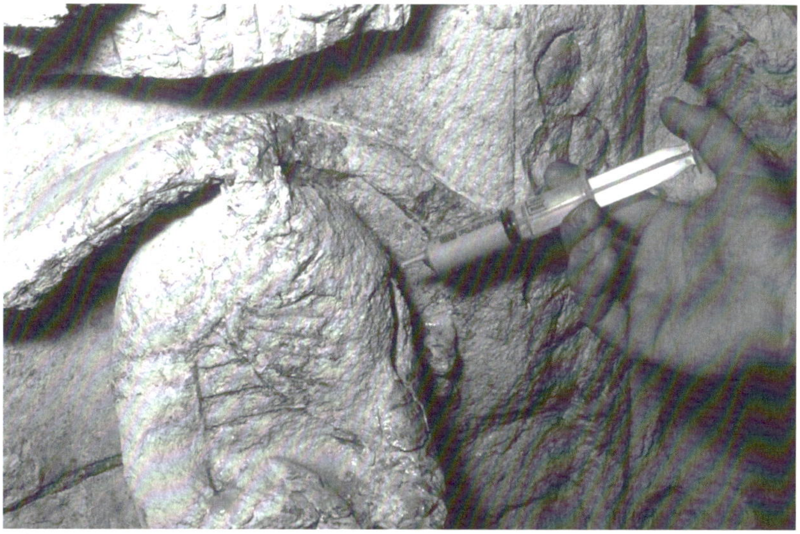

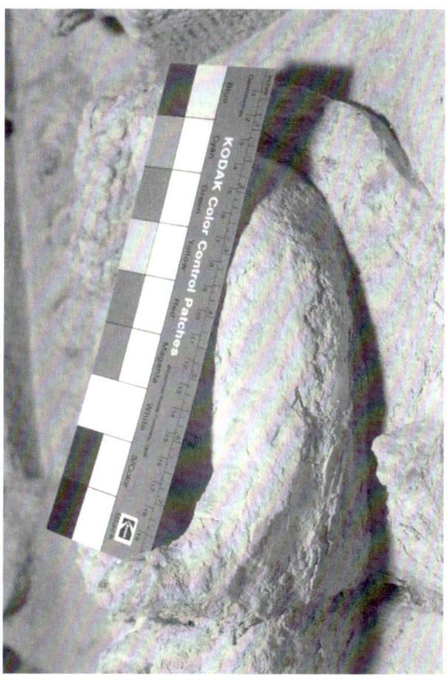

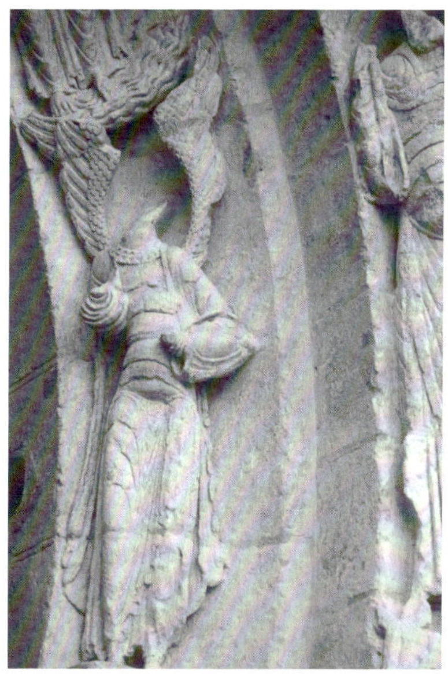

[12] Example of grout injection in a crack. **[13]** Aspect of sculpture after repair showing the filling of the crack. **[14]** Example of sculpture after restoration.

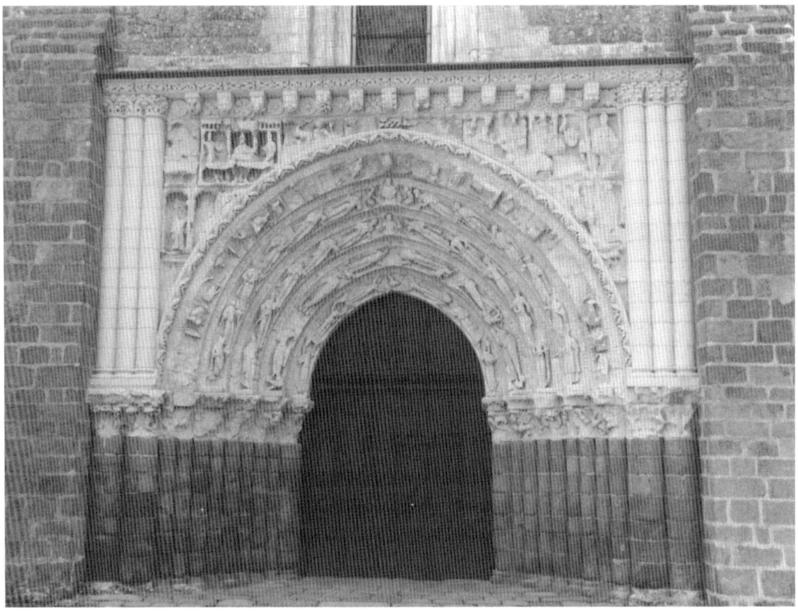

[15] View of Argenton-Château portal after restoration.

These weather fillets are indispensable; otherwise, the grout will "leak" and it will be impossible to maintain it within the cracks.

Some applications reached a thickness of approximately 1 cm.

The lack of wettability of the injection grout makes it difficult to access the bottom of the cracks; an addition of tensio-active may be useful provided it has been verified that it is not harmful to the bacteria.

The climatic conditions are determining factors. For the survival of the bacteria, a minimum of 10°C is essential. In the case of Argenton-Château, work at the site began in summer and continued throughout the winter, with temperatures as low as 0°C. It was therefore necessary to set up an enclosed structure around the zones of intervention with a heating fan system to maintain the stone support at temperatures from 12 to 15°C.

It is necessary to observe a certain number of rules of hygiene, considering that living organisms are being manipulated:

• Preparation and cleaning of the tools and containers with detergents without bactericide agents; and

• Meticulous rinsing after use of the material (atomizer, syringes, etc.); otherwise the ducts might be filled with biological calcium carbonate formation.

Control of the application of these products and the quality of the final results are directly related to the restorer, who must be highly qualified.

CONCLUSIONS AND PERSPECTIVES The experimental application of biological mortar at the Argenton-Château is a first. The use of this range of products required precision and a total respect for the specifications and the "instructions for use," in order to guarantee a good result. Even though the constraints are greater than those for the use of a classic mortar, the final result is outstanding, the replacement material being perfectly integrated with the original material.

Various experiments with the biotechnology company have shown that it is possible to industrially produce and preserve a bacterial paste; this represents significant progress in the development of this type of mortar.

The implementation of this project in real on-site conditions has provided numerous complementary lessons that go beyond laboratory experience in the field of Tuffeau powder based mortars. The synthesis of these data will permit adaptation of the research to other applications for the process (mortar with other types of stone powder, bonding of elements). Scientific investigation will be broadened to assess other characteristics of mortar (elasticity module, capillarity, resistance to compression, etc.) in comparison with traditional mortars.

The application of a high-technology approach fits the minimum-intervention principle and falls perfectly well within the framework of present conservation-restoration ethics in France, which aims at a minimal conservation above all.

Last of all, this project couldn't have been successfully managed without close collaboration between scientists, restorers, and industrial specialists.

ACKNOWLEDGMENTS We wish to thank:

- Mrs. Castanier of Nantes University, Mrs. Levrel of the monument restoration company Quélin, ARD-Soliances in Reims, biotechnology company, and the Laboratoire d'Etudes et de Recherche des Matériaux d'Arles for their active participation in the research and finalization; and
- Mr. Jeanneau, chief architect of the Historical Monuments, and Mr. Blin, regional curator for the Historical Monuments, for their confidence in this project and for making this experimentation possible.

REFERENCES

ANDERSON, D.H., and ROBINSON, R.J., 1946. Rapid electrometric determination of the alkalinity of sea water, using a glass electrode. *Industrial and Engineering Chemistry*, Analytical Ed., vol. 18, p. 767.

BOQUET, E., BORONAT, A., and RAMOS-CORMENZANA, A., 1973. Production of calcite (calcium carbonate) crystals by soil bacteria is a general phenomenon. *Nature* (London), vol. 246, (21/28), pp. 527–529.

BROMBLET, P., 1992. Argenton-Château (Deux-sèvres, 79, France). Église Saint-Gilles. *Portail : état de conservation de la pierre*. L.R.M.H., Champs-sur-Marne (France), 34 f.

CASTANIER, S., 1987. Microbiogéologie : processus et modalités de la carbonatogenèse bactérienne. Thèse: Sci. Biogéol., Nantes, Faculté des sciences et des techniques, 541 pp.

CASTANIER, S., LE MÉTAYER-LEVREL, G., ORIAL, G., LOUBIÈRE J.F., and PERTHUISOT J.P., 1999. Bacterial carbonatogenesis and applications to preservation and restoration of historic property. In : Ciferri, O., Tiano, P., and Mastromei, G., (Eds.), *Of Microbes and Art: The Role of Microbial Communities in the Degradation and Protection of Cultural Heritage*, Florence, Italy, pp. 203–218.

CASTANIER, S., LE MÉTAYER-LEVREL, G., and PERTHUISOT, J.P., 1999. Ca-carbonates precipitation and limestone genesis: The microbiogeologist point of view. *Sedimentary Geology*, no. 126, 9–23.

LEBESGUE, Y., 1979. La fixation des microorganismes sur les supports minéraux. *Journal Français d'Hydrologie*, vol. 10, fasc. I, n° 28, pp. 59–66.

LE MÉTAYER-LEVREL, G., 1993. Biominéralisation de surfaces: Application à la protection des pierres de taille, Diplôme de Recherche Universitaire, Université de Nantes, 127 pp.

LE MÉTAYER-LEVREL, G., 1996. Microbiogéologie du carbonate de calcium : applications industrielles, implications géologiques. Thèse: Sciences de la terre. Microbiogéologie: Nantes, Faculté des sciences et des techniques, 168 f.

LE MÉTAYER-LEVREL, G., CASTANIER, S., ORIAL, G., [ET AL.], 1997. From carbonatogenesis concepts to bacterial regeneration of limestones (microbial lifting). In: *Association des sédimentologiste français*, International Workshop on Microbial mediation in carbonate diagenesis, Chichilianne, France, pp. 41–42.

LE MÉTAYER-LEVREL, G., CASTANIER, S., ORIAL, G., [ET AL.], 1999. Applications of bacterial carbonatogenesis to the protection and regeneration of limestones in buildings and historic patrimony. *Sedimentary Geology*, no. 126, pp. 25–34.

MARTINET, G., 1997. *Caractérisation de mortiers obtenus par biominéralisation*. LERM, Arles, 7 pp.

ORIAL, G., CASTANIER, S., LE MÉTAYER, G., [ET AL.], 1992. The biomineralization: a new process to protect calcareous stone : applied to historic monuments. In: Toishi, K., Arai, H., Kenjo, T., and Yamano, K., (Eds.), *Biodeterioration of Cultural Property* 2: Proceedings of the 2nd International Conference on Biodeterioration of Cultural Property, Yokohama, pp. 98–116.

ORIAL, G., 2000. La biomineralizacion aplicada a la conservacion del patrimonio: Balance de diez anos de experimentacion. *Restaurar la Memoria: Secundo Congreso Internacional*, Valladolid, pp. 81–82.

RAFAI, N., and MARTINET, G., 2002. *Caractérisation de traitements par biominéralisation*. LERM, Arles, 36 pp.

M.P. NUGARI[1]
O. SALVADORI[2]

BIOCIDES AND TREATMENT OF STONE: LIMITATIONS AND FUTURE PROSPECTS

1 Istituto Centrale per il Restauro, Rome, Italy 2 Laboratorio Scientifico, Cannaregio, Venice, Italy

INTRODUCTION Control of biological growth is one of the operations carried out on cultural heritage during restoration. Chemical substances, biocides, are usually used for this purpose; they are applied before and/or after restoration and conservation treatments. Generally, biocides are used to eliminate the macro- or microflora responsible for biodeterioration, although more recently some products have been developed to prevent new microbial colonization of restored surfaces and consequently should be applied at the end of the conservation treatment (Malagodi et al., 2000; Urzì et al., 2000). For the conservation of stone materials, many studies have been carried out on commercial biocides to test their effectiveness, toxicity level, chemical stability, and harmlessness for treated surfaces, in order to identify the most suitable products (Koestler and Salvadori, 1996; Tiano, 1998; Warscheid and Braams, 2000). Unfortunately, the results of laboratory and in-situ tests often provide information on a single case or on a specific problem without taking into account the overall conservation project for the monument in question.

The control of biological growth requires further in-depth analysis, which should fulfill the following aims:

• Evaluate the efficacy of chemical products against macro- and microorganisms deteriorating stone (target biodeteriogens);

- Assess the substrate-biocide interaction;
- Define methods for testing and evaluating biocides;
- Define the best method for applying biocide on stone;
- Evaluate long-term effects of biocide applications and interactions with other products used during restoration; and
- Single out new and alternative methods to disinfect and de-infest.

EVALUATION OF EFFICACY It is well known that a biocide treatment should be recommended only after accurate diagnosis to detect the stone-colonizing macro- and microflora and evaluation of their role in the deterioration process (Koestler et al., 1997). When organisms and/or microorganisms are not responsible for serious damage, or the effectiveness of the treatment is uncertain, and/or the environmental or substrate conditions are favorable to new colonization and cannot be modified, no treatment should be carried out. Ariño and Saiz-Jimenez (1996) warn of the risk of repeated applications of biocides. Monuments in polluted areas can shelter the development of species that adapt to the urban environment and represent an actual refuge for certain species when their natural habitat is threatened. Moreover, new associations seem to have established themselves on artificial stone substrata in some archaeological areas in the Mediterranean Basin. The risk of an irreversible loss of biological diversity should also be taken into consideration.

There are many inorganic and organic products **[TABLE 1]** and commercial formulae adopted in the restoration field for biocidal purposes, and their efficacy should be carefully evaluated. Tests of effectiveness against the deteriogen organisms should be performed in the laboratory and/or in situ. The procedures selected can strongly influence the results, and a discrepancy in the behavior of biocides has frequently been found between laboratory and outdoor conditions, depending on the type of microorganisms as well as on practical conditions. Laboratory tests may therefore have limited predictive value with regard to the microbicidal activity of biocides on stone surfaces (Koestler and Salvadori, 1996). Isolated microorganisms used in efficacy tests are much more sensitive than those present in biofilms, which are far less susceptible to biocides, so the spectrum of activity of chemicals could be significantly affected by practical conditions. Finally, the presence of endolithic microorganisms and their interaction with biocides has only rarely been considered. Krumbein and Gross (1992) indicated that most biocides have to be applied in a far higher concentration than minimum inhibitory concentration (MIC) to be effective on natural stone. Nevertheless, Pinck et al. (2002), in a comparative study on different biocides against mixed biofilms, have demonstrated that two of

MICROBIOCIDES	HERBICIDES
ORGANOMETALLIC COMPOUNDS	DIAZINE
organotin compounds	IMIDAZOLINONES
	PHOSPHOORGANIC COMPOUNDS
PHENOLIC COMPOUNDS AND DERIVATES	PYRIDINES
o-phenyl-phenol	TRIAZINES
p-chloro-m-cresol	UREA DERIVATES
p-hydroxybenzoic acid ethyl ester	
thymol	
QUATERNARY AMMONIIUM COMPOUNDS (QACS)	
benzalkonium chloride	
NITROGEN-CONTAINING COMPOUNDS	
triazine-derivates	
carbamates	
halogenate pyridine derivate	
dodecylamine lactate – dodecylamine salicylate	
UREA DERIVATES	
n-(3,4-dichlorophenyl)-n-n-dimethyl urea	

[TABLE 1] Organic biocides most frequently tested or used in the field of stone conservation

the biocides they tested were highly effective in killing bacteria and fungi on natural stone at MIC while the other products failed to kill nitrifying bacteria, chemoorganothrophic bacteria, and fungi on natural stone at 10 times MIC.

Other factors that may affect the effectiveness of biocides, and which should be taken into account, are:

- The type of substrate (e.g., the stone porosity affects the quantity of the product absorbed, and the presence of certain clay minerals, such as illite and smectite, can increase efficacy, absorbing more biocide) (Grant and Bravery, 1985; Young et al., 1995);
- The presence of organic materials (dust, pollen, etc.) or pollutants on the surface;
- The solvent used (e.g., water hardness can reduce the activity of quaternary ammonium compounds, while alcoholic solutions can increase their penetration power and sometimes their stability);
- The contact time and temperature during the treatment; and
- Weather conditions (wind, rain) and light intensity.

Some of these conditions can be simulated in the laboratory but it is impossible to recreate a complex ecosystem such as is found in colonized outdoor stone.

A further complication is that the chemical compatibility, or incompatibility, of biocides with other chemical substances that may be used on the stone should be well known. A mixture of two chemicals does not necessarily produce an effect that equals the sum of the effects of the two separate components, so in attempting to reach an additive interaction, the opposite effect has often been obtained. For example, BAC products are incompatible with anionic surfactants, nitrates, hydrogen peroxide, and many other substances that may reduce the efficacy of treatment. Yet a positive result has been obtained when adding potassium hydroxide to acidic herbicides. In fact, the neutralization of herbicides decreases the damage caused by acidic substances to limestone without affecting the efficiency of the weed control on monuments (Mouga and Almeida, 1997).

In-situ applications are very useful, as biocide effects on microorganisms and biocide-stone interactions in the short and long term can be studied at the same time. Efficacy tests are in any case carried out in the laboratory, except for some measurements of phototrophic microorganisms that can be performed in situ (e.g., fluorometric analysis).

BIOCIDE-SUBSTRATE INTERACTIONS A careful appraisal of biocide interactions with stone materials in the short and/or long term should be made. Many factors affect the interaction of a biocide with stone. The most significant that affect efficacy of a product are the chemical composition of the product, the conditions of use (concentration, method of application, duration on substrate, etc.), the mineralogical and petrographic characteristics of the stone, and the environmental conditions surrounding the monument (Nugari, 1999).

Sometimes chromatic alterations can be induced by different commercial formulae with the same active ingredient, because of different coformulants or solvents used in each product. Different types of stone react differently with different formulations of the biocide, suggesting the possibility of selecting the product that causes the minimum interference. For instance, the biocide TBTN+BAC caused a darkening and yellowing of travertine, even at low concentrations, while several other products did not interfere with this lithotype (Altieri et al., 1999). Marble and limestone are particularly sensitive to acid action; therefore biocides with a pH of around 7 or those with added neutralizing substances should be used to prevent damage (Mouga and Almeida, 1997). Corrosion of calcite and oxidation of minerals that produce rusty stains when H_2O_2 is used is well known (Nugari et al., 1993a, 1993b; Kumar and Kumar, 1999). Despite this, several products, whose negative interaction with stone has been clearly demonstrated, still appear in many lists of recommended biocides and are still in use.

On the basis of many reports (Nugari et al., 1993a, 1993b; Koestler and Salvadori, 1996; Prieto et al., 1996; Wakefield and Jones, 1996; Altieri et al., 1997; Altieri et al., 1999; Nugari, 1999) analyzing the behavior of biocides on stone materials, the best techniques of verifying the inertness of a biocide treatment seem to be:

- Measuring changes of surface color by reflected-light colorimetry;
- Verifying variation of water absorption by capillarity to evaluate changes in the water affinity;
- Observing surface morphology modification by optical microscopy or SEM; and
- Estimating changes in dry and wet weight.

Further information is obtained through conductivity and pH measurements during treatment of samples by immersion in biocide solution other than the quality and the quantity of released ions in water (by FTIR) or on dried residues (by SEM/EDX) (Quaresima et al., 1995; Wakefield and Jones, 1996; Altieri et al., 1997; Nugari et al., 1993b; Nugari, 1999). Chromatic alteration is the only variation that can be measured in situ, as all the other techniques require that a large quantity of sample be taken to make the measurements.

The biocide-stone interaction is influenced by environmental factors, and reducing the residence time of biocide on stone to the minimum decreases the risk of possible negative interference. For example, exposure to UVA radiation in the laboratory tests caused color changes in biocide-treated stone samples. If the stone was washed after treatment, no alteration was detected (Altieri et al., 1999). For this reason, the stone surface should normally be washed after the biocide has performed its action, although this operation reduces or undoes any long-term effectiveness of the treatment; on the other hand, the stability (i.e., light stability) of the products used for preventive purposes should be attentively evaluated.

METHODS OF TREATMENT The practical procedures in biocide application represent individual steps that must be integrated into a general restoration work scheme. The planning phase must be consonant with the aim of the treatment: to eliminate or prevent biological colonization. Whenever compounds are applied before restoration to remove microbial colonization, these steps should be followed: accurate diagnosis, choice of appropriate commercial biocide, concentration, and solvent, taking into account its efficiency against detected organisms and inertness to the material to be protected; application of the biocide (after a mechanical reduction of biomass, if necessary); time allotted for the product to perform its action; repetition of treatment, if necessary; and cleaning of growth residuals and removal of biocide remains.

As to the need for correct diagnosis before disinfection, microbiological analysis in conservation practice is only occasionally carried out for various reasons, often due to lack of funds, despite the fact that biocide application is a common procedure. Furthermore, only a few laboratory experiments have resulted in practical suggestions, and so an evaluation of the most suitable biocide to use is not carried out and the choice of products and methods is often left to the practical experience of the restorers, and the poor selective efficacy of many biocides unfortunately encourages empirical solutions.

Ideally, restorers should strive to attain better knowledge of chemical and physical properties, such as solubility, stability, incompatibility with other chemicals, etc., of the active ingredients and/or of the commercial products employed. Too frequently the mode of action of the biocide is unknown and consequently the timing of the application is wrong, or recommended concentrations are doubled or tripled in an attempt to improve effectiveness, with negative consequences ensuing. Higher concentrations are sometimes far less effective than diluted solutions (Pinck et al., 2002) and can frequently cause negative collateral effects. For instance, treatments with concentrated solutions of a product based on benzalkonium chloride (BAC), one of the most frequently used biocides, caused yellowing of Carrara marble and turned the color of porous limestone containing iron traces pink or red [FIG. 1], whereas no change of color was detected using the same product at lower (and recommended) concentrations. On the other hand, sometimes the in-situ experience suggests practical strategies useful to increase the effectiveness of biocide treatment; for example, in the case of black patinas due to cyanobacteria and/or fungi colonization, nebulization of water on the surfaces for 6–8 hours before biocide application increases the antimicrobial efficacy of the treatment. This is probably due to an increase of metabolic activity connected with water availability. Likewise, the recommendation of spraying water on lichens before treatment in order to increase treatment effectiveness is already known (Caneva et al., 1996).

The time it takes biocides to act varies depending upon the product and the organisms or microorganisms to be removed. Microbicides can be divided into two main categories: those that act through contact, such as oxidizing agents like hypochlorite or membrane active disinfectants like quaternary ammonium compounds; and those able to inhibit some metabolic activity of target microorganisms (Denyer and Stewart, 1998). Oxidizing agents have a rapid effect but may be ineffective and very aggressive for the substrate. The QACs are the most used because of their efficacy and relative inertness to the substrate. In the case of biocides that inhibit metabolic activity, the efficacy is strictly linked to the duration of the treatment. The time of action, in this case, is normally 3–7 days, depending on the group of microorganisms (bacteria, fungi, or algae) to be eliminated (Gomez-Alarcon et al., 1999; Pietrini et al., 1999). A second application of the product is always necessary when spores or a biofilm formation are present.

Furthermore, better results from biocide treatment could be achieved by improving the methods of application. Besides brushing, paper poultice application, or addition to absorbent clays, the biocide could be added to certain gels that are frequently used as bases for solvents or other cleaning agents in the cleaning of paintings, ensuring better biocide-biofilm contact, and thus reducing the biocide spreading within the stone and consequent negative effects. Instead, in the case of endolithic microorganism colonization, the spreading and penetration of biocide within the stone enhances the effectiveness of the treatment, while crystal formation inside the stone pores may inhibit the penetration of biocide solutions. New investigation techniques, such as SEM or confocal scanning laser microscopy, have yielded information on distribution of biocides on rock surfaces, and on their efficacy (Blazquez et al., 2000).

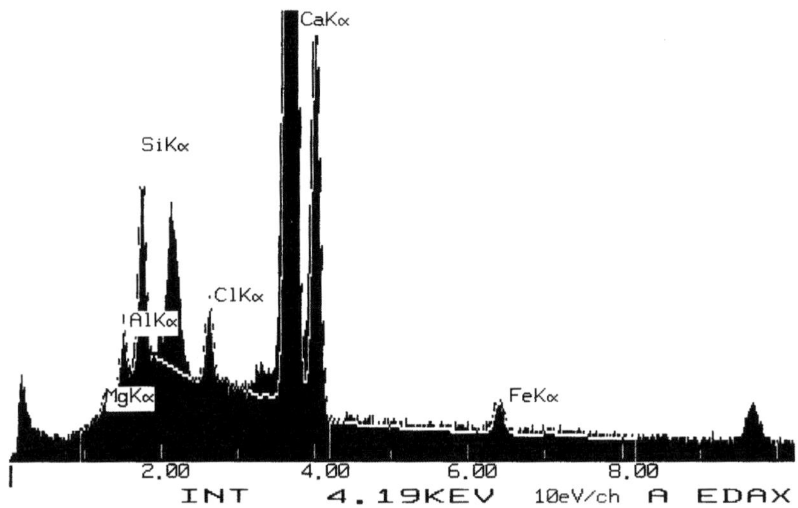

[1] Energy dispersion spectrum referring to limestone surface area after treatment with concentrated solution of BAC causing iron mobilization to the surface and deposition of chlorine.

The durability of the biocide treatment is strongly influenced by different environmental factors. Sunlight radiation and rainfall washing-out can play an important role, so the chemicophysical stability of products should also be considered (Nugari, 1999). Generally when the biocide application is intended for removing existing growth and not to prevent new growth it must be removed by washing with water or other solvents after its intended action has taken place. This practice is normally carried out during surface cleaning of the stone, one of the main restoration phases. With regard to herbicides, several problems may arise from the seasonal repetition of treatment to control or prevent vegetation growth. In addition to these cases, especially in archaeological sites where restoration is not foreseen, biocides are left on surfaces until the rain washes them away. Herbicidal remains could cause problems of interaction with stone and mortars over time.

BIOCIDES AND RESTORATION PRODUCTS During restoration, many operations are carried out for different reasons and with different products. Biocide treatment is only one of them. Combining biocides with cleaning agents, consolidants, and/or water repellents, or applying them one after the other, should be done with extreme caution because of possible negative interactions between the different products with consequent effects on the effectiveness of each. In order to determine compatibility and the most appropriate procedure for applying different products, exhaustive investigations should be carried out (Leznicka, 1992; Tiano et al., 1995; Malagodi et al., 2000; Ariño et al., 2002); also, though, empirical attempts are undertaken.

Recently the application of protective polymers and biocides at the end of restoration projects to waterproof the stone and inhibit recolonization for long periods has become an increasingly common practice. A study on mortars treated either with silicone or with separate applications of silicone and biocide in two successive steps, has demonstrated biocidal inhibition of microbial growth for a period of 5–8 years, except under continuous conditions of rising damp when phototrophic microorganisms developed on the silicone alone (Ariño et al., 2002). However, the application of biocides before applying water repellents seems to modify the characteristics of the latter, thereby diminishing their water-repellent properties. On the other hand, treating the stone initially with protective polymers and then with biocides does not significantly alter the water-repellent characteristics. Moreover, the sequence of applications and the biocide-protective interrelations vary according to the products selected (Balzarotti-Kämmlein et al., 1999; Malagodi et al., 2000).

While restorers often empirically attempt to mix water repellents or consolidants with biocides, the real efficiency of these mixtures has not yet been sufficiently assessed. Commercial products composed of a biocide-laced resin, i.e., polymers with antimicrobial activity, are required for different purposes by industries (Leeming et al., 2002), and the immobilization of poorly water-soluble biocides on resins represents a recent interesting field of research. When the biocide is not immobilized it may interact with the substrate and cause unintended changes; e.g., Tudor et al. (1990) report on an organo-iodide biocide that had been added to a silane consolidant that when applied to a Carrara marble statue produced a bright orange surface on the stone. The aim of immobilization should be a physisorption of the biocide to the hydrophobic compounds and consequently the donation of biocide "on demand" from the supporting resin to the microorganisms [FIG. 2] (Denyer and Stewart, 1998). The immobilization of biocides can be obtained with different methods, depending on the resin and the biocide used (i.e., whether water-soluble or not) and has to be confirmed analytically. It does not consist of simply mixing two different compounds during restoration procedures. Quaresima et al. (1977) polymerized some Cu^{2+}-laced monomers and controlled the copper release over a period of time, demonstrating that these materials could be models for the development of new products for the control of biodeterioration (Corain et al., 1998). Some commercial products with these characteristics (resin + biocide) are marketed for restoration, but preliminary analysis with agar diffusion tests and on limestone slabs have not improved antimicrobial efficiency compared with biocides or resins used separately (Blazquez et al., 2000).

Recently the development of new repair mortars for restoration displaying antimicrobial activity has been studied with the purpose of inhibiting the growth of living organisms. They were obtained by adding to lime mortars (i) 5–10% zeolithized Neapolitan yellow tuff previously subjected to Cu exchange (Ferone et al., 2000) or (ii) sepiolite acting as biocide support (Martinez-Ramirez et al., 1998a, 1988b). The biocidal activity is due, respectively, to Cu^{2+} release or to the presence of pentachlorophenol absorbed to sepiolite. These studies are preliminary reports and require more attentive evaluation (especially since the use of PCP is banned in many countries for toxicological reasons), but represent an interesting new field of research.

DESIGN OF STANDARD TEST METHOD The control of stone biodeterioration is a complex issue, and further research is therefore necessary to develop efficient and safe products. The difficulty of comparing results obtained from different works and of achieving unambiguous results has shown the need to determine and standardize

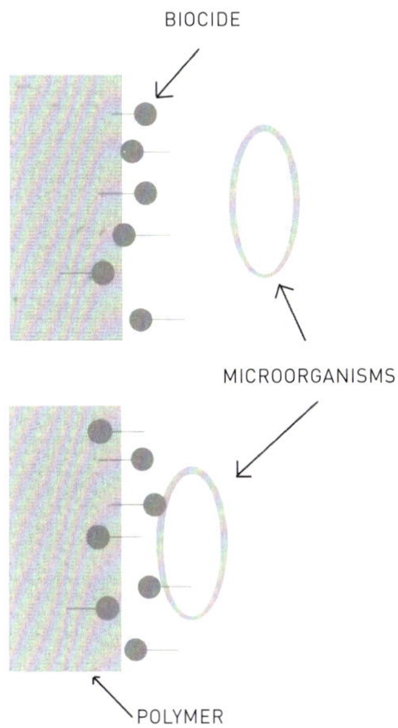

[2] Biocide uptake by microorganisms from the supporting resin (modified from Denyer and Stewart, 1998).

appropriate procedures for test methods (Nugari et al., 1993a,b; Koestler and Salvadori, 1996).

Research should focus on establishing harmonized test protocols, choosing specific target organisms among those affecting monuments, and developing methods for evaluating biocidal efficacy and biocide-stone interaction both in the laboratory and in situ. The final aim should be the standardization of test methods including potential efficacy under in-use conditions, and recommendations for use.

In the late 1980s, the European Committee for Standardization (CEN) conceived the CEN/TC (Technical Committee) 216 with the scope of standardizing terminology, test methods, recommendations for use, and labelling in the whole field of chemical disinfection and antiseptics. Areas of activity include agriculture, domestic service, food hygiene, and other industrial fields. The major issues involve assessing the performance of the tests in practice (Holah, 2001). This activity may help to evaluate the performance of biocides in the restoration field but, until now, not a single method has been produced with this specific aim. A working group responsible for the cultural heritage sector should be established to help manufacturers and users select the appropriate tests for this field.

A working group concerned with cultural heritage has been active in Italy for some years within the Ente Nazionale di Unificazione (UNI). Its purpose is to determine appropriate methodology to check tests that evaluate the effectiveness and possible interaction between biocide and substrate and to standardize test method procedures. However, European standards must be developed within the framework of a testing strategy.

CONCLUSION AND FUTURE PROSPECTS In practice, preventive methods to inhibit or slow down biological colonization should concentrate more on modifying the environmental conditions and physicochemical parameters of the stone in order to avoid or reduce repetition of biocide treatment. Periodic control and maintenance work (i.e., removing the initial phase of "dirt" deposition) should be carried out. This is often the principal, and sometimes the only, way of preventing biological colonization.

In order to guarantee long-term protection of restored monuments against biodeterioration, studies must also be promoted to test biocides for their preservative properties. A few products have shown long-lasting efficacy and it is necessary to verify the duration of residual action and the possible effect on stone over time and under diverse environmental conditions. A pyridine derivative, insoluble in water, has been shown to slow down lichen, algal, cyanobacterial, and fungal recolonization for as long as 8 years in environmental conditions of relatively low humidity, but this substance failed to inhibit bacterial growth (Balzarotti-Kämmlein et al., 1999; Gomez-Bolea et al., 1999; Pietrini et al., 1999; Malagodi et al., 2000; Urzì et al., 2000; Ariño et al., 2002).

More scientific attention should be paid to this field, as careful examination of the literature shows a progressive decrease in research dealing with biocide treatments during recent years. Further studies need to be done on nonbiocide (and nontoxic!) substances such as use of enzymes (selective substances catalyzing the fission processes of specific target molecules and macromolecules) used in a stepwise manner to clean other kinds of materials (paper, polychrome works of art). Experimentation involving the use of enzymes to remove biodeteriogens, although yielding good preliminary results (Capponi and Meucci, 1987), was not followed up, probably due to the difficulty of outdoor application and to high costs. Nowadays, the greater availability of enzymes, together with reduced costs, should encourage new tests with these substances which, being highly specific, should not negatively affect the substrate.

Unfortunately, many nontoxic (e.g., modified atmospheres) or physical (UV radiation and γ-rays) methods applied to control biodeterioration on other kinds of materials or on stone indoors (Bartolini et al., 1999; Koestler, 2000) are unsuitable for outdoor application and consequently cannot be adopted for stone treatment.

Further research into traditional or alternative disinfection treatments should be carried out to provide as much information as possible to be able to establish a "code of practice" for carrying out restoration and/or maintenance work with the best results.

REFERENCES

ALTIERI, A., COLADONATO, M., LONATI, G., MALAGODI, M., NUGARI, M.P., and SALVADORI, O. 1997. Effects of biocidal treatments on some Italian lithotypes samples. In: Moropoulou, A., Zezza, F., and Kollias, E., Papachristodoulou, (Eds.), *Proceedings of the 4th International Symposium on the Conservation of Monuments in the Mediterranean*, Rhodes, vol. 3, pp. 31–40.

ALTIERI, A., LONATI, G., MALAGODI, M., and NUGARI, M.P. 1999. Colorimetric evaluation of biocidal interference with stone in UV-A aging test. In: Marabelli M., and Parisi, C., (Eds), *6th International Conference on "Non Destructive Testing and Microanalysis for the Diagnostics and Conservation of the Cultural and Environmental Heritage,"* Rome, May 17–20, 1999, pp. 107–116.

ARIÑO, X., CANALS, A., GOMEZ-BOLEA, A., and SAIZ-JIMENEZ, C. 2002. Assessment of the performance of a water-repellent/biocide treatment after 8 years. In: Galán, E., and Zezza, F., (Eds.), *Protection and Conservation of the Cultural Heritage of the Mediterranean Cities. Proceeding of the 5th International Symposium on the Conservation of Monuments in the Mediterranean Basin*, Sevilla, pp. 121–125.

ARIÑO, X., and SAIZ-JIMENEZ, C. 1996. Biological diversity. *Aerobiologia*, 12, 279–282.

BALZAROTTI-KÄMMLEIN, R., SANSONI, M., and CASTRONOVO, A. 1999. An innovative water-compatible formulation of Algophase® for treatment of mortars. In: *Extended Abstracts: Of Microbes and Art, an International Conference on Microbiology and Conservation*, Florence, pp. 217–220.

BARTOLINI, M., PIETRINI, A.M., RICCI, S. 1999. Use of UV-C irradiation on artistic stoneworks for control of algae and Cyanobacteria. In: *Extended Abstracts: Of Microbes and Art, an International Conference on Microbiology and Conservation*, Florence, pp. 221–227.

BLAZQUEZ, A.B., LORENZO, J., FLORES M., and GÓMEZ-ALARCÓN, G. 2000. Evaluation of the effect of some biocides against organisms from historic monuments. *Aerobiologia*, 16, 423–428.

CANEVA, G., NUGARI, M.P., PINNA, D., and SALVADORI, O. 1996. Il controllo del degrado biologico. *I biocidi nel restauro dei materiali lapidei*. Nardini Editore, Florence, pp. 1–208.

CAPPONI, G., AND MEUCCI, C. 1987. Il restauro del paramento lapideo della facciata della Chiesa di S.Croce a Lecce. In: *Bollettino d'Arte*, supplement to n. 41, Materiali lapidei, vol. I, pp. 263–282.

CORAIN, B., FILIPPO, R., QUARESIMA, R., SCOCCIA, G., and VOLPE, R. 1998. Impiego di resine metallorganiche innovative per la protezione biologica dei beni culturali. In: Sanna, U., (Ed.), *Atti IV Congresso nazionale AIMAT*, PTM Editrice, pp. 769–779.

DENYER, S.P., and STEWART, G.S.A.B. 1998. Mechanisms of action of disinfectants. *International Biodeterioration and Biodegradation*, 41, 261–268.

FERONE, C., PANSINI, M., MASCOLO, M.C., and VITALE, A. 2000. Preliminary study on the set up of mortars displaying biocidal activity. In: Fassina, V., (Ed.), *Proceedings of the 9th International Congress on Deterioration and Conservation of Stone, Venice*, vol. 2, Elsevier, Amsterdam, pp. 371–378.

GOMEZ-ALARCON, G., BLÁZQUEZ, A.B., and LORENZO, J. 1999. Biocides used in the control of microorganisms on stone monuments. In: *Extended Abstracts: Of Microbes and Art, an International Conference on Microbiology and Conservation*, Florence, pp. 228–232.

GOMEZ-BOLEA, A., ARIÑO, X., BALZAROTTI, R., and SAIZ-JIMENEZ, C. 1999. Surface treatment of stones: consequences on lichenic colonization. In: *Extended Abstracts: Of Microbes and Art, an International Conference on Microbiology and Conservation*, Florence, pp. 233–237.

GRANT, C., and BRAVERY, A.F. 1985. A new method for assessing the resistance of stone to algal disfigurement and the efficacy of chemical inhibitors. In: Felix, G., (Ed.), *Proceedings of the Vth International Congress on Deterioration and Conservation of Stone*, Lausanne, Presses Polytechniques Romandes, pp. 663–674.

HOLAH, J.T. 2001. CEN/TC 216: Its role in producing current and future European disinfectant testing standards. Abstracts from the symposium "Disinfection and Hygiene —Future Prospects" 18–19 October 2001, Wageningen, Netherlands. *International Biodeterioration and Biodegradation*, 47, 243–244.

KOESTLER, R.J. 2000. When bad things happen to good art. *International Biodeterioration and Biodegradation*, 46, 259–260.

KOESTLER, R.J., and SALVADORI, O. 1996. Methods of evaluating biocides for the conservation of porous building materials. *Science and Technology for Cultural Heritage*, 5 (1), 63–68.

KOESTLER, R.J., WARSCHEID, T., and NIETO, F. 1997. Biodeterioration: risk factors and their management. In: Baer, N.S., and Snethlage, R., (Eds.), *Saving Our Architectural Heritage: The Conservation of Historic Stone Structures*, John Wiley & Sons Ltd., pp. 25–36.

KRUMBEIN, W.E., and GROSS, M. 1992. Interactions of biocides with cultures of biodeterioration microbiota in agar diffusion and rock cube test. In: Rodrigues, J.D., Henriques, F., and Jeremias, F.T., (Eds.), *Proceedings of the 7th International Congress on Deterioration and Conservation of Stone*, Lisbon, pp. 501–509.

KUMAR, R., and KUMAR, A.V. 1999. Biodeterioration of stone in tropical environments. *Research in Conservation*, The Getty Conservation Institute, Los Angeles, 85 pp.

LEEMING, K., MOORE, C.P., and DENYER, S.P. 2002. The use of immobilised biocides for process water decontamination. *International Biodeterioration and Biodegradation*, 49, 39–43.

LEZNICKA, S. 1992. Antimicrobial protection of stone monuments with p-hydroxybenzoic acid esters and silicone resin. In: Delgado Rodrigues, J., Henriques, F., and Telmo Jeremias, F., (Eds), *Proceedings of the 7th International Congress on Deterioration and Conservation of Stone, Lisbon*, vol. 1, Laboratorio Nacional de Engenharia Civil, Portugal, pp. 481–490.

MALAGODI, M., NUGARI, M.P., ALTIERI, A., and LONATI, G. 2000. Effects of combined application of biocides and protectives on marble. In: Fassina, V., (Ed.), *Proceedings of the 9th International Congress on Deterioration and Conservation of Stone, Venice*, vol. 2, Elsevier, Amsterdam, pp. 225–233.

MARTINEZ-RAMIREZ, S., PUERTAS, F., BLANCO-VARELA, M.T., and THOMPSON, G.E. 1998a. Effect of dry deposition of pollutants on the degradation of lime mortars with sepiolite. *Cement and Concrete Research*, 28 (1), 125–133.

MARTINEZ-RAMIREZ, S., PUERTAS, F., BLANCO-VARELA, M.T., THOMPSON, G.E., and ALMENDROS, P., 1998b. Behavior of repair lime mortars by wet deposition process. *Cement and Concrete Research*, 28 (2), 221–229.

MOUGA, T.M., and ALMEIDA, M.T. 1997. Neutralisation of herbicides. Effects on wall vegetation. *International Biodeterioration and Biodegradation*, 40, 141–149.

NUGARI, M.P. 1999. Interference of antimicrobial agents on stone. In: *Extended Abstracts: Of Microbes and Art, an International Conference on Microbiology and Conservation*, Florence, pp. 211–214.

NUGARI, M.P., D'URBANO, S., and SALVADORI, O. 1993a. Test methods for comparative evaluation of biocide treatments. In: Thiel, M.J. (Ed.), *Proceedings of the International UNESCO/RILEM Congress on Conservation of Stone and Other Materials*, E & FN Spon, London, pp. 565–572.

NUGARI, M.P., PALLECCHI, P., and PINNA, D. 1993b. Methodological evaluation of biocidal interference with stone materials—preliminary laboratory tests. In: Thiel, M.J., (Ed.), *Proceedings of the International UNESCO/RILEM Congress on Conservation of Stone and Other Materials*, E & FN Spon, London, pp. 295–302.

PIETRINI, A.M., RICCI, S., and BARTOLINI, M. 1999. Long term evaluation of biocide efficacy on algal growth. In: *Extended Abstracts: Of Microbes and Art, an International Conference on Microbiology and Conservation*, Florence, pp. 238–245.

PINCK, C., BALAZAROTTI-KAMMLEIN, R., and MANSCH, R. 2002. Biocidal efficacy of Algophase® against nitrifying bacteria. In: Galán, E., and Zezza, F., (Eds.), *Protection and Conservation of the Cultural Heritage of the Mediterranean Cities*. Proceeding of the 5th International Symposium on the Conservation of Monuments in the Mediterranean Basin, Sevilla, pp. 449–453.

PRIETO, B., RIVAS, M.T., and SILVA, B.M. 1996. Effectiveness of biocide treatments on granite. In: *Proceedings of EC Workshop on Degradation and Conservation of Rocks in Monuments*, European Commission, pp. 361–366.

QUARESIMA, R., DI GIUSEPPE, E., and VOLPE, R. 1995. Impiego di biocidi per la rimozione della microflora algale. In: Biscontin, G., and Driussi, G., (Eds.), *La pulitura delle superfici dell'Architettura. Scienza e Beni Culturali* XI, Bressanone. Libreria Progetto Editore, Padova, pp. 267–275.

QUARESIMA, R., BACCANTE, A., VOLPE, R., and CORAIN, B. 1997. Realisation and possibility of polymeric metalo-organic matrixes with biocides activity. In: Moropoulou, A., Zezza, F., Kollias, E., and Papachristodoulou, I., (Eds.), *Proceedings of the 4th International Symposium on the Conservation of Monuments in the Mediterranean*, Rhodes, vol. 3, pp. 323–335.

TIANO, P. 1998. Biodeterioration of monumental rocks: decay mechanisms and control methods. *Science and Technology for Cultural Heritage*, 7 (2), 19–38.

TIANO P., CAMAITI M., and ACCOLLA P. 1995. Methods for evaluation of products against algal biocenosis of monumental fountains. In: *International Colloquium on Methods of Evaluating Products for the Conservation of Porous Building Materials in Monuments*, ICCROM, Rome, pp. 75–86.

TUDOR, P.B., MATERO, F.G., and KOESTLER, R.J. 1990. Restoration of an outdoor marble sculpture in New Orleans: A case study of the compatibility of biocidal cleaning and consolidation. In: Llewellyn, G.C., and O'Rear, C.E., (Eds.), *Biodeterioration Research* 3. Plenum Press. New York, pp. 525–533.

URZÌ, C., DE LEO, F., GALLETTA, M., and SALAMONE, P. 2000. Efficacy of biocide in "in situ" and "in vitro" treatment. Study case of the "Templete de Mudejar", Guadalupe, Spain. In: Fassina, V., (Ed.), *Proceedings of the 9th International Congress on Deterioration and Conservation of Stone, Venice*, vol. 1, Elsevier, pp. 531–539.

WAKEFIELD, R.D., and JONES, M. 1996. Some effects of masonry biocides on intact and decayed stone. In: *8th Internation Congress on Deterioration and Conservation of Stone*, Berlin, vol. II, pp. 703–716.

WARSCHEID, TH., and BRAAMS, J. 2000. Biodeterioration of stone: A review. *International Biodeterioration and Biodegradation* 46, 343–368.

YOUNG, M.E., WAKEFIELD, R., URQUHART, D.C.M., NICHOLSON, K., and TONG, K. 1995. Assessment in a field setting of various biocides on sandstone. In: *International Colloquium on Methods of Evaluating Products for the Conservation of Porous Building Materials in Monuments*, ICCROM, Rome, pp. 93–99.

D.P. WESSEL

THE USE OF METALLIC OXIDES IN CONTROL OF BIOLOGICAL GROWTH ON OUTDOOR MONUMENTS

Architectural Resources Group, San Francisco, CA

INTRODUCTION The detrimental effect of microorganisms (fungi, lichens, algae, mosses, bacteria) on historic building facades, and methods to control the resultant biodeterioration of outdoor masonry, have been the focus of many studies. The process of biodeterioration on stone has been documented as both chemical and mechanical. For example, lichens can produce organic acids that may damage stone masonry, particularly calcareous rock. An example of mechanical biodeterioration is the process of hyphae penetrating cracks and fissures in stone, where they exert pressure, enlarging these faults and thus further deteriorating the masonry.

The technique commonly identified in the literature for control of microorganisms is the use of commercially available chemical biocides, typically in liquid form. Wakefield (1997) has identified the six most frequently used active ingredients found in masonry chemical biocides—alkyl amine salts, quaternary ammonium salts, inorganic borate/zinc salts, sodium salt, and sodium hypochlorite. Chemical biocide treatment requires periodic reapplication to sustain effectiveness. Reapplication is frequently not possible because of funding constraints. Further, there are possible side effects of chemical treatments to control microorganisms, such as the introduction of soluble salts into the masonry being treated.

Passing reference has been made in the literature to an alternative to chemical biocides for control of microorganisms on buildings, specifically, the strategic positioning of metal strips or wire on building facades (Ashurst and Dimes, 1999). Rainwater, washing over the strips, releases oxides and carbonates, which appear to inhibit biological growth. Building surfaces below the metal strips are kept free of microorganisms.

Field conditions documented in this study indicate that rainwater runoff passing over either zinc or copper is effective in inhibiting or preventing biological growth on a variety of substrates, including wood, architectural concrete, and stone masonry. This technique is most effective on new or freshly cleaned architectural materials. Substrates with established colonies of microorganisms are not readily affected by rainwater laden with either copper or zinc oxides and carbonates.

Little research has been done on this method of microorganism control. Given the increasing concern about health hazards associated with using chemical treatments in building conservation, and also given the cost of repeated application of chemical biocides on buildings, investigation of this topic is warranted.

This paper examines the use of two types of metals for control of microorganisms on buildings: copper and zinc. A field test, in place for five years, using zinc strips will be reviewed. Field conditions in which copper strips on wood and masonry appear to inhibit microorganisms will be examined. Current concern regarding possible adverse environmental effects of the use of copper and zinc oxides in solution will be presented.

BACKGROUND Routine maintenance of historic buildings and monuments is a goal infrequently achieved by either private or public stewards of cultural property. This is despite an ever-increasing awareness of the consequences of deferred maintenance on cultural resources and the cost benefits of cyclical maintenance. Conservation treatment of historic buildings and monuments often occurs only after a crisis situation develops in which significant intervention becomes a necessity. At this juncture, funding becomes available for a one-time project. After the intervention project has been completed, the resource is again left until the next critical situation develops. It is evident that in both the private and public sectors, funding for the care of historic buildings and monuments frequently occurs in one large expenditure rather than incrementally. Therefore, cyclical conservation treatments, such as cleaning and periodic application of chemical biocides, are, in many cases, not possible to implement.

In addition to budget constraints, there is the reality that private organizations such as churches, schools, universities, building owners, and historical groups frequently lack the expertise to provide cyclical maintenance for cultural properties. With public agencies, maintenance budgets are sometimes decided upon based on political considerations rather than a rational assessment of need.

Given the foregoing factors, the control of biological growth on masonry structures can be problematic, as chemical biocides should be renewed periodically to remain effective. Further, there is the possibility of adverse effects resulting from repeated treatments. The use of metallic oxides and carbonates in solution to control biological growth offers a possible alternative to repeated application of chemical biocides. Metallic zinc and copper strips appear to remain effective in controlling biological growth for years, thus eliminating the need for reapplication. In addition, the strips can be installed as part of a major conservation effort for relatively little cost. Since funding for major projects is more likely for routine maintenance, this method of biological control has practical advantages because it represents a one-time installation expense.

ZINC AND COPPER OXIDES AND CARBONATES AS BIOCIDES The efficacy of zinc or copper oxides functioning as biocides is frequently evident on buildings with flashing, gutters, or ornamentation made from these materials. Whether the building cladding is wood, stone, or concrete, field conditions demonstrate that surfaces below galvanized or copper flashing are relatively free of algae and lichens for some distance. Dramatic examples of this condition are produced where runoff affects only a portion of a roof or wall surfaces. [FIG. 1] This photo shows a roof penetration flashed with zinc-coated (galvanized) sheet metal. It is evident that the mineral composition roofing below the flashing is free of chlorella-like algae, as is typically found on roofs in this area.

In figure 2 a concrete building wall is shown. [FIG. 2] Portions of this elevation have copper planters under window openings. The wall surfaces below these planters are clean of algae when compared to adjacent wall surfaces.

In each of the examples shown there was no visual evidence of biological growth on the surfaces below the sheet metal. Also, the use of architectural metalwork was either a component of the waterproofing system or a decorative element of the structure, and not installed to keep the building clean. Yet, when strategically positioned, these metals are effective in keeping the building wall or roof free of biological growth.

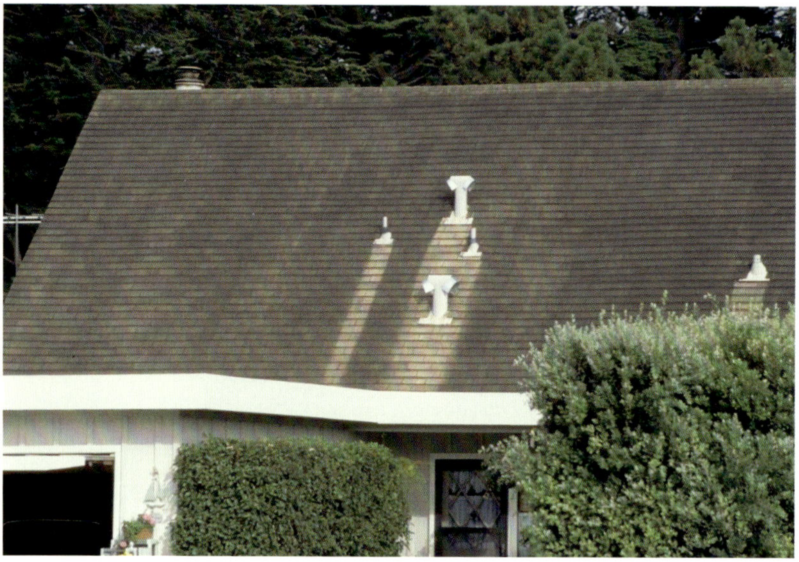

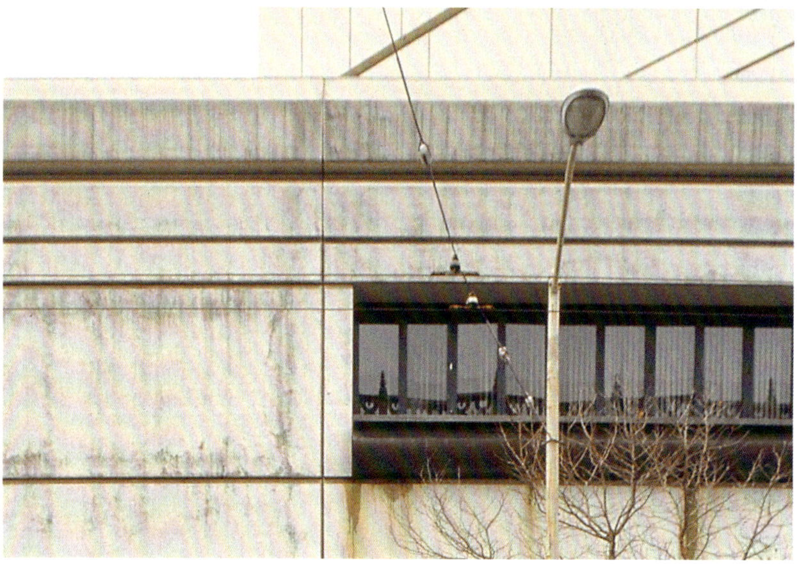

[1] Roof penetration flashed with zinc-coated sheet metal. [2] Copper planters under window openings.

The long history of zinc and copper used in conjunction with masonry and wood building materials suggests that there is no significant deleterious effect produced by oxides or carbonates of these metals. Field observation and testing indicate that this technique is most effective on clean building surfaces as a preventive measure; however, there is evidence that it may be curative over time.

TOXICITY MECHANISMS Copper in various forms has been used for many years as a biocide for plants and fungi. For example, copper carbonate has been used by the wine industry as a fungicide in winemaking equipment. Zinc has been used similarly; zinc is a known fungicide and has been used as an antifungal agent in many products, from paint coatings to carpets.

Both copper and zinc are essential micronutrients, and are utilized by cells of fungi, lichens, algae, and mosses (S. Lebow, personal communication). These cells have mechanisms that allow tight control of uptake of these and other heavy metals; however, when the cells are exposed to high metal levels, their control mechanisms are overwhelmed and the metals begin to disrupt cell processes. In general terms, both metals form complexes with (and sometimes precipitate) various proteins and nucleic acids within the cells, thus disrupting their cell function. In addition, copper is known to form complexes with phosphates, preventing their use by the cells, and is suspected of promoting the conversion of hydrogen peroxide to a destructive oxidant hydroxy radical (OH–).

Algae require trace metals for regular growth, but the optimum amounts are within a limited range. A review of the literature suggests that the actual mechanism of toxicity of metals is not completely understood; it is thought to be by the slowing of the flow of electrons in mitochondrial transfer reactions (Rand et al., 1995). Copper is known to inhibit photosynthesis in algae and nitrogen fixation in blue-green algae (Mikkelsen et al., 2001).

In plants, zinc toxicity can appear similar to copper deficiency, as an excess of zinc will reduce available copper to the plant. Exposure of plants to copper and zinc has been shown to cause micronutrient (iron) deficiency and inhibition of root growth in plants (Ebbs and Kochian, 1997).

In summary, copper and zinc oxides and carbonates, produced by rainwater washing over roof metalwork, inhibit or prevent the growth of fungi, lichens, algae, and mosses because at higher concentrations these compounds disrupt very basic cell processes. However, observation of field conditions suggests that in most cases the carbonates and oxides released by these metals are not of sufficient concentration, nor are they the most effective compounds of these metals to function as biocides. Instead, the mechanism may be more accurately described as that of a fungistat; that is, it inhibits biological growth without destroying it. This helps to explain why it is difficult to eradicate well-established colonies of algae, mosses, and lichens using only this technique without an initial cleaning of the monument or building.

ENVIRONMENTAL CONCERNS An important question about the use of zinc or copper metal to control biological growth on buildings, as well as about their use in general on buildings, is the effect of these metals on the environment. Concentrations of both zinc and copper have been shown to be toxic to freshwater and marine invertebrates and to freshwater fish. This has led water-quality agencies to establish limits for acceptable amounts of heavy metals in water bodies. In addition to industrial pollution and runoff from streets and highways, roof runoff has been identified as a source of metals in water. Until relatively recently, roof runoff was assumed to be clean and not to pose the potential of affecting water quality (U.S. EPA, 1983). This topic has now attracted the attention of urban hydrologists, and several studies have been completed in the past two decades on this topic. There now appears to be disagreement in the literature about the quality of roof runoff water; some researchers have concluded that runoff is polluted by heavy metals while others have found low pollution potential (Good, 1993). This is due to a lack of reliable and comparable data, according to one researcher (Forster, 1999).

Both zinc and copper have been identified as present in roof runoff water. The obvious source is copper- and zinc-plated roofs, flashing, and ornaments. Zinc-rich roof coatings, used primarily to protect flat roofs from ultra-violet degradation, are another source of zinc in roof runoff. One study suggests that galvanized roofing materials are a prime source of zinc in urban area waters (Good, 1993). The amount of copper in roof runoff is less easily quantified.

Roof runoff water quality will be receiving increasing scrutiny because it is now understood that roof surfaces account for approximately half of the total runoff volume from impermeable surfaces in urban areas of industrial countries (Forster, 1996). For industrial areas, roof runoff accounts for 61 percent of the zinc in runoff, and for commercial areas, 17 percent. The percentage is much lower for residential areas, at about 8 percent.

Currently, there is no governmental restriction on the use of galvanized metal, zinc, or copper for roofing or roofing materials such as flashing. However, until recently, commercially produced zinc strips for use on wood or asphalt shingle roofs were classified by the Environmental Protection Agency as a pesticide under federal law. Trade groups petitioning the EPA in 1994 were eventually able to have this product delisted in 2001. Thus, the product is no longer required to comply with federal registration and testing for pesticides. For the purposes of disposal, the federal government does not consider wastes hazardous on the basis of zinc content. However, California and Michigan label wastes as hazardous because of zinc content, although trade groups in both these states are now lobbying to remove zinc from the hazardous waste listing.

The complexities associated with the study of roof water runoff preclude simple answers regarding the environmental effects of using zinc or copper metal to keep buildings clean of biological growth or of their general use in the roofing trade. It should be noted that employing narrow strips of these materials on buildings would represent an infinitesimal increase in exposure when compared to entire roofs constructed of copper or galvanized metal. Consequently, it is reasonable to assume that this method of biological growth control on buildings would have minimal effect on overall water quality for a given area. The most responsible approach is to consult the regional water quality control board to learn if an excess of either of these metals presents a problem for a specific geographic area.

USES IN ARCHITECTURAL CONSERVATION Commonly agreed-upon conservation methodology requires a comprehensive condition assessment prior to any treatment. In practice, though, this frequently does not occur for buildings, and the type or cause of staining or soiling on structures is not identified. In many cases, what appears to be atmospheric soiling or staining is instead biological growth or a mixture of the two. For porous materials such as sandstones, algae and lichens can be very difficult to remove, as they penetrate the interstices of the stone. Aggressive attempts to remove what is mistakenly believed to be soiling, but is actually well established biological growth, can result in significant damage to building materials. Use of metal strips can be effective in preventing biological growth from becoming established. Over many years this method may provide a means to eradicate existing algae, lichens, and mosses that otherwise could not be removed without damaging masonry and wood substrates.

The architectural design of a structure will, in large part, dictate if the technique under discussion can be used successfully. Copper or zinc strips can be placed at the top edge of parapets, cornices, or other projections. Ornate facades with multiple projections or uneven wall surfaces would be difficult to treat. Many historic buildings and monuments fall into this category. However, an increasing number of modernist buildings are becoming recognized for their cultural value.

The designs of these structures, with expanses of flat exterior wall surfaces, are more conducive to the use of this technique of biological control. The Hoover Tower at Stanford University is a historic modernist building. It is shown, in figures 3 and 4, before and after cleaning. [FIGS. 3 AND 4] The cleaning was undertaken approximately five years ago, and biological growth is beginning to reappear. It would have been simple and very inexpensive to have installed metallic strips at the observation level to keep the wall surface below clean of biological growth.

The potential for discoloration of the wall surface to be treated by copper or zinc oxides is a consideration when deciding between these two metals. Copper oxides can be noticeable on lighter-colored stones, and they produce staining. There is also the possibility of zinc oxide producing light-colored streaks; however, the author has not noticed this occurring on any of the examples cited.

The efficacy of this treatment will depend on a number of factors. Slope and exposure are obviously critical. Field observation shows that steep slopes and vertical wall surfaces are more easily kept clean using metal strips, and that walls can be treated at much wider intervals than roofs. North- and east-facing roof and wall surfaces that are kept wet or moist for long periods are more conducive to many types of biological growth and are therefore more difficult to keep clean. It may not be possible to keep surfaces with significant shade from trees or buildings free of biological growth, even if these surfaces are cleaned and strips are installed.

EXAMPLES OF ZINC STRIPS ON STRUCTURES Zinc is a common metal that is brittle at room temperature. It is used in coatings for iron and steel to inhibit corrosion, as a component of fertilizers and animal feed, as an alloy with other metals, and in pharmaceuticals and cosmetics. The ordinary compounds of zinc are considered to be of low toxicity.

The effectiveness of zinc and zinc-plated steel in the control of biological growth—particularly mosses—on wood roofs has been known in the Pacific Northwest for 20 years or more. Mosses and other types of biological growth are particularly prevalent on roofs in the moist Pacific Northwest climate. Several manufacturers produce strips of pure zinc marketed as a method to keep wood and asphalt roofs free of mosses.

For large roofs, manufacturers recommend that the strips be installed in rows horizontally every 15 feet. They also recommend that roofs be cleaned prior to installation of the zinc strips. Anecdotal information suggests that the zinc oxides and carbonate will eventually kill mosses and algae, although it can take as long as eight to ten years for the growth to wither, die, and wash off the building.

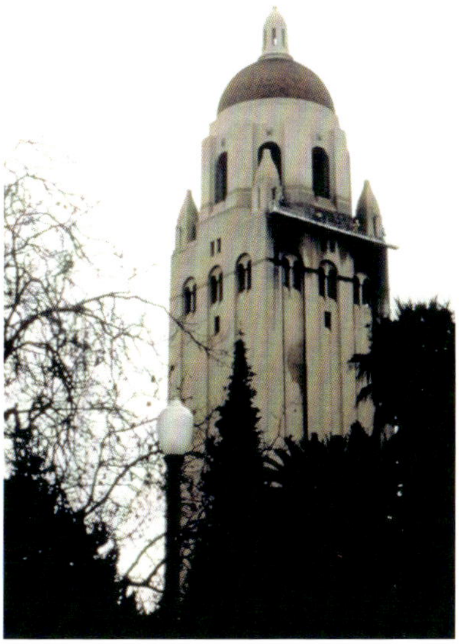

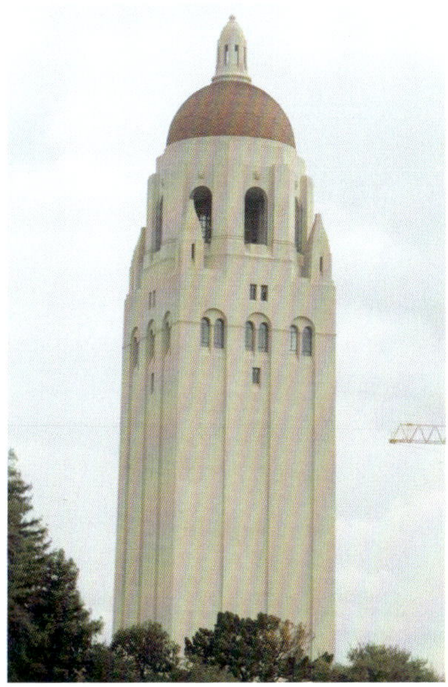

[3] Hoover Tower at Stanford University showing heavy accumulations of biological growth on one elevation before cleaning. [4] The tower after complete cleaning of the structure.

Zinc strips are being tested in Washington, D.C., on the National Holocaust Museum. To keep the exterior concrete masonry walls free of biological growth, zinc strips were installed along the parapet at test locations. After approximately 18 months, results are encouraging because the biological growth is not returning.

Case Study—Stanford Mausoleum The Stanford Mausoleum is a small, one-story neoclassical building measuring 25 by 40 feet in plan. [FIG. 5] It was constructed in 1889 of gray-colored ashlar granite masonry from Barre, Vermont.

In 1995, the university commissioned a conservation study of the monument. It had been many years since the building had had any attention, and its condition had reached a crisis situation with severe water penetration problems. There was significant soiling of the masonry, as well as heavy accumulations of biological growth.

The roof of the structure had very heavy accumulations of lichens, the concentrations of which were equal on both the north and south sides. [FIG. 6] These lichens have been identified as being both foliose and crustose lichens. The foliose are parmelioid, probably genus *Xanthoparmelia,* which is a common lichen found in the San Francisco Bay area.

Based upon previous experience it was unlikely that the structure would receive cyclical maintenance in the future. There would be a one-time expenditure for this project, and then the monument would be left for another significant length of time without routine maintenance.

Use of copper wire or flashing at the peak of the mausoleum roof as a means to keep copper oxides washing over the roof slab was considered. This was rejected because of the potential of staining the light-colored granite. Based upon these considerations, zinc was selected instead of copper.

Test strips of zinc were installed in 1995 on both cleaned and unclean portions of stone to observe any effect on the masonry, including staining. [FIG. 6] The strips have a slight texture to provide a greater exposure of surface area. After approximately 20 months no unsightly staining or other deleterious effects were observed, and there was no evidence of algae or lichen. The condition of the lichen in the uncleaned portion of stone below the zinc was unaffected.

In August 1997, the entire roof was cleaned using a pressure washer and a dilute masonry cleaning product containing hydrofluoric acid. Zinc strips were installed in the same manner as the test panels. [FIG. 7] They were bent along their lengths to conform to the angle of the roof peak, and kept in place with a urethane sealant as an adhesive.

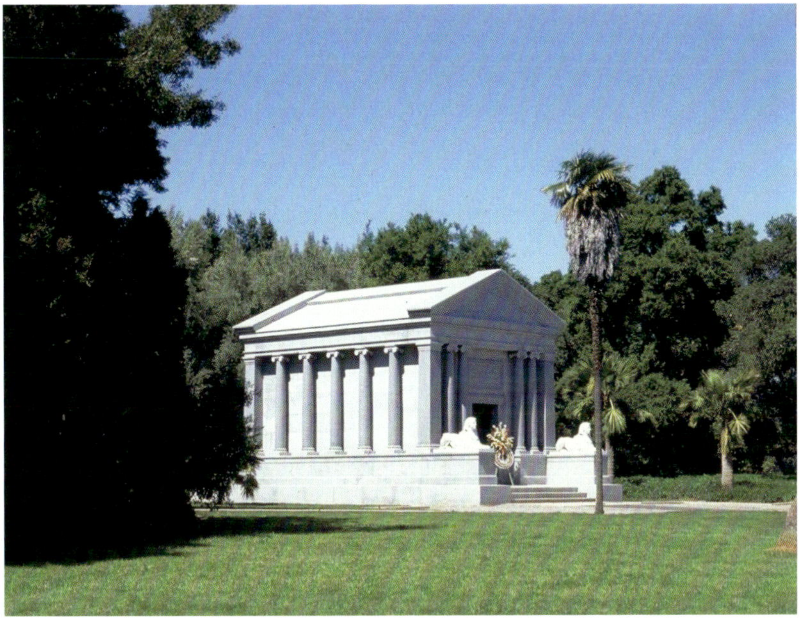

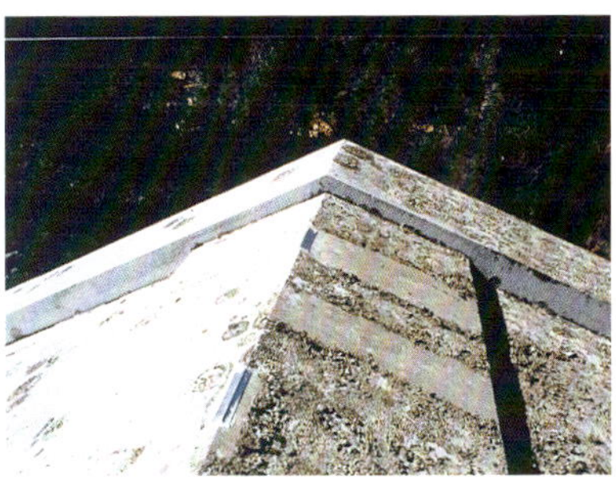

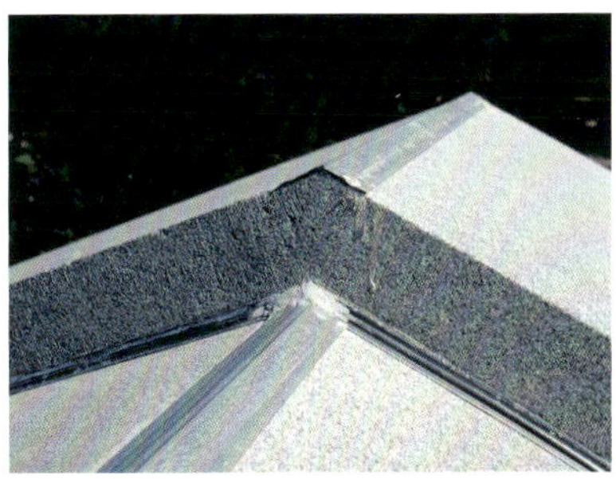

[5] Stanford Mausoleum. [6] Roof structure of the mausoleum showing heavy accumulations of lichen. Strips of zinc have been installed at the roof peak. [7] Roof structure of the mausoleum showing zinc strips installed.

The structure was examined in May 2002. There was no evidence of algae or fungus on the roof surfaces, and the strips have remained well-adhered to the masonry substrate.

EXAMPLES OF COPPER STRIPS ON STRUCTURES Copper formed into sheet metal is a pliable and malleable metal that has a bright reddish metallic luster. In addition to use in roofing building materials, it is used in marine paints, plumbing fixtures, and pipes. Copper compounds are commonly used in wood preservatives and fungicides.

Figure 8 shows the asphalt shingle roof of a historic brick structure that was replaced in 1994. [FIG. 8] Copper strips were installed under the ridge shingles along the roof peak to inhibit biological growth on the shingles below. Recent inspection of the roof indicated that the roof remains clean of algae or other lower plant forms.

Green algae are common on wood buildings, as well as on masonry buildings in northern California. The structure shown in figure 9 is located on the coast, approximately 20 miles from San Francisco. [FIG. 9] It was originally constructed with copper flashing that has kept wood shingles below effectively clean of algae and lichen. Above the copper flashing there is a significant amount of algae.

In one location on the east side of the building shaded by a tree, algae are present even under the copper flashing. This is an example of how algae with moist conditions can survive even when washed by copper carbonate and oxides.

CONCLUSION Oxides and carbonates produced from rainwater flowing over zinc and copper strips have been shown to inhibit the growth of mosses, lichens, and algae on clean building surfaces. This use should be considered as having the qualities of a fungistat rather than a fungicide because eradication of existing biological growth on buildings could take years. Factors that will affect how well this technique works are exposure, shade, and slope.

The architectural design of a given structure will determine if the technique can be used effectively. Zinc may be a better choice on light-colored masonry materials, while copper can be used on darker masonry or wood.

There is currently no restriction on the use of copper and zinc as building materials. The responsible practitioner should check with the regional water-quality control board to determine if specific concerns regarding zinc or copper exist in the municipality where the building or monument is located.

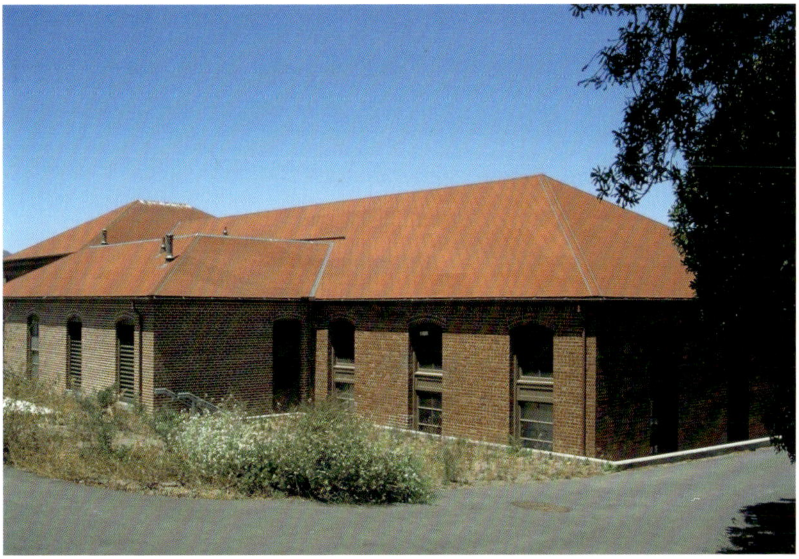

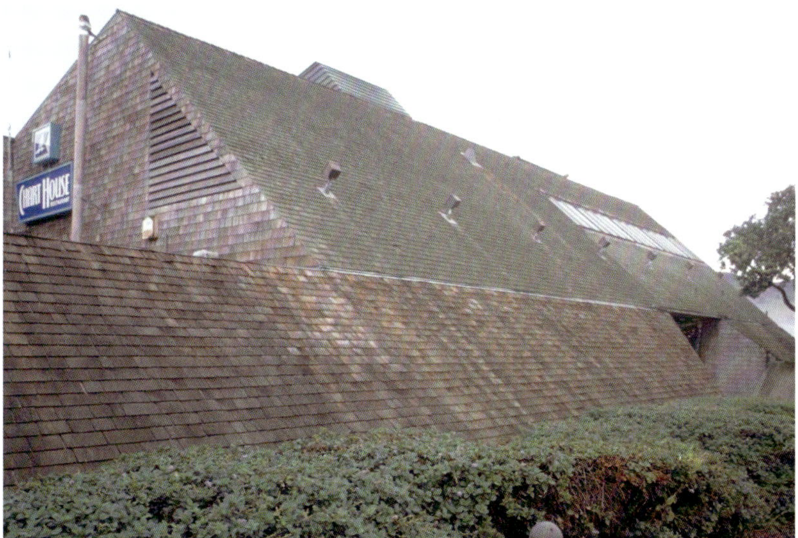

[8] Historic brick structure with copper strips installed under ridge shingles. [9] Copper strips installed under ridge.

The appeal of this technique for inhibiting biological growth is its simplicity and relatively low cost. It can be installed as part of a major conservation effort and minimize the need for future cleanings. For tenaciously adhered algae and lichens, these techniques may be useful over time for eradication of the growths in a manner that does not damage substrates of wood or masonry. Further research and field-testing of this technique are warranted to develop practical guidelines for its use.

REFERENCES

ASHURST, J., DIMES, F.G., 1999. *Conservation of Building & Decorative Stone*, Butterworth-Heinemann. Part 2, pp. 136–137.

BITTON, G., KOOPMAN, B., 1982. Tretrazolium reduction-malachite green method for assessing the viability of filamentous bacteria in activated sludge. *Applied Environmental Microbiology* 43 (4), 964–966.

FORSTER, J. 1999. Variability of Roof Runoff Quality. *Water Science Technology* 39 (5), 137–144.

FORSTER, J. 1996. Patterns of Roof Runoff Contamination and Their Potential Implication on Practice and Regulation of Treatment and Local Infiltration. *Water Science Technology* 33 (6), 39–48.

GOOD, J.C. 1993. Roof Runoff as a Diffuse Source of Metals and Aquatic Toxicity in Stormwater. *Water Science Technology* 43 (3–5), 317–322.

LEBOW, S., 2002. Conversation with author. USDA-Forest Service, Forest Products Laboratory, 1 Gifford Pinchot Drive, Madison, WI 53705–2398.

RAND, G.M., PETROCELLI, S.R. (Ed.), 1985. *Fundamental of Aquatic Toxicology*, Taylor and Francis, Inc. pp. 374–415.

U.S. EPA, 1983. *Results of the Nationwide Urban Runoff Program*. Vol. 1, Final Report. Washington, DC, p. 200.

WAKEFIELD, R., 1997. Masonry Biocides—Assessment of Efficacy and Effects on Stone. *Scottish Society for Conservation and Restoration (SSCR) Journal* 8 (2), 5–10.

SUPPLIERS

Shingle Shield
Chicago Metallic Corporation
4849 South Austin Avenue
Chicago, IL 60638

Z-Stop
WesPac
4015 164th Street SW
Lynwood, WA 98037–6924

ENDNOTE

A.R. CAVALIERE

IN DEFENSE OF THE FUNGI

Gettysburg College, Gettysburg, PA

For three days, during the ABC 2002 symposium and in the dozens of papers written and presented at these proceedings, we've heard a multitude of negative accounts about those "terrible" fungi around us. Time and time again, the destruction they cause to our magnificent historical, cultural, and artistic pieces has been emphasized with vigor. They have been labeled as nasty, rotten, foul, dirty, putrefying, destructive culprits whose aim is to undermine all the treasures of cultural heritage we have so creatively and carefully produced throughout the ages. It saddens me to have heard all these dreadful things attributed to the fungi, for they are, without doubt, one of the most important forms of life on earth. True, they are seasoned animal and plant pathogens and masters at biodegradation, but that is what they are supposed to do! We would surely be in a sorry state if fungi found other ways of eating and left our organic matter to pile up around us. They truly are amazing, for they nourish us, they medicate us, they titillate our taste buds, they inebriate us, they make us hallucinate, and in the end, they dispose of us. In the next few pages, I'd like to tell you some things about these wondrous fungi—what they are, where they come from, how they live, and why we should be grateful to them.

WHAT ARE THESE DREADFUL THINGS? In the classification scheme, fungi are organisms that are neither plant (Kingdom Plantae) nor animal (Kingdom Animalia); they are in a Kingdom of their own, (Kingdom Eumycota). Recent genetic studies of their DNA reveal that they are more closely related to animals, and therefore humankind, than they are to plants. In scientific terminology, the Kingdom Eumycota has an elaborate taxonomic series of levels, but to simplify things all the fungi may be divided into three groups—molds, mushrooms, and yeasts. Of the three, molds and yeasts are of primary concern to conservators, while the mushrooms include some that are more of a culinary delight, and others that are not so delightful!

Molds are filamentous in structure; their body is composed of threads called "hyphae." The hyphae are themselves composed of microscopic units or cells, which, when attached in long chains, are visible to the naked eye. A mass of hyphae, whether growing on a ripened fruit or an exquisite piece of moisture-soaked artwork, usually forms a disc-shaped, more or less flattened matting termed a "mycelium." A hyphal strand placed in the center of a suitable medium in a culture dish will produce a mycelium that will grow and increase in size circumferentially to the edge of the dish. It is the hyphal strands that form the substance of the fungi and are the "culprits" that digest away and deteriorate the substrate from which it draws nourishment.

Mushrooms, the most recognized and probably the most treasured of the fungi, are also composed of filaments of hyphae. Most of the mushroom hyphae are underground, decomposing the roots of some tree, the litter of fallen leaves or branches, or the remains of some long-dead insect or animal. The mushroom itself is composed of a massive number of intertwining hyphae, and, actually, the mushroom is to the fungus what an apple is to a tree. The mushroom forms only when the fungus has had its fill of nutrients and is ready to extend itself above ground and disseminate spores. The so-called "fairy ring" is a result of the circumferential growth of the hyphae underground, similar to the mycelial mat growing in a culture dish. The mushrooms form along the periphery of the colony. As the colony ages, the older, center portion dies out and what we see above ground is a ring of living fungus tissue.

Yeasts are so different from either molds or mushrooms it is a wonder that they are called fungi. Their body plan is very simple, and they most closely resemble bacteria in their structure, except they are much larger. These fungi are spherical cells and form short or long chains that are excellent at digesting living or dead plant material and animal remains, as well as finding their way into ears, eyes, and other human crevices that are best left unmentioned.

THEY'RE EATING AWAY OUR WORLD! One of the most astonishing attributes of the fungi is the variety of materials they can eat. While green, chlorophyll-bearing plants are capable of manufacturing their own food source (carbon compounds) from carbon dioxide and water, animals and fungi must rely on complex carbon compounds already formed. They must live on the products directly or indirectly made by plants. Fungi first claim an area upon which to grow, and then begin to digest their surrounding by using a huge storehouse of enzymes. These masters of digestion have evolved an enzyme system that rivals that of humans, and in some ways surpasses it—consider that they can eat and digest wood.

Fungi differ from animals in the way they obtain their nutrients. Animals take food (a large complex of molecules) into their bodies and by the use of selective enzymes digest the material into smaller, usable molecules that are then absorbed into the cells that make up the digestive system. Fungi have not evolved a cavity or mouth; therefore they cannot absorb large complexes of molecules. Their method of obtaining nutrients or eating is unique: They secrete digestive enzymes from their hyphal cells into their immediate environment. The enzymes that are released digest the material upon which the fungus is growing, and then the hyphae absorb the smaller, simpler molecules into their cells. It is this process of fungal digestion that frustrates so many, especially the art conservator, for the nutrients used by the fungus may be the deteriorated remains of an historical document or cultural piece. With the exception of plastic, fungi can eat and digest almost any kind of material—anything containing cellulose, most proteins, including the chitin of insects, and an array of lipids. Pages can be filled with the specific kinds of materials fungi can digest. And when they partner with some algae and form lichens, these hungry organisms are capable of chipping away at stone, marble, and granite.

THEY START FROM A TINY BEGINNING Each new fungus, regardless of the size of the grown colony, has a simple beginning: It starts from a spore. There are meager requirements for spores to begin their "reign of terror"; they germinate in the presence of a little moisture and a broad range of temperature, the optimum being between 25 and 30°c. Although it's a much less complex machine than a seed, the fungus spore behaves the same way a seed does. It absorbs water, swells, kicks its enzymes into gear, and grows a little germ tube, the forerunner of literally thousands of genetically identical hyphae that will eventually make up the visible colony. But the little spore cannot land just anywhere and expect to flourish; the landing pad must be, in most cases, composed of some variety of organic matter. Once penetration of the surrounding material has been achieved, the hyphae will begin to elongate, produce side branches, and proliferate. As the hyphae grow they release both enzymes and metabolic byproducts we term "mycotoxins." The fungi are no fools—they never release all their digestive enzymes, as that is too costly in terms of energy loss. Instead, the enzymes they secrete are specific for the meal at hand—amylases for starch digestion, cellulases for wood, paper, and other cellulose-containing material, cutinases for plant leaf and young stem surfaces, and an array of other enzymes capable of taking apart the most complex structural design of animal or plant parts. Because there is fierce competition for nutrients out there, the fungus's best strategy is to keep freeloaders, such as bacteria and other microbes, from moving in on their territory and eating the goods. And the fungi do it quite effectively; they have evolved an elaborate assortment of metabolic byproducts that are released into the surroundings and are toxic to other fungi, microbes, and even small insects and animals.

GRANDMOTHER'S PORTRAIT WASN'T MOLDY YESTERDAY! The big question on everyone's minds, I'm sure, is "What makes mold grow so fast?" It follows that if a fungus intends to live from one generation to another, the mother colony must find a way to produce offspring and spread them out a bit. And the colony does just that—as the hyphae continue to grow and mature, they begin to generate spores, called conidia. But, really, the fungi overdo it! Colonies may produce hundreds of thousands, and even millions of spores over and over again. If they all found a suitable place to land and reproduce, we'd be totally swamped by mold. But they don't and we aren't. Only a very few land on a suitable substrate with the right environmental conditions to allow them to germinate.

Another amazing attribute of fungi is that they can produce these conidia with or without sex. Day-to-day production of conidia is a rather simple task accomplished by their proliferation from the mother colony. Terminal hyphae develop simple or elaborate stalks, branches, or heads upon which spores are formed and pinched off. Because most of the time the conidia are dry and powdery, they are easily whisked away by the slightest breeze, or attached and carried away by insects that may have inadvertently come in contact with the colony. The massive populations of fungus spores are generated without sex, that is, without the intervention or cooperation of another colony of the same fungus. They are derived from a single lineage; therefore all the millions of spores that are produced, with the exception of a spore mutation or two, are genetically identical to those of the parent colony and to each other. Learned mycologists often refer to this massive reproductive strategy as "fungus amplification" because it produces so much so fast. However, like every other living creature, the fungi are sexual as well. Books have been written about the why, ways, and when of fungal sexual encounters, and these matters are much too complicated and confusing for discussion here. The essential thing to know is that when the environmental conditions of temperature, nutrient level, and moisture are conducive, and a particular mold finds itself in the company of another sex, a union of the two strains will ensue. The exchange of genetic material between them will result in the production of hundreds or thousands of conidia that are now genetically different from both parent colonies and, often, from each other.

Many people may still be convinced that this world would be a much better place to live if we were free of, at least, snakes, mosquitoes, and fungi. I can't speak for snakes and mosquitoes, but I can assure you that without the fungi life would be nonexistent for some, miserable for many, and a lot less exciting for most. The attributes of fungi, both good and bad—mostly bad—are well documented in hundreds of volumes and thousands of scholarly papers. The subjects discussed during this symposium focused on the evil that fungi do, and, believe me, they commit evils much more serious than the biodeterioration of artworks, documents, monuments, and other historical materials. But because many of you will read the papers presented in this volume and leave with a one-sided view of the fungi, I think it only fair to pay tribute to some of the good they do.

MUSHROOMS, A CULINARY DELIGHT Who among us has not enjoyed the succulent taste of the oyster mushroom (*Pleurotus ostreatus*), or shiitake (*Lentinus edodes*), morels (*Morchella esculenta*), chanterelles (*Cantharellus cibarius*), the common meadow mushroom (*Agaricus bisporus*), or, if you can afford them, black or white truffles (*Tuber melanosporum* and *T. magnatum*)? Some 20 edible species are now grown worldwide and enjoyed by almost every culture. Steak, lamb, pork, sausages, and pizza would surely be lacking without these succulent morsels, and where would mushroom soup be without the "'shrooms?" I clearly remember as a child, during the spring and fall months, being taken by my father, along with several other of his seven sons, into the forests of Connecticut to collect mushrooms. It was a ritual, every Sunday after attending church, or often instead of church. We never collected all of the edible species, only the ones my father knew for certain were edible. Later in the day, mom would take over in the kitchen and prepare some dishes that were so delectable, they are among my favorites to this day—mushrooms pickled in wine vinegar with garlic and herbs; mushrooms fried with Italian sausage, bell peppers, and onions; mushrooms with a variety of pastas and sauces. It is a wonder I ever left home!

For much of our history, while European and Asian cultures regularly gathered and consumed fleshy fungi, most of the English-speaking world looked upon mushrooms or "toadstools" as evil, mysterious, and poisonous creations of elves, witches, or devils. Mushrooms, after all, grew only where lightening strikes the forest floor, or on the very spot wetted by the semen of mating deer. Or even worse, it was a common belief that mushrooms grew wherever witches and warlocks urinated, expectorated, or defecated upon the ground. It's a blessing that we've at least passed that historical period!

Mushroom tissue contains a number of chemical compounds that can be nutritional, hallucinogenic, poisonous, or a combination of the three. The nutritional value of mushrooms has been well established in recent years. Although not considered a health food by most nutritionists, mushrooms do contain several amino acids, as well as carbohydrates and high levels of minerals and vitamins. They are free of cholesterol, low in calories, and high in fiber. They contain an insignificant amount of fat (although of course the fat content will increase substantially when mushrooms are prepared the way a lot of people like them—sautéed in olive oil or butter, or smothered with gravy!)

With about 10,000 species of mushrooms growing in the wild, only a handful are poisonous, and, of those, an even smaller number are deadly. How, then, do we account for the significant number of deaths that occur each year from mushroom poisonings? The answer is that the deadly ones are common, appear often during the growing season, usually smell great, and are very good looking! *Amanita virosa*, the "death angel," is a perfect example; it is almost pure white and really looks angelic—but looks can kill! Even though I'm Italian, I offer rather conservative advice for the would-be mushroom pickers who are inexperienced at distinguishing between safe and poisonous mushrooms—buy your mushrooms at the grocery store. Many guides and texts have been written that deal with mushroom poisonings, how to recognize edible and poisonous varieties, and how to identify mushrooms and other fleshy fungi in the field. In addition to Lincoff and Mitchel (1977), Arora (1986), Phillips (1991), Schaechter (1997), and Barron (1999), you can find over 550 publications dealing with the subject of mushrooms listed on the Internet.

A DIFFERENT KIND OF TRIP "The sacred mushrooms of Mexico," "magic mushrooms of Russia and Siberia," "soma of the gods," "Teonanacatl," and "black 'shrooms from California" all are designations for extraordinary mushrooms that are hallucinogenic. *Psilocybe cubensis, P. mexicana, Amanita muscaria*, and species of *Conocybe, Gymnopilis*, and *Stropharia* are the ones most often associated with mind-altering chemicals. The chemicals tend to work on the central nervous system and result in reactions that have been described as feelings of happiness, liberation, or religious awakening, or as having spiritual overtones. Rage, violence, and disorientation may also result, depending on the amount of mushroom consumed and the mental state of the individual, as well as on his or her physiological condition. Whether the "trip" is a good one or a bad one, it is not uncommon for the individual to conclude the experience with what I call NDS—nausea, diarrhea, and sleep. While several excellent texts discuss this topic in detail, the classical work must be credited to Wasson and Wasson (1957). There is also an authoritative treatment by Lincoff and Mitchel (1977), an informative and readable update by Benjamin (1995), and one devoted entirely to *Psilocybe* and related genera, the most notable of hallucinogenic mushrooms (Stamets, 1996).

GIVE US THIS DAY OUR DAILY BREAD AND WINE It is often difficult to believe that yeasts (*Saccharomyces cereviseae* and several other species) are absolutely essential for the very existence of bread and wine. Both substances depend on the biochemical activities and byproducts of this productive fungus, a process often referred to as alcoholic fermentation. When yeasts are supplied with even a simple source of food and some moisture, they flourish. Provide them with a little sugar from grapes or starch from flour, and they can do marvelous things! As the yeasts consume the nourishment, the cells begin to grow and divide prolifically. Two byproducts are always generated—carbon dioxide and ethyl alcohol. Which of the byproducts is desired depends entirely upon whether the fermentation is done by a brewer for beer, by a distiller for wine and other forms of alcohol, or by a baker for bread. Note the use of the word "byproducts" rather than "waste products." On earth nothing organic is ever wasted, for one organism's byproduct will most certainly be another organism's sustenance. The brewer and distiller are interested primarily in the alcohol byproduct, while the baker is interested in the carbon dioxide gas that is generated by the respiring yeasts. The purpose of promoting alcoholic fermentation in baking is to provide gas production, which is responsible for the rising process in the bread; the alcohol burns off in the oven and releases that heavenly aroma around the kitchen. The various flavors and textures of bread result from the various grains used, the flavors added, and the degree to which yeasts are allowed to ferment the starch.

Although wine is usually fermented grape juice, it can be made by using several kinds of fruits, herbs, or even flowers. Once the plant parts and yeasts are added and some liquid is supplied, the yeasts begin to do their thing; the organic base is slowly consumed, gas is produced and released, and the sugars are converted to alcohol—a miraculous evolutionary process! In wine, the alcohol concentration reaches about 12% before the yeasts begin to die. Some wines, such as sherry, are fortified; that is, alcohol is added to them after the fermenting process is complete. Wines vary tremendously in color and taste. The flavor of red wines comes from the grapes used, the strain of yeasts, and the specific fermentation process. When finished, wines come in all shades of red, pink, and white. White wine is fermented from skinned red grapes, white grapes, or apples. For the serious drinker, "hard liquor" is also credited to fermentation by yeasts. Scotch is the product of fermented barley, rye comes from fermented rye, and we get bourbon from fermented corn.

SOME MOLDS ARE GOOD FOR YOUR HEALTH! For better than 50 years fungi have played a major role in our medical well-being. Although it has only been a short time that fungus compounds have been processed or manufactured for medicinal purposes, their use as aids to combat infection dates back to the Egyptians. In fact, when the remains of a Stone Age man who had been traveling some 5,000 years ago in the Tyrolean Alps was discovered in 1991, he was carrying with him tissue from three different mushrooms. Current research suggests that some of the mushrooms may have served as a source of medicine. Ancient humans most assuredly gathered mushrooms not only to eat but also as a means of curing many ailments. In many Asian cultures, especially China and Japan, mushrooms have been used for centuries for medicinal purposes. Certain mushrooms are believed to fight infection, enhance vitality, and promote good health. Shiitake (*Lentinus edodes*), an edible and delicious mushroom, not only tantalizes our palate, but also contains lentinan, a chemical that enhances the activity of the immune system. This substance has been shown to be extremely useful in boosting the activity of a cancer patient's natural interferon and interleukin production, which is then capable of fighting tumors. Today, in the western world, while mushrooms are enjoyed over steaks and in numerous other dishes, they are less commonly used to heal our bodies, but other fungi abound that are able to fight infectious microbes that invade us.

It began with the discovery of penicillin, from the fungus *Penicillium chrysogenum*, by Alexander Fleming in 1928. Today, our doctors can prescribe a plethora of pills made of fungus derivatives that we take casually to help cure us of bacterial as well as fungal infections. Until recent times, these microbes would have killed us. Griseofulvin, derived from *Penicillium griseofulvum*, is an antifungal agent rather than one that is antibacterial. It is one of the major drugs used today for fighting mycoses (fungal infections of the nails, toes, and several other epidermal and subdermal infestations). Within recent years many new fungus-derived drugs have made their way into general use. Cyclosporin, derived from *Cylindrocarpon lucidum* and *Tolypocladium inflatum,* suppresses the activity of the immune system and is, therefore, widely used in organ transplants to inhibit tissue rejection by the recipient. Even in the lowering of cholesterol and the treatment of heart disease and arteriosclerosis, fungi are playing a positive role. Compactin and Lovastatin are only two of the commercial drugs produced by several species of fungi, among which are *Penicillium, Aspergillus, Trichoderma,* and *Hypomyces.*

In addition to blemishing thousands of pieces of artwork, *Aspergillus niger* grown under the proper culture conditions produces an enzyme that can lower methane gas production. What a tremendous advantage for some human digestive systems! The enzyme is now incorporated into a commercial product that can be taken by the more gaseous individuals in our society. The remedy, called Beano, which is now advertised frequently in magazines and seen on national TV in the evening when we are trying to enjoy dinner, is available in drugstores throughout the U.S. and Canada. It's just another small measure of the good molds do!

Some fungi team up with other organisms, resulting in the formation of some interesting partnerships. One of these is known as Kombucha, a gelatinous, symbiotic growth composed of yeasts enclosed in a matrix of bacteria. Kombucha tea, whose reputation has become widespread, is a drink derived by culturing the two organisms on a mixture of tealeaves, water, and sugar. The symbionts begin to grow on the surface of the tea and secrete a number of amino acids into the liquid. After a week or two of mellowing, the liquid is filtered and consumed at a quantity of 150 ml or more per day. One has to be brave to drink the mixture after seeing it; I was! The bouquet is sweet-smelling and the flavor is subtle, with a lightly acidic aftertaste. The drink, which has been popular in various parts of the world for over a thousand years, has recently become very fashionable in the west; my batch came from California. Although its medicinal properties are questionable, people who drink it are convinced of its health benefits, which they claim include, among many others, cure or suppression of AIDS, curing of various types of cancer and arteriosclerosis, lowering of blood pressure, enhancement of the sex drive, and the reversal from graying to natural hair color in mature adults.

PLEASE PASS THE ROQUEFORT CHEESE Cheese is a protein-lipid, semisolid or solid material manufactured from milk. While bacterial fermentation is primarily responsible for the basic production of cheese, some of the world's finest cheeses are indebted to fungi for their distinctive tastes and textures. Whether they be mild, sharp, or pungent, their flavor, texture, and aroma result from the activity of yeasts or molds. In a very organized and systematic way, the molds feed on the milk products by secreting proteolytic and lipolytic enzymes. The enzymes break down the protein and lipid components of the milk and absorb the reduced molecules into their hyphae, which then release a number of chemicals into the cheese. Even when the cheese is freshly purchased from the market and we hurry home for a delightful morsel accompanied by a glass of red wine, keep in mind that fungi may have started eating that cheese months ago! The fungal hyphae are responsible for the blue "veining" often found in cheese, while their released byproducts offer a multitude of flavors and textures. The flavor of many varieties of cheese, then, is a result of the delightful and universal activity we often call "rotting." *Penicillium roqueforti* is the mold responsible for rotting (ripening might be a more tasteful word) the blue-veined variety of cheeses like Roquefort, blue cheese, Stilton, and that delightful Italian treasure, Gorgonzola. *Penicillium camemberti* plays the crucial role in the maturation of Camembert and Brie. While *P. roqueforti* works on the inside of the cheese, *P. camemberti* starts digesting the milk product from the outside and works in, resulting in a cheese that is hard on the outside and soft (slightly digested) on the inside.

WHEN ALL IS SAID AND DONE Fungi do indeed threaten humankind when one considers the tremendous damage they can to our crops, our livestock, our historic treasures, and our health. On the other hand, through their continuous process of decomposition, they help our world renew its resources by returning chemical elements back into the fabric of the earth for their eventual regeneration. Moreover, fungi not only benefit humans directly with food, medicine, and an array of products, their association with other animals and plants can enhance our lives as well. When we weigh all of the good against all of the not so good, we can't help but declare that fungi are fantastic!

REFERENCES

For well over 25 years Constantine Alexopoulos and his coworkers have produced the most authoritative textbook in the field of mycology; it is now in its fourth edition (Alexopoulos et al., 1996). Several more, excellent ones have appeared within recent years, the most current of which is Kendrick (2000). Throughout the past half-century numerous classical books have been published that cover the history, folklore, and general nature of mushrooms and fungi; some of the best are Rolfe and Rolfe (1925), Large (1940), Ramsbottom (1953), Gray (1959), and Christiansen (1961). More recently several extremely readable and delightful books on fungi have been published; two in particular are Hudler (1998) and Moore (2001).

ALEXOPOULOS, C.J., MIMS, C.W., and BLACKWELL, M., 1996. *Introductory Mycology* (4th edition). John Wiley and Sons, Inc. New York, NY.

ARORA, D., 1986. *Mushrooms Demystified* (2nd edition). Ten Speed Press. Berkeley, CA.

BARRON, G., 1999. *Mushrooms of Northeastern North America*. Lone Pine Publ. Edmonton, Alberta, Canada.

BENJAMIN, D.R., 1995. *Mushrooms, Poisons and Panaceas*. W.H. Freeman Publ. New York, NY.

CHRISTENSEN, C.M., 1961. *The Molds and Man*. University of Minnesota Press. Minneapolis, MN.

GRAY, W.D., 1959. *The Relation of Fungi to Human Affairs*. Holt & Company. New York, NY.

HUDLER, G.W., 1998. *Magical Mushrooms, Mischievous Molds*. Princeton University Press. Princeton, NJ.

KENDRICK, B., 2000. *The Fifth Kingdom* (3rd edition). Focus Publishing. Newburyport, MA.

LARGE, E.C., 1940. *Advance of the Fungi*. Dover Publ. New York, NY. (Reprinted 1966 by Jonathan Capc Limited. London, UK)

LINCOFF, G., and MITCHEL, D.H., 1977. *Toxic and Hallucinogenic Mushroom Poisoning*. Van Nostrand Reinhold. New York, NY.

MOORE, D., 2001. *Slayers, Saviors, Servants, and Sex. An Expose of the Kingdom Fungi*. Springer Publishing. New York, NY.

PHILLIPS, R., 1991. *Mushrooms of North America*. Little, Brown & Co. Boston, MA.

RAMSBOTTOM, J., 1953. *Mushrooms and Toadstools*. Collins Press. London, UK.

ROLFE, R.T., and ROLFE, F.W., 1925. *The Romance of the Fungus World*. Chapman & Hall. New York, NY. (Reprinted 1966 by Johnson Reprint Corp. New York, NY.)

SCHAECHTER, E., 1997. *In the Company of Mushrooms*. Harvard University Press. Cambridge, MA.

STAMETS, P., 1996. *Psilocybin Mushrooms of the World*. Ten Speed Press. Berkeley, CA.

WASSON, R.G., and WASSON, V.P., 1957. *Mushroom, Russia and History*. Pantheon Publ. New York, NY.

Abbruscato, P., 226
Abdulla, H., 452
Albertano, P., 302
Baldwin, A.M., 82, 94, 110
Barabesi, C., 472
Becker, M.B., 190
Berovič, M., 50
Blanchette, R.A., 328, 370, 390
Bracci, S., 486
Breuker, M., 248
Bullock, V., 440
Carlson, D.N., 390
Cavaliere, A.R., 128, 554
Centeno, S.A., 110
Ciferri, O., 190
De Leo, F., 316
De Montebello, P., 9
De Rossi, E., 190
DeSalle, R., 94
Di Bonaventura, M.P., 94, 110
Donato, P., 316
Duncan, S., 370
El-Easely, A.-S.M., 72
Erhardt, D., 72
Eveleigh, D.E., 94
Farrell, R.L., 370
Florian, M.-L., 172
Gonzalez, J.M., 210
Held, B.W., 370
Hernandez-Duque, G., 248
Hoffmann, P., 348
Jones, M., 28
Jurgens, J.A., 370, 390
Karbowska-Berent, J., 266
Koestler, R.J., 10, 94, 110, 410, 426
La Cono, V., 316
Laiz, L., 210
Loubiere, J.-F., 498
Mastromei, G., 472
May, E., 452
McNamara, C.J., 248
Mitchell, R., 248
Mouzouras, R., 28
Müeller, R., 248
Nieto-Fernandez, F., 110
Nugari, M.P., 518
Onelli, E., 226
Orial, G., 498
Papida, S., 452
Patrignani, G., 226
Perito, B., 472
Perry, T.D., 248
Pitman, A.J., 28
Pohleven, F., 426
Pohleven, J., 426
Pointing, S.B., 28
Realini, M., 226
Rescic, S., 486
Rosa, H.E., 60
Saiz-Jimenez, C., 210
Salvadori, O., 518
Scaramelli, L., 226
Simpson, E., 358
Sorlini, C., 226
Strzelczyk, A., 60
Suzuki, J., 410
Szczepanowska, H., 128
Tamburini, E., 472
Tavzes, C., 426
Tiano, P., 486
Tumosa, C.S., 72
Urzì, C., 316
Van Dyke, Y., 152
Vieweger, T., 498
von Endt, D., 72
Warscheid, Th., 14
Wessel, D.P., 536
Wypyski, M.T., 110
Zanardini, E., 226
Zinn, M., 248

CONTACT LIST

PROF. PATRIZIA ALBERTANO
Department of Biology
University of Rome "Tor Vergata"
via della Ricerca scientifica
00133 Rome, Italy
Tel: +39 06 7259 4859 / 4332
Fax: +39 06 202 3500
Email: albertano@uniroma2.it

MS. ANN M. BALDWIN
The Sherman Fairchild Center for Works on Paper and Photograph Conservation
The Metropolitan Museum of Art
1000 Fifth Avenue
New York, NY 10028
Tel: 212-650-2165
Email: ann.baldwin@metmuseum.org

PROF. MARIN BEROVIČ
Faculty of Chemistry & Chemical Technology, Dept. Chem, Biochem and Ecol. Engineering
Hajdrihova 19
1000 Ljubljana, Slovenia
Tel: +386 61 1760 438
Fax: +386 61 1259 244
Email: marin.berovic@ki.si

DR. ROBERT A. BLANCHETTE
Department of Plant Pathology
495 Borlaug Hall
1991 Upper Buford Circle
University of Minnesota
St. Paul, MN 55108-6030
Email: robertb@umn.edu

MS. VERONICA BULLOCK
2300 Arapahoe Ave., Apt. 129
Boulder, CO 80302
Email: bullconserves@hotmail.com

PROF. A. RALPH CAVALIERE
Department of Biology
Gettysburg College
Gettysburg, PA 17325-1484
Tel: 717-337-6151
Fax: 717-337-6157
Email: rcavalie@gettysburg.edu

DR. A. ELENA CHAROLA
Email: charola@worldnet.att.net

PROF. ORIO CIFERRI
Department of Genetics and Microbiology
Univ. of Pavia
Via Abbiategrasso, 207
27100 Pavia, Italy
Tel: +39 0382 505 576/7
Fax: +39 0382 528 496
Email: ociferri@ipvgen.unipv.it

DR. MARIA PIA DI BONAVENTURA
Objects Conservation
The Metropolitan Museum of Art
1000 Fifth Avenue
New York, NY 10028
Tel: 212-396-5344
Email: mariapia1@juno.com

DR. DAVID ERHARDT
Smithsonian Center for Materials Research and Education
Smithsonian Institution
4210 Silver Hill Road
Suitland, MD 20746-2863
Tel: 301-238-3700 ext. 116
Fax: 301-238-3709
Email: ErhardtD@SCMRE.si.edu

PROF. DOUGLAS E. EVELEIGH
Chair of Applied Microbiology
Douglas E. Eveleigh
Rutgers University, Cook College
Dept. of Biochemistry & Microbiology
76 Lipman Drive
New Brunswick, NJ 08901-8525
Tel: 732-932-9763 ext. 328
Fax: 732-932-8965
Email: eveleigh@aesop.rutgers.edu

MS. MARY-LOU FLORIAN
133 Simcoe Street
Victoria, British Columbia
Canada, V8V 1K5
Tel/Fax: 250-385-8263
Email: mflorian@telus.net

DR. JUAN M. GONZALEZ
IRNAS
P.O. Box 1052
41080 Sevilla, Spain
Fax: +34 95 462 4002
Email: jmgrau@irnase.csic.es

MR. BRIAN W. HELD
Department of Plant Pathology
495 Borlaug Hall
1991 Upper Buford Circle
University of Minnesota
St. Paul, MN 55108-6030
Email: bwheld@umn.edu

DR. PER HOFFMANN
Deutsches Schiffahrtsmuseum
Hans-Scharoun Platz 1
D-27568 Bremerhaven, Germany
Tel: +49 471 4820762
Fax: +49 471 482074
Email: phoffmann@dsm.de

DR. MARK JONES
The Mary Rose Trust
College Road, HM Naval Base
Portsmouth PO1 3LX, UK
Email: maryrose@cix.co.uk
mark-jones@maryrose-conservation.
freeserve.co.uk

MR. JOEL A. JURGENS
Department of Plant Pathology
495 Borlaug Hall, 1991 Upper Buford Circle
University of Minnesota
St. Paul, MN 55108-6030
Email: jajurg@umn.edu

PROF. J. KARBOWSKA-BERENT
Institute of Conservation &
Restoration of Cultural Property
Nicolaus Copernicus Univ.
ul. Sienkiewicza 30/32
87-100 Toruń, Poland
Tel: +48 56 6540512
Email: karber1@wp.pl

DR. ROBERT J. KOESTLER
The Metropolitan Museum of Art
The Sherman Fairchild Center
for Objects Conservation
1000 Fifth Avenue
New York, NY 10028
Tel: 212-396-5390
Fax: 212-570-3859
Email: Robert.Koestler@metmuseum.org

PROF. GIORGIO MASTROMEI
Dept. Animal Biology and Genetics
Via Romana 17
50125 Florence, Italy
Tel: +39 055 228 8240
Fax: +39 055 228 8250
Email: mastromei@dbag.unifi.it

DR. ERIC MAY
Reader in Microbiology
School of Biological Sciences
University of Portsmouth
King Henry Building
King Henry I Street
Portsmouth PO1 2DY, UK
Tel: +44 023 9284 2025
Fax: +44 023 9284 2070
Email: eric.may@port.ac.uk

DR. CHRISTOPHER J. MCNAMARA
Harvard University
Div. Engineering and Applied Science
29 Oxford Street
Cambridge, MA 02138
Email: mcnamara@deas.harvard.edu

PROF. RALPH MITCHELL
Harvard University
Div. Engineering and Applied Science
29 Oxford Street
Cambridge, MA 02138
Tel: 617-495-2846
Fax: 617-496-1471
Email: Mitchell@deas.harvard.edu

PROF. FERNANDO NIETO-FERNANDEZ
Biology Dept.
Old Westbury College
Old Westbury, NY 11568
Tel: 516-876-2729
Fax: 516-876-2749
Email: nieto@oldwestbury.edu

PROF. MARIA PIA NUGARI
I.C.R.-Lab Indagini Biologiche
Piazza S. Francesco di Paola 9
00184 Rome, Italy
Tel: +39 06 4889 6410
Email: mariapia.nugari@tiscalinet.it

DR. GENEVIÈVE ORIAL
Responsable de la section microbiologie
Laboratoire de Recherche des
Monuments Historiques
29, rue de Paris
77420 Champs sur Marne, France
Tel: +33 1 60 37 77 80
Fax: +33 1 64 37 77 99
Email: Genevieve.orial@culture.gouv.fr

PROF. FRANC POHLEVEN
Department of Wood Science
and Technology
Biotechnical Faculty
University of Ljubljana
Rozna dolina, Cesta VIII/34
SI - 1000, Ljubljana, Slovenia
Tel: +38 661 1831167
Fax: +38 661 1235035
Email: franc.pohleven@bf.uni-lj.si

DR. HALINA ROSA
Institute of Conservation &
Restoration of Cultural Property
Nicholas Copernicus Univ.
Ul Sienkiewicza 30/32-87100
Toruń, Poland
Tel: +48 56 6540512
Email: alstrze@art.uni.torun.pl
Note: this is the email for her colleague
and co-author Prof. Strzelczyk

DR. ORNELLA SALVADORI
BBAASS
Laboratorio Scientifico
Cannareggio 3553
30131 Venice, Italy
Tel: +39 041 720 661
Email: orsalva@tin.it

DR. ELIZABETH SIMPSON
2 Hudson Street
Ossining, NY 10562
Tel: 914-944-0161

PROF. CLAUDIA SORLINI
DISTAM-Fac. Agraria
Univ. Milan
Via G. Celoria 2
20133 Milan, Italy
Tel: +39 02 2395 5822
Email: claudi.sorlini@unimi.it

PROF. ALICIA STRZELCZYK
Institute of Conservation &
Restoration of Cultural Property
Nicholas Copernicus Univ.
Ul Sienkiewicza 30/32-87100
Toruń, Poland
Tel: +48 56 6227051 (w)
Tel: +48 56 6227527 (h)
Email: alstrze@art.uni.torun.pl

MR. JUN SUZUKI
Paintings Conservation Dept.
Museum of Fine Arts
465 Huntington Ave.
Boston, MA 02115
Email: juns@kdd.net

MS. HANNA SZCZEPANOWSKA
Department of Anthropology
Anthropological Conservation Laboratory
National Museum of Natural History
Washington, DC 20560
Email: hennas@gwu.edu

DR. CRTOMIR TAVZES
Department of Wood Science
and Technology
Biotechnical Faculty
University of Ljubljana
Rozna dolina, Cesta VIII/34
SI-1000, Ljubljana, Slovenia
Tel: +386 (0)1 423-11-61
Fax: +386 (0)1 423-50-35
Email: crtomir.tavzes@bf.uni-lj.si

DR. PIERO TIANO
C.N.R. Cause di Deperimento e
Metodi di Conservazione Opere d'Arte
Via degli Alfani 74
50125 Florence, Italy
Tel: +39 055 214 777
Email: tiano@cscoa.fl.cnr.it

PROF. CLARA URZÌ
Dipartimento di Scienze Microbiologiche,
Genetiche e Molecolari
Università di Messina
Department of Microbiological,
Genetic and Molecular Sciences
University of Messina
Salita Sperone 31
I-98166 Messina, Italy
Tel: +39 090 676 5196 / +39 090 393 398
Fax: +39 090 392 733
E-mail: urzicl@unime.it
Web page: http://ww2.unime.it/mbio/
clara.htm

MS. YANA VAN DYKE
Paper Conservation Dept.
The Metropolitan Museum of Art
1000 Fifth Avenue
New York, NY 10028
Tel: 212-650-2276
Email: Yana.VanDyke@metmuseum.org

DR. THOMAS WARSCHEID
LBW – Bioconsult
Am Sportplatz 21
26215 Wiefelstede, Germany
Email: lbw.bb@gmx.de

MR. DAVID WESSEL
Architectural Resources Group
Pier 9, The Embarcadero
San Francisco, CA 94111
Tel: 415-421-1680
Fax: 415-421-0127
Email: David@argsf.com

DR. ELIZABETTA ZANARDINI
DISTAM – Fac. Agraria
Univ. Milan
Via G. Celoria 2
20133 Milan, Italy
Tel: +39 02 2395 5825
Fax: +39 02 7063 0829
Email: elisabetta.zanardini@unimi.it